Effects of Lifestyle on Men's Health

Effects of Lifestyle on Men's Health

Editors

Faysal A. Yafi

Department of Urology, University of California Irvine,
Newport Beach, CA, United States

Natalie R. Yafi

Independent Registered Dietitian,
Costa Mesa, CA, United States

Academic Press is an imprint of Elsevier
125 London Wall, London EC2Y 5AS, United Kingdom
525 B Street, Suite 1650, San Diego, CA 92101, United States
50 Hampshire Street, 5th Floor, Cambridge, MA 02139, United States
The Boulevard, Langford Lane, Kidlington, Oxford OX5 1GB, United Kingdom

© 2019 Elsevier Inc. All rights reserved.

No part of this publication may be reproduced or transmitted in any form or by any means, electronic or mechanical, including photocopying, recording, or any information storage and retrieval system, without permission in writing from the publisher. Details on how to seek permission, further information about the Publisher's permissions policies and our arrangements with organizations such as the Copyright Clearance Center and the Copyright Licensing Agency, can be found at our website: www.elsevier.com/permissions.

This book and the individual contributions contained in it are protected under copyright by the Publisher (other than as may be noted herein).

Notices
Knowledge and best practice in this field are constantly changing. As new research and experience broaden our understanding, changes in research methods, professional practices, or medical treatment may become necessary.

Practitioners and researchers must always rely on their own experience and knowledge in evaluating and using any information, methods, compounds, or experiments described herein. In using such information or methods they should be mindful of their own safety and the safety of others, including parties for whom they have a professional responsibility.

To the fullest extent of the law, neither the Publisher nor the authors, contributors, or editors, assume any liability for any injury and/or damage to persons or property as a matter of products liability, negligence or otherwise, or from any use or operation of any methods, products, instructions, or ideas contained in the material herein.

Library of Congress Cataloging-in-Publication Data
A catalog record for this book is available from the Library of Congress

British Library Cataloguing-in-Publication Data
A catalogue record for this book is available from the British Library

ISBN 978-0-12-816665-9

For information on all Academic Press publications
visit our website at https://www.elsevier.com/books-and-journals

Publisher: Stacy Masucci
Acquisition Editor: Stacy Masucci
Editorial Project Manager: Megan Ashdown
Production Project Manager: Maria Bernard
Cover Designer: Matthew Limbert

Typeset by SPi Global, India

Contents

CONTRIBUTORS xvii
AUTHOR BIOGRAPHY xxi

PART 1 Diet

CHAPTER 1.1 Diet and Sexual Health 3
Maxwell M. Towe, Faysal A. Yafi, and Natalie R. Yafi

Introduction 3
Diet and Health 3
Sexual Health and Overall Health 3
Diet and Erectile Function 4
Food Groups and Erectile Function 4
Specific Diets 6
Obesity and ED 7
Diet vs Weight Loss 8
Diet and Ejaculatory/Orgasmic Dysfunction 9
Folate 10
Serotonin/Tryptophan 10
Other Nutrients 11
Delayed/Anejeculation 11
Diet and Peyronie's Disease 12
Vitamin E 12
Blueberries 12
Other Foods 13
Diet and Hypogonadism 14
Obesity and Hypogonadism 14
Weight Loss and Hypogonadism 15
Diets and Dietary Patterns 15
Foods That Affect Male Hormones 16
Androgen Suppressing 16
Androgen Inducing 17

v

	Future Perspectives	19
	Childhood Diet	19
	Organic Food	19
	Ketogenic Diet	19
	Take Home Messages	19
	References	20
CHAPTER 1.2	**Effects of Lifestyle on Urinary Health**	**27**
	Joseph Mahon, and Kevin T. McVary	
	Introduction	27
	Diet and Urological Health	27
	Summary	36
	References	38
	Further Reading	40
CHAPTER 1.3	**Diet and Fertility in Men: Are Sperm What Men Eat?**	**41**
	Feiby L. Nassan, and Jorge E. Chavarro	
	Introduction	41
	Diet and the Building Blocks for Sperm	42
	Dietary Fats, the Sperm Cell Membrane and Men's Fertility	42
	Antioxidants: Defense Against Oxidative Stress	44
	The One-Carbon Metabolism and Spermatogenesis	46
	Food as a Vehicle for Environmental Toxins	47
	Meats, Dairy Products and Hormone Residues	48
	Fish and Methylmercury	49
	Fruits, Vegetables, and Pesticide Residues	50
	Soy Products and Xenoestrogens	51
	Diet Patterns	51
	Conclusions	52
	References	53
CHAPTER 1.4	**The Gut Microbiome**	**61**
	Sarah Ashman, and Hari Krishnamurthy	
	Introduction	61
	Technological Approaches to Studying the Gut Microbiome	61
	DNA Approaches—What Is Present?	61
	RNA Approaches—What Is the Functional Pathway?	62
	Differentiation of the Intestinal Microbiome	62
	Protein and Metabolite Approaches—Host Interaction and Outcome	62

Broad Functions 63
Microbes and Immunity 63
Homeostasis Is Maintained by Two-Way Communication
    Between the Microbiota and Intestinal Immune System 64
Microbial Diversity 66
Keystone Species 67
Metabolic Crossfeeding 67
Suppressing Diversity Through Prescription Antibiotic Use 68
Colon Health and Colorectal Cancer 68
Men's Health and the Gut Microbiome 69
Gut Microbiome and Testosterone 69
Gut Microbiome and Prostate Health 70
Cardiovascular Health 70
Dietary Influence 71
Probiotic Foods 71
Probiotic Supplements 72
Prebiotics 75
Additional Beneficial Nutrients 80
Conclusion 80
References 84
Further Reading 98

PART 2   Metabolic Health

CHAPTER 2.1   Metabolic Health: Inflammation and Men's Health 101
Brent M. Hanson, and James M. Hotaling

Introduction 101
Genitourinary Conditions, Pain Syndromes, and Sexual
    Dysfunction 103
Prostatitis, Chronic Pelvic Pain, and Inflammation 103
Erectile Dysfunction and Inflammation 104
Hypogonadism and Inflammation 104
Male Infertility 105
Varicocele, Infertility, and Inflammation 106
Cryptorchidism, Infertility, and Inflammation 107
Inflammation and Testicular Torsion 108
Genitourinary Infections and Infertility 109
Environmental Exposures, Inflammation, and Infertility 110
The Effects of Inflammation on Assisted Reproductive
    Technology 111

Urologic Malignancy 111
Inflammation and Prostate Cancer 111
Inflammation and Bladder Cancer 113
Implications 114
References 114

CHAPTER 2.2 Diabetes and Men's Health 121
Carolyn A. Salter, and John P. Mulhall

Functional Anatomy 121
Erectile Tissue 121
Nerves 121
Blood Vessels 122
Ejaculatory Apparatus 123
Functional Physiology 123
Erectile Physiology 123
Physiology of Ejaculation 124
Physiology of Testosterone Production 125
Erectile Dysfunction 126
Overview 126
Pathophysiology of ED in Diabetics 127
Patient Evaluation 129
Management 130
Ejaculatory Dysfunction 135
Overview 135
Anejaculation Evaluation 136
Treatment of Anejaculation 136
Orgasmic Dysfunction 137
Pathophysiology 137
Patient Evaluation 137
Treatment 138
Testosterone Deficiency 139
Overview 139
Pathophysiology 140
Treatment of Testosterone Deficiency and the
    Impact on DM 141
Patient Evaluation 141
Treatment 141
Complications 144
Conclusion 144
References 144

CHAPTER 2.3 Obesity and Men's Health 149
Ahmet Tevfik Albayrak, and Ege Can Serefoglu

Overview 149
Obesity and BPH/LUTS 149
Introduction 149
Underlying Mechanisms 150
Behavioral Modifications 151
Obesity and Erectile Dysfunction 152
Introduction 152
Underlying Mechanisms 152
Obesity, ED and CVD Risk 153
Treatment of Obesity-Related ED 153
Obesity and Hypogonadism 154
Introduction 154
Underlying Mechanism 154
Treatment of Obesity-Related
 Hypogonadism 155
Obesity and Infertility 155
Introduction 155
Underlying Mechanisms 156
Treatment of Obesity-Related Infertility 157
Obesity and Prostate Cancer 158
Introduction 158
Underlying Mechanisms 158
Possible Interventions 159
Key Points: Obesity and Men's Health 159
References 160

CHAPTER 2.4 Cardiovascular Disease and Men's
Health 169
*İyimser Üre, John M. Masterson, and
Ranjith Ramasamy*

Cardiovascular Disease and Erectile
 Dysfunction 169
Pathophysiology 169
ED as a Predictor of CVD 170
Sexual Health in CVD Patients 171
CVD Drugs and ED 173
Hypogonadism and CVD 174
References 175

PART 3 Mental Health

CHAPTER 3.1 Sleep, Shift Work, and Men's Health 181
Jorge Rivera Mirabal, Mohit Khera and Alexander W. Pastuszak

Introduction 181
Sleep, Shift Work, Insomnia and Occupational Health 182
Sleep, Shift Work and Chronic Illness 183
Sleep Disorders, Shift Work and Erectile Dysfunction 186
Sleep Disorders, Shift Work and Hypogonadism 188
Sleep Disorders, Shift Work and Lower Urinary Tract
    Symptoms 192
Sleep Disorders, Shift Work, Occupational Exposures
    and Infertility 195
Sleep Disorders, Shift Work and Prostate Cancer 198
Conclusions 200
References 200

CHAPTER 3.2 Stress, Depression, Mental Illness, and Men's Health 207
Nnenaya Q. Agochukwu, and Daniela Wittmann

Introduction 207
Stress and the Contemporary Male Role 208
Contribution of Class, Race/Ethnicity, and Sexual Minority
    Status to Men's Experience of Stress 208
Stress and Physical Health 210
Depression, Mental Illness, and Men's Health 210
Masculinity and Sexual Health 212
Men and Coping 213
Help-Seeking, Athleticism, and Risk Taking Behaviors 213
Healthcare Providers and Men's Health 214
Summary 214
References 216

CHAPTER 3.3 Meditation, Yoga, and Men's Health 223
Claire Postl, and Lawrence C. Jenkins

Introduction 223
Stress 223
Socialized for Stress 224
Meditative Practices 224
Health Benefits of Meditative Practices 225
Sexual Health 225

Fertility 225
Weight and Diabetes Management 226
Cardiovascular Disease 226
Anxiety and Depression 227
Recommendations 227
Conclusion 228
References 228
Further Reading 231

PART 4 Hormones

CHAPTER 4.1 Testosterone and Men's Health 235
James Anaissie, Alexander W. Pastuszak, and Mohit Khera

Introduction 235
Diagnosis of Hypogonadism 236
Treatment of Hypogonadism 237
Efficacy of Testosterone Therapy 239
Erectile Dysfunction 239
Urinary Function 240
Metabolic Syndrome 240
Safety of Testosterone Therapy 241
Conclusion 242
References 242

CHAPTER 4.2 Other Hormonal Therapies and Men's Health 253
Dorota J. Hawksworth, and Arthur L. Burnett, II

Introduction 253
Human Chorionic Gonadotropin 255
Introduction 256
Mechanism of Action 256
Indications 256
Dosing 256
Side Effects and Safety Concerns 256
Benefits 257
Selective Estrogen Receptor Modulators 258
Introduction 258
Mechanism of Action 258
Indications 259
Dosing 259
Side Effects and Safety Concerns 259
Benefits 260

Aromatase Inhibitors 261
Introduction 261
Mechanism of Action 261
Indications 262
Dosing 262
Side Effects and Safety Concerns 262
Benefits 263
SERM and AI Combination Therapy (Clomiphene Citrate and Anastrozole) 263
Conclusions 264
References 264

PART 5   Supplements

CHAPTER 5.1 Chinese Medicine and Men's Health 271
Barbara M. Chubak, and Jillian Capodice

References 278

CHAPTER 5.2 Over-the-Counter Supplements and Men's Health 281
Farouk M. El-Khatib, Natalie R. Yafi, and Faysal A. Yafi

Introduction 281
Scope of OTC 281
Market Size of OTC 281
Naturopathic Medicine 282
OTC and Sexual Function 282
Erectile Dysfunction 282
Ejaculatory and Orgasmic Function 286
OTC and Hormones 288
Testosterone 288
Estrogen 290
OTC and Urinary function 291
Benign Prostatic Hyperplasia 291
Prostatitis 293
Risks of OTC 295
Take Home Messages 295
References 295
Further Reading 300

PART 6   Recreational Habits

CHAPTER 6.1 Smoking and Men's Health 303
U. Milenkovic, and M. Albersen

Introduction 303
Smoking and Male Fertility 304

Smoking and Testosterone 306
Smoking and Erectile Dysfunction...307
Peyronie's Disease...309
Nephrolithiasis 310
Urological Cancers 311
Bladder Cancer..311
Prostate Cancer...311
Kidney Cancer (Renal Cell Carcinoma [RCC]) 312
References...312

CHAPTER 6.2 The Opioid Epidemic and Men's Sexual Health 321
Hossein Sadeghi-Nejad, and Radhika Ragam
Introduction...321
Background and Epidemiology..321
Pathophysiology of Opioid-Induced Hypogonadism 322
Diagnosis 323
Treatment 325
Treatment Options..326
Intramuscular Testosterone Enanthate or Cypionate,
    75–100 mg Every Week or 150–200 mg Every 2 Weeks 326
Transdermal Testosterone Patches, One or Two 5 mg
    Applied Nightly..326
1.625% Testosterone Gel 20.25–81 mg Daily 326
Subcutaneous Testosterone Pellets, Varied Dosing,
    Interval Injection of 3–6 Months 326
Conclusion 328
References...329

CHAPTER 6.3 Alcohol and Men's Health 333
*Brian Dick, Scott Brimley, Peter Tsambarlis
and Wayne Hellstrom*
Introduction...333
Acute Effects of Alcohol 333
Erectile Dysfunction (ED)...333
Ejaculatory Dysfunction 335
Chronic Effects of Alcohol...335
Erectile Dysfunction 336
Ejaculatory Dysfunction 341
Hypogonadism..342
Potential Confounding Factors 342
Assessing Erectile Dysfunction 342
Societal Difference in Alcohol Consumption..........................343

	Conclusion	343
	References	344
CHAPTER 6.4	Exercise, Sports, and Men's Health	349
	Joshua T. Randolph, Lindsey K. Burleson, Alyssa Sheffield and Johanna L. Hannan	
	Introduction	349
	Exercise and Men's Health	349
	Exercise Benefits Erectile Function	349
	Can Erectile Function be Recovered With Exercise?	350
	Exercise and Erectile Function in Prostate Cancer Survivors	351
	Pelvic Floor Muscle Rehabilitation and Men's Sexual Health	352
	Exercise and Fertility in Men	352
	Exercise and Libido in Men	354
	Sports and Men's Health	354
	Sports and Genitourinary Trauma	354
	Continued Controversy—Does Cycling Lead to ED?	355
	Sex Before Sports—Does It Impact Athletic Performance?	355
	Overall Conclusions	356
	References	357

PART 7 External factors

CHAPTER 7.1	Environmental Toxins and Men's Health	363
	J. Marinaro, and C. Tanrikut	
	Environmental Toxins and Men's Health	363
	Air Pollution	363
	Water Pollution	366
	Heat	368
	Radiation	375
	Hypoxia	378
	Physical and Mental Stress	379
	Endocrine Disruptors and Men's Health	380
	What Are Endocrine Disrupting Chemicals?	380
	Types of Endocrine Disrupting Compounds	380
	Conclusions	390
	References	390
	Further Reading	401
CHAPTER 7.2	Endocrine Disruptors and Men's Health	403
	Christian Fuglesang S. Jensen, Ulla N. Joensen, Zainab G. Nagras, Dana A. Ohl, and Jens Sønksen	
	Endocrine Disruptors	403
	Gonadal Development and Male Differentiation	404

Endocrine Disruptors and Associated Clinical
  Conditions—The "Testicular Dysgenesis Syndrome"............406
Cryptorchidism..407
Hypospadias..407
Testis Cancer..408
Male Infertility...409
Hypogonadism..410
Evidence Synthesis and Conclusion..............................410
References..411

INDEX..413

Contributors

Nnenaya Q. Agochukwu Department of Urology, University of Michigan, Ann Arbor, MI, United States

Ahmet Tevfik Albayrak Health Science University, Sisli Etfal Training & Research Hospital, Istanbul, Turkey

M. Albersen Department of Urology, University Hospitals Leuven, Leuven, Belgium

James Anaissie Scott Department of Urology, Baylor College of Medicine, Houston, TX, United States

Sarah Ashman Clinical Education, Vibrant America Clinical Laboratory, San Carlos, CA, United States

Scott Brimley Tulane University School of Medicine, Department of Urology, New Orleans, LA, United States

Lindsey K. Burleson Department of Physiology, Brody School of Medicine, East Carolina University, Greenville, NC, United States

Arthur L. Burnett, II The James Buchanan Brady Urological Institute and Department of Urology, Johns Hopkins University School of Medicine, Baltimore, MD, United States

Jillian Capodice Department of Urology, Icahn School of Medicine at Mount Sinai, New York, NY, United States

Jorge E. Chavarro Department of Nutrition; Department of Epidemiology, Harvard T. H. Chan School of Public Health; Channing Division of Network Medicine, Harvard Medical School, and Brigham and Women's Hospital, Boston, MA, United States

Barbara M. Chubak Department of Urology, Icahn School of Medicine at Mount Sinai, New York, NY, United States

Brian Dick Tulane University School of Medicine, Department of Urology, New Orleans, LA, United States

Farouk M. El-Khatib Department of Urology, University of California Irvine, Newport Beach, CA, United States

Johanna L. Hannan Department of Physiology, Brody School of Medicine, East Carolina University, Greenville, NC, United States

Brent M. Hanson RMA New Jersey, Sidney Kimmel Medical College at Thomas Jefferson University, Basking Ridge, NJ, United States

Dorota J. Hawksworth The James Buchanan Brady Urological Institute and Department of Urology, Johns Hopkins University School of Medicine, Baltimore, MD, United States

Wayne Hellstrom Tulane University School of Medicine, Department of Urology, New Orleans, LA, United States

James M. Hotaling University of Utah Center for Reconstructive Urology and Men's Health, Department of Surgery—Urology, Salt Lake City, UT, United States

Lawrence C. Jenkins Men's Health Program, Department of Urology, The James Comprehensive Cancer Center, The Ohio State University Wexner Medical Center, Columbus, OH, United States

Christian Fuglesang S. Jensen Department of Urology, Herlev and Gentofte Hospital, University of Copenhagen, Copenhagen, Denmark

Ulla N. Joensen Department of Urology, Rigshospitalet, University of Copenhagen, Copenhagen, Denmark

Mohit Khera Scott Department of Urology, Baylor College of Medicine, Houston, TX, United States

Hari Krishnamurthy Technology Development, Vibrant Sciences, San Carlos, CA, United States

Joseph Mahon Department of Urology, Stritch School of Medicine, Loyola University Medical Center, Maywood, IL, United States

J. Marinaro Department of Urology, Georgetown University School of Medicine, Washington, DC, United States

John M. Masterson University of Miami, Miller School of Medicine, Department of Urology, Miami, FL, United States

Kevin T. McVary Department of Urology, Stritch School of Medicine, Loyola University Medical Center, Maywood, IL, United States

U. Milenkovic Department of Urology, University Hospitals Leuven, Leuven, Belgium

Jorge Rivera Mirabal Scott Department of Urology, Baylor College of Medicine, Houston, TX, United States

John P. Mulhall Department of Urology, Memorial Sloan-Kettering Cancer Center, New York, NY, United States

Zainab G. Nagras Department of Urology, Herlev and Gentofte Hospital, University of Copenhagen, Copenhagen, Denmark

Feiby L. Nassan Department of Environmental Health; Department of Nutrition, Harvard T. H. Chan School of Public Health, Boston, MA, United States

Dana A. Ohl Department of Urology, University of Michigan, Ann Arbor, MI, United States

Alexander W. Pastuszak Division of Urology, Department of Surgery, University of Utah School of Medicine, Salt Lake City, UT, United States

Claire Postl Men's Health Program, Department of Urology, The James Comprehensive Cancer Center, The Ohio State University Wexner Medical Center, Columbus, OH, United States

Radhika Ragam Rutgers New Jersey Medical School, Newark, NJ, United States

Ranjith Ramasamy University of Miami, Miller School of Medicine, Department of Urology, Miami, FL, United States

Joshua T. Randolph Department of Physiology, Brody School of Medicine, East Carolina University, Greenville, NC, United States

Hossein Sadeghi-Nejad Professor of Surgery, Rutgers New Jersey Medical School and Hackensack University Medical Center, Hackensack, NJ, United States

Carolyn A. Salter Department of Urology, Memorial Sloan-Kettering Cancer Center, New York, NY, United States

Ege Can Serefoglu Bahceci Health Group, Istanbul, Turkey

Alyssa Sheffield Department of Public Health, Brody School of Medicine, East Carolina University, Greenville, NC, United States

Jens Sønksen Department of Urology, Herlev and Gentofte Hospital, University of Copenhagen, Copenhagen, Denmark

C. Tanrikut Department of Urology, Georgetown University School of Medicine, Washington, DC, United States

Maxwell M. Towe Department of Urology, University of California Irvine, Newport Beach, CA, United States

Peter Tsambarlis Tulane University School of Medicine, Department of Urology, New Orleans, LA, United States

İyimser Üre Eskişehir Osmangazi University, Faculty of Medicine, Department of Urology, Eskişehir, Turkey

Daniela Wittmann Department of Urology, University of Michigan, Ann Arbor, MI, United States

Faysal A. Yafi Department of Urology, University of California Irvine, Newport Beach, CA, United States

Natalie R. Yafi Independent Registered Dietitian, Costa Mesa, CA, United States

Author Biography

Faysal A. Yafi, MD FRCSC, is an assistant professor of urology, chief of andrology, and director of Men's Health at the University of California, Irvine. He earned his medical degree from the American University of Beirut. He then served his internship in general surgery at the Mayo Clinic in Rochester, Minnesota, and subsequently completed his residency in urology at McGill University in Montreal, Quebec. He then completed a 2-year fellowship in andrology, sexual medicine, and prosthetic urology at Tulane University. He is certified by the American Board of Urology and is a fellow of the Royal College of Surgeons of Canada. He serves on numerous national and international urology and sexual medicine society committees as both member and chair. He has an avid interest in both basic and clinical research. He has written more than 100 peer-reviewed publications and book chapters, has been an invited speaker to multiple national and international venues, and has received multiple national and international awards and recognitions.

Natalie R. Yafi, RD, is a registered dietitian specializing in the prevention and treatment of cardiovascular disease, diabetes, food allergies, and gut imbalances. She received her degree in nutrition followed by a clinical graduate program at Mississippi State University. She has since held multiple positions as both a dietitian and physician consultant. She has been published multiple times in peer-reviewed journals. Her main areas of interest revolve around incorporating her expertise in nutrition and advanced metabolic testing to promote healthy living, behavioral changes, and weight loss and to improve all elements of men's health.

PART 1

Diet

CHAPTER 1.1

Diet and Sexual Health

Maxwell M. Towe*, Faysal A. Yafi*, and Natalie R. Yafi[†]

*Department of Urology, University of California Irvine, Newport Beach, CA, United States,
[†]Independent Registered Dietitian, Costa Mesa, CA, United States

INTRODUCTION

Diet and Health

Our understanding of how diet affects our health is constantly evolving. Nutrition guidelines are changed year to year by the Food and Drug Administration (FDA) and other governing bodies as more research is put out regarding what exactly constitutes a healthy diet. Processed meat was officially classified as a carcinogen in 2015 with red meat being classified as "probably carcinogenic" at the same time [1]. Twenty years ago, it was thought that a proper diet should avoid all fat and include complex carbohydrates as a substantial portion of what we consume [2]. More recently, we have come to realize that refined carbohydrates like white bread and pasta raise blood sugar more than whole grains and that the type of fat (monounsaturated vs polyunsaturated vs saturated) matters more when looking at disease development [3]. Diets such as the Mediterranean diet incorporate many of the nutrients that we now consider to be health-promoting such as fish, fruits, vegetables, nuts, monounsaturated fats, and whole grains, and research has shown adherence to this diet correlates with a lower risk of cardiovascular disease (CVD), diabetes, obesity, cancer, and neurodegenerative disorders [4–7]. The Western diet on the other hand involves processed meat, refined carbohydrates, dairy, and salt, and its health detriments have been documented by numerous studies [8–10]. Data are constantly being produced on this topic to help guide us make the right decisions when deciding what to eat for dinner or what type of snack to have. The fact that we keep adjusting our attitudes toward diet reemphasizes the importance it has for our overall health.

Sexual Health and Overall Health

Normal sexual function is dictated by the interaction of multiple systems including genitourinary, endocrine, neurological, psychological, and cardiovascular [11, 12]. Disturbances of any one of these systems can interrupt various parts of the sexual response cycle and negatively affect sexual health. Rarely then will men present in clinic solely with a sexual complaint, and usually, there is an associated health issue to go along with it. Many times, the two are interrelated, and it can be difficult to distinguish which one causes the other. For example, erectile dysfunction (ED) can be the result of numerous comorbidities such as a diabetes-induced neuropathy [13] or atherosclerosis from CVD [14]. On the other hand, sexual dysfunction can also exacerbate and even be the cause of certain diseases. For example, sexual dysfunction can cause low self-esteem and anxiety that may lead to depression in some men [15, 16]. Additionally, men who are hypogonadal are at an increased risk for developing metabolic syndrome and CVD [17]. Regardless of which one came first, it is important to look at a man's sexual health in the context of his overall health, as one can usually serve as a window into the other.

DIET AND ERECTILE FUNCTION

ED is defined as the consistent or recurrent inability to attain and/or maintain penile erection sufficient for sexual satisfaction, including satisfactory sexual performance, and its prevalence increases with age with a worldwide prevalence of ED predicted to reach 322 million cases by 2025 [18]. The pathophysiology of ED is complex and likely variable depending on the context in which it appears. The ED disease process is closely related to that of CVD and diabetes [19], and thus, they share many of the same risk factors such as smoking, obesity, alcohol use, sedentary lifestyle, and psychological distress. The mechanisms that are believed to play a role in ED include endothelial dysfunction, impaired nitric oxide (NO) signaling, oxidative stress, impaired blood flow, and an increase in general inflammation. As diet has been shown to influence the progression or prevention of CVD, diabetes, and other elements of metabolic syndrome, most likely by affecting those mentioned parameters, diet should also play a role in the development of ED.

Food Groups and Erectile Function

Oxidative stress has been proposed as a pathological mechanism behind ED [20, 21]. It stands to reason then that foods that contain a high amount of antioxidant compounds should be protective for sexual health. Fruits and vegetables have largely been studied in this regard and overall have been shown to have this effect [20, 21]. A study in Iranian men found that daily consumption of fruits was associated with a lower risk of ED compared with a diet with only weekly or seldom consumption of fruits and vegetables [22]. Similarly, a study in diabetic Canadian men found an inverse relationship between fruit/vegetable consumption and ED, and this relationship held true while controlling for sociodemographic factors, lifestyle factors, duration of diabetes, and diabetic complications [23]. The authors of this study were able to quantify this beneficial effect further, reporting a 10% decrease in ED risk for each additional daily serving of fruits/vegetables. These results were replicated in a cohort of Albanian men, where the authors reported a 13% decrease in ED risk per each additional serving [24].

Other food types that seem to be associated with lower rates of ED are nuts and fiber. A study looking at the representation of food intake in both ED and no ED groups found nut and fruit intake to be significantly higher in the group without ED [25]. This difference existed even when controlling for BMI, waist circumference, physical activity, and total energy intake. A diet with high fiber content was also found to correlate with normal erectile function in a randomized clinical trial [26].

Fat intake is also important to consider when looking at ED risk, and specifically, it is the type of fat that seems to matter the most. Monounsaturated fat intake is higher in men without ED, and diet interventions with increases in monounsaturated fat lead to fewer incidences of ED [26]. Saturated fats on the other hand are more correlated with ED development. A study on Chinese men found dairy products, which are high in saturated fat, to be negatively correlated with International Index of Erectile Function (IIEF-5) scores [27].

Polyunsaturated fats like omega-3 and omega-6 fatty acids are known to play a role in cardiovascular health [28] and likely are beneficial when considering ED risk. Moreover, the ratio of omega-6 to omega-3 fatty acid intake is what's thought to be important when considering a healthy diet [29]. The Western diet is characterized by a higher amount of omega-6 fatty acids and a larger omega-6/omega-3 ratio, which has been linked to a higher prevalence of CVD, metabolic syndrome, Alzheimer's disease, and inflammatory bowel disease, among others [30, 31]. This is likely due to the fact that the Western diet lacks plants and fish, which are foods that contain some of the highest levels of omega-3s. Similarly, the Western diet is characterized by high quantities of processed meat that conversely contains high levels of omega-6 fatty acids [32]. Grass-fed red meat on the other hand contains a high level of omega-3 fat and a smaller omega-6/omega-3 ratio. Consumers who switched their diet to primarily grass-fed red meat saw an increase in plasma and platelet concentrations of omega-3 fatty acids than those fed red meat with a cereal concentrated diet [33]. These types of dietary modifications may have important implications on cardiovascular and overall general health, which could translate into better erectile functioning by decreasing atherosclerotic plaque buildup. This effect has been documented in atherosclerotic rat models, whereby significant improvement in endothelial and subsequent erectile function was seen following omega-3 fatty acid administration [34].

Specific Diets

The previous data might suggest that a diet that incorporates a higher number of servings of these foods might also be associated with better erectile function. The Mediterranean diet, which is rich in fruits, vegetables, nuts, fish, and monounsaturated fats, has been the most studied regarding its effect on ED. Many have reported its favorable effects on erectile function [25, 26, 35, 36]. Indeed, lower rates of ED were seen in men who had stricter adherence to the Mediterranean diet compared with men who did not [36]. Additionally, the severity of ED was worse in the group with the lowest adherence to the Mediterranean diet. The benefits of diets like this one have been reported in other cultures as well. Men living in the Amazon rainforest, whose diet is similar to the Mediterranean diet, also seem to have low rates of ED and other metrics of sexual health like ejaculatory function [37]. The Amazonian diet consists of river fish, Brazilian nuts, fruits, and vegetables [37]. One such diet that is not as well represented in the literature for this topic is vegetarianism/veganism. Because these diets are based around nonanimal products, fruit and vegetables make up a substantial proportion of the diet, and one would expect relatively low rates of ED in this population. On the other hand, a diet lacking in animal products may be compensated for by an equivalent increase in refined carbohydrates similar to the Western diet and may be detrimental for erectile health. There are only a few anecdotal reports of vegetarian diets in this respect, and in these cases, this dietary pattern improved erectile function in men [38], but it has not been systematically studied on a large scale.

Other diets that require more attention are the Paleolithic "Paleo" diet and ketogenic diet. The Paleo diet follows that of the hunter-gatherer era, with emphasis on foods that could be foraged for in the wild and do not require industrial production [39]. In this regard, it is similar to the Mediterranean diet and consists of lean meats, plant-based foods, fruits, and nuts. Its beneficial effects on decreasing CVD and metabolic syndrome risk have been documented in the literature [40], but a comparable risk reduction in ED has yet to be discovered. Like the Paleolithic diet, the ketogenic diet has been shown to reduce the risk of metabolic syndrome, possibly by restoring insulin sensitivity through benign dietary ketosis [41]. In a study looking at dietary effects on common metrics of metabolic state, men consuming a ketogenic diet for 8 weeks had a significantly lower BMI, hemoglobin A1c, serum triglycerides,

and body fat percentage compared with men eating a standard American diet. Even more revealing, an additional group that included exercise with their diet intervention still did not see the same level of reduction in these parameters compared with the ketogenic group [41].

Obesity and ED

Many studies have investigated the link between increasing body mass and its effect on erectile function [42–45]. In the Massachusetts Male Aging Study, obesity (defined as body mass index (BMI) >28) emerged as an independent risk factor associated with ED prevalence [46]. Other large-scale male cohorts such as the osteoporotic risk in men study have corroborated the same finding [47]. This finding was significant even when controlling for potential confounders such as smoking, alcohol consumption, hypertension, amount of physical activity, and cholesterol level. Moreover, other studies have shown that weight loss from both diet changes and physical exercise can improve erectile function both from baseline and over control groups [48]. Specifically, by restricting caloric intake to 900 kcal/day, patients who were diabetic and nondiabetic were able to significantly lose weight and improve IIEF-5 scores [49]. And in a group of obese men, ED was only shown to improve after bariatric surgery despite adopting a new diet for 16 weeks beforehand [50]. This evidence suggests that the buildup of adipose tissue strongly exerts a negative influence over erectile function [51]. From a physiological standpoint, there are a few reasons why this may be the case. First, fat cells contain the aromatase enzyme that converts testosterone to estrogen and decreases testosterone's ability to positively affect sexual health [52]. Second, a decrease in nitric oxide and resulting endothelial dysfunction has also been implicated as a potential consequence of obesity. Vascular generation of tumor necrosis factor (TNF)-alpha decreases the bioavailability of NO in small arteries, which is required for normal erectile function [53]. Furthermore, obesity decreases the concentration of NO synthase in endothelial and nervous tissue [54], impairing the signaling system from the brain to the vasculature required for a normal erection. Taken together, these disturbances in the natural communication and coordination of the erectile response point to a strong argument for obesity's pathological role in ED development.

Diet vs Weight Loss

It is not entirely clear which is more important for the beneficial effects certain diets have on erectile function—either the actual components of the diet or the accompanying weight loss that usually occurs from adherence to a healthy diet. For example, obesity is a known risk factor for ED, and many studies have looked at dietary intervention on ED rates, specifically in obese or overweight men. In the Mediterranean diet and type 2 diabetes (MEDITA) trial, a group of diabetic men with a BMI > 25 were randomized to either a Mediterranean diet or a low-fat diet to compare the effects of these two diets on sexual functioning. The men in the Mediterranean diet group had lesser deterioration of erectile function (as measured by the IIEF-5) over time, but they also experienced a significantly greater weight loss than the low-fat diet group [55]. However, another study investigating the Mediterranean diet in patients with metabolic syndrome found an independent effect of diet adherence to an increase in erectile function [56]. In this study, the recruited men were not necessarily overweight, and the difference in weight loss between the two groups was minimal. Another study that compared a high-protein (HP), low-fat diet with a high-carbohydrate (HC), low-fat diet in obese men found a significant difference in weight loss and improvement in erectile function within both groups during the weight loss phase (weeks 0–12), but no significant difference of these outcomes between groups [57]. Additionally, in the weight-maintenance phase of their study (weeks 12–52), there were no purported changes in erectile function in either group despite adherence to their diets. It is possible that no differences were seen between the HP and low-carbohydrate (LC) group because the meat given to the HP group had a mix of processed and unprocessed meat. Supplementing with grass-fed meat might have shown a between groups difference. A very similar study found that a high-protein diet increased IIEF-5 score up to week 52 even after there was no reported weight loss [58], pointing to the idea that it was the diet that made a stronger impact on erectile health.

Whether or not diet by itself or the accompanying weight loss of diet is the main factor predicting erectile function has not entirely been teased out, but it remains a topic that is extensively studied. Most likely, it is a combination of effects from these two approaches, possibly from different physiological mechanisms, that has the greatest impact on preventing ED and restoring a man's sexual health.

DIET AND EJACULATORY/ORGASMIC DYSFUNCTION

Ejaculatory function is a continuum with premature ejaculation (PE) anchoring at one end, normal ejaculation in the center and difficulties with delayed or anejaculation at the opposite end. Premature ejaculation (PE) is characterized by ejaculation that always or nearly always occurs prior to or within about 1 min of vaginal penetration; the inability to delay ejaculation on all or nearly all vaginal penetrations; and negative personal consequences, such as distress, bother, frustration, and/or the avoidance of sexual intimacy [59]. PE is the most common male sexual dysfunction in young men [60] and is most often an embarrassing topic for patients to talk about with physicians. Delayed ejaculation (DE) is not as common as PE, but it can still cause much distress to the patient and affect his confidence. Additionally, the patient's partner can be even more upset by the problem by misinterpreting the prolonged ejaculatory latency as a lack of attraction to the partner.

The pathophysiology behind these ejaculatory disorders is not well understood, and much of current thinking is extrapolated on from what is already known about the normal ejaculatory cycle.

Folate

The process of ejaculation is orchestrated by the central nervous system and involves multiple neurotransmitter systems such as norepinephrine, dopamine, serotonin, and GABA [61]. Serotonin has been demonstrated to play an inhibitory role in this process both pre- and postsynaptically. Animal studies have shown that administration of serotonin delays ejaculation time in male rats [62]. Indeed, this pathway is the basis behind the well-known sexual side effects of serotonin reuptake inhibitors (SSRIs) and the use of SSRIs for treating PE. SSRIs block the reuptake of serotonin at the presynapse, extending the time with which serotonin can act on postsynaptic receptors [63]. Similarly to SSRIs, dietary intake of folic acid may also prolong ejaculatory time by enhancing serotonergic neurotransmission [64]. Additionally, low serum levels of folate are associated with a lesser response to SSRIs, and folate deficiency is associated with decreased serotonin activity [65]. In men with PE, folic acid levels were found to be significantly lower compared with normal controls [66]. Serum levels of folate were also found to be positively correlated with intravaginal ejaculation latency time (IELT), a common measure for evaluating PE [66]. To date, there are no randomized, placebo-controlled studies to assess the effectiveness of treating PE with folate. This may serve as a safer and cheaper alternative to SSRI use in patients with an already low folate level or in those who may be unresponsive to SSRI therapy. Therefore, a diet that is deficient in folate may raise one's risk for PE.

Serotonin/Tryptophan

Due to the importance of serotonin's inhibitory role on ejaculation, one might postulate then that foods that contain serotonin in them, such as bananas, could raise serotonin levels in the brain to the point that would delay ejaculation time in men. This myth however turns out to be false, as the serotonin in bananas is not able to cross the blood-brain barrier [67]. Another idea is that foods that contain the precursor to serotonin—tryptophan—specifically in a higher proportion to other amino acids, will allow it to cross the blood-brain barrier and affect serotonin levels. Such foods that contain a higher proportion of tryptophan include chickpeas and corn, and some studies have looked at their beneficial effect on serotonin levels in the realm of improving cognition [68] and preventing pellagra [69], respectively. Whether or not higher tryptophan containing foods would have a meaningful effect on ejaculatory time and possibly be implicated in delayed/anejaculation remains to be seen.

Other Nutrients

Besides modulation through the serotonergic pathway, a few other dietary components have been explored as potential moderators of PE that don't necessarily take advantage of the serotonergic pathway. Biotin and zinc have been shown to be important for normal sexual functioning and a healthy ejaculatory reflex, respectively, and their incorporation into a supplement for men with PE was shown to improve ejaculatory control and increase IELT [70]. Magnesium has also been investigated, and men with PE had decreased magnesium levels in their semen as compared with controls [71]. The authors theorized that hypomagnesemia resulting from either dietary deficiency or excessive alcohol intake could account for symptoms of PE in men. Finally, caffeine has recently been proposed as a possible treatment for premature ejaculation. In a double-blinded, placebo-controlled study, men with PE who received caffeine 2 h before sexual intercourse had improved sexual satisfaction and prolonged IELT scores [72]. However, the mechanisms behind caffeine's therapeutic effect on men with PE have yet to be elucidated, and more work needs to be done on investigating the practicality of this treatment to see if coffee or energy drinks can be substituted for caffeine pills.

Delayed/Anejeculation

There is less published research relating diet or dietary intervention to delayed ejaculation or anejaculation. The earlier referenced study on men in the Amazon noted a low prevalence of DE as part of their outcomes measured [37], but no other studies examine similar types of diets on DE in particular. As mentioned previously, these disorders along with PE lie on two different ends on the same continuum. It is therefore reasonable to assume that the same mechanisms are at play in both but in reverse. For example, serotonin activity may be upregulated as opposed to downregulated in DE, prolonging ejaculation time and causing sexual dysfunction. It stands to reason that some of the treatments proposed for PE might actually cause DE if taken in excess (e.g., excessive consumption of folic acid-containing foods or caffeinated drinks). However, this idea has not been systematically studied. Delayed ejaculation can also be a symptom of low testosterone and hypogonadism in general [73], and a diet that affects sex hormones can also play a causal role. Alcohol use may be a precipitating or even causative factor for DE, presumably due to its general inhibitory effects. Indeed, there is evidence that alcohol has an acute ejaculatory delaying effect, and some young men report using alcohol with the intention of delaying time to ejaculation during sexual encounters [74]. Additionally, in studies investigating sexual dysfunction, there is generally a higher amount of alcohol consumption in males that have DE and lower rates of alcohol use in males with PE. Taken together, this suggests that the inhibitory action of alcohol has similar action on the ejaculatory cycle as serotonin, albeit possibly through a different mechanism.

DIET AND PEYRONIE'S DISEASE

Vitamin E

There is a paucity of literature investigating a link between diet and the risk of developing Peyronie's disease. While there are numerous nonpharmacological oral therapies that exist for men with PD, the American Urological Association (AUA) currently recommends against the use of this type of therapy for PD management. One such oral therapy is vitamin E, and there is a lack of consensus on its efficacy for improving outcomes like penile curvature, penile pain, or plaque size [75] in these patients. It seems that vitamin E administration alone does not confer much benefit but may have some use when combined with other antioxidant and antiinflammatory therapies [76].

Blueberries

If the pathophysiology behind PD involves oxidative stress and the therapies serve to reduce that issue, it stands to reason that foods with antioxidant properties should be protective against the development of PD or at least in treating it. Blueberries contain flavonoids, anthocyanins, and polyphenols and contain one of the highest concentrations of antioxidants in terms of fruits and vegetables [77]. The anthocyanins in blueberries have also been shown to have antifibrotic effects and inhibit some of the inflammatory mediators (NF-kB, iNOS, and COX2) involved with PD [78]. Certain studies have added blueberries to their treatment modality due to these properties and systematically studied their effect on improving PD outcomes. In studies that include blueberries as part of their treatment regimens, plaque size is reduced, and degree of curvature lessens [79]. Like with vitamin E, the combination of blueberries with other antioxidants seems to produce the largest effect than any one component alone.

Diet and Peyronie's Disease 13

Other Foods

Although not necessarily studied, other foods with a high antioxidant component should also serve a role for patients with PD. Citrus fruits that are high in ascorbic acid or soybeans that contain anthocyanin are foods that contain these chemicals. Diets that incorporate these and other antioxidant-rich foods (e.g., Mediterranean and vegetarian diets) could be protective for PD and need to be studied in this patient population.

DIET AND HYPOGONADISM
Obesity and Hypogonadism
Hypogonadism in men is classified as biochemically low testosterone levels in the setting of a cluster of symptoms, which may include reduced sexual desire (libido) and activity, decreased spontaneous erections, decreased energy, and depressed mood [80]. Obesity is a known risk factor for developing hypogonadism and is associated with a decrease in both total testosterone and sex hormone binding globulin (SHBG) [81]. In fact, having a BMI > 25 is listed as a risk factor for developing hypogonadism in certain reviews [82]. Hypogonadism itself is also a risk factor for obesity, and it seems that the pathophysiological mechanisms behind these conditions exacerbate one another [83]. This concept can be easily understood when one considers the way in which a buildup of adipose tissue affects body function. For one, adipose tissue is pro-inflammatory and contributes to the production of cytokines like TNF-alpha, IL-6, and CRP [84]. This increase in inflammation from adipose tissue upregulates the hypothalamic-pituitary-adrenal (HPA) axis [85] and downregulates the hypothalamic-pituitary-gonadal (HPG) axis, leading to a generalized hypogonadal state. As mentioned earlier, adipose tissue also contains the enzyme aromatase, which converts testosterone to estradiol. Studies have shown that glucocorticoids induce the aromatase enzyme in adipose tissue [86], causing an increase in the conversion of testosterone to estradiol. Release of glucocorticoids due to the HPA axis upregulation from inflammation may further make an obese man hypogonadal and impair his sexual functioning. Dietary interventions that result in weight loss have been shown to decrease inflammatory mediators such as IL-6 and CRP [87], presumably through decreasing central adiposity. This reduction in inflammation is paralleled by improvements in sexual functioning as well [87].

Diets that predispose toward developing obesity such as the Western Diet [88–90] should be a risk factor for developing hypogonadism. However, the data seem to suggest that it is not necessarily the Western Diet itself but rather the obesity that can result from it that determines hypogonadal development. An early study comparing a Western diet and vegetarian diet among Black men found no significant differences in sexual function or total testosterone levels based on diet [91]. This could be because all the men in the study were healthy and had a normal BMI. Moreover, obese men that are given a low-calorie, low-carbohydrate diet lose weight and see an associated increase in testosterone levels and sexual function [92]. Healthy men, however, when placed on a similar low-carbohydrate and high-fiber diet experience a decrease in testosterone levels in the short term [93]. Furthermore, diets that have a high carbohydrate-to-protein ratio are reported to increase serum testosterone in certain men [94, 95]. Refined carbohydrates are a central component of the

Western diet. A study in obese bariatric surgery patients found that 16 weeks of a low-saturated-fat, high-monounsaturated-fat, and fiber diet did not improve IIEF-5 scores or testosterone values. Instead, these patients only saw a significant increase in these parameters after bariatric surgery [50].

Weight Loss and Hypogonadism

Many studies looking at the effect of weight loss on sex hormones have found mixed results. Numerous studies report an increase in SHBG with either no change or a decrease in total testosterone levels when participants lost weight [57, 96]. This is especially true in healthy nonobese individuals. However, many of these studies only have short-term follow-up. There could also be a discrepancy in the latency time for some sex hormones to be affected by weight loss. One study found that SHBG rose significantly during the calorie-restriction phase but leveled out during the weight-maintenance phase [57]. However, the increases in total testosterone were maximal during the weight-maintenance phase, and free testosterone only increased during that phase as well. This could be explained by the idea that the body needs time to adjust after a drastic decrease in dietary intake is detected. When it first detects this change, it interprets the subsequent weight loss as a stressful stimulus and upregulates the HPA axis. The subsequent release of glucocorticoids exerts a negative effect on sex hormone production, potentially leaving some men hypogonadal. Then, after the body adjusts and reaches a new homeostasis, the effects of a reduction in adipose tissue and its associated aromatizing properties are manifested by increases in total and free testosterone. This effect may be minimal or irrelevant in obese individuals due to the fact that they have increased energy reserves from fat. Their body may not respond and have the same stress response that a nonobese male has, and as a result, they may see an earlier increase in both total and free testosterone.

Diets and Dietary Patterns

There is not much research that explores established dietary patterns or specific foods related to hypogonadism. Vitamin D has been linked to hypogonadism in a few studies [97, 98], and vitamin D deficiencies were associated with lower testosterone levels and hip fractures in elderly men [98]. It may be inferred that diets with vitamin D-rich foods (e.g., fatty fish and fortified cereals) such as the Mediterranean would protect against becoming hypogonadal and this should be a topic for future study. There are a couple of isolated case reports of hypogonadism being induced from unusual dietary habits. One case of hypogonadism occurred in a young healthy male from overconsumption of soy products [99]. He presented with ED and a loss of libido and had recently started a new vegan diet with mainly soy products. The amount of soy he was consuming daily contained close to 360 mg of isoflavones (normal amount for a Western diet = 2 mg/day), and these hypogonadal effects were reversed after cessation of soy consumption. Another report describes a young male athlete who had also started a new diet that was rich in carotenoids (carrots, broccoli, lettuce, tomatoes, cabbage, and garlic). The authors of this report classified it as a case of hypothalamic hypogonadism due to his associated low levels of LH and FSH. His testosterone levels and sexual functioning also returned to normal after cessation of his diet [100].

FOODS THAT AFFECT MALE HORMONES

Much of the chapter has focused on dietary patterns and not necessarily specific foods when considering diet and sexual health. However, several individual foods have been referenced in the literature as having a direct influence on the hormonal milieu in an adult male (Table 1). Usually, these foods affect testosterone because they contain unique compounds that act to either alter production in some way or by altering testosterone metabolism. Many of these foods also have an effect on SHBG and consequently free testosterone as well.

Androgen Suppressing

Flax Seed

Flax seed, a food that has been renowned due to its high omega-3 fatty acid content, has been shown to decrease total testosterone levels in men [101, 102]. This is primarily due to the high concentration of lignans in flax seed, a compound that is suggested to bind testosterone in the enterohepatic circulation and lead to its excretion [102]. Lignans have also been cited to increase SHBG in some studies leading to a consequent decrease in free testosterone [103].

Table 1 Foods Affecting Hormones

Food	Active Compounds	Effect on Hormones
Flaxseed	Lignans	↓TT, ↑SHBG, ↓FT
Licorice	Glycyrrhizic acid	↓TT, (↓FT)
Soy	Isoflavones, equol	↓TT, ↑DHEA
Ginger	Gingerol	↑TT, ↑LH, ↑FSH
Coffee	Caffeine?	↑TT, ↑SHBG
Pomegranates	Polyphenols	↑Salivary testosterone

DHEA, *dehydroepiandrosterone;* FSH, *follicle-stimulating hormone;* FT, *free testosterone;* LH, *luteinizing hormone;* SHBG, *sex hormone binding globulin;* TT, *total testosterone.*

Current research has focused on using flax seed as a potential treatment for polycystic ovarian syndrome in women due to its androgen-suppressing effects.

Licorice
Licorice contains a compound called glycyrrhizic acid, which inhibits 17β-hydroxysteroid dehydrogenase, the enzyme responsible for converting androstenedione to testosterone, in vitro [104]. Licorice consumption in men led to a significant decrease in testosterone and a buildup of 17-hydroxyprogesterone, the precursor of androstenedione [105]. A follow-up study found similar results after just 1 week of licorice consumption and also a slight but nonsignificant decrease in free testosterone levels [106].

Soy
As previously discussed, ingestion of soy in extreme quantities can lead to a hypogonadal state through dysregulation of the HPG axis. This is due to the isoflavones contained in soy products, which are chemically and functionally similar to estradiol [107]. In the case report, decreases in total testosterone were noted along with an increase in dehydroepiandrosterone (DHEA). In the gut, the isoflavones are further broken down into a compound called equol [108]. Whereas isoflavones are mainly estrogenic, equol has a strong antiandrogenic property and will bind androgens such as dihydrotestosterone (DHT) preventing them from working at their target tissues [109].

Androgen Inducing
Ginger
Unlike the previous foods, ginger may play a positive role in androgenic regulation. It has antiinflammatory properties due to its main constituent, gingerol [110], which will enhance HPG axis activity. In a study of Iraqi men, supplementation with ginger was shown to increase testosterone by 17% from baseline [111]. Other components of the HPG axis like LH and FSH increased significantly as well.

Coffee
Many people drink coffee, but not many people know that regular coffee consumption can signal increased sex hormone levels. In a cross-sectional study, coffee was found to be positively associated with total testosterone levels and SHBG [112]. The authors theorized that the caffeine may affect hepatic metabolism leading to these results.

Pomegranates
Pomegranate juice has a number of reported benefits on cardiac health, including decreasing plaque size in atherosclerosis and reducing LDL cholesterol oxidation [113]. It has also been shown to increase levels of salivary testosterone by 24% after 1 week of consumption and reduce both systolic and diastolic blood pressure [114]. These purported effects are most likely due to the high concentration of polyphenolic compounds in pomegranates, which have strong antioxidant action. In test tube studies, pomegranate extract was shown to have antiestrogenic and antiaromatase activity as well [115].

FUTURE PERSPECTIVES

Childhood Diet

More research needs to be done on the way in which early in life dietary patterns affect later sexual health. Much of the research at this point is focused on how dietary interventions can ameliorate sexual dysfunction that has already occurred. However, if diet is to affect sexual health in the same way that it affects the development of comorbidities like CVD and diabetes, then earlier identification and implementation of healthy dietary habits is necessary to prevent dysfunction from ever occurring. In that way, diet can be seen as a type of preventative care and not just a treatment for problems later on down the road.

Organic Food

An interesting direction to look further at is how organic foods affect sexual health. Modern-day food culture places a huge emphasis on the importance of organically grown food for our health, which is evidenced by the prevalence of organic grocery stores like Whole Foods. Furthermore, pesticide exposure has been linked to ED [116] and hypogonadism [117], so it's possible that regular consumption of organic foods would decrease a man's risk of developing these two disorders.

Ketogenic Diet

The ketogenic diet is a relatively new concept that may start being studied given its growing popularity. It is a low-carbohydrate but high-fat diet that cuts out many of the components of the Mediterranean diet that have shown to be beneficial for sexual health and overall health. For example, the ketogenic diet excludes whole grains, legumes, cereals, and most fruits. It also does not differentiate between monounsaturated and saturated fats, and encourages consumption of red meat, butter, cream, and eggs. One would think that a diet with these foods would contribute to atherosclerotic plaque buildup, which is seen in ED associated with CVD. Since much of the current literature seems to support the Mediterranean diet for its favorable effect on erectile function, it will be interesting to see how the ketogenic diet affects men's sexual health given that it encourages consumption of apparently disease-propagating food groups.

TAKE HOME MESSAGES

- Diets such as the Mediterranean diet, which cut out processed/refined foods and incorporate high amounts of fruits, vegetables, and foods with

omega-3 fatty acids, have a positive effect on erectile function, most likely by improving cardiovascular health.
- Ejaculation is coordinated by specific pathways in the brain; overactivation of these pathways can lead to PE, while their inhibition can cause DE. Consumption of certain foods that take advantage of this circuitry such as coffee or alcohol may be able to influence ejaculatory speed and health.
- Obesity negatively affects sexual function by creating a generalized inflammatory state in the body and dysregulating the hormonal milieu. These effects may manifest as erectile dysfunction, delayed ejaculation, or hypogonadism.

References

[1] Domingo JL, Nadal M. Carcinogenicity of consumption of red meat and processed meat: a review of scientific news since the IARC decision. Food Chem Toxicol 2017;105:256–61.

[2] Willett WC. Overview and perspective in human nutrition. Asia Pac J Clin Nutr 2008;17.

[3] Srinath Reddy K, Katan MB. Diet, nutrition and the prevention of hypertension and cardiovascular diseases. Public Health Nutr 2004;7:167–86.

[4] Fliss-Isakov N, et al. Mediterranean dietary components are inversely associated with advanced colorectal polyps: a case-control study. World J Gastroenterol 2018;24(24):2617–27.

[5] Trichopoulou A, Costacou T, Bamia C, et al. Adherence to a Mediterranean diet and survival in a Greek population. N Engl J Med 2003;348:2599–608.

[6] Salas-Salvadó J, Bulló M, Babio N, et al. Reduction in the incidence of type 2 diabetes with the Mediterranean diet: results of the PREDIMED-Reus nutrition intervention randomized trial. Diabetes Care 2011;34:14–9.

[7] Estruch R, Ros E, Salas-Salvadó J, et al. Primary prevention of cardiovascular disease with a Mediterranean diet. N Engl J Med 2013;368:1279–90.

[8] Odermatt A. The Western-style diet: a major risk factor for impaired kidney function and chronic kidney disease. Am J Physiol Ren Physiol 2011;301:F919–31.

[9] Sherzai A, Heim LT, Boothby C, et al. Stroke, food groups, and dietary patterns: a systematic review. Nutr Rev 2012;70:423–35.

[10] Fabiani R, Minelli L, Bertarelli G, et al. A western dietary pattern increases prostate cancer risk: a systematic review and meta-analysis. Nutrients 2016;8.

[11] De Tejada IS, Angulo J, Cellek S, et al. Pathophysiology of erectile dysfunction. J Sex Med 2005;2:26–39.

[12] Althof SE, Leiblum SR, Chevret-Measson M, Hartmann U, Levine SB, McCabe M, et al. Psychological and interpersonal dimensions of sexual function and dysfunction. J Sex Med 2005;2:793–800.

[13] Morano S. Pathophysiology of diabetic sexual dysfunction. J Endocrinol Investig 2003;26(3):65–9.

[14] Billups KL. Erectile dysfunction as an early sign of cardiovascular disease. Int J Impot Res 2005;17:S19–24.

[15] Quilter M, Hodges L, von Hurst P, et al. Male sexual function in New Zealand: a population-based cross-sectional survey of the prevalence of erectile dysfunction in men aged 40–70 years. J Sex Med 2017;14:928–36.

[16] Atlantis E, Sullivan T. Bidirectional association between depression and sexual dysfunction: a systematic review and meta-analysis. J Sex Med 2012;9:1497–507.

[17] Corona G, Rastrelli G, Monami M, et al. Hypogonadism as a risk factor for cardiovascular mortality in men: a meta-analytic study. Eur J Endocrinol 2011;165(5):687–701.

[18] Ayta IA, McKinlay JB, Krane RJ. The likely worldwide increase in erectile dysfunction between 1995 and 2025 and some possible policy consequences. BJU Int 1999;84:50–6.

[19] Sullivan ME, Thompson CS, Dashwood MR, et al. Nitric oxide and penile erection: is erectile dysfunction another manifestation of vascular disease? Cardiovasc Res 1999;43:658–65.

[20] Azadzoi KM, Schulman RN, Aviram M, Siroky MB. Oxidative stress in arteriogenic erectile dysfunction, prophylactic role of antioxidants. J Urol 2005;174:386–93.

[21] De Young L, et al. Oxidative stress and antioxidant therapy: their impact in diabetes-associated erectile dysfunction. J Androl 2004;25(5):830–6.

[22] Shiri R, et al. Association between comorbidity and erectile dysfunction in patients with diabetes. Int J Impot Res 2005;18(4):348–53.

[23] Wang F, et al. Erectile dysfunction and fruit/vegetable consumption among diabetic Canadian men. Urology 2013;82(6):1330–5.

[24] Kim DM, Bruka L. Benefits from fruit/vegetable consumption on erectile dysfunction among diabetic Albanian men. Int J Sci Res 2016;5(4):1037–9.

[25] Esposito K, et al. Dietary factors in erectile dysfunction. Int J Impot Res 2006;18(4):370–4.

[26] Esposito K, et al. Effects of intensive lifestyle changes on erectile dysfunction in men. J Sex Med 2009;6(1):243–50.

[27] Chen Y, et al. Relationship among diet habit and lower urinary tract symptoms and sexual function in outpatient-based males with LUTS/BPH: a multiregional and cross-sectional study in China. BMJ Open 2016;6(8).

[28] Patterson E, et al. Health implications of high dietary omega-6 polyunsaturated fatty acids. J Nutr Metab 2012;2012.

[29] Allayee H, Roth N, Hodis HN. Polyunsaturated fatty acids and cardiovascular disease: implications for nutrigenetics. Lifestyle Genomics 2009;2(3):140–8.

[30] Calder PC. Polyunsaturated fatty acids, inflammatory processes and inflammatory bowel diseases. Mol Nutr Food Res 2008;52(8):885–97.

[31] Calder PC. n-3 Polyunsaturated fatty acids, inflammation, and inflammatory diseases. Am J Clin Nutr 2006;83(6):1505S–19S.

[32] Simopoulos AP. Evolutionary aspects of diet, the omega-6/omega-3 ratio and genetic variation: nutritional implications for chronic diseases. Biomed Pharmacother 2006;60(9):502–7.

[33] McAfee AJ, et al. Red meat from animals offered a grass diet increases plasma and platelet n-3 PUFA in healthy consumers. Br J Nutr 2011;105(1):80–9.

[34] Shim JS, et al. Effects of omega-3 fatty acids on erectile dysfunction in a rat model of atherosclerosis-induced chronic pelvic ischemia. J Korean Med Sci 2016;31(4):585–9.

[35] Ramírez R, et al. Erectile dysfunction and cardiovascular risk factors in a Mediterranean diet cohort. Intern Med J 2016;46(1):52–6.

[36] Giugliano F, et al. Adherence to Mediterranean diet and erectile dysfunction in men with type 2 diabetes. J Sex Med 2010;7(5):1911–7.

[37] Teixeira T, et al. Male sexual quality of life is maintained satisfactorily throughout life in the Amazon rainforest. Sex Med 2018;6(2):90–6.

[38] Trapp CB, Barnard ND. Usefulness of vegetarian and vegan diets for treating type 2 diabetes. Curr Diab Rep 2010;10(2):152–8.

[39] Frassetto LA, et al. Metabolic and physiologic improvements from consuming a paleolithic, hunter-gatherer type diet. Eur J Clin Nutr 2009;63(8):947.

[40] Jönsson T, et al. Beneficial effects of a Paleolithic diet on cardiovascular risk factors in type 2 diabetes: a randomized cross-over pilot study. Cardiovasc Diabetol 2009;8(1):35.

[41] Gibas MK, Gibas KJ. Induced and controlled dietary ketosis as a regulator of obesity and metabolic syndrome pathologies. Diabetes Metab Syndr Clin Res Rev 2017;11:S385–90.

[42] Kolotkin RL, et al. Sexual functioning and obesity: a review. Obesity 2012;20(12):2325–33.

[43] Diaz-Arjonilla M, et al. Obesity, low testosterone levels and erectile dysfunction. Int J Impot Res 2008;21(2):89–98.

[44] Paulos M, Miner M. The metabolic syndrome and ED. Male Sex Dysfunct 2016;109–19.

[45] Tamler R. Diabetes, obesity, and erectile dysfunction. Gend Med 2009;6:4–16.

[46] Feldman HA, et al. Erectile dysfunction and coronary risk factors: prospective results from the Massachusetts Male Aging Study. Prev Med 2000;30(4):328–38.

[47] Garimella PS, et al. Association between body size and composition and erectile dysfunction in older men: osteoporotic fractures in men study. J Am Geriatr Soc 2013;61(1):46–54.

[48] Wing RR, et al. Effects of weight loss intervention on erectile function in older men with type 2 diabetes in the look AHEAD trial. J Sex Med 2010;7(1):156–65.

[49] Khoo J, et al. Effects of a low-energy diet on sexual function and lower urinary tract symptoms in obese men. Int J Obes 2010;34(9):1396–403.

[50] Reis LO, et al. Erectile dysfunction and hormonal imbalance in morbidly obese male is reversed after gastric bypass surgery: a prospective randomized controlled trial. Int J Androl 2010;33(5):736–44.

[51] Chitaley K, et al. Diabetes, obesity and erectile dysfunction: field overview and research priorities. J Urol 2009;182(6):S45–50.

[52] Simpson ER, Mendelson CR. Effect of aging and obesity on aromatase activity of human adipose cells. Am J Clin Nutr 1987;45(1):290–5.

[53] Agarwal A, et al. Role of oxidative stress in the pathophysiological mechanism of erectile dysfunction. J Androl 2006;27(3):335–47.

[54] Rodríguez-Hernández H, et al. Obesity and inflammation: epidemiology, risk factors, and markers of inflammation. Int J Endocrinol 2013;2013.

[55] Maiorino MI, et al. Effects of Mediterranean diet on sexual function in people with newly diagnosed type 2 diabetes: the MÈDITA trial. J Diabetes Complicat 2016;30(8):1519–24.

[56] Esposito K, et al. Mediterranean diet improves erectile function in subjects with the metabolic syndrome. Int J Impot Res 2006;18(4):405.

[57] Moran LJ, et al. Long-term effects of a randomised controlled trial comparing high protein or high carbohydrate weight loss diets on testosterone, SHBG, erectile and urinary function in overweight and obese men. PLoS One 2016;11(9):e0161297.

[58] Khoo J, et al. Comparing effects of a low-energy diet and a high-protein low-fat diet on sexual and endothelial function, urinary tract symptoms, and inflammation in obese diabetic men. J Sex Med 2011;8(10):2868–75.

[59] McMahon CG, et al. An evidence-based definition of lifelong premature ejaculation: Report of the International Society for Sexual Medicine Ad Hoc Committee for the Definition of Premature Ejaculation. BJU Int 2008;102(3):338–50.

[60] Papaharitou S, et al. Erectile dysfunction and premature ejaculation are the most frequently self-reported sexual concerns: profiles of 9,536 men calling a helpline. Eur Urol 2006;49(3):557–63.

[61] Buvat J. Pathophysiology of premature ejaculation. J Sex Med 2011;8:316–27.

[62] Matuszcyk JV, Larsson K, Eriksson E. The selective serotonin reuptake inhibitor fluoxetine reduces sexual motivation in male rats. Pharmacol Biochem Behav 1998;60(2):527–32.

[63] Stahl SM. Mechanism of action of serotonin selective reuptake inhibitors: serotonin receptors and pathways mediate therapeutic effects and side effects. J Affect Disord 1998;51(3):215–35.

[64] Yin T-L, et al. Folic acid supplementation as adjunctive treatment premature ejaculation. Med Hypotheses 2011;76(3):414–6.

[65] Botez MI, et al. Folate deficiency and decreased brain 5-hydroxytryptamine synthesis in man and rat. Nature 1979;278(5700):182–3.

[66] Yan W-J, et al. A new potential risk factor in patients with erectile dysfunction and premature ejaculation: folate deficiency. Asian J Androl 2014;16(6):902.

[67] Young SN. How to increase serotonin in the human brain without drugs. J Psychiatry Neurosci 2007;32(6):394.

[68] Kerem Z, et al. Chickpea domestication in the Neolithic Levant through the nutritional perspective. J Archaeol Sci 2007;34(8):1289–93.

[69] Katz SH, Hediger ML, Valleroy LA. Traditional maize processing techniques in the New World. Science 1974;184(4138):765–73.

[70] Cai T, et al. Rhodiola rosea, folic acid, zinc and biotin (EndEP®) is able to improve ejaculatory control in patients affected by lifelong premature ejaculation: results from a phase I-II study. Exp Ther Med 2016;12(4):2083–7.

[71] Aloosh M, Hassani M, Nikoobakht M. Seminal plasma magnesium and premature ejaculation: a case-control study. BJU Int 2006;98(2):402–4.

[72] Saadat H, Seyyed KA, Panahi Y. The effect of on-demand caffeine consumption on treating patients with premature ejaculation: a double-blind randomized clinical trial. Curr Pharm Biotechnol 2015;16(3):281–7.

[73] Corona G, et al. Different testosterone levels are associated with ejaculatory dysfunction. J Sex Med 2008;5(8):1991–8.

[74] Santtila P, Kenneth Sandnabba N, Jern P. Prevalence and determinants of male sexual dysfunctions during first intercourse. J Sex Marital Ther 2009;35(2):86–105.

[75] Pryor JP, Farell CR. Controlled clinical trial of vitamin E in Peyronie's disease. Prog Reprod Biol Med 1983;9(9):41–5.

[76] Prieto Castro RM, et al. Combined treatment with vitamin E and colchicine in the early stages of Peyronie's disease. BJU Int 2003;91(6):522–4.

[77] Wu X, et al. Characterization of anthocyanins and proanthocyanidins in some cultivars of Ribes, Aronia, and Sambucus and their antioxidant capacity. J Agric Food Chem 2004;52(26):7846–56.

[78] Choi JH, et al. Anti-fibrotic effects of the anthocyanins isolated from the purple-fleshed sweet potato on hepatic fibrosis induced by dimethylnitrosamine administration in rats. Food Chem Toxicol 2010;48(11):3137–43.

[79] Paulis G, et al. Effectiveness of antioxidants (propolis, blueberry, vitamin E) associated with verapamil in the medical management of Peyronie's disease: a study of 151 cases. Int J Androl 2012;35(4):521–7.

[80] Mulhall JP, et al. Evaluation and management of testosterone deficiency: AUA guideline. J Urol 2018;200(2):423–32.

[81] Kapoor D, et al. Testosterone replacement therapy improves insulin resistance, glycaemic control, visceral adiposity and hypercholesterolaemia in hypogonadal men with type 2 diabetes. Eur J Endocrinol 2006;154(6):899–906.

[82] Zarotsky V, et al. Systematic literature review of the risk factors, comorbidities, and consequences of hypogonadism in men. Andrology 2014;2(6):819–34.

[83] Cohen PG. The hypogonadal–obesity cycle: role of aromatase in modulating the testosterone–estradiol shunt–a major factor in the genesis of morbid obesity. Med Hypotheses 1999;52(1):49–51.

[84] Park HS, Park JY, Rina Y. Relationship of obesity and visceral adiposity with serum concentrations of CRP, TNF-α and IL-6. Diabetes Res Clin Pract 2005;69(1):29–35.

[85] Silverman MN, Sternberg EM. Glucocorticoid regulation of inflammation and its functional correlates: from HPA axis to glucocorticoid receptor dysfunction. Ann N Y Acad Sci 2012;1261(1):55–63.

[86] Zhao Y, Mendelson CR, Simpson ER. Characterization of the sequences of the human CYP19 (aromatase) gene that mediate regulation by glucocorticoids in adipose stromal cells and fetal hepatocytes. Mol Endocrinol 1995;9(3):340–9.

[87] Esposito K, et al. Effect of lifestyle changes on erectile dysfunction in obese men: a randomized controlled trial. JAMA 2004;291(24):2978–84.

[88] Astrup A, et al. Obesity as an adaptation to a high-fat diet: evidence from a cross-sectional study. Am J Clin Nutr 1994;59(2):350–5.

[89] Fung TT, et al. Association between dietary patterns and plasma biomarkers of obesity and cardiovascular disease risk. Am J Clin Nutr 2001;73(1):61–7.

[90] Manzel A, et al. Role of "Western diet" in inflammatory autoimmune diseases. Curr Allergy Asthma Rep 2014;14(1):404.

[91] Hill P, et al. Diet and urinary steroids in black and white North American men and black South African men. Cancer Res 1979;39(12):5101–5.

[92] Schulte DM, et al. Caloric restriction increases serum testosterone concentrations in obese male subjects by two distinct mechanisms. Horm Metab Res 2014;46(4):283–6.

[93] Hämäläinen E, et al. Diet and serum sex hormones in healthy men. J Steroid Biochem 1984;20(1):459–64.

[94] Hoffer LJ, et al. Effects of severe dietary restriction on male reproductive hormones. J Clin Endocrinol Metab 1986;62(2):288–92.

[95] Anderson KE, et al. Diet-hormone interactions: protein/carbohydrate ratio alters reciprocally the plasma levels of testosterone and cortisol and their respective binding globulins in man. Life Sci 1987;40(18):1761–8.

[96] Reed MJ, et al. Dietary lipids: an additional regulator of plasma levels of sex hormone binding globulin. J Clin Endocrinol Metab 1987;64(5):1083–5.

[97] Lee DM, et al. Association of hypogonadism with vitamin D status: the European Male Ageing Study. Eur J Endocrinol 2012;166(1):77–85.

[98] Diamond T, et al. Hip fracture in elderly men: the importance of subclinical vitamin D deficiency and hypogonadism. Med J Aust 1998;169(3):138–41.

[99] Siepmann T, et al. Hypogonadism and erectile dysfunction associated with soy product consumption. Nutrition 2011;27(7–8):859–62.

[100] Adamopoulos D, et al. Association of carotene rich diet with hypogonadism in a male athlete. Asian J Androl 2006;8(4):488–92.

[101] Demark-Wahnefried W, et al. Pilot study of dietary fat restriction and flaxseed supplementation in men with prostate cancer before surgery: exploring the effects on hormonal levels, prostate-specific antigen, and histopathologic features. Urology 2001;58(1):47–52.

[102] Adlercreutz H, et al. Effect of dietary components, including lignans and phytoestrogens, on enterohepatic circulation and liver metabolism of estrogens and on sex hormone binding globulin (SHBG). J Steroid Biochem 1987;27(4–6):1135–44.

[103] Martin ME, et al. Interactions between phytoestrogens and human sex steroid binding protein. Life Sci 1995;58(5):429–36.

[104] Sakamoto K, Wakabayashi K. Inhibitory effect of glycyrrhetinic acid on testosterone production in rat gonads. Endocrinol Jpn 1988;35(2):333–42.

[105] Armanini D, Bonanni G, Palermo M. Reduction of serum testosterone in men by licorice. N Engl J Med 1999;341(15):1158.

[106] Armanini D, et al. Licorice consumption and serum testosterone in healthy man. Exp Clin Endocrinol Diabetes 2003;111(06):341–3.

[107] Kurzer MS, Xia X. Dietary phytoestrogens. Annu Rev Nutr 1997;17(1):353–81.

[108] Setchell KDR, Brown NM, Lydeking-Olsen E. The clinical importance of the metabolite equol—a clue to the effectiveness of soy and its isoflavones. J Nutr 2002;132(12):3577–84.

[109] Lund TD, et al. Equol is a novel anti-androgen that inhibits prostate growth and hormone feedback. Biol Reprod 2004;70(4):1188–95.

[110] Grzanna R, Lindmark L, Frondoza CG. Ginger—an herbal medicinal product with broad anti-inflammatory actions. J Med Food 2005;8(2):125–32.

[111] Mares WAA-K, Najam WS. The effect of ginger on semen parameters and serum FSH, LH & testosterone of infertile men. Med J Tikrit 2012;18(182):322–9.

[112] Svartberg J, et al. The associations of age, lifestyle factors and chronic disease with testosterone in men: the Tromso study. Eur J Endocrinol 2003;149(2):145–52.

[113] Zarfeshany A, Asgary S, Javanmard SH. Potent health effects of pomegranate. Adv Biomed Res 2014;3.

[114] Al-Dujaili E, Smail N. Pomegranate juice intake enhances salivary testosterone levels and improves mood and well being in healthy men and women. In: Society for Endocrinology BES 2012., vol. 28. BioScientifica; 2012.

[115] Adams LS, et al. Pomegranate ellagitannin-derived compounds exhibit antiproliferative and antiaromatase activity in breast cancer cells in vitro. Cancer Prev Res 2010;3(1):108–13.

[116] Oliva A, Giami A, Multigner L. Environmental agents and erectile dysfunction: a study in a consulting population. J Androl 2002;23(4):546–50.

[117] Roychoudhury S, Bhattacharjee R. Environmental issues resulting in andropause and hypogonadism. In: Bioenvironmental issues affecting men's reproductive and sexual health; Elsevier-Academic Press; 2018, pp. 261–73.

CHAPTER 1.2

Effects of Lifestyle on Urinary Health

Joseph Mahon, Kevin T. McVary

Department of Urology, Stritch School of Medicine, Loyola University Medical Center, Maywood, IL, United States

INTRODUCTION

Diet and Urological Health

Benign prostatic hyperplasia (BPH) represents the most common neoplastic entity afflicting aging men. The clinical manifestation of BPH represents a spectrum; while many men will not experience any urinary symptoms, others experience significant symptomatic distress (referred to as lower urinary tract symptoms or LUTS) resulting in a significant impact to their quality of life. The interplay between BPH and LUTS has proved far more complex than simple obstruction of the prostatic urethra. The underlying pathophysiology of LUTS may include any combination of outlet obstruction, oxidative stress, acute and/or chronic inflammation, and sympathetic hyperactivity, all factors that may manifest in the prostate, bladder, spinal cord control of pelvic function, or even more central processing of signal from the male pelvis [1–3]. It is likely through these physiological changes that diet and lifestyle alterations may have the greatest effect on BPH/LUTS and general urological health. While these topics are discussed individually below, it is important to remember that the complexity of prostatic health remains to be completely understood. These topics represent individual twines in a large web of forces affecting prostate and bladder function, each influencing the others. While we may never fully understand the complete pathophysiology of BPH and LUTS, fundamental knowledge of the various forces at play will aid any clinician treating men afflicted with the disease.

   a. **Sources of nutrition and their effects on LUTS/BPH**
      i. **Protein sources**
         1. The role of dietary protein in the pathogenesis and progression of BPH and LUTS remains to be fully elucidated. What is clear is that intake of these macronutrients plays a complex role in male

urinary health. One of the earliest inquiries into this relationship was a 1983 examination of 100 Japanese men with BPH compared with 100 age-matched controls. Men with BPH were more likely to consume meat and dairy products regularly [4]. Daily meat consumption elicited a relative risk (RR) increase of 3.18, while regular milk consumption elicited a RR of 2.25. Suzuki et al., in an 8-year study of over 3500 men with BPH, found that high-total-protein diet was associated with BPH [5]. Animal-based protein exhibited a stronger association than cold-water fish-based and plant-based protein. Chen and colleagues noted a higher incidence of BPH in men with higher monthly meat, fish, and egg consumption across three regions of China [6]. A similar association was also described by Raheem and Parsons in their 2014 study [7]. However, in a better designed study, a 2008 cross-sectional study of over 1500 men in the Boston Area Community Health survey compared patient completed food frequency questionnaires with in-person interview data related to urinary symptoms. While total energy intake was associated with higher symptom scores, men who consumed more protein were less likely to report LUTS [8].

2. In a large case-control study by Bravi and colleagues, both eggs (odds ratios (OR) 1.43) and poultry (OR 1.39) exhibited a significant association with BPH [9, 10]. This was also supported by Lagiou and colleagues evaluating a Greek population, in which dairy products, meats, fish, and eggs were all associated with increased BPH risk [11]. A cohort of Japanese-American men associated increased beef intake with increased obstructive uropathy [12].

3. One of the possible roles that dietary proteins play in the maintenance of prostate health may be related to their contributions to the maintenance of various minerals and micro- and macronutrients. Zinc levels of prostatic tissue are higher than most other tissue in the human body, which has led to significant interest in its interplay with prostate health [13]. Whereas excessive zinc intake (>100 mg/day) has been suggested to increase the risk of prostate cancer, moderate zinc intake may serve a protective role in both prostate cancer risk and BPH [14]. In the Prostate Cancer Prevention Trial (PCPT), in which 4770 men were followed over a 7-year time period, a protective role of zinc with BPH was suggested [15]. Critics of this trial point out that the questionnaire utilized (Food Frequency Questionnaire) has not performed well in evaluating associations between diet and micro- and macromolecules with disease. While all meat products serve as

great sources of dietary zinc, red meat in particular is high in the mineral. One hundred grams of raw ground beef contains 4.8 mg of zinc, which represents roughly 40% of recommended daily intake. Other sources rich in dietary zinc include shellfish, legumes (e.g., chickpeas, lentils, and beans), nuts (e.g., peanuts, cashews, and almonds), dairy products, and eggs.

4. In a 2017 systematic review by Bradley and colleagues, in which a total of 28 publications met criteria for inclusion of diet and LUTS, the authors concluded that red meat is only weakly associated with BPH risk [16]. For general urinary symptoms, dietary protein intake may decrease risk in men, though storage symptoms in women may increase. In summary, dietary protein serves as a rich source of zinc, a mineral found in great quantities in the prostate and may influence prostatic health. Total dietary protein intake has been associated with a decreased incidence of LUTS; however, when individual protein sources have been examined, weak associations have been noted for red meat, eggs and poultry, and fish when compared with plant-based protein sources. As such, further clarity of the role of dietary proteins and urinary health is needed.

ii. **Fruits and vegetables**
1. Fruits and vegetables are significant sources of vitamins, minerals, fiber, antioxidants, and polyphenols. The benefits of these substances were shown in the PCPT where Kristal and colleagues showed that men who consumed at least four daily servings of vegetables exhibited a significantly lower risk of developing BPH compared with men who consumed one or less serving per day [15]. In 2007, Rohrmann and colleagues examined responders of the Health Professionals Follow-up Study looking at vegetable consumption in men either who reported having surgery for BPH or who exhibited an International Prostate Symptom Score (IPSS) of 15–35 ($n=6092$), as compared with those who had not previously undergone surgery and had never scored ≥ 7 on the IPSS. Men with IPSS scores of 8–14 were excluded from analysis [17]. They found that vegetable intake was inversely associated with BPH (OR 0.89; 95% CI: 0.80 and 0.99; $P=.03$), though no significant association was revealed for fruits. Further analysis did reveal that fruits and vegetables rich in β-carotene, lutein, or vitamin C all had significant inverse associations with BPH. Interestingly, those whose diets were rich in β-carotene and/or lutein were also less likely to undergo surgery for BPH; though, this trend was not observed for vitamin C. The identified foods rich in β-carotene, lutein, or both included cooked spinach, raw spinach, brussels sprouts, peas, and peaches. Looking at overall

BPH incidence (regardless of symptoms), they found fruits and vegetables of the botanical groups Rutaceae, legumes, cruciferous vegetables, and "other vegetables" were all associated with a lower OR for incidental BPH. Being a large population-based study, the study group carried significant confounders. Indeed, as the authors identified, those men who consumed larger amounts of fruits and vegetables tended to be older, were less likely to be active smokers, consumed less alcohol, were more physically active, had lower body mass indexes, and were more likely to utilize supplementation. Liu and colleagues examined the role of fruits and vegetables in the maintenance of prostate health. In a 4-year prospective study in Hong Kong, high intake of dark and leafy vegetables significantly reduced the risk of LUTS progression by 37.2% and symptomatic BPH by 34.3% after 4 years compared with those with moderate intake [18].

2. With the establishment of oxidative stress as one of the driving forces of BPH pathophysiology, much attention has been given to the role of dietary antioxidants on urinary health. Attempts to identify specific sources of fruit-/vegetable-derived antioxidants and polyphenols have shown mixed results. In a well-constructed randomized constructed trial by Spettel and colleagues evaluating the role of grape juice (rich in polyphenol components) in men with LUTS, no significant changes in symptom scores were appreciated between the grape juice and placebo arms [19]. Cranberry represents a natural source of phenolic acids, flavonoids, and other micro- and macronutrients believed to be of benefit to prostate health. In a 2010 study by Vidlar and colleagues, the use of dried cranberries in men with LUTS was shown to improve LUTS symptom scores, quality-of-life responses, urine flow rate, postvoid residual (PVR) volume, and serum prostate-specific antigen (PSA) levels [20].

3. Tomatoes, rich in lycopene (a natural source of dietary antioxidants), have gained particular interest for men with BPH. In a randomized, double-blind, placebo-controlled trial by Schwarz and colleagues, men consuming 15 mg/day of lycopene for 6 months were found to have both a decreased incidence of BPH progression and symptomatic improvement of LUTS [21]. Additionally, it has been hypothesized that lycopene may play a significant role in influencing cell growth and differentiation, key processes in the natural history of BPH [22].

4. With knowledge of the beneficial role of garlic and onion in cardiovascular disease, effects of aging, and their antiproliferative action on human cancers, Galeone and colleagues utilized the multicenter case-control series in Italy to assess any association

intake of these vegetables may have on BPH [23]. Compared with nonusers, the highest intake group of onion and garlic produced ORs of 0.41 (95% CI: 0.24–0.72) and 0.72 (95% CI: 0.57–0.91), respectively. They postulate that the phytochemical components and their antioxidant effect may be responsible for their benefit, given the known role of oxidative stress in BPH/LUTS.

5. Additionally, the nutritional benefits of fruits and vegetables may demonstrate a beneficial effect on BPH risk factors—specifically, Barnard and Aronson found a protective shift in testosterone to estradiol ratios and insulin sensitivity (known risk factors for BPH) in men who consumed a low-fat, high-fiber diet consisting of abundant fruits, vegetables, and whole grains [24]. These effects may, at least in part, be attributed to plant-based estrogens such as isoflavones and flaxseed. These components contain lignans that have been shown to inhibit 5-α-reductase, thereby inhibiting prostatic testosterone to dihydrotestosterone conversion. Additionally, their role as sources of dietary fiber has also shown potential in curbing the development/progression of BPH [25].

6. In summary, increased intake of fruits and vegetables has been shown to positively influence prostate and urinary health. As rich sources of antioxidants, such as lycopene, β-carotene, and lutein, fruits and vegetables are able to reduce oxidative stress on the prostate. Additionally, their intake has been shown to augment the hormonal milieu of the prostate to provide a protective role. For these reasons and more, dietary intake of fruits and vegetables aids in the maintenance of male urinary health.

iii. **Dairy**

1. The relationship of dairy intake and BPH has been met with mixed results. As mentioned earlier, Araki and colleagues found that regular milk consumption held a relative risk (RR) of 2.25 for symptomatic BPH among Japanese study participants [4]. In a Greek population-based study, Lagiou and colleagues noted a weak association between dairy products and BPH that did not meet statistical significance [11]. Meanwhile, the PCPT trial noted no association with increasing daily dairy consumption [15]. In a recent case-control series by Chen and colleagues, 4208 men across China were examined, of which 2584 (61.4%) suffered from LUTS/BPH [6]. In the evaluation of dietary milk and dairy intake, they too found a weak but positive association with IPSS scores.

2. The role of dairy product intake in the maintenance of prostatic health remains unclear. While small studies have noted a detrimental effect, larger studies failed to identify any association.

As such, further inquiry is required before any conclusions may be drawn.

    iv. **Grains/starch**
1. In a large case-control study of 2820 men under the age of 75 taking place in Italy from 1991 to 2002, the intake of 78 food and beverage items was evaluated for association with BPH [9, 10]. On multivariate analysis, several food groups were identified to exhibit elevated OR for BPH. Significant trends of increasing risk with increased consumption of bread (OR 1.69 for highest quantile vs lowest quantile), cereals (OR 1.55), eggs (OR 1.43), and poultry (OR 1.39) were observed. Additionally, a nonlinear relation was observed for pastas and rice. Starch intake itself was associated with an OR 1.51. They identified that the refined cereals and bread commonly consumed in Italy represent a high glycemic load. They then postulated that this glycemic load leads to increased circulating insulin and release of insulin-like growth factors, which may lead to worsening of clinical BPH [26, 27].
2. This positive relationship between cereal consumption and BPH, however, was not observed in a cross-sectional study by Lagiou and colleagues of 184 men with histologically confirmed BPH compared with 246 controls with no clinical evidence of BPH in Greece. In their study, cereal and starchy roots exhibited no increased risk of BPH on logistic regression. This may reflect the different glycemic loads of various bread products common to different regions/countries.
3. The role of grains/starch in the maintenance of urinary health remains unclear. The available studies have suggested both detrimental and beneficial roles. This contradiction may be more a reflection of the glycemic load their intake provides. Given the significant effect on insulin resistance and increased insulin-like growth factor exhibit on prostatic health, it is not difficult to conclude that the high glycemic load of refined grains and starch may be detrimental to prostatic health, while natural grains and starch rich in fiber may prove beneficial. However, until sufficient evidence is available, no definitive role of grains and starch can be identified.

b. **Caloric load/obesity and the effects on LUTS/BPH**
    i. When examining the literature for dietary caloric load and obesity's relationship with BPH, we see an alarming trend. While evaluation of caloric intake alone fails to predict the incidence of BPH, evaluation of the derivation of those calories yields clearer relationships [28]. High caloric diets with significant fat content are associated with increased prevalence of BPH [15]. Previously, a high intake of

polyunsaturated fatty acids (PUFAs) was observed to be associated with a higher risk of BPH in the Health Professionals Follow-Up Study [5]. In fact, a dose-response effect of +4.5% per 100 kcal of total fat intake can be seen for the incidence of clinical BPH. As previously discussed, the role of caloric intake on insulin sensitivity and circulating insulin-like growth factor exhibit several effects on urinary and prostatic health. Furthermore, the role of insulin sensitivity has been at the forefront of understanding the role of the metabolic syndrome (MetS). With these striking results, much attention has been turned to the understanding of the MetS as it continues to increase in prevalence at an alarming rate among present-day populations.

ii. MetS represents a constellation of abnormalities in one's metabolic health, which has been increasing in prevalence within the United States and worldwide [29]. This increased risk of MetS represents a significant epidemic for the health care of most countries. The abnormalities most commonly cited include abdominal obesity, hypertension, insulin resistance, and dyslipidemia. Obesity and the constellation that comprises MetS have been found to be detrimental throughout a number of body systems, and recent research reveals that prostatic health is no exception. MetS has also been associated with hyperactivity of the autonomic nervous system, a connection that may serve as the foundation for its association with urinary health [30, 31]. McVary and colleagues used a rat model to demonstrate an association between autonomic activity and prostatic growth [32]. When this signaling was dampened, regression of the prostate was observed. Epidemiological studies reflected this relationship in humans, with elevated autonomic tone being associated with a higher incidence of clinical BPH, finding significantly larger prostate volumes with faster growth rates in men who met criteria for MetS [33, 34]. This relationship between the autonomic nervous system and BPH/LUTS was best detailed in a 2005 single-center substudy to the medical therapy of prostatic symptoms (MTOPS) trial by McVary and colleagues [3]. In this prospective study, assessment of the functional state of the autonomic nervous system in BPH patients with significant untreated LUTS, they postulated that increased sympathetic activity may serve a causal role in both development and progression of symptomatic BPH as measured by IPSS scores, quality-of-life measures, and BPH impact scores. This autonomic hyperactivity likely affects detrusor instability as well. With both the outlet and the detrusor under the influence of autonomic hyperactivity, it is no wonder that BPH/LUTS shows such a strong association with MetS.

iii. To summarize, caloric load and obesity play a paramount role in the maintenance of urinary and prostatic health. Both caloric load and obesity have been associated with progression of BPH/LUTS. The role of MetS may prove to be at the heart of this association. With increased sympathetic activity, decreased insulin sensitivity, and increased circulating insulin-like growth factors, MetS proves detrimental to urinary and prostatic health from several fronts. Knowing the alarming population trends, the role of dietary intake in combating the development of MetS deserves significant attention (Fig. 1).

c. **Inflammation and the effects on LUTS/BPH**
   i. The role of prostatic inflammation in BPH was first suggested by Nickel in 1994, when he suggested that inflammation represented the third leg in the etiology and pathogenesis of BPH (along with prostatic enlargement and smooth muscle tension) leading to prostatic enlargement and LUTS [35]. Through the results of several histological studies on BPH, the presence of infiltrative lymphocytes and macrophages supports this hypothesis that there is a role for chronic inflammation in the pathogenesis of the disease. Indeed, baseline prostate biopsies of over 1000 men in the MTOPS study predicted not only LUTS symptom progression but also risk for acute urinary retention and future need for surgery [36]. Additional studies have further correlated the presence of histological inflammation with urinary/prostatic health—indicating a higher prevalence in men with acute urinary retention, increased risk for medical treatment failure and risk for BPH-related surgery [37–39].
   ii. In a 2008 study, Nickel and colleagues utilized baseline prostate biopsy samples from the REduction by DUtasteride of prostate Cancer Events (REDUCE) trial to assess the association of histological

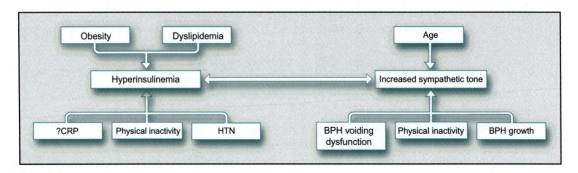

FIG. 1
The interplay of MetS and urinary/prostatic health *From Kasturi S, Russell S, McVary KT. Metabolic syndrome and lower urinary tract symptoms secondary to benign prostatic hyperplasia. Curr Urol Rep 2006;7:288–92.*

evidence of inflammation with LUTS presence and severity via IPSS scores [40, 41]. Average acute and chronic inflammation was rated on a four-point scale (0 = none, 1 = mild, 2 = moderate, and 3 = severe), then these findings were correlated with total IPSS scores and grouped IPSS scores for subcategories of irritative symptoms, obstructive symptoms, and nocturia. Of the 8224 men evaluated, 15.4% had evidence of acute inflammation, 77.6% had evidence of chronic inflammation, but 21.6% had no evidence of inflammation. Total IPSS scores and subscores were higher in the group with evidence of chronic inflammation compared with those with no evidence of chronic inflammation. A weak relationship was noted between the severity of chronic inflammation and LUTS symptom scores; however, this may be limited by the exclusion of men with clinical prostatitis or severe LUTS from the original REDUCE trial recruitment. A similar relationship was also appreciated in a subsequent trial by Descazeaud and colleagues [42]. In a 2016 follow-up longitudinal evaluation of REDUCE, Nickel and colleagues confirmed that chronic inflammation is associated with higher IPSS scores, larger prostatic volume, and higher risk of acute urinary retention [43].

iii. C-reactive protein (CRP) has utility as a marker for systemic inflammation. Elevations in CRP are predictive of development of noninsulin-dependent diabetes mellitus and cardiovascular disease—two of the criteria for diagnosis of MetS. Elevations in CRP have been shown to be associated with increasing severity of LUTS [44]. Indeed, multiple population-based studies have linked elevations in CRP with various LUTS symptoms. In an evaluation of the third National Health and Nutrition Examination Survey (NHANES III) database, men with elevated CRP were more likely to report three or four urinary symptoms (nocturia, incomplete emptying, hesitancy, and weak stream) than men with undetectable CRP levels. Similarly, data from the Olmsted County study reported an association between CRP levels and progression of irritative (frequency, urgency, and nocturia) LUTS in men; though, no association was appreciated with overall symptom scores or obstructive (incomplete emptying, intermittency, weak stream, and straining) symptoms [45]. In a 2009 study using the Boston Area Community Health Survey (BACH), Kupelian and colleagues further investigated the association of CRP levels and LUTS [46]. The BACH survey is a population-based epidemiological survey of urological symptoms and risk factors of 5502 individuals (2301 men and 3201 women). During an in-person interview, subjects completed the American Urological Association Symptom Index (AUA-SI, aka IPSS) (IPSS) among other health interview topics, and venous blood samples were collected. The AUA-SI consists of seven question prompts to which the patient selects a quantitative score (0–5) corresponding to the frequency with which they experienced the symptom designated in the prompt. Symptom score totals 1–7 represent mild, 8–19 moderate, and 20–35 severe LUTS. An additional prompt measures the degree of intrusion or disruption the LUTS have on the individual and is referred to as its degree of bother or effect on quality of life (QOL). When this is included, the AUA-SI is then referred to as the International Prostate Symptom Score (IPSS). Among the male subjects, there was a statistically significant association between overall IPSS scores and CRP levels OR 1.21 (95% CI: 1.04 and 1.39). After multivariate analysis, CRP levels were associated with straining (OR 1.71; 95% CI: 1.22 and 2.39) and nocturia (OR 1.36; 95% CI: 1.15 and 1.59) (Fig. 2).

iv. The association of chronic prostatic inflammation with urinary and prostatic health cannot be denied; histological evidence has correlated with the serum marker CRP. How inflammation fits into the complex mesh work that is BPH/LUTS remains to be elucidated. In the web of prostatic health, inflammation likely influences and is influenced by factors in the pathogenesis of BPH/LUTS. With the knowledge we have gained concerning the roles of diet, oxidative stress, obesity, and MetS, the interplay of prostatic inflammation with these other factors deserves investigation.

SUMMARY

Epidemiological and cross-sectional observation studies of BPH/LUTS have proliferated in the past 15 years, leading to the accumulation of data regarding the prevalence and risk factors for LUTS. There are no population-based studies of histological BPH or urodynamic variables, however, and the current estimates are based on symptoms, usually recorded by means of the AUA symptom index or comparable measures. Despite this limitation, evidence has emerged of the role of BPH/LUTS as a marker or harbinger of men's health. This chapter reviews the growing evidence of association between BPH/LUTS and diet and between BPH/LUTS and other indicators of systemic illness or inflammation. The influence of dietary intake on systemic health cannot be denied. Through evaluation of specific caloric sources, we have identified a significant association between the intakes of certain foods; however, most remain to be clearly defined. Similarly, the role of caloric intake, obesity, metabolic syndrome, and prostatic inflammation clearly hold heavy influence in the maintenance of urinary and prostatic health. Further research is needed to elucidate the underlying mechanisms of these associations, though the

Dependent variable	CRP	Men Unadjusted OR (95%CI)	Adjusted* OR (95%CI)
AUA-SI ≥8	log(CRP)	**1.38 (1.20, 1.57)**	**1.21 (1.04, 1.39)**
	<1 mg/L	1.00	1.00
	1-3 mg/L	1.31 (0.85, 2.00)	1.21 (0.76, 1.92)
	>3 mg/L	**2.05 (1.33, 3.16)**	1.43 (0.91, 2.24)
Incomplete emptying	log(CRP)	**1.50 (1.04, 2.17)**	1.30 (0.91, 1.86)
	<1 mg/L	1.00	1.00
	1-3 mg/L	1.89 (0.91, 3.94)	1.56 (0.79, 3.07)
	>3 mg/L	**3.27 (1.06, 10.11)**	2.20 (0.78, 6.17)
Intermittency	log(CRP)	1.11 (0.89, 1.39)	0.93 (0.74, 1.17)
	<1 mg/L	1.00	1.00
	1-3 mg/L	1.25 (0.56, 2.77)	1.07 (0.49, 2.35)
	>3 mg/L	0.72 (0.30, 1.70)	0.42 (0.18, 1.00)
Weak stream	log(CRP)	1.09 (0.90, 1.33)	1.00 (0.82, 1.22)
	<1 mg/L	1.00	1.00
	1-3 mg/L	1.18 (0.52, 2.71)	1.13 (0.49, 2.58)
	>3 mg/L	1.06 (0.52, 2.14)	0.87 (0.42, 1.79)
Straining	log(CRP)	**1.78 (1.20, 2.64)**	**1.71 (1.22, 2.39)**
	<1 mg/L	1.00	1.00
	1-3 mg/L	**4.29 (1.13, 16.37)**	3.52 (0.88, 14.13)
	>3 mg/L	**4.55 (1.43, 14.47)**	3.36 (0.98, 11.45)
Frequency	log(CRP)	1.09 (0.93, 1.27)	1.04 (0.89, 1.22)
	<1 mg/L	1.00	1.00
	1-3 mg/L	1.09 (0.72, 1.66)	1.10 (0.73, 1.66)
	>3 mg/L	1.28 (0.80, 2.05)	1.18 (0.73, 1.90)
Urgency	log(CRP)	**1.44 (1.11, 1.87)**	1.21 (0.94, 1.56)
	<1 mg/L	1.00	1.00
	1-3 mg/L	0.85 (0.45, 1.54)	0.74 (0.38, 1.43)
	>3 mg/L	**2.52 (1.32, 4.80)**	1.62 (0.83, 3.18)
Nocturia	log(CRP)	**1.54 (1.33, 1.78)**	**1.36 (1.15, 1.59)**
	<1 mg/L	1.00	1.00
	1-3 mg/L	1.25 (0.87, 1.80)	1.09 (0.75, 1.6)
	>3 mg/L	**3.04 (1.95, 4.72)**	**2.09 (1.30, 3.38)**

FIG. 2
Association of CRP levels and urological symptoms assessed using odds ratios (ORs) and 95% confidence intervals (95% CI) for continuous CRP levels (natural log transformation). Bold=statistically significant ORs. *Adjusted for age, race/ethnicity, SES, heart disease, type 2 diabetes, depression, antiinflammatory, and LUTS medication use. *From Kupelian V, McVary KT, Barry MJ, et al. Association of C-reactive protein and lower urinary tract symptoms in men and women. Results from the Boston Area Community Health (BACH) Survey. Urology 2009;73(5):950–57.*

interplay of these factors likely represents an intricate web of influence. The hope for the future is that if the mechanisms of action for the influence of lifestyle and dietary factors were known, then prospective trials using diet for LUTS treatment could be done. Additionally, if these mechanisms were known, then preventative strategies could be organized with great impact on health-care policy.

References

[1] Azadzoi KM, Yalla SV, Siroky MB. Oxidative stress and neurodegeneration in the ischemic overactive bladder. J Urol 2007;178:710–5.

[2] Kramer G, Mitteregger D, Marberger M. Is benign prostatic hyperplasia (BPH) an immune inflammatory disease? Eur Urol 2007;51:1202–16.

[3] McVary KT, Rademaker A, Lloyd GL, Gann P. Autonomic nervous system overactivity in men with lower urinary tract symptoms secondary to benign prostatic hyperplasia. J Urol 2005;174:1327–433.

[4] Araki H, Watanabe H, Mishina T, Nakao M. High-risk group for benign prostatic hypertrophy. Prostate 1983;4(3):253–64.

[5] Suzuki S, Platz EA, Kawachi I, et al. Intakes of energy and macronutrients and the risk of benign prostatic hyperplasia. Am J Clin Nutr 2002;75:689–97.

[6] Chen Y, Yu L, Wu S, et al. Relationship among diet habit and lower urinary tract symptoms and sexual function in outpatient-based males with LUTS/BPH: a multiregional and cross-sectional study in China. BMJ Open 2016;6.

[7] Raheem OA, Parsons JK. Associations of obesity, physical activity and diet with benign prostatic hyperplasia and lower urinary tract symptoms. Curr Opin Urol 2014;24(1):10–4.

[8] Maserejian NN, Giovannucci EL, McKinlay JB. Dietary macronutrients, cholesterol, and sodium and lower urinary tract symptoms in men. Eur Urol 2009;55:1179–89.

[9] Bravi F, Bosetti C, Dal Maso L, Talamini R, Montella M, Negri E, et al. Food groups and risk of benign prostatic hyperplasia. Urology 2006;67:73.

[10] Bravi F, Bosetti C, Dal Maso L, Talamini R, Montella M, Negri E, et al. Macronutrients, fatty acids, cholesterol, and risk of benign prostatic hyperplasia. Urology 2006;67:1205.

[11] Lagiou P, Wuu J, Trichopoulou A, Hsieh CC, Adami HO, Trichopoulos D. Diet and benign prostatic hyperplasia: a study in Greece. Urology 1999;54:284.

[12] Chyou PH, Nomura AM, Stemmermann GN, Hankin JH. A prospective study of alcohol, diet, and other lifestyle factors in relation to obstructive uropathy. Prostate 1993;22:253.

[13] Li XM, Zhang L, Li J, et al. Measurement of serum zinc improves prostate cancer detection efficiency in patients with PSA levels between 4 ng/mL and 10ng/m. Asian J Androl 2005;7:323–8.

[14] Leitzmann MF, Stampfer MJ, Wu K, et al. Zinc supplement use and risk of prostate cancer. J Natl Cancer Inst 2002;95:1004–7.

[15] Kristal AR, Arnold KB, Schenk JM, et al. Dietary patterns, supplement use, and the risk of symptomatic benign prostatic hyperplasia: results from the prostate cancer prevention trial. Am J Epidemiol 2008;167:925–34.

[16] Bradley CS, Erickson BA, Messersmith EE, et al. Evidence of the impact of diet, fluid intake, caffeine, alcohol, and tobacco on lower urinary tract symptoms: a systematic review. J Urol 2017;198:1010–20.

[17] Rohrmann S, Giovannucci E, Willett WC, Platz EA. Fruit and vegetable consumption, intake of micronutrients, and benign prostatic hyperplasia in US men. Am J Clin Nutr 2007;85:523–9.

[18] Liu ZM, Wong CK, Chan D, et al. Fruit and vegetable intake in relation to lower urinary tract symptoms and erectile dysfunction among Southern Chinese elderly men: a 4-year prospective study of Mr OS Hong Kong. Medicine 2016;95(4):e2557.

[19] Spettel S, Chughtai B, Feustel P, Kaufman A, Levin R, De E. A prospective randomized double-blind trial of grape juice antioxidants in men with lower urinary tract symptoms. Neurourol Urodyn 2013;32:261–5.

[20] Vidlar A, et al. The effectiveness of dried cranberries (Vaccinium macrocarpon) in men with lower urinary tract symptoms. Br J Nutr 2010;104:1181–9.

[21] Schwarz S, Obermüller-Jevic UC, Hellmis E, et al. Lycopene inhibits disease progression in patients with benign prostate hyperplasia. J Nutr 2008;138:49–53.

[22] Ilic D. Lycopene for the prevention and treatment of prostate disease. Recent Results Cancer Res 2014;202:109–14.

[23] Galeone C, Pelucchi C, Talamini R, et al. Onion and garlic intake and the odds of benign prostatic hyperplasia. Urology 2007;70(4):672–6.

[24] Barnard RJ, Aronson WJ. Benign prostatic hyperplasia: does lifestyle play a role? Phys Sportsmed 2009;37:141–6.

[25] Moyad MA, Lowe FC. Educating patients about lifestyle modifications for prostate health. Am J Med 2008;121(8):S34–42.

[26] Denis L, Morton MS, Griffiths K. Diet and its preventive role in prostatic disease. Eur Urol 1999;35:377–87.

[27] Monti S, Di Silverio F, Lanzara S, et al. Insulin-like growth factor-I and -II in human benign prostatic hyperplasia relationship with binding proteins 2 and 3 and androgens. Steroids 1998;63:362–6.

[28] Poon KS, McVary KT. Dietary patterns, supplemental use and the risk of benign prostatic hyperplasia. Curr Urol Rep 2009;10:279–86.

[29] Ford ES, Giles WH, Dietz WH. Prevalence of the metabolic syndrome among US adults: findings from the third National Health and Nutrition Examination Survey. JAMA 2002;287:356–9.

[30] Reaven GM, Lithell H, Landsberg L. Hypertension and associated metabolic abnormalities: the role of insulin resistance and the sympathoadrenal system. N Engl J Med 1996;334:374–81.

[31] Landsberg L. Insulin-mediated sympathetic stimulation: role in the pathogenesis of obesity-related hypertension (or, how insulin affects blood pressure, and why). J Hypertens 2001;19:523–8.

[32] McVary KT, Razzaq A, Lee C, et al. Growth of the rat prostate gland is facilitated by the autonomic nervous system. Biol Reprod 1994;51:99–107.

[33] Hammarsten J, Hogstedt B. Clinical, anthropometric, metabolic and insulin profile of men with fast annual growth rates of benign prostatic hyperplasia. Blood Press 1999;8:29–36.

[34] Glynn RJ, Campion EW, Bouchard GR, Silbert JE. The development of benign prostatic hyperplasia among volunteers in the Normative Aging Study. Am J Epidemiol 1985;121:78–90.

[35] Nickel JC. Prostatic inflammation in benign prostatic hyperplasia the third component? Can J Urol 1994;1(1).

[36] Roehrborn CG, Kaplan SA, Noble WD, et al. The impact of acute or chronic inflammation in baseline biopsy on the risk of clinical progression of BPH: results from the MTOPS study. J Urol 2005;173:346.

[37] Tuncel A, Uzun B, Eruyar T, et al. Do prostatic infarction, prostatic inflammation and prostate morphology play a role in acute urinary retention? Eur Urol 2005;48:277.

[38] Mishra VC, Allen DJ, Nicolaou C, et al. Does intraprostatic inflammation have a role in the pathogenesis and progression of benign prostatic hyperplasia? BJU Int 2007;100:327.

[39] Kwon YK, Choe MS, Seo KW, et al. The effect of intraprostatic chronic inflammation on benign prostatic hyperplasia treatment. Korean J Urol 2010;51:266.

[40] Nickel JC, Roehrborn CG, O'Leary MP, et al. The relationship between prostate inflammation and lower urinary tract symptoms: examination of baseline data from the REDUCE trial. Eur Urol 2008;54(6):1379–84.

[41] Andriole G, Bostwick D, Brawley O, et al. Chemoprevention of prostate cancer in men at high risk: rationale and design of the reduction by dutasteride of prostate cancer events (REDUCE) trial. J Urol 2004;172:1314–7, 15371831.

[42] Robert G, Descazeaud A, Nicolaïew N, et al. Inflammation in benign prostatic hyperplasia: a 282 patients' immunohistochemical analysis. Prostate 2009;69:1774.

[43] Nickel CJ, Roehrborn CG, Castro-Santamaria R, et al. Chronic prostate inflammation is associated with severity and progression of benign prostatic hyperplasia, lower urinary tract symptoms and risk of acute urinary retention. J Urol 2016;196:1493–8.

[44] Rohrmann S, De Marzo AM, Smit E, et al. Serum C-reactive protein concentration and lower urinary tract symptoms in older men in the Third National Health and Nutrition Examination Survey (NHANES III). Prostate 2005;62:27–33.

[45] St Sauver JL, Sarma AV, Jacobson DJ, McGree ME, Lieber MM, Girman CJ, Nehra A, Jacobsen SJ. Association between C-reactive protein levels and longitudinal changes in urologic measures. J Urol 2008;179:S30.

[46] Kupelian V, McVary KT, Barry MJ, et al. Association of C-reactive protein and lower urinary tract symptoms in men and women. Results from the Boston Area Community Health (BACH) Survey. Urology 2009;73(5):950–7.

Further Reading

[47] Parsons JK, Sarma AV, McVary KT, Wei JT. Obesity and benign prostatic hyperplasia: clinical connections, emerging etiological paradigms and future directions. J Urol 2013;189(1 Suppl):S102–6.

[48] Maserejian NN, Wager CG, Giovannucci EL, Curto TM, McVary KT, McKinlay JB. Intake of caffeinated, carbonated, or citrus beverage types and development of lower urinary tract symptoms in men and women. Am J Epidmiol 2013;177(12):1399–410. https://doi.org/10.1093/aje/kws411. Epub 2013 June.

[49] Gacci M, Sebastianelli A, Salvi M, De Nunzio C, Tubaro A, Vignozzi L, Corona G, McVary KT, Kaplan SA, Maggi M, Carini M, Serni S. Central obesity is predictive of persistent storage LUTS after surgery for benign prostatic enlargement: results of a multicenter prospective study. BJU Int 2015;116(2):271–7.

CHAPTER 1.3

Diet and Fertility in Men: Are Sperm What Men Eat?

Feiby L. Nassan[*,†], Jorge E. Chavarro[†,‡,§]

[*]Department of Environmental Health, Harvard T. H. Chan School of Public Health, Boston, MA, United States, [†]Department of Nutrition, Harvard T. H. Chan School of Public Health, Boston, MA, United States, [‡]Department of Epidemiology, Harvard T. H. Chan School of Public Health, Boston, MA, United States, [§]Channing Division of Network Medicine, Harvard Medical School, and Brigham and Women's Hospital, Boston, MA, United States

INTRODUCTION

In 1826, Jean Anthelme Brillat-Savarin (French writer, (1755–1826)) wrote, Dis-moi ce que tu manges, je te dirai ce que tu es. "Tell me what you eat, and I will tell you what you are [1]." Since then, Brillat-Savarin's quote had been commonly appropriated today. It is also cited as the origin of the adage "You are what you eat," which was not literally found in English until in the 1920s when nutritionist Victor Lindlahr used it. Lindlahr was then quoted in a 1923 ad for United Meat Market. If men are what they eat, are their sperm so too?

Approximately one in six couples who try to become pregnant fails to do so within a year and receive the infertility diagnosis [2]. Although male factor infertility is identified in approximately 50% of couples seeking medical help for fertility problems [3, 4], infertility research has traditionally focused on women. Yet, sperm counts have declined in the Western countries by 50%–60% between 1973 and 2011 [5–7]. While the causes of this downward trend in semen quality are still being studied, concurrently, shifts to worsening diet quality [8] and increasing obesity [9–11] could explain, to some extent, these alarming trends.

Although the evidence that overweight and obesity play a significant role in compromising men's semen quality [11] and couples' fertility [12] is strong and consistent, there are still no clear dietary guidelines for men in couples seeking to become pregnant. In this chapter, we aim to summarize the growing literature about the role of diet on men's fertility. Although far from a perfect proxy of men's fertility [13–18], conventional semen analysis is the cornerstone of fertility assessment [19] and does shed insight into testicular function and men's reproductive potential. In addition, most of the literature has focused on the association between diet and semen quality as a proxy of men's fertility status. Therefore, we will also focus on the relation between diet and semen quality and to a less extent the profile of the serum reproductive hormones and the emerging literature on men's diet and couple-based outcomes, including measures of fecundity in studies of pregnancy planners and studies of pregnancy outcomes after assisted reproductive technologies (ART).

DIET AND THE BUILDING BLOCKS FOR SPERM

The transformation from spermatogonia to mature sperm involves several complex steps, some of which are susceptible to influence from environmental factors including diet. In studying the relation between nutritional factors and semen quality as a marker of spermatogenesis, it is convenient to highlight three key characteristics of spermatogenesis. First, as sperm mature, the fatty acid composition of the cell membrane changes dramatically with a preferential accumulation of long-chain polyunsaturated fatty acids (PUFAs) [20–23], a highly specialized feature that appears to be crucial for sperm function [24]. Second, as sperm mature, they lose most of their cytoplasm. In addition, condensation of the sperm chromatin occurs after they progress from the caput epididymis in transit to the cauda epididymis. These two processes include repeated oxidation reactions [24] during which the shuttle systems for removing and transferring reducing equivalents into the mitochondria are not operational. Therefore, during these reactions, sperm are left without intracellular defense mechanisms against oxidative damage. Moreover, the preferentially incorporation of PUFAs, a highly oxidizable substrate, into the cell membrane makes the sperm more susceptible to oxidative stress. Third, spermatogenesis requires nearly constant production of DNA and, therefore, an adequate supply of substrates for DNA production. These three characteristics of spermatogenesis—membrane specialization, the loss of cytoplasm, and high demand for DNA substrates—can be influenced by diet.

Dietary Fats, the Sperm Cell Membrane and Men's Fertility

The composition of the sperm cell membrane with rich fatty acid content is critical for proper sperm function. The sperm cell membrane plays an important role in the key fertilization events such as capacitation, acrosome reaction, and sperm-oocyte fusion [24]. The amount of PUFAs in the sperm cell membrane, particularly of docosahexaenoic acid (DHA), a crucial omega-3 fatty acid, increases as the sperm matures [25]. These changes appear to be the combined result of local metabolism and dietary input [26–30]. DHA represents approximately 20% of the fatty acid content in the mature sperm, as compared with approximately 4% in immature germ cells [25]. For example, in mice, DHA content is relatively higher in epididymal than in testicular sperm [20]. Similarly, in human orchiectomy specimens, the proportion of PUFAs in cell membranes is higher in sperm recovered from the cauda than from sperm recovered from the caput epididymis [21]. A similar pattern has also been observed in other mammals [22, 23].

The fatty acid composition in the sperm cell membrane is partly explained by a highly specialized local metabolism. Sertoli cells express Δ6-desaturase (the rate-limiting enzyme in the PUFA metabolism) and Δ5-desaturase at concentrations comparable with those in the liver [31]. Moreover, enzymes involved in the elongation of PUFAs (ELOVL2, ELOVL4, and ELOVL5) are expressed in a limited number of tissues but are highly expressed in the testis [32–35]. In addition to the presence of specialized enzymatic machinery, Sertoli cells can actively convert the 18-carbon PUFAs and 20-carbon PUFAs into their 22- and 24-carbon metabolites more efficiently than the hepatocytes [36–38]. Furthermore, the enzymes involved in this pathway in Sertoli cells prefer the conversion of omega-3 fatty acids into their 22- and 24-carbon metabolites over converting omega-6 fatty acids [36–38]. This could explain to some extent the high concentration of DHA in the sperm relative to other tissues.

Fatty acid intake also influences the fatty acid composition of the testis and sperm function. PUFAs cannot be endogenously synthesized by humans and must, therefore, be obtained from consuming seafood in the case of longer-chain omega-3 PUFAs such as eicosapentaenoic acid (EPA) and DHA or seeds, nuts, and vegetable oils in the case of the 18-carbon linoleic (LA) and α-linolenic acids (ALA). Diets rich in ALA or fish oil, which is rich in EPA and DHA, increase testicular levels of DHA in rodent models [28–30] and in the sperm membrane DHA in humans [39]. Higher DHA content in the sperm membrane has in turn been associated with higher sperm motility [26, 39–44], sperm normal morphology [39, 44], and sperm concentration [39, 40, 43–46]. Moreover, among men presenting to fertility clinics, higher intake of omega-3 PUFAs was related to a higher proportion of morphologically normal sperm [47], and fish intake was positively related to total sperm count and normal sperm morphology [46]. In a small trial (n=28) among asthenospermic men, 3 months of DHA supplementation did not improve sperm motility [48]. This trial, however, is an outlier in the literature. For example, a trial of long-chain omega-3 fatty acid (DHA+EPA) supplementation (1.84g/day for 32 weeks) among 211 men with idiopathic oligoasthenoteratospermia resulted in a significant increase in total sperm count, sperm concentration, and the percentages of motile and morphologically normal sperm [39]. In a randomized controlled trial of young healthy men consuming a typical Western-style diet, men randomized to walnut supplementation (rich in ALA) of 75g/day for 12 weeks had improvements in the sperm vitality, motility, and morphology compared with controls [49]. Furthermore, recent studies suggested that the benefit may extend beyond semen quality. In a prospective cohort of couples trying to conceive, men's fish intake was related to shorter time to pregnancy and lower risk of infertility independently of their partner's fish intake [50].

On the other hand, *trans*-fatty acids and saturated fats appear to have the opposite effect on spermatogenesis from PUFAs. Like PUFAs, *trans* fats—which are primarily found in commercially baked and fried foods—accumulate in the testis [51, 52]. Unlike PUFAs, however, sperm membrane levels and intake of these fatty acids have been consistently associated with poor semen quality, particularly lower sperm counts [45, 47, 53, 54]. Animal models suggest that diets supplemented with *trans* fats not only result in decreased spermatogenesis but also can induce decreased production of testosterone and decreased testicular mass and testicular degeneration, in a dose-dependent manner [51, 55–57]. Of note, as of June 2018, the United States Food and Drug Administration (FDA) decided to exclude *trans* fats from industrial origin from the list of substances generally recognized as safe (GRAS). This decision is expected to effectively eliminate that concern from the United States once fully implemented. Nevertheless, given the widespread of *trans* fats in the global food supply and in particular in Africa and South Asia, this will remain a global public health concern. Moreover, saturated fats will not disappear from the food supply. Despite the scarcity of the evidence, two observational studies have found that intake of saturated fats was associated with lower total sperm count and concentration among fertility patients in the United States [47] and among healthy young men in Denmark [53].

Antioxidants: Defense Against Oxidative Stress

The combination of significant cytoplasm loss, the generation of ROS by the mitochondria, and the preferential accumulation of the highly oxidizable substrates in the cell membrane make the sperm highly susceptible to the oxidative damage. This high susceptibility of the sperm has led to a big interest in the role of antioxidants in spermatogenesis especially among subfertile men. The most compelling evidence supporting a beneficial role of antioxidants improving semen quality and male fertility comes from a Cochrane review of randomized clinical trials (RCTs) in couples undergoing infertility treatment [58]. The review found that antioxidant supplementation of the male partner over the course of infertility treatment improves sperm motility as well as clinical pregnancy and live birth rates during infertility treatment [58]. However, Cochrane reviewers also indicated important gaps, which complicate the interpretation of these data. Most importantly, few trials included in the meta-analysis evaluated the same intervention (i.e., the same dose and combination of antioxidants) against the same comparator (i.e., placebo or same active comparator). In addition, the expansive definition of "antioxidant" used in this meta-analysis resulted in the inclusion of trials evaluating the effects of agents without antioxidant capacity, making it difficult to attribute any observed effect to protection against oxidative stress. Moreover, three of the four trials evaluating live birth as an outcome were published in the mid-1990s and reported live birth rates of 10% or lower [59–61], calling into question their relevance to the current practice of reproductive medicine. Nevertheless, the biological plausibility and clear signal from the RCTs warrant further attention to the role antioxidants may play on spermatogenesis and male fertility.

The damaging effects of oxidative stress may contribute to male factor infertility [62]. Oxidative stress occurs when reactive oxygen species (ROS) overcome the semen's natural antioxidant defenses and cause cellular damage to the sperm [62]. Oxidative parameters are higher in the semen of idiopathic infertile men than in fertile men, and a high correlation is found between oxidative parameters, sperm ROS formation, and DNA fragmentation levels [63]. Antioxidants such as vitamin C and carotenoids are naturally found in semen [62] and can act as free radical scavengers that help to overcome ROS [64]. Other antioxidants, like vitamin E, can directly neutralize the ROS in sperm plasma membranes [65]. The direct antioxidant capacity of these nutrients appears to translate into clinically observable benefits. Vitamin C and β-carotene have also been related to semen quality among young healthy men [66], and as mentioned above, antioxidant supplementation improves sperm motility among subfertile men [58]. The effect of antioxidants may extend beyond their ability to prevent oxidation or reduce the ROS. For example, in observational studies, other antioxidant intakes such as β-carotene have been associated with a lower prevalence of disomy in the sperm X chromosome [67].

The evidence for a benefit on pregnancy and live birth rates is exceedingly intriguing but not conclusive. As mentioned above, a meta-analysis of RCTs concluded that antioxidant supplementation to men in couples undergoing infertility treatment may improve pregnancy and live birth rates [58]. This conclusion is based on the pooled data from four RCTs [59–61, 68], three of which were published during the 1990s and may not be relevant to current practice. The most recent one conducted in 2007 found that a combination multivitamin/multimineral supplement improved pregnancy and live birth rates during in vitro fertilization/intracytoplasmic sperm injection (IVF-ICSI) treatment [68]. This particular intervention makes it difficult to discern whether the benefit is due to the specific combination of nutrients included in the study or due to one or a limited number of them. Subgroup analyses suggested that vitamin E may be a particularly important antioxidant for further study. This subgroup analysis included two studies [59, 60], and the beneficial effect was mainly due to one of them [60] in which men were treated with a high dose of vitamin E (300 mg/day). Nonrandomized studies, however, also suggest a benefit of vitamin E alone [60, 69] or in combination with vitamin C [70]. Clearly, additional prospective studies and in particular RCTs sufficiently powered to evaluate the independent role of antioxidant supplemental of men in couples seeking fertility treatment are needed to clarify their clinical utility.

The One-Carbon Metabolism and Spermatogenesis

The one-carbon metabolism is a metabolic pathway with a series of related pathways in which one-carbon moieties are transferred from donors to intermediate carriers and ultimately used in methylation reactions or in the synthesis of purines and thymidine, which are used in DNA building blocks [71, 72]. Specific nutrients such as folates play an important role in the one-carbon metabolism, serving as either substrates or cofactors, and thus play an important role in spermatogenesis. One-carbon metabolism is highly active in the testes during spermatogenesis [73–75]. In addition, disruption of this metabolic pathway, including genetic [76] or pharmacological [77–79] disruption, has detrimental effects on spermatogenesis.

In a model of genetic deficiency, male 5,10-methylenetetrahydrofolate reductase null (MTHFR$^{-/-}$) mice have altered spermatogenesis evidenced by drastic reduction in the number of germ cells at 6 days postnatal that continues into adult life and results extremely abnormal testicular histology with most tubules devoid of germ cells, an equally drastic reduction of testicular sperm counts and infertility [76]. Interestingly, this phenotype can be partially reversed by lifetime supplementation with betaine [76], a methyl donor. Pharmacological inhibition of this pathway has similar effects. Administration of pyrimethamine, a dihydrofolate reductase (DHFR) inhibitor, to sexually mature mice causes a reversible arrest of spermatogenesis in a dose-dependent manner showing marked reduction in sperm counts after 30 days and complete infertility after 50 days of administration [77, 78]. Likewise, administration of etoprine, also a DHFR inhibitor, causes spermatogenic arrest and infertility [79].

Decreased methylenetetrahydrofolate reductase (MTHFR) activity in humans also appears to be detrimental to male fertility. Among men presenting to a fertility center, hypermethylation in the promoter region of *MTHFR* was found in 53% of testis biopsies of men with nonobstructive azoospermia, compared with none of the testis biopsies of men with obstructive azoospermia but normal spermatogenesis, suggesting that deregulation of *MTHFR* might impair spermatogenesis [80]. Several human studies have examined the potential role on male fertility of variation in the genes encoding for the enzymes involved in this pathway, in particular for the *C677T* SNP in *MTHFR*. A meta-analysis on the association between the C677T single nucleotide polymorphism (SNP) in MTHFR and male infertility reported pooled statistically significant odd ratios (OR) (95% confidence interval (95%CI)) of 1.39 (1.15–1.69) for TT homozygotes and 1.23 (1.08–1.41) for T allele carriers with male factor infertility [81]. Another large study carried out in Korea has reported an association between homozygosity for the variant G allele in methionine synthase A2756G (MTR A2756G) and nonobstructive azoospermia (OR (95% CI) = 4.63 (1.40–15.31)) and an association between being a carrier (OR (95%CI) = 1.75 (1.07–2.86)) and homozygote (OR (95%CI) = 2.96 (1.51–5.82)) for the variant G allele in methionine synthase reductase A66G (MTRR A66G) and oligoasthenoteratospermia [82].

Folic acid intake also influences spermatogenesis. In a randomized trial, folic acid supplementation of 15 mg per day for 90 days has led to a 53% increase in sperm concentration and a higher proportion of motile sperm [83]. In another randomized trial in which subfertile men were assigned to either 5 mg of folate per day for 182 days, zinc, folate+zinc, or placebo, the total normal sperm count in the subfertile men assigned to the folate+zinc arm increased with a 74% compared with preintervention and with a 41% increase (did not reach statistical significance though) when compared with men who were assigned to the placebo [84]. Neither folate alone nor folate+zinc had an effect on serum concentrations of the follicle-stimulating hormone (FSH), testosterone, or inhibin B [85]. In a case-control study, men from a fertility clinic in Spain who were in the highest tertile of folate intake had lower odds (87%) of oligoteratospermia than men in the lowest tertile of folate intake [86]. Other two cross-sectional studies found a positive association between seminal plasma folate and vitamin B12 and sperm concentration [87, 88]. Among men who have previously fathered a pregnancy and have sperm concentrations above 20 million/mL, seminal plasma folate was inversely related to sperm DNA fragmentation, another marker of sperm integrity [89]. Furthermore, folate intake has been associated with a lower prevalence of sperm aneuploidy [67]. In a study of healthy nonsmoking men, men who had the highest folate intake (722–1150 μg/day) had a lower incidence of sex nullisomy, disomy X, disomy 21, and aggregate sperm aneuploidy [67]. Similarly, in rodent models, diet that was deficient in folate resulted in differential sperm DNA methylation at sites that is known to be associated with decreased pregnancy rates, increased postimplantation embryo loss, differences in placental vasculature development birth weight, and increased gross anatomical abnormalities in their offspring, independently of maternal folate status [90–93]. Whether the same is true in humans is unclear as is also whether the apparent benefits of folate and other nutrients of the one-carbon metabolism on semen parameters translate into clinically relevant differences in pregnancy or live births. Ongoing randomized trials (folic acid and zinc supplementation trial; NCT01857310) may answer this question.

FOOD AS A VEHICLE FOR ENVIRONMENTAL TOXINS

There has been focus on studying the different environmental toxicants as potentially modifiable risk factors on men infertility particularly semen quality, for decades. In addition to nutrients, food also acts as a potential vehicle to several environmental toxins. Therefore, research has started to study the diet as a vehicle for those environmental chemicals.

Meats, Dairy Products and Hormone Residues

Specific modern dairy farming practices [94] and livestock production [95] may result in dairy products and meat carrying environmental estrogens. Commercial milk is derived from pregnant cows in most of the world [94, 96]. Therefore, milk contains the naturally occurring placental hormones such as estradiol and progesterone in measurable concentrations [97, 98]. This raises concerns about the reproductive effects of milk and dairy products on consumers. It has been estimated that dairy accounts for 60%–80% of dietary estrogen intake in the Western countries [99]. For example, intake of milk and other dairy products in boys has been associated with increased excretion of estrone, estriol, estradiol, and pregnanediol [100]; higher concentrations of prepubertal growth hormone (GH), insulin-like growth factor 1 (IGF-1), and the ratio of IGF-1 to insulin-like growth factor-binding protein 3 [101]; and higher frequency of teenage acne [102]. In healthy young men, intake of dairy products has been associated with lower serum concentrations of testosterone, FSH, and luteinizing hormone (LH) [100].

However, the evidence of the relation between intake of the dairy product and semen quality is still inconclusive. Some studies have suggested that dairy products are possible risk factors for poorer semen parameters [94, 103–106]; however, other studies did not support this conclusion [103, 104, 107]. A case-control study comparing dietary habits of oligoasthenoteratospermic versus normospermic men in Spain found that cases had higher intake of full-fat dairy products (whole milk, semi-skim milk, yogurt, and cheese) and lower intakes of skim milk than controls [103]. In another case–control study of asthenozoospermic men in Iran, men with greater intake of total dairy products had marginally higher odds of asthenozoospermia, and men with greater intake of skim milk had significantly lower odds of asthenozoospermia [104]. Furthermore, intake of the low-fat dairy products was associated with higher sperm concentration and better motility among 155 men attending a fertility clinic in Boston, Massachusetts, in a longitudinal cohort study [106]. Intake of full-fat dairy products, specifically cheese, was associated with lower normal sperm morphology and progressive sperm motility among physically active young men in a cross-sectional study [105], whereas, in another cross-sectional study among men attending a fertility clinic in the Netherlands, dairy intake was not related to semen quality [107]. Although the evidence largely supports the benefit of low-fat versus the harmful effects of full-fat dairy products, there is still a need for more studies, especially randomized trials, for a well-supported conclusion.

In the United States and other countries, 60–90 days prior to slaughter of the meat-producing animals; anabolic sex steroids including combinations of estrogen, progesterone, and testosterone; and any of three synthetic hormones (zeranol, melengestrol acetate, and trenbolone acetate) are commonly administered for growth promotion [95, 108]. Therefore, the resultant meat products contain some hormonal residues in meat and meat-product consumers [108, 109], with the potential reproductive consequences [98, 110, 111]. Studies, however, are inconsistent about the relation between meat intake and semen quality. Male offspring of the women in the United States with high beef consumption during pregnancy had lower sperm concentration in adulthood [110]. In a cross-sectional study, young college men in the United States who consumed higher amounts of processed meat had lower total sperm counts and total progressive motile counts [112]. In Spain, in a cross-sectional study, oligoasthenoteratospermic men had an approximately 31% higher intake of processed red meat than controls, but there was no difference in unprocessed red meat intake between the two groups [103]. A study in Iran reported that men in the highest intake tertile of processed red meat had twice of the odds of asthenozoospermia compared with those in the lowest tertile of consumption, but red meat intake was not related to asthenozoospermia [104]. Another study in fertility clinic in the Netherlands reported that intake of meat products was not related to semen quality parameters [107]. A longitudinal study of men attending a fertility clinic in the United States reported that processed meat intake was inversely related to sperm morphology [46]. It is important to point out that the Spanish and Dutch studies were conducted after the European Union has banned the steroid hormones for beef cattle [108, 113]. This may suggest that the similar findings in the Spanish and US studies are probably not due to hormonal residues but may instead reflect a nutritional characteristic of meat itself or of dietary patterns where meat is preferentially consumed (see dietary patterns, below).

Fish and Methylmercury

Another important protein source is seafood. Fish intake may have beneficial effects on semen quality and couples' fecundity, as we discussed above. However, contaminated fish and shellfish are the main source of methylmercury exposure, the most common organic mercury compound found in the environment [114]. Both animal [115, 116] and in vitro [117, 118] studies have shown detrimental effects of methylmercury on male reproductive health including impaired spermatogenesis [115], lower testicular weight and decreased sperm count [116], decreased sperm motility, and increased abnormal tail morphology [119]. The fact that fish itself has higher contents of omega-3 fatty acids with its potential beneficial effects and also could be simultaneously contaminated with methylmercury could explain these apparently contradictory findings. The reason for that is the fact that studies hardly ever consider simultaneously mercury and fish consumption and the association with semen parameters, which leads to residual confounding. Only few studies have addressed these issues concurrently. For example, in a recent study, hair mercury that is the best biological biomarker was positively associated with total sperm count, sperm concentration, and progressive motility in a cohort of men attending a fertility clinic. This association was attenuated after further adjustment for fish intake [120]. Specifically, men in the highest quartile of hair mercury concentrations had higher sperm concentration (50%), total sperm count (46%), and progressive motility (31%) compared with men in the lowest quartile. Men with fish intake above the study population median had even stronger associations [120]. These results that confirm about the methylmercury exposure from fish intake show the important role of diet while assessing the associations between heavy metals and semen parameters among men of couples seeking fertility care. These results suggest that the beneficial effects of fish intake (at least the fish types this population have consumed) on spermatogenesis may outweigh the potential negative effects of methylmercury. These findings warrant that further investigation and local fish inventories need to be consulted.

Fruits, Vegetables, and Pesticide Residues

According to the World Health Organization, fruits and vegetables are universally recommended as an essential component of a healthy diet [121]. However, they are still the main source of exposure to pesticide residues from diet and the most important source of exposure to pesticides in the general population [122]. Fruits and vegetables with low to moderate pesticide residues, such as avocado, onions, and beans, are positively associated with semen parameters in young healthy men [123]. Total fruit and vegetable intake was unrelated to semen parameters, in men attending a fertility clinic. However, in the same study population, intake of high-pesticide-residue fruits and vegetables, such as spinach, strawberries, and apples, was associated with poorer semen quality [123]. On average, men who consumed 1.5 or more servings per day of high-pesticide-residue fruit and vegetable (the highest quartile of intake) had 49% lower total sperm count and 32% lower percentage of morphologically normal sperm than men who consumed <0.5 servings per day (the lowest quartile of intake). However, low- to moderate-pesticide-residue fruit and vegetable intake was associated with a higher percentage of morphologically normal sperm [124]. Concerns regarding exposure to pesticide residues on health have extended beyond fertility and have recently been suggested to contribute to greater risk of some cancers [125]. At this point, evidence linking intake of pesticide residues from food with health in general and male fertility in particular is nascent and far from definitive but certainly merits additional inquiry [126].

Soy Products and Xenoestrogens

Isoflavones are known to be weakly estrogenic plant-derived polyphenolic compounds. They are present in soybeans and soy-derived products [127–132] that can bind to estrogen receptors [133]. In animal models, isoflavones were associated with smaller testes in rats [134] and nongenomic adverse effects on sperm capacitation and acrosome reaction [135]. However, the literature in humans on soy or soy-derived products and male fertility is still scarce and inconsistent. Semen quality and reproductive hormone concentrations did not change after supplementation of 40 mg/day of isoflavones for 2 months in 14 men [136]. In another study among 48 men with abnormal semen parameters and 10 fertile control men, however, dietary isoflavone intake was associated with higher sperm count and motility and lower sperm DNA damage [137]. On the other hand, another study among 99 men attending a fertility clinic showed that dietary intake of soy-derived foods was associated with lower sperm concentrations [138]. Although one may argue that Asian diets, for example, include high amounts of phytoestrogens from soy foods in general and they do not appear to have any apparent deleterious effects on fertility, however, in a study in China among 609 men with idiopathic infertility and 469 fertile controls, higher urinary concentrations of isoflavones were associated with lower sperm concentration, total sperm count, sperm motility, and higher odds of idiopathic male infertility [139]. In addition, men's soy intake was not associated with the probability of live birth after assisted reproductive technologies (ART) in couples attending fertility clinics [140], and men's urinary isoflavone concentrations were not associated with fecundity in a prospective cohort of pregnancy planners [141].

DIET PATTERNS

Healthier dietary patterns have the most consistent association with better semen quality. In recent reviews [142–144], a healthy dietary pattern like the Mediterranean pattern or patterns with high intakes of seafood, poultry, whole grains, legumes, fruits and vegetables, and skim milk have been consistently associated with better semen parameters in studies in North America, Europe, Middle East, and East Asia [104, 107, 142, 145, 146]. Unhealthy dietary patterns that are high in fats, red and processed meats, refined grains, sweets, and sweetened beverages were associated with poorer semen quality. A more recent study in Israel has concluded that adherence to any of the four dietary indexes including healthy eating index (HEI), dietary approaches to stop hypertension (DASH), alternate Mediterranean diet (aMED) score, and alternative healthy eating index (AHEI) was associated with better overall sperm quality, with AHEI best associated [147]. Also, in Greece, a recent study has reported that men who adhered to the Mediterranean diet had lower likelihood of abnormal sperm concentration, total sperm count, and sperm motility [148]. Also recently in the Netherlands, men who adhered to a healthy diet pattern had higher sperm concentrations, sperm count, and motile sperm [149]. This beneficial association was even more prominent among men that had lower total motile sperm count <10 million semen quality to start with [149]. They suggested that the preconceptional nutritional counseling and coaching of couples who are trying to conceive need to be tailored [149]. In addition, a healthy dietary pattern was associated with lower fragmentation of the sperm DNA [107]. Following a healthy diet pattern as an overall healthy behavior is consistently associated with better semen quality overall in many studies; however, more research is still needed to determine the effect of individual food categories on men's reproduction.

CONCLUSIONS

In an age where couples seek pregnancy in older age and with alarming decline in men's semen quality, there is a big need to know the modifiable risk factors such as diet that couples particularly men can change to improve their reproductive potentials. Although the relation between diet and men's fertility is still under research and the evidence is no always confirmed, there are many general patterns that can be shared with patients and clinicians as potential means to modify and hopefully improve their reproductive potentials. First, increased intake of omega-3 fatty acids, whether from foods (either nuts or fish) or from supplements, appears to have a positive effect on spermatogenesis. Supplementation with antioxidants including vitamins C and E and with nutrients involved in the one-carbon metabolism including folate, B12, and zinc pathway also appears to be beneficial. On the other hand, evidence that environmental toxicants carried by diet, including xenoestrogens from soy, dairy products, and meat, are harmful to men's reproductive potential is still, at best, questionable. Lastly and most importantly, a robust body of evidence from studies spanning the globe has emerged suggesting that dietary patterns generally consistent with those already promoted for the prevention of cardiovascular disease and other chronic conditions may also be beneficial for men's fertility. Does this mean that sperm is what men eat? More research is needed to more clearly understand not only how diet can affect semen parameters and other proxies for men's fertility but also how diet impacts couple-based outcomes including natural fecundity and those who undertake fertility treatments.

Conflicts of Interest

The authors report no conflicts of interest.

References

[1] Brillat S. Physiologie du goût, ou, Méditations de gastronomie transcendante: ouvrage théorique, historique et à l'ordre du jour. Paris; 1826.

[2] Thoma ME, McLain AC, Louis JF, et al. Prevalence of infertility in the United States as estimated by the current duration approach and a traditional constructed approach. Fertil Steril 2013;99(5):1324–1331.e1321.

[3] Legare C, Droit A, Fournier F, et al. Investigation of male infertility using quantitative comparative proteomics. J Proteome Res 2014;13(12):5403–14.

[4] Thonneau P, Marchand S, Tallec A, et al. Incidence and main causes of infertility in a resident population (1 850 000) of three French regions (1988-1989)*. Hum Reprod 1991;6(6):811–6.

[5] Levine H, Jørgensen N, Martino-Andrade A, et al. Temporal trends in sperm count: a systematic review and meta-regression analysis. Hum Reprod Update 2017;1–14.

[6] Carlsen E, Giwercman A, Keiding N, Skakkebaek NE. Evidence for decreasing quality of semen during past 50 years. BMJ 1992;305:609–13.

[7] Swan SH, Elkin EP, Fenster L. Have sperm densities declined? A reanalysis of global trend data. Environ Health Perspect 1997;105(11):1228–32.

[8] U.S. Department of Agriculture. Profiling food consumption in America. In: Agriculture fact book 2001–2002. Washington, DC: United States Government Printing Office; 2003. p. 13–21.

[9] Finucane MM, Stevens GA, Cowan MJ, et al. National, regional, and global trends in body-mass index since 1980: systematic analysis of health examination surveys and epidemiological studies with 960 country-years and 9.1 million participants. Lancet 2011;377(9765):557–67.

[10] Sermondade N, Faure C, Fezeu L, et al. Obesity and increased risk for oligozoospermia and azoospermia. Arch Intern Med 2012;172(5):440–2.

[11] Sermondade N, Faure C, Fezeu L, et al. BMI in relation to sperm count: an updated systematic review and collaborative meta-analysis. Hum Reprod Update 2013;19(3):221–31.

[12] Sundaram R, Mumford SL, Buck Louis GM. Couples' body composition and time-to-pregnancy. Hum Reprod 2017;32(3):662–8.

[13] Buck Louis GM, Sundaram R, Schisterman EF, et al. Semen quality and time to pregnancy: the longitudinal investigation of fertility and the environment study. Fertil Steril 2014;101(2):453–62.

[14] Patel CJ, Sundaram R, Buck Louis GM. A data-driven search for semen-related phenotypes in conception delay. Andrology 2017;5(1):95–102.

[15] Sripada S, Townend J, Campbell D, Murdoch L, Mathers E, Bhattacharya S. Relationship between semen parameters and spontaneous pregnancy. Fertil Steril 2010;94(2):624–30.

[16] Jedrzejczak P, Taszarek-Hauke G, Hauke J, Pawelczyk L, Duleba AJ. Prediction of spontaneous conception based on semen parameters. Int J Androl 2008;31(5):499–507.

[17] Nallella KP, Sharma RK, Aziz N, Agarwal A. Significance of sperm characteristics in the evaluation of male infertility. Fertil Steril 2006;85(3):629–34.

[18] Guzick DS, Overstreet JW, Factor-Litvak P, et al. Sperm morphology, motility, and concentration in fertile and infertile men. N Engl J Med 2001;345(19):1388–93.

[19] Practice Committee of American Society for Reproductive Medicine. Diagnostic evaluation of the infertile female: a committee opinion. Fertil Steril 2015;103(6):e44–50.

[20] Ollero M, Powers RD, Alvarez JG. Variation of docosahexaenoic acid content in subsets of human spermatozoa at different stages of maturation: implications for sperm lipoperoxidative damage. Mol Reprod Dev 2000;55(3):326–34.

[21] Haidl G, Opper C. Changes in lipids and membrane anisotropy in human spermatozoa during epididymal maturation. Hum Reprod 1997;12(12):2720–3.

[22] Hall JC, Hadley J, Doman T. Correlation between changes in rat sperm membrane lipids, protein, and the membrane physical state during epididymal maturation. J Androl 1991;12(1):76–87.

[23] Rana APS, Majumder GC, Misra S, Ghosh A. Lipid changes of goat sperm plasma membrane during epididymal maturation. Biochim Biophys Acta 1991;1061(2):185–96.

[24] Flesch FM, Gadella BM. Dynamics of the mammalian sperm plasma membrane in the process of fertilization. Biochim Biophys Acta 2000;1469(3):197–235.

[25] Lenzi A, Gandini L, Maresca V, et al. Fatty acid composition of spermatozoa and immature germ cells. Mol Hum Reprod 2000;6(3):226–31.

[26] Conquer JA, Martin JB, Tummon I, Watson L, Tekpetey F. Fatty acid analysis of blood serum, seminal plasma and spermatozoa of normozoospermic vs. asthenozoospermic males. Lipids 1999;34:793–9.

[27] Jeong B-Y, Jeong W-G, Moon S-K, Ohshima T. Preferential accumulation of fatty acids in the testis and ovary of cultured and wild sweet smelt Plecoglossus altivelis. Comp Biochem Physiol B Biochem Mol Biol 2002;131(2):251–9.

[28] Sebokova E, Garg ML, Wierzbicki A, Thomson ABR. Alteration of the lipid composition of rat testicular plasma membranes by dietary (n-3) fatty acids changes the responsiveness of Leydig cells and testosterone synthesis. J Nutr 1990;120:610–8.

[29] Ayala S, Brenner RR. Effect of polyunsaturated fatty acids of the alpha-linolenic series in the lipid composition of rat testicles during development. Acta Physiol Lat Am 1980;30(3):147–52.

[30] Ayala S, Brenner RR, Dumm C. Effect of polyunsaturated fatty acids of the α-linolenic series on the development of rat testicles. Lipids 1977;12(12):1017–24.

[31] Saether T, Tran TN, Rootwelt H, Christophersen BO, Haugen TB. Expression and regulation of {delta}5-desaturase, {delta}6-desaturase, stearoyl-coenzyme a (CoA) desaturase 1, and stearoyl-CoA desaturase 2 in rat testis. Biol Reprod 2003;69(1):117–24.

[32] Tvrdik P, Westerberg R, Silve S, et al. Role of a new mammalian gene family in the biosynthesis of very long chain fatty acids and sphingolipids. J Cell Biol 2000;149(3):707–18.

[33] Mandal MNA, Ambasudhan R, Wong PW, Gage PJ, Sieving PA, Ayyagari R. Characterization of mouse orthologue of ELOVL4: genomic organization and spatial and temporal expression. Genomics 2004;83(4):626–35.

[34] Zhang K, Kniazeva M, Han M, et al. A 5-bp deletion in ELOVL4 is associated with two related forms of autosomal dominant macular dystrophy. Nat Med 2001;27(1):89–93.

[35] Leonard AE, Bobik EG, Dorado J, et al. Cloning of a human cDNA encoding a novel enzyme involved in the elongation of long-chain polyunsaturated fatty acids. Biochem J 2000;350(3):765–70.

[36] Retterstøl K, Haugen TB, Woldseth B, Christophersen BO. A comparative study of the metabolism of n-9, n-6 and n-3 fatty acids in testicular cells from immature rat. Biochim Biophys Acta 1998;1392(1):59–72.

[37] Retterstøl K, Haugen TB, Christophersen BO. The pathway from arachidonic to docosapentaenoic acid (20:4n-6 to 22:5n-6) and from eicosapentaenoic to docosahexaenoic acid (20:5n-3 to 22:6n-3) studied in testicular cells from immature rats. Biochim Biophys Acta 2000;1483(1):119–31.

[38] Christophersen BO, Hagve TA, Christensen E, Johansen Y, Tverdal S. Eicosapentaenoic- and arachidonic acid metabolism in isolated liver cells. Scand J Clin Lab Investig Suppl 1986;184:55–60.

[39] Safarinejad MR. Effect of omega-3 polyunsaturated fatty acid supplementation on semen profile and enzymatic anti-oxidant capacity of seminal plasma in infertile men with idiopathic oligoasthenoteratospermia: a double-blind, placebo-controlled, randomised study. Andrologia 2011;43(1):38–47.

[40] Zalata A, Christophe A, Depuydt C, Schoonjans F, Comhaire F. The fatty acid composition of phospholipids of spermatozoa from infertile patients. Mol Hum Reprod 1998;4(2):111–8.

[41] Gulaya NM, Margitich VM, Govseeva NM, Klimashevsky VM, Gorpynchenko II, Boyko MI. Phospholipid composition of human sperm and seminal plasma in relation to sperm fertility. Arch Androl 2001;46(3):169–75.

[42] Tavilani H, Doosti M, Abdi K, Vaisiraygani A, Joshaghani HR. Decreased polyunsaturated and increased saturated fatty acid concentration in spermatozoa from asthenozoospermic males as compared with normozoospermic males. Andrologia 2006;38(5):173–8.

[43] Aksoy Y, Aksoy H, Altinkaynak K, Aydin HR, Ozkan A. Sperm fatty acid composition in subfertile men. Prostaglandins Leukot Essent Fatty Acids 2006;75(2):75–9.

[44] Tavilani H, Doosti M, Nourmohammadi I, et al. Lipid composition of spermatozoa in normozoospermic and asthenozoospermic males. Prostaglandins Leukot Essent Fatty Acids 2007;77(1):45–50.

[45] Chavarro JE, Furtado J, Toth TL, et al. Trans-fatty acid levels in sperm are associated with sperm concentration among men from an infertility clinic. Fertil Steril 2011;95(5):1794–7.

[46] Afeiche MC, Gaskins AJ, Williams PL, et al. Processed meat intake is unfavorably and fish intake favorably associated with semen quality indicators among men attending a fertility clinic. J Nutr 2014;144(7):1091–8.

[47] Attaman JA, Toth TL, Furtado J, Campos H, Hauser R, Chavarro JE. Dietary fat and semen quality among men attending a fertility clinic. Hum Reprod 2012;27(5):1466–74.

[48] Conquer JA, Martin JB, Tummon I, Watson L, Tekpetey F. Effect of DHA supplementation on DHA status and sperm motilty in asthenozoospermic males. Lipids 2000;35:149–54.

[49] Robbins WA, Xun L, FitzGerald LZ, Esguerra S, Henning SM, Carpenter CL. Walnuts improve semen quality in men consuming a Western-style diet: randomized control dietary intervention trial. Biol Reprod 2012;87(4):101–8.

[50] Gaskins AJ, Sundaram R, Buck Louis GM, Chavarro JE. Seafood intake, sexual activity, and time to pregnancy. J Clin Edocrinol Metab 2018;103(7):2680–8.

[51] Jensen B. Rat testicular lipids and dietary isomeric fatty acids in essential fatty acid deficiency. Lipids 1976;11(3):179–88.

[52] Privett OS, Phillips F, Shimasaki H, Nozawa T, Nickell EC. Studies of effects of trans fatty acids in the diet on lipid metabolism in essential fatty acid deficient rats. Am J Clin Nutr 1977;30:1009–17.

[53] Jensen TK, Heitmann BL, Jensen MB, et al. High dietary intake of saturated fat is associated with reduced semen quality among 701 young Danish men from the general population. Am J Clin Nutr 2013;97(2):411–8.

[54] Chavarro JE, Minguez-Alarcon L, Mendiola J, Cutillas-Tolin A, Lopez-Espin JJ, Torres-Cantero AM. Trans fatty acid intake is inversely related to total sperm count in young healthy men. Hum Reprod 2014;29(3):429–40.

[55] Hanis T, Zidek V, Sachova J, Klir P, Deyl Z. Effects of dietary trans-fatty acids on reproductive performance of Wistar rats. Br J Nutr 1989;61(03):519–29.

[56] Veaute C, Andreoli MF, Racca A, et al. Effects of isomeric fatty acids on reproductive parameters in mice. Am J Reprod Immunol 2007;58(6):487–96.

[57] MInguez-Alarcón L, Chavarro JE, Mendiola J, et al. Fatty acid intake in relation to reproductive hormones and testicular volume among young healthy men. Asian J Androl 2017;19(2):184–90.

[58] Showell MG, Mackenzie-Proctor R, Brown J, Yazdani A, Stankiewicz MT, Hart RJ. Antioxidants for male subfertility. Cochrane Database Syst Rev 2014;(12).

[59] Kessopoulou E, Powers HJ, Sharma KK, et al. A double-blind randomized placebo cross-over controlled trial using the antioxidant vitamin E to treat reactive oxygen species associated male infertility. Fertil Steril 1995;64(4):825–31.

[60] Suleiman SA, Ali ME, Zaki ZM, el-Malik EM, Nasr MA. Lipid peroxidation and human sperm motility: protective role of vitamin E. J Androl 1996;17(5):530–7.

[61] Omu AE, Dashti H, Al-Othman S. Treatment of asthenozoospermia with zinc sulphate: andrological, immunological and obstetric outcome. Eur J Obstet Gynecol Reprod Biol 1998;79(2):179–84.

[62] Tremellen K. Oxidative stress and male infertility–a clinical perspective. Hum Reprod Update 2008;14(3):243–58.

[63] Aktan G, Dogru-Abbasoglu S, Kucukgergin C, Kadioglu A, Ozdemirler-Erata G, Kocak-Toker N. Mystery of idiopathic male infertility: Is oxidative stress an actual risk? Fertil Steril 2013;99(5):1211–5.

[64] Talevi R, Barbato V, Fiorentino I, Braun S, Longobardi S, Gualtieri R. Protective effects of in vitro treatment with zinc, d-aspartate and coenzyme q10 on human sperm motility, lipid peroxidation and DNA fragmentation. Reprod Biol Endocrinol 2013;11:81.

[65] Agarwal A, Nallella KP, Allamaneni SSR, Said TM. Role of antioxidants in treatment of male infertility. Reprod BioMed Online 2004;8(6):616–27.

[66] Minguez-Alarcon L, Mendiola J, Lopez-Espin JJ, et al. Dietary intake of antioxidant nutrients is associated with semen quality in young university students. Hum Reprod 2012;27(9):2807–14.

[67] Young SS, Eskenazi B, Marchetti FM, Block G, Wyrobek AJ. The association of folate, zinc and antioxidant intake with sperm aneuploidy in healthy non-smoking men. Hum Reprod 2008;23(5):1014–22.

[68] Tremellen K, Miari G, Froiland D, Thompson J. A randomised control trial examining the effect of an antioxidant (Menevit) on pregnancy outcome during IVF-ICSI treatment. Aust N Z J Obstet Gynaecol 2007;47(3):216–21.

[69] Geva E, Bartoov B, Zabludovsky N, Lessing JB, Lerner-Geva L, Amit A. The effect of antioxidant treatment on human spermatozoa and fertilization rate in an in vitro fertilization program. Fertil Steril 1996;66(3):430–4.

[70] Greco E, Romano S, Iacobelli M, et al. ICSI in cases of sperm DNA damage: beneficial effect of oral antioxidant treatment. Hum Reprod 2005;20(9):2590–4.

[71] Bailey LB, Gregory IIIJF. Folate. In: Bowman BA, Russell RM, editors. Present knowledge in nutrition. 9th ed. vol. 1. Washington, DC: ILSI Press; 2006. p. 125–37.

[72] Lucock M. Folic acid: nutritional biochemistry, molecular biology, and role in disease processes. Mol Genet Metab 2000;71(1–2):121–38.

[73] Holm J, Hansen SI, Høier-Madsen M. A high-affinity folate binding protein in human semen. Biosci Rep 1991;11(5):237–42.

[74] Holm J, Hansen SI, Høier-Madsen M, Christensen TB, Nichols CW. Characterization of a high-affinity folate receptor in normal and malignant human testicular tissue. Biosci Rep 1999;19(6):571–80.

[75] Chen Z, Karaplis AC, Ackerman SL, et al. Mice deficient in methylenetetrahydrofolate reductase exhibit hyperhomocysteinemia and decreased methylation capacity, with neuropathology and aortic lipid deposition. Hum Mol Genet 2001;10(5):433–43.

[76] Kelly TLJ, Neaga OR, Schwahn BC, Rozen R, Trasler JM. Infertility in 5,10-methylenetetrahydrofolate reductase (MTHFR)-deficient male mice is partially alleviated by lifetime dietary betaine supplementation. Biol Reprod 2005;72(3):667–77.

[77] Cosentino MJ, Pakyz RE, Fried J. Pyrimethamine: an approach to the development of a male contraceptive. Proc Natl Acad Sci U S A 1990;87(4):1431–5.

[78] Kalla NR, Saggar SK, Puri R, Mehta U. Regulation of male fertility by pyrimethamine in adult mice. Res Exp Med (Berl) 1997;197(1):45–52.

[79] Malik NS, Matlin SA, Fried J, Pakyz RE, Consentino MJ. The contraceptive effects of etoprine on male mice and rats. J Androl 1995;16(2):169–74.

[80] Khazamipour N, Noruzinia M, Fatehmanesh P, Keyhanee M, Pujol P. MTHFR promoter hypermethylation in testicular biopsies of patients with non-obstructive azoospermia: the role of epigenetics in male infertility. Hum Reprod 2009;24(9):2361–4.

[81] Tüttelmann F, Rajpert-De Meyts E, Nieschlag E, Simoni M. Gene polymorphisms and male infertility a meta-analysis and literature review. Reprod BioMed Online 2007;15:643–58.

[82] Lee H-C, Jeong Y-M, Lee SH, et al. Association study of four polymorphisms in three folate-related enzyme genes with non-obstructive male infertility. Hum Reprod 2006;21(12):3162–70.

[83] Bentivoglio G, Melica F, Cristoforoni P. Folinic acid in the treatment of human male infertility. Fertil Steril 1993;60(4):698–701.

[84] Wong WY, Merkus HMWM, Thomas CMG, Menkveld R, Zielhuis GA, Steegers-Theunissen RPM. Effects of folic acid and zinc sulfate on male factor subfertility: a double-blind, randomized, placebo-controlled trial. Fertil Steril 2002;77(3):491–8.

[85] Ebisch IMW, Pierik FH, Jong FHD, Thomas CMG, Steegers-Theunissen RPM. Does folic acid and zinc sulphate intervention affect endocrine parameters and sperm characteristics in men? Int J Androl 2006;29(2):339–45.

[86] Mendiola J, Torres-Cantero AM, Vioque J, et al. A low intake of antioxidant nutrients is associated with poor semen quality in patients attending fertility clinics. Fertil Steril 2010;93(4):1128–33.

[87] Wallock LM, Tamura T, Mayr CA, Johnston KE, Ames BN, Jacob RA. Low seminal plasma folate concentrations are associated with low sperm density and count in male smokers and nonsmokers. Fertil Steril 2001;75(2):252–9.

[88] Boxmeer JC, Smit M, Weber RF, et al. Seminal plasma cobalamin significantly correlates with sperm concentration in men undergoing IVF or ICSI procedures. J Androl 2007;28:521–7.

[89] Boxmeer JC, Smit M, Utomo E, et al. Low folate in seminal plasma is associated with increased sperm DNA damage. Fertil Steril 2009;92(2):548–56.

[90] Ly L, Chan D, Aarabi M, et al. Intergenerational impact of paternal lifetime exposures to both folic acid deficiency and supplementation on reproductive outcomes and imprinted gene methylation. Mol Hum Reprod 2017;23(7):461–77.

[91] Lambrot R, Xu C, Saint-Phar S, et al. Low paternal dietary folate alters the mouse sperm epigenome and is associated with negative pregnancy outcomes. Nat Commun 2013;4:2889.

[92] Kim HW, Kim KN, Choi YJ, Chang N. Effects of paternal folate deficiency on the expression of insulin-like growth factor-2 and global DNA methylation in the fetal brain. Mol Nutr Food Res 2013;57(4):671–6.

[93] Kim HW, Choi YJ, Kim KN, Tamura T, Chang N. Effect of paternal folate deficiency on placental folate content and folate receptor alpha expression in rats. Nutr Res Pract 2011; 5(2):112–6.

[94] Ganmaa D, Wang PY, Qin LQ, Hoshi K, Sato A. Is milk responsible for male reproductive disorders? Med Hypotheses 2001;57(4):510–4.

[95] Willingham EJ. Environmental review: trenbolone and other cattle growth promoters: need for a new risk-assessment framework. Environ Pract 2006;8(01):58–65.

[96] Ganmaa D, Qin L-Q, Wang P-Y, Tezuka H, Teramoto S, Sato A. A two-generation reproduction study to assess the effects of cows' milk on reproductive development in male and female rats. Fertil Steril 2004;82(Suppl. 3):1106–14.

[97] Pape-Zambito DA, Roberts RF, Kensinger RS. Estrone and 17β-estradiol concentrations in pasteurized-homogenized milk and commercial dairy products. J Dairy Sci 2010;93 (6):2533–40.

[98] Daxenberger A, Ibarreta D, Meyer HHD. Possible health impact of animal oestrogens in food. Hum Reprod Update 2001;7(3):340–55.

[99] Hartmann S, Lacorn M, Steinhart H. Natural occurrence of steroid hormones in food. Food Chem 1998;62(1):7–20.

[100] Maruyama K, Oshima T, Ohyama K. Exposure to exogenous estrogen through intake of commercial milk produced from pregnant cows. Pediatr Int 2010;52(1):33–8.

[101] Rich-Edwards J, Ganmaa D, Pollak M, et al. Milk consumption and the prepubertal somatotropic axis. Nutr J 2007;6(1):28.

[102] Adebamowo CA, Spiegelman D, Berkey CS, et al. Milk consumption and acne in teenaged boys. J Am Acad Dermatol 2008;58(5):787–93.

[103] Mendiola J, Torres-Cantero AM, Moreno-Grau JM, et al. Food intake and its relationship with semen quality: a case-control study. Fertil Steril 2009;91(3):812–8.

[104] Eslamian G, Amirjannati N, Rashidkhani B, Sadeghi M-R, Hekmatdoost A. Intake of food groups and idiopathic asthenozoospermia: a case-control study. Hum Reprod 2012;27 (11):3328–36.

[105] Afeiche MC, Williams P, Mendiola J, et al. Dairy food intake in relation to semen quality and reproductive hormone levels among physically active young men. Hum Reprod 2013;28 (8):2265–75.

[106] Afeiche MC, Bridges ND, Williams PL, et al. Dairy intake and semen quality among men attending a fertility clinic. Fertil Steril 2014;101:1280–7.

[107] Vujkovic M, de Vries JH, Dohle GR, et al. Associations between dietary patterns and semen quality in men undergoing IVF/ICSI treatment. Hum Reprod 2009;24(6):1304–12.

[108] Andersson A, Skakkebaek N. Exposure to exogenous estrogens in food: possible impact on human development and health. Eur J Endocrinol 1999;140(6):477–85.

[109] Henricks DM, Gray SL, Owenby JJ, Lackey BR. Residues from anabolic preparations after good veterinary practice. APMIS 2001;109(4):273–83.

[110] Swan SH, Liu F, Overstreet JW, Brazil C, Skakkebaek NE. Semen quality of fertile US males in relation to their mothers' beef consumption during pregnancy. Hum Reprod 2007;22 (6):1497–502.

[111] Sharpe RM. Environmental/lifestyle effects on spermatogenesis. Philos Trans R Soc B Biol Sci 2010;365(1546):1697–712.

[112] Afeiche MC, Williams PL, Mendiola J, et al. Meat intake and reproductive parameters among young men. Epidemiology 2014;25:323–30.

[113] European Commission. Opinion of the scientific committee on veterinary measures relating to public health. Assessment of potential risks to human health from hormone residues in bovine meat and meat products. In: Directorate General XXIV Cpachp, 1999.

[114] EPA-FDA. EPA-FDA advice about eating fish and shellfish. 2017, https://www.epa.gov/choose-fish-and-shellfish-wisely; 2017. Accessed 16 April 2018.

[115] Homma-Takeda S, Kugenuma Y, Iwamuro T, Kumagai Y, Shimojo N. Impairment of spermatogenesis in rats by methylmercury: involvement of stage- and cell- specific germ cell apoptosis. Toxicology 2001;169(1):25–35.

[116] Orisakwe OE, Afonne OJ, Nwobodo E, Asomugha L, Dioka CE. Low-dose mercury induces testicular damage protected by zinc in mice. Eur J Obstet Gynecol Reprod Biol 2001;95(1):92–6.

[117] Arabi M, Heydarnejad MS. In vitro mercury exposure on spermatozoa from normospermic individuals. Pak J Biol Sci 2007;10(15):2448–53.

[118] Ernst E, Lauritsen JG. Effect of organic and inorganic mercury on human sperm motility. Pharmacol Toxicol 1991;68(6):440–4.

[119] Mohamed MK, Burbacher TM, Mottet NK. Effects of methyl mercury on testicular functions in Macaca fascicularis monkeys. Pharmacol Toxicol 1987;60(1):29–36.

[120] Minguez-Alarcon L, Afeiche MC, Williams PL, et al. Hair mercury (Hg) levels, fish consumption and semen parameters among men attending a fertility center. Int J Hyg Environ Health 2018;221(2):174–82.

[121] WHO. Promoting fruit and vegetable consumption around the world, http://www.who.int/dietphysicalactivity/fruit/en/index2.html; 2018. Accessed 4 May 2018.

[122] FDA. Pesticide monitoring program fiscal year 2012 pesticide report, https://www.ams.usda.gov/datasets/pdp/pdpdata; 2012. Accessed 4 May 2018.

[123] Chiu YH, Gaskins AJ, Williams PL, et al. Intake of fruits and vegetables with low-to-moderate pesticide residues is positively associated with semen-quality parameters among young healthy men. J Nutr 2016;146(5):1084–92.

[124] Chiu YH, Afeiche MC, Gaskins AJ, et al. Fruit and vegetable intake and their pesticide residues in relation to semen quality among men from a fertility clinic. Hum Reprod 2015;30(6):1342–51.

[125] Baudry J, Assmann KE, Touvier M, et al. Association of frequency of organic food consumption with cancer risk: findings from the NutriNet-Sante prospective cohort study. JAMA Intern Med 2018;178(12):1597–606.

[126] Hemler EC, Chavarro JE, Hu FB. Organic foods for cancer prevention-worth the investment? JAMA Intern Med 2018;178(12):1606–7.

[127] Kuiper GGJM, Lemmen JG, Carlsson B, et al. Interaction of estrogenic chemicals and phytoestrogens with estrogen receptor β. Endocrinology 1998;139(10):4252–63.

[128] Branham WS, Dial SL, Moland CL, et al. Phytoestrogens and mycoestrogens bind to the rat uterine estrogen receptor. J Nutr 2002;132:658–64.

[129] Miksicek RJ. Interaction of naturally occurring nonsteroidal estrogens with expressed recombinant human estrogen receptor. J Steroid Biochem Mol Biol 1994;49:153–60.

[130] Matthews J, Celius T, Halgren R, Zacharewski T. Differential estrogen receptor binding of estrogenic substances: a species comparison. J Steroid Biochem Mol Biol 2000;74:223–34.

[131] Harris HA, Bapat AR, Gonder DS, Frail DE. The ligand binding profiles of estrogen receptors α and β are species dependent. Steroids 2002;67:379–84.

[132] Song TT, Hendrich S, Murphy PA. Estrogenic activity of glycitein, a soy isoflavone. J Agric Food Chem 1999;47:1607–10.

[133] Thomas P, Dong J. Binding and activation of the seven-transmembrane estrogen receptor GPR30 by environmental estrogens: a potential novel mechanism of endocrine disruption. J Steroid Biochem Mol Biol 2006;102:175–9.

[134] Atanassova N, McKinnell C, Turner KJ, et al. Comparative effects of neonatal exposure of male rats to potent and weak (environmental) estrogens on spermatogenesis at puberty and the relationship to adult testis size and fertility: evidence for stimulatory effects of low estrogen levels. Endocrinology 2000;141:3898–907.

[135] Fraser LR, Beyret E, Milligan SR, Adeoya-Osiguwa SA. Effects of estrogenic xenobiotics on human and mouse spermatozoa. Hum Reprod 2006;21:1184–93.

[136] Mitchell JH, Cawood E, Kinniburgh D, Provan A, Collins AR, Irvine DS. Effect of a phytoestrogen food supplement on reproductive health in normal males. Clin Sci (Lond) 2001;100:613–8.

[137] Song G, Kochman L, Andolina E, Herko RC, Brewer KJ, Lewis V. Beneficial effects of dietary intake of plant phytoestrogens on semen parameters and sperm DNA integrity in infertile men [abstract]. Fertil Steril 2006;86:S49.

[138] Chavarro JE, Toth TL, Sadio SM, Hauser R. Soy food and soy isoflavone intake in relation to semen quality parameters among men from an infertility clinic. Hum Reprod 2008;23(11):2584–90.

[139] Xia Y, Chen M, Zhu P, et al. Urinary phytoestrogen levels related to idiopathic male infertility in Chinese men. Environ Int 2013;59(0):161–7.

[140] Minguez-Alarcon L, Afeiche MC, Chiu YH, et al. Male soy food intake was not associated with in vitro fertilization outcomes among couples attending a fertility center. Andrology 2015;3(4):702–8.

[141] Mumford SL, Sundaram R, Schisterman EF, et al. Higher urinary lignan concentrations in women but not men are positively associated with shorter time to pregnancy. J Nutr 2014;144(3):352–8.

[142] Salas-Huetos A, Bullo M, Salas-Salvado J. Dietary patterns, foods and nutrients in male fertility parameters and fecundability: a systematic review of observational studies. Hum Reprod Update 2017;1–19.

[143] Giahi L, Mohammadmoradi S, Javidan A, Sadeghi MR. Nutritional modifications in male infertility: a systematic review covering 2 decades. Nutr Rev 2016;74(2):118–30.

[144] Li Y, Lin H, Li Y, Cao J. Association between socio-psycho-behavioral factors and male semen quality: systematic review and meta-analyses. Fertil Steril 2011;95(1):116–23.

[145] Liu CY, Chou YC, Chao JC, Hsu CY, Cha TL, Tsao CW. The association between dietary patterns and semen quality in a general Asian population of 7282 males. PLoS ONE 2015;10(7).

[146] Gaskins AJ, Colaci DS, Mendiola J, Swan SH, Chavarro JE. Dietary patterns and semen quality in young men. Hum Reprod 2012;27(10):2899–907.

[147] Efrat M, Stein A, Pinkas H, Unger R, Birk R. Dietary patterns are positively associated with semen quality. Fertil Steril 2018;109(5):809–16.

[148] Karayiannis D, Kontogianni MD, Mendorou C, Douka L, Mastrominas M, Yiannakouris N. Association between adherence to the Mediterranean diet and semen quality parameters in male partners of couples attempting fertility. Hum Reprod 2017;32(1):215–22.

[149] Oostingh EC, Steegers-Theunissen RP, de Vries JH, Laven JS, Koster MP. Strong adherence to a healthy dietary pattern is associated with better semen quality, especially in men with poor semen quality. Fertil Steril 2017;107(4):916–923.e912.

CHAPTER 1.4

The Gut Microbiome

Sarah Ashman*, Hari Krishnamurthy[†]

*Clinical Education, Vibrant America Clinical Laboratory, San Carlos, CA, United States,
[†]Technology Development, Vibrant Sciences, San Carlos, CA, United States

INTRODUCTION

The human superorganism is a collective term for our human bodies and the microbes they host, mostly symbiotic but occasionally pathogenic. The superorganism can be thought of as an interdependent ecosystem in which microbes rely on the host for necessary sustenance in order to thrive and in turn provide the human host with critical support functions such as immunomodulation.

Technological and computational advances in genome studies have enabled the study of the human microbiome, which is an order of magnitude higher in complexity compared with the human genome by itself. Much of the work in this field has been accomplished since the Human Microbiome Project started in 2007 by the National Institutes of Health. The first phase of the project involved the basic characterizing of human microbial flora. The second phase, called the Integrative Human Microbiome Project (iHMP), was started in 2014 with the aim of studying the role of microbes in health and disease states and is expected to lead to new discoveries in diagnosis, prognosis, and treatment of a variety of human diseases [1].

TECHNOLOGICAL APPROACHES TO STUDYING THE GUT MICROBIOME

DNA Approaches—What Is Present?

The complexity of the gut microbiome with its variety of species is challenging to study. A culture-independent method of evaluation is critical, because many of these organisms have never been cultured and it is unknown if novel growth conditions are necessary to culture some [2].

Initial studies in this space with bacteria involved targeted sequencing of specific regions such as 5S and 16S rRNA genes [3]. Though this method contributed to significant understanding of the microbiome, large-scale studies were not possible until the advent of next-generation sequencing. Sequenced organisms were grouped into operational taxonomic units and referred to as a single bacterial species when the sequence similarity was 97% [4]. Although more progress has been made in the study of bacteria than viruses and eukaryotic microbes, considerable efforts are now being made to reduce this imbalance. Viruses are primarily studied using shotgun sequencing and microarrays [5], while eukaryotic organisms are studied using 18S rRNA and specific signature sequences [6, 7].

RNA Approaches—What Is the Functional Pathway?

Apart from DNA-based studies, studies have also been done based on the RNA of these organisms that are referred to as metatranscriptomics. It is estimated that the microbiome expresses 100 times more genes than the human host. The activity of the microbiome can be gauged by studying microbial mRNA via sequencing. The main challenge in this space has been the stability of the RNA, which is being addressed in novel methods to extract and study it [8].

DIFFERENTIATION OF THE INTESTINAL MICROBIOME

Protein and Metabolite Approaches—Host Interaction and Outcome

Metaproteomics is the study of the microbial proteome, which is expected to give us a better understanding of how the gut microbiome interacts with the host. The complexity of looking for multiple proteins enables researchers to look at the functional characteristics and activity of the gut microbiome [9]. Another aspect of the gut microbiome is the study of the metabolites produced by them. These small molecules have been shown to affect the physiology and even modify the intestinal permeability of the host [10]. Both metaproteomics and metabolite studies of the gut microbiome are relatively new but hold promise to a better understanding of the human superorganism.

Within the superorganism, multiple microbiomes exist. An integumentary microbiome interacts with the external environment; a salivary microbiome can be either protective or destructive to oral health [11]; the intestinal microbiome is probably one of the most well known and studied of late; and the genital microbiome(s) [12], according to some animal studies and depending on gender, might influence reproductive health and possibly even epigenetics of offspring [13].

The intestinal microbiome has gained notoriety recently due to its pivotal role in human health. While current research into the intestinal microbiome is merely scratching the surface of what humans can know about their microscopic symbionts, much has already been elucidated in regard to this complex relationship. The majority of research has occurred and still does occur in animals; however, within the last decade, a significant number of human trials, both in vivo and in vitro, have begun to fortify the knowledge base and further explore the interplay of humans and microbes as they relate to a variety of disease states, nutritional outcomes, and epigenetic influences.

Broad Functions

The intestinal microbiome plays a number of roles in human health, from very complex immunomodulatory functions to broad influence on nutritional status. Microorganisms included in the intestinal microbiome include bacteria, yeasts/fungi, and sometimes parasites. Bacteria can be classified into three main groups: **commensal** microorganisms (those acquired through exposure over the life span and that help the host symbiotically), **pathobionts** (bacteria normally found in the intestinal tract, perform some beneficial functions, and whose populations may also be opportunistic without adequate commensal abundance), and **pathogens** (bacteria that cause disease and should not be found in any normal intestinal microbial ecosystem).

The main functions of the intestinal microbiome consist of immunomodulation, fiber fermentation, vitamin production, inflammatory response, competitive inhibition of pathogens, and mucosal barrier fortification. Refer to Table 1 for greater detail on each of these functions.

MICROBES AND IMMUNITY

T-cell differentiation is one of the primary immunomodulatory influences that microbes have on the host's immune system. T-cell differentiation is the act of T helper cells (Th) becoming regulatory T cells, or Tregs, and preventing unchecked Th17 immune responses. This occurs partly through the production of the short-chain fatty acids (SCFAs), butyrate, propionate, and acetate, which are the by-products of fiber fermentation by microbes in the colon. SCFAs bind to G-protein-coupled receptors (GPCRs) on intestinal epithelial cells and influence the activation or differentiation of T cells, in a dose-dependent manner [33]. SCFAs have differing mechanisms of influence on T-cell differentiation depending on which SCFA is acting. Refer to Fig. 1 for an illustration of specific mechanisms of SCFAs on T cells.

Table 1 Broad Functions of the Intestinal Microbiome

Major Function	Central Concepts
Immunomodulation	1. Recognition of immune-stimulatory bacterial molecules and microbial metabolites shapes the innate immune system [14]
	2. The absence of sufficient microbial populations and their microbial metabolites has downstream effects on immune function in the host [15]
Fiber fermentation	1. The human host's survival depends on its microscopic intestinal symbionts, as they depend on the host for both real estate and sustenance
	2. The majority of microbiota are capable of fermenting fiber, simple carbohydrates, and even complex carbohydrates
	3. Microbial abundance relies on the abundance of metabolites produced by other microbes [16], and the amount and types of fiber being fermented are just as relevant as the consistency or frequency of intake [17]
Vitamin and nutrient metabolism	1. The primary vitamins that intestinal microbes are known to liberate or synthesize are the B vitamins (primarily B12, B6, and folate) and vitamin K2 [18, 19]
	2. Humans lack B vitamin absorption mechanisms in the large bowel where the majority of vitamins are synthesized [20, 21]. Animal studies, however, have shown that some B vitamins may be absorbed into circulation [18]
	3. Oxalate-degrading microbes in the intestinal tract affect the absorption of oxalate as evidenced by urinary excretion rates and reduced incidence of renal calculi [22–24]
Inflammatory response	1. Intestinal microbiota modulate inflammatory responses, both locally in the intestines and systemically outside the gastrointestinal tract [15, 25, 26]
	2. If certain groups of commensal microorganisms are not acquired or cannot thrive, the host lacks some degree of immune development [27–29]
Competitive inhibition of pathogens	1. Commensals crowd out opportunistic and pathogenic microorganisms, preventing them from adhering to luminal surfaces [30]
	2. Commensal bacteria also perform interference inhibition by producing antimicrobial peptides (AMPs) that limit or suppress the growth of pathobionts and pathogens [30]
Mucosal barrier fortification	1. Commensal microbes in the mucosa provide metabolites that fortify the epithelial barrier through paracellular tight junction modulation, cellular cytoskeleton fortification, and short-chain fatty acid (SCFA) production [31, 32]

Homeostasis Is Maintained by Two-Way Communication Between the Microbiota and Intestinal Immune System

All biological systems and organisms will seek to maintain or return to homeostasis as a primary goal of survival. Maintenance of homeostasis relies upon delicate systems of communication between the host and the intestinal microbiome. The mutual survival of both parties is the ultimate goal. The primary methods of communication occur through metabolites produced by the microbes [34], antimicrobial peptides produced by the host and the microbes [30], and lipopolysaccharide (LPS) or lipoteichoic acid (LTA) on the outer membrane of microbes that allow the host immune system and other microbes to easily recognize friend from foe [35].

Microbes and Immunity

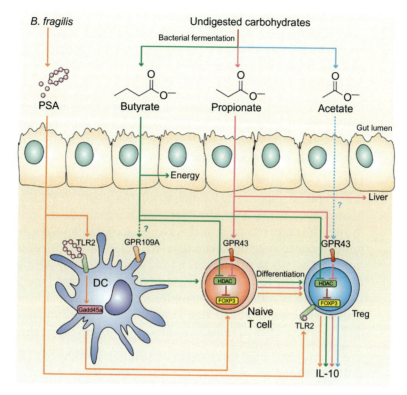

FIG. 1
Direct and indirect influence of short-chain fatty acids on immunomodulation through T-cell differentiation. *From: Hoeppli RE, Wu D, Cook L, Levings MK. The environment of regulatory T cell biology: cytokines, metabolites, and the microbiome. Front Immunol 2015;6:61. https://doi.org/10.3389/fimmu.2015.00061.*

Metabolites are able to modulate host immune responses [15], influence metabolism of nutrients [36], and even affect neurological processes such as appetite and mood [37]. These metabolites produced by microbes include SCFAs, such as butyrate, propionate, and acetate; lactic acid; and vitamins, such as K2 and biotin [38]. SCFAs from fermentable carbohydrates specifically modulate immune responses such as T-cell differentiation [33], promote satiety [39], and even modulate host gene expression through histone deacetylase activity [40].

The aforementioned antimicrobial peptides (AMPs) produced by intestinal microbes are capable of inhibiting the growth of pathobiont and pathogenic microorganisms and are released by commensal organisms when invasive microbes are sensed in their specific niche of the intestinal ecosystem. AMPs such as defensins and cathelicidins are also produced by the host to shape the intestinal microbiome of commensals and pathogens and are produced by cells in the Peyer's patches in response to changes in concentration and location of both commensal and invasive microorganisms [41]. The main mechanism of action of AMPs against various bacterial species is their ability to bind to bacterial membranes and disrupt them, leading to cell death [42]. Without adequate abundance of commensal microorganisms in the intestinal mucosa, production of AMPs for competitive inhibition of opportunists and pathogens is reduced.

One last category of metabolites that are produced by microorganisms and participate in modulation of the microbiome and host immune response is bacterial endotoxins: LPS produced by gram-negative bacteria and LTA produced by gram-positive bacteria.

LPS is a potent bacterial endotoxin involved in inflammatory stimulation of the host immune response by binding to toll-like receptors (TLRs) [35] in the local luminal sites, where it causes an increase in intestinal tight junction permeability, which allows relatively unchecked paracellular flow of substances from the luminal side to systemic circulation [43]. If LPS translocates across an impaired intestinal barrier [44], it has been linked to the development of the inflammatory precursors to metabolic syndrome, obesity, and type 2 diabetes [45]; propagation of inflammatory conditions such as osteoarthritis [46] and even neurological symptoms in vivo in mice and in vitro human studies [47, 48]; and, in severe cases of impaired immune function, septic shock [35].

LTA has a very similar mechanism of action in binding to TLRs, which initiate inflammatory responses from the host. Once it has bound to its ligand, the same types of inflammatory cytokine responses result from the host [35, 49].

MICROBIAL DIVERSITY

The key to a healthy intestinal microbiome in humans is diversity [50]. Taxa of microbiota rely on each other for sustenance and defense of their niches. With the loss of abundance of one group, others also decline. Thus, the symbiotic nature of the microbiome is not only as it relates to the host and microbes but also as microbes relate to one another.

The concept of diversity of the intestinal microbiome can be summarized as follows: the greater the number of species, the greater the entire system is at adapting and surviving. Low diversity is correlated with a number of chronic diseases including irritable bowel syndrome (IBS) [51], inflammatory bowel disease (IBD) [50], colorectal cancer [52], celiac disease [53], obesity [54], and even autism [55]. While human studies are merely beginning to elucidate cause-and-effect relationships between alterations in the intestinal microbiome and host disease, some have already shown that these microbial alterations, or dysbiosis, are at least not the result of the disease, but more likely to have been present before disease onset and, therefore, involved in its progression or exacerbation [56–58]. While long-term human studies are still pending, animal studies have shown so far that over four generations of nutritional changes that affect microbiome diversity, the low diversity of the intestinal microbiome becomes permanent in the host [59].

Keystone Species

Exposure to a variety of microbes throughout the earliest years of life is one of the critical components of the microbiota's development and maintenance. The concept of keystone species refers to those species that, in humans, are considered important, if not critical, to balanced immune responses, interconnected interactions with other taxa, proper homeostasis of the intestinal tract, and the host's long-term health. Taxa that have been identified as keystone groups include *Lactobacillus, Bifidobacterium, Eubacterium, Clostridia, Butyrivibrio, Roseburia, Akkermansia, Faecalibacterium, Bacillus, Prevotella, Lachnospiraceae, Ruminococcus, Oxalobacter,* and *Blautia* [60].

Diversity is acquired early in life from birth to as late as 5 years of age but does continue to evolve over the life span of the host in response to environmental influences [61]. Infants born via vaginal delivery acquire a microbiome that differs from those born via cesarean delivery [62]. Vaginally delivered infants acquire the first inhabitants of their internal microbiome within hours of birth, via vaginal and fecal exposure from their mother [63]. Infants fed primarily breastmilk tend to have much richer *Bifidobacterium* and some *Lactobacillus* abundance from these microbes naturally present in breastmilk [64]. As infants begin to add solid foods more indicative of adult eating patterns, such as complex carbohydrates, their microbiome fluctuates to resemble that of an adult pattern more closely [63].

While birth and early diet initially impart diversity, environment influences the development of diversity in the microbiome. The host's continued exposure to its surroundings is also a major contributory factor in the development of microbial diversity. The hygiene hypothesis suggests that children (and adults) who experience a greater diversity of microbes earlier in life have stronger immune systems, which can adapt to a greater number of threats without long-term harm to the host, and less incidence of chronic inflammatory diseases later in life [27–29, 65, 66]. Both early-life exposures to diverse environments through birth and breastfeeding provide important exposure to microbes, while environmental influences contribute as humans age.

Metabolic Crossfeeding

Another central concept to the microbiome's balance and adaptability is metabolic cross-feeding between taxa. Bifidobacteria can cross-feed with certain butyrate producers [16], for instance. This means that one bacterium produces a by-product or metabolite that feeds another. The *Lactobacillus* genus is well known for its production of lactic acid, which is used by other species, such as *Anaerostipes caccae*, to produce butyrate [67]. When levels of the primary genus fall, the secondary genera that rely on its metabolites are affected as well [67, 68]. In order to bring balance and homeostasis, consideration must be given to the interrelated nature of the ecosystem and not just one or two keystone species.

Suppressing Diversity Through Prescription Antibiotic Use

Use of prescription antibiotics is a critical and lifesaving tool for health-care providers. Because of widespread overuse in past decades, however, many species of microorganisms have grown tolerant or resistant to many antibiotics [69, 70]. Additionally, because recommending a probiotic supplement alongside an antibiotic drug is still not common in many health-care providers' protocols [71], research is discovering that individuals with multiple courses of antibiotics are often lacking sufficient abundance of keystone microorganisms, which have been almost permanently suppressed through that antibiotic use [72].

The use of probiotics concurrently or immediately following a course of antibiotics has been shown to lessen the side effects of antibiotics and may prevent some loss of microorganism diversity typically seen with standard and routine antibiotic use [71, 73–75].

Antibiotic effects on diversity are variable. A study conducted with 21 patients treated with broad-spectrum antibiotics found that patients saw an average of 25% reduction in microbial diversity found in fecal samples after antibiotic administration, and the number of taxa present fell from 29 to 12; furthermore, analysis revealed that taxa in the gram-negative groups grew in abundance in relation to the gram-positive microbes, causing a shift in gram-positive to gram-negative balance, which may explain inflammatory symptoms such as diarrhea [76]. Abundance of commensal microorganisms, particularly some SCFA producers, may be affected between 40 days [77] and 4 years [72] post treatment, while the *Firmicutes-Bacteroidetes* ratio increases in favor of Firmicutes, which has been linked to inflammatory conditions such as obesity [78, 79] and a lower competitive inhibition of pathogens [80].

COLON HEALTH AND COLORECTAL CANCER

Perhaps, one of the most beneficial discoveries of microbiome research in human health is the connection between SCFA levels in the colon and T-cell differentiation, immunomodulation, colitis, and colorectal cancer. The effect of butyrate levels on T-cell differentiation has been extensively researched in the pursuit of better understanding of chronic inflammatory diseases.

Butyrate's main contribution to immunomodulation in the intestinal lumen and colonocyte is its influence on T-cell differentiation through GPCRs, which influences those T cells' response to antigens, from sources such as diet, microbes, and environmental chemicals [33]. In two in vitro studies, one pathway through which butyrate's immunomodulation was shown was the inhibition of NF-κB [81] and modulating gene expression to reduce the risk of colorectal cancer [40]. It is this latter effect that is of particular importance to many health-care providers, as the early detection of colorectal cancer has contributed to a significant reduction (20%) of new colorectal cancer cases [82]. The risk for developing colorectal cancer is also slightly higher in men (4.49% over one's lifetime) than in women (4.15% over one's lifetime) [83].

Butyrate is not the only SCFA produced by colonic bacteria that benefits the colonocytes and intestinal immunity. Acetate and propionate are also by-products of fiber fermentation in the colon and have broad influence on T-cell differentiation through different mechanisms but yield the same overall result: greater immunomodulation to reduce aberrant immune responses, in addition to increases in *Lactobacillus* genus, *Lachnospiraceae* family, and an abundance of *Akkermansia* [84].

Abundance of SCFAs is correlated with lower risk of colitis [85]. SCFAs bind to GPCRs to stimulate production of Tregs, the abundance of which controls or influences inflammatory responses [86]. Increased inflammatory responses are associated with increased incidence of colitis [87], which, if left untreated, has a higher incidence of tumorigenesis and colorectal cancer [85].

MEN'S HEALTH AND THE GUT MICROBIOME

The gender of the host has been shown to affect the gut microbiota composition [88]. Specifically, some studies have also shown that gender-specific immune system modulation could be attributed to the gut microbiota [89, 90].

Gut Microbiome and Testosterone

Male obesity has been associated with decreased testosterone production and lowered fertility. A high-fat- or high-calorie-based diet affects the gut microbiome and could result in intestinal permeability leading to the circulation of LPS, which in turn could result in the body lowering testosterone production to help fight infections [91]. Testosterone is a known immunosuppressant, and it has been postulated that lowering testosterone could have resulted in evolutionary benefits to fight infection [92, 93].

Gut Microbiome and Prostate Health

Several studies have been conducted on the role of the gut microbiome in the pathogenesis of prostate cancer. One such study observed higher abundance of *Bacteroides massiliensis* in prostate cancer cases as compared with controls [94]. Meta studies have indicated multiple anatomic sites involved in prostate health and disease [95]. Dysbiosis of the microbiome could lead to inflammatory response that increases the chances of diseases at different anatomical sites. A novel microbiome-derived risk factor for prostate cancer based on 10 aberrant metabolic pathways has also been proposed [96]. Overall studies in this space are still preliminary, and more extensive studies are needed to understand the underlying relationships.

Cardiovascular Health

Due to technological limitations, the connection between the gut microbiome and cardiovascular inflammation has not been fully elucidated. The model of microbial influence on atherosclerosis has been studied in great depth in animals; however, human trials are still ongoing. Previous findings show that there is a direct and indirect influence from the periodontal microbiome on immune dysregulation and inflammation that precipitate atherosclerosis through disruption of endothelial cell function [97].

Making a leap to the contribution of inflammation from intestinal microbiota to atherosclerosis is much more complex. There appear to be mechanisms with potential influence on the development and propagation of inflammatory arterial plaque. Of most recent import is the involvement of intestinal bacteria that produce trimethylamine, a by-product of carnitine metabolism by microbes in the gut and that is converted to trimethylamine N-oxide (TMAO) in the liver. In animal models, TMAO causes atherosclerosis [98, 98a]. In human models, TMAO increases platelet production and reactivity and the formation of foam cells within plaques. There is also significantly higher relative risk of atherosclerosis in patients with the highest TMAO levels compared with the lowest levels [99].

In human studies of the microbiota/TMAO model, vegans fed carnitine did not have greater TMAO levels, which led researchers to theorize that this population lacks sufficient meat-degrading microbes due to dietary differences (Koeth et al., 2013). While carnitine is found most abundantly in red meat and eggs, other food sources of carnitine include poultry and fish. In studies comparing dietary carnitine sources, consumption of fish high in carnitine was shown to be protective against atherosclerosis, and therefore, carnitine alone cannot be the single high-risk variable in atherosclerosis [99]. Other studies in humans have shown that individuals with impaired renal function are most at risk for accumulation of TMAO and, therefore, see greater risk of atherosclerosis alongside intestinal microbiome dysbiosis [100].

Bogiatzi et al. [101] found, in a study of 316 at risk patients, that patients with the highest levels of atherosclerotic plaque had higher levels of TMAO compared with those in the lowest plaque group, independent of renal function. This study also stated that no differences between the microbiota of sample and control existed; however, no explanation of microbiome screening was given, nor were data provided on intestinal microbial composition of participants.

Kasselman et al. [102] posit that the intestinal microbiome plays a role in development of atherosclerotic plaques through inflammatory mechanisms originating in the gut, including the activation of NF-κB, which affects gene expression by activating inflammatory immune cells and downstream by-products such as cytokines, nitric oxide synthase (NOS), and leukocyte adhesion. The proposed mechanism of NF-κB relates to the overproduction of metabolites by microbes in the *Bacteroidetes phyla*, which is seen in higher abundance in individuals with obesity and metabolic syndrome. This increased binding of metabolites such as acetate and propionate, SCFAs favored in a *Bacteroidetes*-dominant microbiome type, to TLR-4 stimulates NF-κB, which stimulates production of inflammatory cytokines by adipocytes [103].

DIETARY INFLUENCE

Probiotic Foods

Perhaps, one of the most ubiquitous nutritional trends of the last decade has been the resurgence of interest in probiotics. Both probiotic foods and supplements work through modifying host immune responses and competitive inhibition of pathogens, with indirect benefits to nutrient digestion and absorption [104]. Many health claims are made about the powers of probiotics, but only a few have been substantiated in human or animal trials.

Fermented foods in the human diet have been found to date back at least 9000 years. Humans have fermented everything from fruits and vegetables, dairy, and alcoholic and nonalcoholic beverages [105]. An important distinction is that while all probiotic foods are fermented, not all fermented foods are probiotic. In order to be probiotic, a food or drink must contain live microbial cultures at the time of ingestion. Some fermentation processes do not yield live cultures, such as alcoholic beverages. The most popular traditional ferments found across broad cultures include yogurt, kombucha, sauerkraut, kimchi, and miso.

Probiotic foods have been found to positively influence the intestinal microbiome in both growth of commensal microorganisms and suppression of pathogens. Zou et al. [106] found that probiotic foods and supplements containing *Lactobacillus* species can prevent and even eradicate *Helicobacter pylori* infections, reducing the rates of stomach cancer. Chiu et al. [107] found that a fermented plant extract drink lowered body weight, body fat, and body mass index, while increasing total phenolic compounds in the plasma of individuals. The same drink also reduced total cholesterol and low-density lipoprotein cholesterol (LDL-C). Subjects were found to have increased abundance of *Bifidobacterium* and *Lactobacillus* genus, with reduced abundance of Escherichia coli and Clostridium perfringens. Another study in children using a probiotic drink containing *Lactobacillus casei* found that levels of *Bifidobacterium* and *Lactobacillus* increased, while levels of *Enterobacteriaceae* and *Staphylococcus* decreased [108]. A probiotic drink containing *Bifidobacterium animalis* was shown to potentiate colonic SCFA production and decrease abundance of pathobiont *Bilophila wadsworthia* [109].

Debate exists whether probiotic supplements confer benefit for the same period of time as probiotic foods. Elli et al. [110] suggest that bacteria in fermented foods last about as long as those found in supplements, approximately 5–7 days, based on stool sample collections of study participants. Bezkorovainy [111] suggests probiotic foods may confer more benefit or yield a greater number of surviving microbes due to the presence of naturally existing prebiotic compounds such as fiber and lactose.

Probiotic Supplements

Despite significant research into the field of probiotics, only a limited number of species have been studied in human trials. Most studies use probiotic strains that are dosed at much higher concentrations than what would be found in probiotic foods [112]. Probiotic supplements usually do not have prebiotics included as traditional fermented foods would; however, some findings suggest that prebiotics included with probiotics increase survivability in harsh gastrointestinal conditions [113].

Probiotics are commonly recommended by health-care providers for patients suffering from IBS; however, due to limited training and education of health-care professionals, recommendations vary over a wide range of products, as do the results of the patient. Most probiotics contain either *Lactobacilli* or *Bifidobacteria* or a combination of strains from both genera. A commonly used yeast-based probiotic strain is *Saccharomyces boulardii*.

Table 2 summarizes the most well-studied clinical benefits of probiotics by specific species association based primarily on human studies, but animal studies were noted.

Table 2 Probiotic Microorganisms With Known Clinical Benefits Based on Human Trials

Clinical Associations and Benefits	Probiotic Microorganisms Associated	Source
Antibiotic-associated diarrhea	*Saccharomyces boulardii* *Lactobacillus rhamnosus* LGG *Lactobacillus acidophilus* *Lactobacillus bulgaricus* *Bifidobacterium longum* *Bifidobacterium lactis*	[73, 104, 114]
Celiac disease	*Bifidobacterium breve*	[115, 116]
Clostridium difficile prevention or treatment	*Lactobacillus plantarum* 299V *Saccharomyces boulardii*	[117, 118]
Constipation	*Lactobacillus casei* *Lactobacillus plantarum* *Lactobacillus rhamnosus* LGG *Lactobacillus reuteri* *Bifidobacterium bifidum* *Bifidobacterium lactis* *Bifidobacterium infantis* *Bifidobacterium longum*	[114]
Depression	Multistrain product (*L. acidophilus*, *L. casei*, and *B. bifidum*) Multistrain product (*B. bifidum*, *B. lactis*, *L. acidophilus*, *L. brevis*, *L. casei*, *L. salivarius*, and L. *lactis*) *Lactobacillus helveticus* *Bifidobacterium longum*	[119–121]
Dermatological disorders	*Lactobacillus rhamnosus* LGG *Lactobacillus salivarius*	[114, 122–124]
Fortification of intestinal barrier (reducing permeability)	*Lactobacillus plantarum* 299V *Bifidobacterium*	[68, 117, 125, 126]
Glycemic control	Multispecies supplement (*L. acidophilus*, *L. casei*, *L. rhamnosus*, *L. bulgaricus*, *B. breve*, *B. longum*, and *Streptococcus thermophilus*)	[127]
Helicobacter pylori infection reduction	Bifidobacteria—multiple strains (BIR-0304, BIR-0307, BIR-0312, BIR-0324, BIR-0326, BIR-0349)	[128]
Inflammatory bowel disease (IBD)—Crohn's disease	*Saccharomyces boulardii* *Bifidobacterium*	[104, 129]
Inflammatory bowel disease (IBD)—ulcerative colitis (UC)	*Saccharomyces boulardii* *Bifidobacterium* VSL#3[a] Multiple strain product (*B. breve*, *B. bifidum*, and *L. acidophilus*)	[104, 130, 130a]
Intestinal hyperpermeability	*Lactobacillus plantarum* *Lactobacillus rhamnosus* LGG *Lactobacillus acidophilus* Multistrain product (*L. acidophilus*, *L. rhamnosus*, *L. casei*, *L. plantarum*, *L. fermentum*, *B. lactis*, *B. breve*, *B. bifidum*, and *S. thermophilus*)	[125, 131, 132]
Irritable bowel syndrome (diarrhea-predominant)	VSL#3[a] *Bifidobacterium longum* spp. *infantis* *Lactobacillus rhamnosus* LGG	[104, 133–135]

Continued

Table 2 Probiotic Microorganisms With Known Clinical Benefits Based on Human Trials *Continued*

Clinical Associations and Benefits	Probiotic Microorganisms Associated	Source
Kidney stones	*Oxalobacter formigenes*	[24]
Nonsteroidal antiinflammatory drug (NSAID) enteropathy	*Lactobacillus acidophilus*	[136]
Obesity	*Lactobacillus gasseri* *Lactobacillus acidophilus*	[114, 137]
Respiratory tract infections	*Lactobacillus rhamnosus* LGG *Lactobacillus plantarum* *Lactobacillus paracasei*	[138, 139]
Rheumatoid arthritis	*Lactobacillus casei* Multiple strain product (*L. acidophilus*, *L. casei*, and *B. bifidum*)	[140, 141]
Seasonal allergies	VSL#3 *Lactobacillus casei* Combination of *L. rhamnosus* LGG + *L. gasseri* *Bifidobacterium longum*	[142]
Travelers' diarrhea	*Lactobacillus rhamnosus* LGG *Saccharomyces boulardii* *Lactobacillus acidophilus* *Lactobacillus casei*	[104, 114]

[a]*VSL#3 is a multistrain probiotic product containing* L. casei, L. plantarum, L. acidophilus, L. bulgaricus, B. longum, B. breve, B. infantis, *and* S. thermophilus.

In addition to the disease- or symptom-specific benefits in Table 2, general benefits of specific probiotic supplements include reduction of inflammation caused by pathogenic or inflammatory microbes [143]; competitive inhibition of pathogens and prevention of pathogenic adhesion to mucosal surfaces [126]; and even contributing to the production and absorption of antioxidants from both diet and endogenous production, such as glutathione [127].

When selecting a probiotic product, consideration should be given to important features such as its colony-forming unit (CFU) count and its ability to deliver live microorganisms to the colon, where they are able to thrive and produce benefit to the host. In vitro studies have claimed that probiotic organisms do not survive digestion [144]; however, microencapsulated or enteric coated products are often more effective than products in standard capsules or powders due to allowing for better survival of extreme pH changes in the upper GI tract [104]. Probiotic supplements can be detected in the stool of the host for about 5–7 days after consumption and, therefore, must be consumed regularly to provide significant long-term benefit [110, 111]. Some concerns exist as to the contents of probiotic supplements being accurate based on labeling, as some have been found not to contain any live organisms or not contain the organisms listed on their labels [145, 146].

One last category of probiotics of relevance is that of spore-forming bacteria. Currently, only products containing *Bacillus* species are available commercially with spore-forming microorganisms, and some concerns exist about their safety in humans [147]. Limited research available suggests that species without toxin-producing genes are considered safe for human consumption [148]. Several human pathogens exist in the same genus (B. anthracis and B. cereus) but are unrelated to those commonly used in probiotic products (B. subtilis and B. coagulans) [149]. Spore formers are antibiotic-resistant while in their spore form. *Bacillus*, a genus traditionally introduced to humans through soil content or consumption, is more resilient in the human GI tract than nonspore formers due to its ability to survive more extreme pH and temperature [147]. *Bacillus* strains have been reported to display antimicrobial, antioxidative, and immune-modulatory activity in the host, due to their ability to produce AMPs [149].

Prebiotics

Fiber Variety and Diversity

Prebiotics are nutritional components that, when consumed by the host, impart benefit to commensal microbes in the GI tract and increase abundance of one or more of those microbes. Prebiotics can be found in plant sources and come from either fibrous components or phenolic components. Fibrous components of nutrients are not available for digestion by the human host due to a lack of appropriate fiber-degrading enzymes; however, commensal microbes within the host's GI tract possess such enzymes. SCFAs are just one by-product of that microbial digestion, and diets low in fiber and high in animal protein have been shown to decrease the abundance of SCFA (butyrate and acetate) in the colon [17]. Animal studies have also shown that, in addition to balancing the *Firmicutes-Bacteroidetes* ratio, they also dose-dependently increase satiety hormones in the host [150].

The main categories of fiber include fructooligosaccharides (FOS), galactooligosaccharides (GOS), xylooligosaccharides (XOS), resistant starch, and pectin. Each type of fiber has different effects on the microbiome and on the host's metabolism and health, due to differing chemical structures and metabolites of its fermentation.

FOS has a strong stimulatory effect on the abundance of *Bifidobacteria* and on levels of SCFAs found in the colon. This fiber group also increases the abundance of *Lactobacilli* in both the small and large intestines [151, 152]. In addition, the increase in SCFAs appears to reduce plasma levels of free fatty acids (FFA), which not only improves glucose uptake but also influences gut hormones such as GLP-1 and PYY, which influence glycemic control through stimulation of insulin production [151].

GOS also has stimulatory effects on the abundance of *Bifidobacteria* [153] and has been shown to significantly improve clinical outcomes for individuals with reduced lactose digestion and tolerance through increasing abundance of lactose-fermenting *Bifidobacterium*, *Faecalibacterium*, and *Lactobacillus* [154]. In addition, GOS has also shown to increase fecal butyrate concentrations [155].

XOS stimulates growth of *Bifidobacteria* [156–158] and has been shown to suppress overgrowths of pathogens in the GI tract [158]. A trial conducted by Van den Abbeele et al. [159] demonstrated that a XOS and green coffee extract blend increased *Bifidobacteria* and *Akkermansia muciniphila*, an important mucus-degrading commensal in the colon, and modulated the intestinal immune system to favor antiinflammatory responses.

Resistant starch (RS) is so named because its molecular structure prevents its digestion by host enzymes; therefore, it is resistant to digestion by the human host, but not by commensal microbes in the colon. RS reduces the abundance of pathogenic *Proteobacteria* such as E. coli/*Shigella* and significantly increases the abundance of *Bifidobacteria*, which increases SCFA levels in the colon [160]. RS also attenuates postprandial (but not fasting) insulin and glucose responses in insulin-resistant individuals, while increasing abundance of *Faecalibacterium*, *Roseburia*, *Akkermansia*, and *Ruminococcus*, which increases colonic SCFAs [84, 161–163].

Major dietary sources of prebiotic fibers are listed in Table 3 to assist in nutritional selection and variety for optimal intake of prebiotic foods.

Of note, prebiotics should be used with caution if one suspects the presence of small intestinal bacterial overgrowth (SIBO), as they may exacerbate overgrowths of microorganisms in the small bowel and contribute to further inflammatory symptoms [165–169].

Phenolic Compounds From Diet and Diversity
Phenolic compounds are chemicals in plants that impart some beneficial action either on the microbiota, which consume or convert them, or on the human host, who absorbs them. Phenolic compounds are usually pigment-associated,

Table 3 Sources of Prebiotic Fibers From Diet

Type of Fiber	Food Sources
Fructooligosaccharides (FOS) (primarily feed *Lactobacillus* and *Bifidobacteria*)	Chicory root, agave, bananas, inulin (onions, leeks, and garlic), asparagus, wheat, barley, nuts
Galactooligosaccharides (GOS) (primarily feed *Lactobacillus* and *Bifidobacteria*) [153]	Jerusalem artichokes, black beans, kidney beans, lima beans, beet roots, broccoli, chickpeas, lentils
Xylooligosaccharides (XOS) (primarily feed *Lactobacillus* and *Bifidobacteria*)	Milk, honey, vegetables with a high cellulose content (celery, Brussels sprouts, cabbage, kale, squash, and sprouts), rice bran, soybeans, bamboo shoots
Resistant starch (can be fermented to yield butyrate) [161]	Banana flour, cooked oats, lentils, green bananas, white beans, barley, green peas, whole wheat, nuts
Pectin (can be fermented to yield butyrate and increases intestinal epithelial cell proliferation [164])	Apples, pears, guavas, citrus fruits, plums, gooseberries

and the amount and diversity of them differ by the hue of the plant, the quality of the soil in which the plant grows, and factors such as the environmental conditions under which the plant is grown [170].

The chemical structure of phenolic compounds is what classifies each compound. The main mechanisms of benefit in the gastrointestinal tract appear to be the stimulation of growth of commensal microbes and the antimicrobial properties of phenolic compounds toward pathobiont and pathogenic microorganisms [171].

The most commonly used and potent antimicrobial phenolic compounds are presented in Table 4.

Phenolic compounds are known to ameliorate inflammation through a variety of pathways, including locally reducing oxidative stress in the intestinal tract; altering gene expression related to inflammatory response, which includes NF-κB; and suppression of downstream cytokine responses [182].

Microbes in the gastrointestinal tract act on phenolic compounds by performing deglycosylation, the hydrolysis of esters and amides, and deglucuronidation of excreted mammalian metabolites; aromatic dehydroxylation, demethoxylation, and demethylation; and hydrogenation, α-oxidation, and β-oxidation [189].

There exists some debate as to whether the majority of phenolic compounds are even available to the host, due to limited absorption in the gastrointestinal tract. Williamson and Clifford [189] suggest that there is some partial absorption of most phenolic compounds, while Kahle et al. [190] found that around 20% of phenolic compounds are absorbed by the host in the small intestine,

Table 4 Most Commonly Used and Potent Phenolic Compounds From Nutritional Sources

Plant Source/Phenolic Compound	Microbes Impacted	Main Attributes
Berberine (found most abundantly in goldenseal, Oregon grape, barberry, and Chinese goldthread plants)	Gram-negative microbes producing LPS	Improves intestinal barrier integrity in rats with endotoxemia (LPS); injections of LPS inhibited the expression of TJ proteins but was attenuated by berberine administration [172]
	Staphylococcus aureus	Berberine contains a weak acid that inhibits bacterial defenses against berberine alkaloids and enhances the antimicrobial properties of the polyphenol against S. aureus [173]
	Escherichia coli, Bacillus subtilis, S. aureus, and Enterococcus faecalis (methicillin-resistant S. aureus (MRSA) and vancomycin-resistant Enterococcus (VRE))	Inhibition of both gram-positive and gram-negative bacteria [171]
	Decreases Firmicutes and increases B. phyla, reducing overall ratio; decrease in Ruminococcus species	Berberine is an alkaloid and considered an "antibiotic with broad spectrum" [174]
Carvacrol (from oregano)	Escherichia coli O157:H7, Salmonella typhimurium, Listeria monocytogenes	Inhibition of pathogenic microorganisms [171]
	Mycobacterium tuberculosis, S. aureus, L. monocytogenes, E. coli, E. faecium, S. enterica, Pseudomonas aeruginosa	Inhibition of pathogenic microorganisms [171]
	Bacillus subtilis	Effective antimicrobial against soil-based bacteria [175]
	Klebsiella oxytoca, K. pneumoniae, E. coli	Inhibits pathogenic microorganisms [176]
	Enterohemorrhagic E. coli O157:H7 (EHEC)	Reduces EHEC motility and attachment to human intestinal epithelial cells and decreased Shiga-like toxin synthesis [177]
Thymol (from thyme)	Salmonella typhimurium, L. monocytogenes, M. tuberculosis	Inhibition of pathogenic microorganisms [171]
	Escherichia coli, K. pneumoniae, methicillin-resistant S. aureus (MRSA); extremely strong activity against 120 strains of Staphylococcus, Enterococcus, Escherichia, and P. genera	Inhibits growth of pathogenic microorganisms [178]
	Klebsiella oxytoca, K. pneumoniae, E. coli	Inhibits pathogenic microorganisms [176]
	Enterohemorrhagic E. coli O157:H7 (EHEC)	Reduces EHEC motility and attachment to human intestinal epithelial cells and decreased Shiga-like toxin synthesis [177]

Dietary Influence

Table 4 Most Commonly Used and Potent Phenolic Compounds From Nutritional Sources *Continued*

Plant Source/Phenolic Compound	Microbes Impacted	Main Attributes
Cinnamic acid and cinnamaldehyde (from cinnamon)	*Mycobacterium tuberculosis, E. coli* O157:H7, *S. typhimurium, L. monocytogenes*	Inhibit growth of pathogenic microorganisms [171]
	Pseudomonas aeruginosa	Strongly inhibit pathogenic microorganisms [179]
	Enterohemorrhagic *E. coli* O157:H7 (EHEC)	Reduce EHEC motility and attachment to human intestinal epithelial cells and decreased Shiga-like toxin synthesis [177]
Resveratrol (most concentrated in red wine)	*Helicobacter pylori*	May reduce abundance of pathogen and inflammation resulting in gastritis [180]
	Bifidobacteria, Akkermansia	Promotes abundance of commensal microorganisms [174]
	Enterococcus, Prevotella, Bacteroides, Bifidobacterium, B. uniformis, Eggerthella lenta, and *Blautia coccoides, Eubacterium rectale*	Daily consumption of red wine polyphenols for 4 weeks significantly increased commensal microorganisms [181]
	Vibrio cholerae, **Proteus mirabilis**	Inhibits growth of pathogenic microorganisms [171]
	Gram-negative LPS-producing bacteria	Reduces abundance of endotoxin-producing microorganisms [182]
Curcumin (from turmeric)	Decrease in *Enterobacteria* and Enterococci, increase in *Lactobacillus* and *Bifidobacterium*	Animal study: decreases abundance of pathogenic and pathobiont taxa and reduces translocation of microbes and endotoxins through intestinal epithelial barrier into systemic circulation [183]
	Increased the abundance of butyrate-producing bacteria	Animal study: suppression of NF-κB activation in colonic epithelial cells, increased fecal butyrate level, increased expansion of Treg cells regulatory dendritic cells [184]
	Reduction in *Prevotella*; significant increase in *Alistipes*; abundance of *Bacteroides* was significantly higher; *Lactobacillus* increased, *Ruminococcus* decreased	Animal study: significant beneficial effects on increases in *Lactobacillus*, and reducing *Firmicutes*-to-*Bacteroidetes* ratio and reducing abundance of opportunists [185]
	Nonpathogenic *E. coli*, may stimulate growth of beneficial strains	Exhibits the highest curcumin-converting ability [186]
Catechins/epigallocatechins (EGCG) (from tea)	*Helicobacter pylori* and *E. coli*	Inhibits growth of pathogenic microorganisms [187]
Quercetin	*Escherichia coli, Serratia*, and *K. pneumoniae*	Inhibits growth of pathogenic microorganisms [188]
	Bacteroides vulgatus, A. muciniphila	Increased abundance of commensal microorganisms [174]

while the majority arrive in tact in the colon and are degraded by colonic microbes, stimulating growth of certain taxa.

Examples of those phenolic compounds that stimulate growth of beneficial bacteria include hydroxycinnamic acid and chlorogenic acid in coffee (both also found in fruits), which stimulate butyrate producers in the colon [191], and anthocyanins in blueberries and other blue, purple, and red produce that have shown stimulatory effects on the growth of commensal microorganisms such as *Bifidobacteria* and *Akkermansia* [192, 193].

Thousands of plant-based phenolic compounds exist in the human diet, and the most optimal intake of those compounds will be a diet inclusive of a high degree of variety from plant sources. Table 5 lists 50 of the highest phenolic content foods commonly consumed in the human diet with the most prevalent phenolic compounds found in those foods.

ADDITIONAL BENEFICIAL NUTRIENTS

Beyond probiotics, prebiotics, and phenolic compounds, other micronutrients common to the human diet impart benefit to both the microbiota and the host. Table 6 lists and describes the most well studied of those, which include the fat-soluble vitamins A and D, omega-3 fatty acids, and the amino acid L-glutamine. All of these nutrients also impart benefit to the intestinal epithelial barrier through actions of immunomodulation and/or tight junction modulation [34, 84, 195, 196, 198, 203, 205, 210, 213, 214].

CONCLUSION

While the intestinal microbiome obviously holds influence over broad health functions in the host, there is still much to be studied in regard to its role in disease. Technological advances have allowed for more accurate sequencing of the microbiome and accelerated speed and reach of research, which have allowed for greater understanding of how the human microbiome differs from animal models in studies. The functions of the gut microbiome in human health and disease include immunomodulation, fiber fermentation, vitamin and nutrient metabolism, inflammatory response modulation, competitive inhibition of pathogens, and mucosal barrier fortification.

Communication between the microbiota and the human host is bidirectional and takes place via chemical messages from metabolites such as SCFAs, AMPs, and microbial membrane peptides. Of key importance to this communication is the diversity and abundance of species, which are affected by a number of

Table 5 Commonly Consumed High-Polyphenol Foods and Drinks by Phenolic Content and Associated Phenolic Compounds

	Food	Major Phenolic Compounds Found		Food	Major Phenolic Compounds Found
1	Cilantro	Caffeic acid, protocatechuic acid, glycitin, and vanillic acid [194]	26	Roasted soybean seed	Isoflavonoids: daidzein, glycitein, genistein, and glucosides
2	Cloves (spice)	Hydroxhenylpropenes: eugenol, acetyl eugenol	27	Milk chocolate	Flavanols: EC
3	Peppermint, dried (herb)	Flavonoids, eriocitrin; hydroxycinnamic acids, rosmarinic acid	28	Strawberry	Anthocyanins, flavanols, hydroxybenzoic acids, hydroxycinnamic acids, stilbenes
4	Celery seed	Apigenin, luteolin	29	Red raspberry	Anthocyanins, flavanols, hydroxycinnamic acids
5	Cocoa powder	Flavanols: epicatechin (EC)	30	Coffee	Phenolic acids: chlorogenic acid
6	Mexican oregano, dried (herb)	Dihydroquercetin; naringenin; luteolin; flavonols: galangin, quercetin	31	Ginger, dried (root)	Hydroxycinnamic acids, other: hydroxyphenylpropenes
7	Dark chocolate (70% or higher)	Flavanols, epicatechin (EC); hydroxycinnamic acid, ferulic acid	32	Whole grain wheat flour	Phenolic acids: hydroxybenzoic acids, hydroxycinnamic acids
8	Flaxseed meal	Hydroxycinnamic acids, lignans	33	Prune	Flavonols, hydroxycinnamic acids
9	Black elderberry	Anthocyanins; flavonols: quercetin	34	Almond	Flavonols: kaempferol, quercetin, hydroxybenzoic acids
10	Chestnut	Hydroxybenzoic acids: gallic acid, ellagic acid, tannins	35	Black grape	Anthocyanins, flavanols, stilbenes
11	Sage, dried (herb)	Hydroxybenzoic acids: gallic acid, vanillic acid; hydroxycinnamic acids: caffeic acid, rosmarinic acid	36	Red onion	Anthocyanins, flavonols
12	Rosemary, dried (herb)	Flavonols; hydroxycinnamic acids: rosmarinic acid, caffeic acid	37	Thyme, fresh (herb)	Flavones; hydroxycinnamic acids: rosmarinic acid, caffeic acid
13	Thyme, dried (herb)	Hydroxybenzoic acids; hydroxycinnamic acids, rosmarinic acid	38	Refined maize flour	Hydroxycinnamic acids
14	Blueberry	Anthocyanins; flavonols, quercetin; phenolic acids, chlorogenic acid	39	Soy, tempeh	Isoflavonoids: daidzein, glycitein, genistein, and glucosides
15	Capers (herb/seasoning)	Flavonols, kaempferol; quercetin	40	Whole grain rye flour	Alkylphenols
16	Curcumin	Curcuminoids, flavonoids, phenolic acids	41	Apple	Phlorizin; phenolic acids: chlorogenic acid, quercetin

Continued

Table 5 Commonly Consumed High-Polyphenol Foods and Drinks by Phenolic Content and Associated Phenolic Compounds *Continued*

	Food	Major Phenolic Compounds Found		Food	Major Phenolic Compounds Found
17	Black olive	Anthocyanins; flavones, luteolin; flavonols; hydroxycinnamic acids; tyrosols: hydroxytyrosol, oleuropein	42	Spinach	Flavonols
18	Hazelnut	Flavonols: epigallocatechin (EGCG)	43	Black tea	Flavanols: catechin, EGCG, procyanidin; flavonols: kaempferol, quercetin; hydroxybenzoic acids
19	Pecan nut	Flavonols: catechin, EGCG	44	Red wine	Phenolic acids, anthocyanins, tannins, stilbenes (resveratrol)
20	Plum	Phenolic acids: chlorogenic acid; procyanidins, anthocyanins	45	Green tea	Flavanols: EC, EGCG
21	Green olive	Hydroxycinnamic acids; tyrosols, oleuropein	46	Yellow onion	Flavonols: quercetin
22	Sweet basil, dried (herb)	Hydroxycinnamic acids	47	Pure apple juice	Dihydrochalcones; flavanols: catechin, procyanidin; flavonols: kaempferol, quercetin; hydroxycinnamic acids
23	Curry powder (spice)	Curcuminoids	48	Pure pomegranate juice	Punicalagin (an ellagitannin)
24	Sweet cherry	Anthocyanins, flavonols, hydroxycinnamic acids	49	Extra virgin olive oil	Tyrosols, lignans: pinoresinol; phenolic acids, hydrolysable tannins
25	Blackberry	Anthocyanins; flavanols, EC; phenolic acid, ellagic acid	50	Peaches (whole, including peel)	Flavanols, catechin; hydroxycinnamic acids

Data reproduced from (unless noted otherwise) Phenol Explorer. Rothwell JA, Pérez-Jiménez J, Neveu V, Medina-Ramon A, M'Hiri N, Garcia Lobato P, Manach C, Knox K, Eisner R, Wishart D, Scalbert A. Phenol-Explorer 3.0: a major update of the Phenol-Explorer database to incorporate data on the effects of food processing on polyphenol content. Database 2013. https://doi.org/10.1093/database/bat070.

influences from birth through adulthood, including environmental exposures and the use of antibiotics. Low diversity is associated with a number of chronic diseases in humans. Restoring balance to this ecosystem therapeutically is possible through a combination of prebiotic foods, probiotic foods and supplements, and plants high in phenolic compounds.

Table 6 Additional Beneficial Nutrients With Immune and Microbial Modulatory Properties

Nutrient	Associations With Immunity	Associations With Microbiota
Vitamin A	Dietary intake of fiber can alter the populations of SCFA-producing microbes favoring growth of *Lactobacillus* genus, *Bacteroidetes* phylum, and *Akkermansia* genus [84] The conversion of vitamin A to retinoic acid, its active form, influences dendritic cell communication with T cells to increase differentiation into Tregs. Without adequate levels of retinoic acid, tolerogenic dendritic cells have reduced T-cell differentiation [84] Retinoic acid influences dendritic cell signaling to promote Treg production [34] Trans-retinoic acid prevents the overconversion of T cells into Th-like cells, which can lead to autoimmunity; adequate retinoic acid selects for Treg production and reduces autoimmune responses [196]	Retinoic acid/vitamin A levels are associated with amelioration of pathogen inflammation and increased levels of zonulin (ZO-1) and occludin proteins into the cellular tight junctions [195]
Vitamin D	Vitamin D influences both FOXp3- and IL-10-producing Tregs and inhibits Th17 pro-inflammatory pathways [34] Defensins are produced within human macrophages and dendritic cells, influenced by vitamin D; vitamin D induces host response to microbes [198] Vitamin D stimulates production of pattern recognition receptors, antimicrobial peptides, and cytokines and affects microbe sensing and inhibition of pathogens [198]	5000 IU of vitamin D3 per day for 90 days increased the abundance of *Akkermansia*, which promotes immune tolerance and increased butyrate producers *Faecalibacterium* and *Coprococcus* [197] Vitamin D3 treatment caused beneficial changes in *Firmicutes, Actinobacteria,* and *Proteobacteria* levels in MS patients and an increase in *Enterobacteria* [199] Animal study showed that high-dose vitamin D3 supplementation is associated with a shift to a more inflammatory fecal microbiome and increased susceptibility to colitis in animals with genetic predisposition to colitis, with a fall in circulating vitamin D occurring as a secondary event in response to the inflammatory process [200] Vitamin D deficiency changes the intestinal microbiome and reduces microbial B vitamin production in the gut. A lack of pantothenic acid results and produces a pro-inflammatory state [201] Vitamin D3 supplementation changed the gut microbiome in the upper GI tract to a less inflammatory state [202]
Omega-3 fatty acids	Eicosapentaenoic acid reinforces cellular tight junctions [203]	Higher omega-3 (specifically DHA) intake is highly correlated with microbiome diversity [204] Increase is seen in butyrate-producing *Lachnospiracea* species: *Eubacterium,*

Continued

Table 6 Additional Beneficial Nutrients With Immune and Microbial Modulatory Properties *Continued*

Nutrient	Associations With Immunity	Associations With Microbiota
	An in vitro study found that omega-3s alone decreased epithelial permeability and improved tight junction stability [205]	*Roseburia, Anaerostipes*, and *Coprococcus* [206]
		Essential fatty acid double bonds being hydrolyzed in the large intestine has antimicrobial properties and increases *Bifidobacteria* while decreasing intestinal permeability [206]
		Omega-3s from plant sources may lower *Bacteroidetes*; in an animal study, a significant decrease in the proportion of phylum *Bacteroidetes* species was observed; a saturated fatty acid-rich diet group showed a significantly greater decrease in *Bacteroidetes* proportion (resulting in a higher *Firmicutes*-to-*Bacteroidetes* ratio) [207]
		In an animal study, fish sources rich in omega-3 fatty acids appear to bring *Firmicutes*-to-*Bacteroidetes* ratio back into balance [208]
		Omega-3 fatty acids decreased *Firmicutes*-to-*Bacteroidetes* ratio [209]
L-Glutamine	An animal study in rats suggests that glutamine alone, but not glutamine with arginine, improves intestinal barrier permeability that resulted from chemotherapy [210]	After 14 days of supplementation, subjects in the glutamine group had significant differences in the *Firmicutes* and *Actinobacteria* phyla compared with those in the control group [211]
	In vitro glutamine and arginine supplementation improve methotrexate-induced barrier permeability [212]	
	Glutamine is used by rapidly dividing epithelial cells, enhances expression of tight junction protein genes and production of tight junction proteins to fortify the barrier [213]	
	Glutamine supplementation resulted in increased secretory IgA (SIgA) and a shift in the *Firmicutes*-to-*Bacteroidetes* ratio to favor *Bacteroidetes* in the ileum and increased the abundance of *Streptococcus* and *Bifidobacterium* in the jejunum [214]	

References

[1] Turnbaugh PJ, Ley RE, Hamady M, Fraser-Liggett C, Knight R, Gordon JI. The human microbiome project: exploring the microbial part of ourselves in a changing world. Nature 2007;449(7164):804–10. https://doi.org/10.1038/nature06244.

[2] Weinstock GM. Genomic approaches to studying the human microbiota. Nature 2012;489 (7415):250–6. https://doi.org/10.1038/nature11553.

[3] Olsen GJ, Lane DJ, Giovannoni SJ, Pace NR, Stahl DA. Microbial ecology and evolution: a ribosomal RNA approach. Annu Rev Microbiol 1986;40(1):337–65. https://doi.org/10.1146/annurev.mi.40.100186.002005.

References

[4] Mysara M, Vandamme P, Props R, Kerckhof F-M, Leys N, Boon N, Monsieurs P. Reconciliation between operational taxonomic units and species boundaries. FEMS Microbiol Ecol 2017;93(4):fix029. https://doi.org/10.1093/femsec/fix029.

[5] Rampelli S, Turroni S. From whole-genome shotgun sequencing to viral community profiling: the ViromeScan tool. In: Pantaleo V, Chiumenti M, editors. Viral metagenomics. Methods in molecular biology. vol. 1746. New York, NY: Humana Press; 2018.

[6] Gupta RS. Protein phylogenies and signature sequences: a reappraisal of evolutionary relationships among archaebacteria, eubacteria, and eukaryotes. Microbiol Mol Biol Rev 1998;62(4):1435–91.

[7] Wang Y, Tian RM, Gao ZM, Bougouffa S, Qian P. Optimal eukaryotic 18S and universal 16S/18S ribosomal RNA primers and their application in a study of symbiosis. PLoS ONE 2014;9(3). https://doi.org/10.1371/journal.pone.0090053.

[8] Bashiardes S, Zilberman-Schapira G, Elinav E. Use of metatranscriptomics in microbiome research. Bioinform Biol Insights 2016;2016(10):19–25. https://doi.org/10.4137/BBI.S34610.

[9] Lee PY, Chin S, Neoh H, Jamal R. Metaproteomic analysis of human gut microbiota: where are we heading? J Biomed Sci 2017;24(1):1–8. https://doi.org/10.1186/s12929-017-0342-z.

[10] Hofer U. Microbiome: precision engineering of gut metabolites. Nat Rev Microbiol 2017;16(1):2–3. https://doi.org/10.1038/nrmicro.2017.159.

[11] Fábián T, Fejérdy P, Csermely P. Salivary genomics, transcriptomics and proteomics: the emerging concept of the oral ecosystem and their use in the early diagnosis of cancer and other diseases. Curr Genomics 2008;9(1):11–21. https://doi.org/10.2174/138920208783884900.

[12] Mändar R, Punab M, Borovkova N, Lapp E, Kiiker R, Korrovits P, Truu J. Complementary seminovaginal microbiome in couples. Res Microbiol 2015;166(5):440–7. https://doi.org/10.1016/j.resmic.2015.03.009.

[13] Javurek AB, Spollen WG, Ali AMM, Johnson SA, Lubahn DB, Bivens NJ, Rosenfeld CS. Discovery of a novel seminal fluid microbiome and influence of estrogen receptor alpha genetic status. Sci Rep 2016;6(1):23027. https://doi.org/10.1038/srep23027.

[14] Shapiro H, Thaiss CA, Levy M, Elinav E. The cross talk between microbiota and the immune system: metabolites take center stage. Curr Opin Immunol 2014;30:54–62. https://doi.org/10.1016/j.coi.2014.07.003.

[15] Sharon G, Garg N, Debelius J, Knight R, Dorrestein P, Mazmanian S. Specialized metabolites from the microbiome in health and disease. Cell Metab 2014;20(5):719–30. https://doi.org/10.1016/j.cmet.2014.10.016.

[16] Belenguer A, Duncan SH, Calder AG, Holtrop G, Louis P, Lobley GE, Flint HJ. Two routes of metabolic cross-feeding between *Bifidobacterium adolescentis* and butyrate-producing anaerobes from the human gut. Appl Environ Microbiol 2006;72(5):3593–9. https://doi.org/10.1128/AEM.72.5.3593-3599.2006.

[17] Levy R, Borenstein E. Metagenomic systems biology and metabolic modeling of the human microbiome: from species composition to community assembly rules. Gut Microbes 2014;5(2):265–70. https://doi.org/10.4161/gmic.28261.

[18] Rossi M, Amaretti A, Raimondi S. Folate production by probiotic bacteria. Nutrients 2011;3(1):118–34. https://doi.org/10.3390/nu3010118.

[19] Sato T, Yamada Y, Ohtani Y, Mitsui N, Murasawa H, Araki S. Production of menaquinone (vitamin K 2)-7 by Bacillus subtilis. J Biosci Bioeng 2001;91(1):16–20. https://doi.org/10.1016/S1389-1723(01)80104-3.

[20] Degnan PH, Taga ME, Goodman AL. Vitamin B12 as a modulator of gut microbial ecology. Cell Metab 2014;20(5):769–78. https://doi.org/10.1016/j.cmet.2014.10.002.

[21] Magnúsdóttir S, Ravcheev D, de Crécy-Lagard V, Thiele I. Systematic genome assessment of B-vitamin biosynthesis suggests co-operation among gut microbes. Front Genet 2015;6:148. https://doi.org/10.3389/fgene.2015.00148.

[22] Mehta M, Goldfarb DS, Nazzal L. The role of the microbiome in kidney stone formation. Int J Surg 2016;https://doi.org/10.1016/j.ijsu.2016.11.024.

[23] Sadaf H, Raza SI, Hassan SW. Role of gut microbiota against calcium oxalate. Microb Pathog 2017;109:287–91. https://doi.org/10.1016/j.micpath.2017.06.009.

[24] Sikora P, Niedźwiadek J, Mazur E, Paluch-Oleś J, Zajączkowska M, Kozioł-Montewka M. Intestinal colonization with oxalobacter formigenes and its relation to urinary oxalate excretion in pediatric patients with idiopathic calcium urolithiasis. Arch Med Res 2009;40(5):369–73. https://doi.org/10.1016/j.arcmed.2009.05.004.

[25] Parks OB, Pociask DA, Hodzic Z, Kolls JK, Good M. Interleukin-22 signaling in the regulation of intestinal health and disease. Front Cell Dev Biol 2015;3:85.

[26] Thaiss CA, Zmora N, Levy M, Elinav E. The microbiome and innate immunity. Nature 2016;535(7610):65–74. https://doi.org/10.1038/nature18847.

[27] Ege MJ, Mayer M, Normand A, Genuneit J, Cookson WOCM, Braun-Fahrländer C, GABRIELA Transregio 22 Study Group. Exposure to environmental microorganisms and childhood asthma. N Engl J Med 2011;364(8):701–9. https://doi.org/10.1056/NEJMoa1007302.

[28] Neu J, Rushing J. Cesarean versus vaginal delivery: long term infant outcomes and the hygiene hypothesis. Clin Perinatol 2011;38(2):321–31. https://doi.org/10.1016/j.clp.2011.03.008.

[29] von Mutius E, Vercelli D. Farm living: effects on childhood asthma and allergy. Nat Rev Immunol 2010;10(12):861–8. https://doi.org/10.1038/nri2871.

[30] Umu ÖCO, Rudi K, Diep DB. Modulation of the gut microbiota by prebiotic fibres and bacteriocins. Microb Ecol Health Dis 2017;28(1). https://doi.org/10.1080/16512235.2017.1348886. 1348886-11.

[31] Noble EE, Hsu TM, Kanoski SE. Gut to brain dysbiosis: mechanisms linking western diet consumption, the microbiome, and cognitive impairment. Front Behav Neurosci 2017;11. https://doi.org/10.3389/fnbeh.2017.00009.

[32] Riordan S, McIver C, Thomas D, Dunscombe V, Bolin T, Thomas M. Luminal bacteria and small-intestinal permeability. Scand J Gastroenterol 1997;32:556–63.

[33] Bollrath J, Powrie FM. Controlling the frontier: regulatory T-cells and intestinal homeostasis. Semin Immunol 2013;25(5):352–7. https://doi.org/10.1016/j.smim.2013.09.002.

[34] Hoeppli RE, Wu D, Cook L, Levings MK. The environment of regulatory T cell biology: cytokines, metabolites, and the microbiome. Front Immunol 2015;6:61. https://doi.org/10.3389/fimmu.2015.00061.

[35] Warshakoon HJ, Burns MR, David SA. Structure-activity relationships of antimicrobial and lipoteichoic acid-sequestering properties in polyamine sulfonamides. Antimicrob Agents Chemother 2009;53(1):57–62. https://doi.org/10.1128/AAC.00812-08.

[36] Maurice C, Haiser H, Turnbaugh P. Xenobiotics shape the physiology and gene expression of the active human gut microbiome. Cell 2013;152(1–2):39–50. https://doi.org/10.1016/j.cell.2012.10.052.

[37] Hsiao E, McBride S, Hsien S, Sharon G, Hyde E, McCue T, Mazmanian S. Microbiota modulate behavioral and physiological abnormalities associated with neurodevelopmental disorders. Cell 2013;155(7):1451–63. https://doi.org/10.1016/j.cell.2013.11.024.

[38] Said HM. Intestinal absorption of water-soluble vitamins in health and disease. Biochem J 2011;437(3):357–72. https://doi.org/10.1042/BJ20110326.

[39] Havenaar R. Intestinal health functions of colonic microbial metabolites: a review. Benefic Microbes 2011;2(2):103–14. https://doi.org/10.3920/BM2011.0003.

[40] Waldecker M, Kautenburger T, Daumann H, Busch C, Schrenk D. Inhibition of histone-deacetylase activity by short-chain fatty acids and some polyphenol metabolites formed in the colon. J Nutr Biochem 2008;19(9):587–93. https://doi.org/10.1016/j.jnutbio.2007.08.002.

[41] Ostaff MJ, Stange EF, Wehkamp J. Antimicrobial peptides and gut microbiota in homeostasis and pathology. EMBO Mol Med 2013;5(10):1465–83. https://doi.org/10.1002/emmm.201201773.

[42] Zhang L, Gallo R. Antimicrobial peptides. Curr Biol 2015;0960-9822. 26(1):R14–9. https://doi.org/10.1016/j.cub.2015.11.017.

[43] Guo S, Al-Sadi R, Said HM, Ma TY. Lipopolysaccharide causes an increase in intestinal tight junction permeability in vitro and in vivo by inducing enterocyte membrane expression and localization of TLR-4 and CD14. Am J Pathol 2013;182(2):375–87. https://doi.org/10.1016/j.ajpath.2012.10.014.

[44] Cani PD, Bibiloni R, Knauf C, Waget A, Neyrinck AM, Delzenne NM, Burcelin R. Changes in gut microbiota control metabolic endotoxemia-induced inflammation in high-fat diet-induced obesity and diabetes in mice. Diabetes 2008;57(6):1470–81. https://doi.org/10.2337/db07-1403-1403.

[45] Monte S, Caruana J, Ghanim H, Sia CL, Korzeniewski K, Schentag J, Dandona P. Reduction in endotoxemia, oxidative and inflammatory stress, and insulin resistance after Roux-en-Y gastric bypass surgery in patients with morbid obesity and type 2 diabetes mellitus. Surgery 2012;151(4):587–93. https://doi.org/10.1016/j.surg.2011.09.038.

[46] Huang Z, Kraus VB. Does lipopolysaccharide-mediated inflammation have a role in OA? Nat Rev Rheumatol 2016;12(2):123–9. https://doi.org/10.1038/nrrheum.2015.158.

[47] Jeong H-K, Jou I, Joe E. Systemic LPS administration induces brain inflammation but not dopaminergic neuronal death in the substantia nigra. Exp Mol Med 2010;42(12):823–32. https://doi.org/10.3858/emm.2010.42.12.085.

[48] Qin L, Wu X, Block ML, Liu Y, Breese GR, Hong J-S, Crews FT. Systemic LPS causes chronic neuroinflammation and progressive neurodegeneration. Glia 2007;55(5):453–62. https://doi.org/10.1002/glia.20467.

[49] Ginsburg I. Role of lipoteichoic acid in infection and inflammation. Lancet Infect Dis 2002;2(3):171–9. https://doi.org/10.1016/S1473-3099(02)00226-8.

[50] Gong D, Gong X, Wang L, Yu X, Dong Q. Involvement of reduced microbial diversity in inflammatory bowel disease. Gastroenterol Res Pract 2016;2016:1–7. https://doi.org/10.1155/2016/6951091.

[51] Carroll IM, Ringel-Kulka T, Siddle JP, Ringel Y. Alterations in composition and diversity of the intestinal microbiota in patients with diarrhea-predominant irritable bowel syndrome. Neurogastroenterol Motil 2012;24:521–30. https://doi.org/10.1111/j.1365-2982.2012.01891.x.

[52] Ahn J, Sinha R, Pei Z, Dominianni C, Goedert JJ, Hayes RB, Yang L. Abstract 2290: Human gut microbiome and risk of colorectal cancer, a case-control study. Cancer Res 2013;73 (8 Suppl):2290. https://doi.org/10.1158/1538-7445.AM2013-2290.

[53] Schippa S, Iebba V, Barbato M, Di Nardo G, Totino V, Checchi MP, Conte MP. A distinctive "microbial signature" in celiac pediatric patients. BMC Microbiol 2010;10:175. https://doi.org/10.1186/1471-2180-10-175.

[54] Turnbaugh PJ, Hamady M, Yatsunenko T, Cantarel BL, Duncan A, Ley RE, Gordon JI. A core gut microbiome in obese and lean twins. Nature 2009;457(7228):480–4. https://doi.org/10.1038/nature07540.

[55] Kang D-W, Park JG, Ilhan ZE, Wallstrom G, LaBaer J, Adams JB, Krajmalnik-Brown R. Reduced incidence of Prevotella and other fermenters in intestinal microflora of autistic children. PLoS ONE 2013;8(7). https://doi.org/10.1371/journal.pone.0068322.

[56] Abrahamsson TR, Jakobsson HE, Andersson AF, Björkstén B, Engstrand L, Jenmalm MC. Low gut microbiota diversity in early infancy precedes asthma at school age. Clin Exp Allergy 2014;44:842–50. https://doi.org/10.1111/cea.12253.

[57] De Goffau MC, Luopajärvi K, Knip M, Ilonen J, Ruohtula T, Härkönen T, Vaarala O. Fecal microbiota composition differs between children with β-cell autoimmunity and those without. Diabetes 2013;62(4):1238–44. https://doi.org/10.2337/db12-0526.

[58] Wang M, Karlsson C, Olsson C, Adlerberth I, Wold AE, Strachan DP, Institutionen för kliniska vetenskaper, sektionen för kvinnors och barns hälsa. Reduced diversity in the early fecal microbiota of infants with atopic eczema. J Allergy Clin Immunol 2008;121(1):129–34. https://doi.org/10.1016/j.jaci.2007.09.011.

[59] Sonnenburg ED, Smits SA, Tikhonov M, Higginbottom SK, Wingreen NS, Sonnenburg JL. Diet-induced extinctions in the gut microbiota compound over generations. Nature 2016;529(7585):212–5. https://doi.org/10.1038/nature16504.

[60] Trosvik P, de Muinck EJ. Ecology of bacteria in the human gastrointestinal tract—identification of keystone and foundation taxa. Microbiome 2015;3. https://doi.org/10.1186/s40168-015-0107-4.

[61] Koenig JE, Spor A, Scalfone N, Fricker AD, Stombaugh J, Knight R, Ley RE. Succession of microbial consortia in the developing infant gut microbiome. Proc Natl Acad Sci U S A 2011;108(Suppl. 1):4578–85. https://doi.org/10.1073/pnas.1000081107.

[62] Salminen S, Gibson GR, McCartney AL, Isolauri E. Influence of mode of delivery on gut microbiota composition in seven year old children. Gut 2004;53(9):1388–9. https://doi.org/10.1136/gut.2004.041640.

[63] Palmer C, Bik E, DiGiulio D, Relman D, Brown P. Development of the human infant intestinal microbiota. PLoS Biol 2007;5:e177. https://doi.org/10.1371/journal.pbio.0050177.

[64] Soto A, Martín V, Jiménez E, Mader I, Rodríguez JM, Fernández L. Lactobacilli and Bifidobacteria in human breast milk: influence of antibiotherapy and other host and clinical factors. J Pediatr Gastroenterol Nutr 2014;59(1):78–88. https://doi.org/10.1097/MPG.0000000000000347.

[65] Correale J, Farez MF. The impact of parasite infections on the course of multiple sclerosis. J Neuroimmunol 2011;233(1):6–11. https://doi.org/10.1016/j.jneuroim.2011.01.002.

[66] Strachan DP. Hay fever, hygiene, and household size. Br Med J 1989;299(6710):1259–60.

[67] Moens F, Verce M, De Vuyst L. Lactate- and acetate-based cross-feeding interactions between selected strains of lactobacilli, bifidobacteria and colon bacteria in the presence of inulin-type fructans. Int J Food Microbiol 2017;241:225–36. https://doi.org/10.1016/j.ijfoodmicro.2016.10.019.

[68] Bottacini F, Ventura M, van Sinderen D, Motherway MO. Diversity, ecology and intestinal function of bifidobacteria. Microb Cell Fact 2014;13. https://doi.org/10.1186/1475-2859-13-S1-S4.

[69] Edelsberg J, Weycker D, Barron R, Li X, Wu H, Oster G, Weber DJ. Prevalence of antibiotic resistance in US hospitals. Diagn Microbiol Infect Dis 2014;78(3):255–62. https://doi.org/10.1016/j.diagmicrobio.2013.11.011.

[70] Lautenbach E, Synnestvedt M, Weiner M, Bilker W, Vo L, Schein J, Kim M. Epidemiology and impact of imipenem resistance in Acinetobacter baumannii. Infect Control Hosp Epidemiol 2009;30(12):1186–92. https://doi.org/10.1086/648450.

[71] Johnston BC, Goldenberg JZ, Vandvik PO, Sun X, Guyatt GH. Probiotics for the prevention of pediatric antibiotic-associated diarrhea. Cochrane Database Syst Rev 2011;11.

[72] Jakobsson HE, Jernberg C, Andersson AF, Sjölund-Karlsson M, Jansson JK, Engstrand L. Short-term antibiotic treatment has differing long-term impacts on the human throat and gut microbiome. PLoS ONE 2010;5(3). https://doi.org/10.1371/journal.pone.0009836.

[73] Blaabjerg S, Artzi DM, Aabenhus R. Probiotics for the prevention of antibiotic-associated diarrhea in outpatients-a systematic review and meta-analysis. Antibiotics (Basel, Switzerland) 2017;6(4):21. https://doi.org/10.3390/antibiotics6040021.

[74] Goldenberg JZ, Lytvyn L, Steurich J, Parkin P, Mahant S, Johnston BC. Probiotics for the prevention of pediatric antibiotic-associated diarrhea. Cochrane Database Syst Rev 2015;12. https://doi.org/10.1002/14651858.CD004827.pub4.

[75] Xu H, Jiang R, Sheng H. Meta-analysis of the effects of bifidobacterium preparations for the prevention and treatment of pediatric antibiotic-associated diarrhea in China. Complement Ther Med 2017;33:105–13. https://doi.org/10.1016/j.ctim.2017.07.001.

[76] Panda S, El khader I, Casellas F, López Vivancos J, García Cors M, Santiago A, Manichanh C. Short-term effect of antibiotics on human gut microbiota. PLoS ONE 2014;9(4):e95476. https://doi.org/10.1371/journal.pone.0095476.

[77] Pérez-Cobas AE, Gosalbes MJ, Frieichs A, Knecht H, Martins Dos Santos VAP. Gut microbiota disturbance during antibiotic therapy: a multi-omic approach. Gut 2013;62(11):1591–601. https://doi.org/10.1136/gutjnl-2012-303184.

[78] Koliada A, Syzenko G, Moseiko V, Budovska L, Puchkov K, Perederiy V, Vaiserman A. Association between body mass index and Firmicutes/Bacteroidetes ratio in an adult Ukrainian population. BMC Microbiol 2017;17. https://doi.org/10.1186/s12866-017-1027-1.

[79] Ley RE, Turnbaugh PJ, Klein S, Gordon JI. Microbial ecology: human gut microbes associated with obesity. Nature 2006;444(7122):1022–3. https://doi.org/10.1038/4441022a.

[80] Lange K, Buerger M, Stallmach A, Bruns T. Effects of antibiotics on gut microbiota. Dig Dis 2016;34(3):260–8. https://doi.org/10.1159/000443360.

[81] Kinoshita M, Suzuki Y, Saito Y. Butyrate reduces colonic paracellular permeability by enhancing PPARγ activation. Biochem Biophys Res Commun 2002;293(2):827–31. https://doi.org/10.1016/S0006-291X(02)00294-2.

[82] National Institutes of Health. Colorectal cancer fact sheet. Retrieved from, https://report.nih.gov/NIHfactsheets/ViewFactSheet.aspx?csid=84; 2018. Accessed 1 July 2018.

[83] American Cancer Society. Cancer facts & figures 2018. Atlanta, GA: American Cancer Society; 2018.

[84] Goverse G, Molenaar R, Macia L, Tan J, Erkelens MN, Konijn T, Mebius RE. Diet-derived short chain fatty acids stimulate intestinal epithelial cells to induce mucosal tolerogenic dendritic cells. J Immunol 2017;198(5):2172–81. https://doi.org/10.4049/jimmunol.1600165.

[85] Wong J, de Souza R, Kendall C, Emam A, Jenkins D. Colonic health: fermentation and short chain fatty acids. J Clin Gastroenterol 2006;40(3):235–43.

[86] Le Poul EL, Loison C, Struyf S, Springael J, Lannoy V, Decobecq M, Detheux M. Functional characterization of human receptors for short chain fatty acids and their role in polymorphonuclear cell activation. J Biol Chem 2003;278(28):25481–9. https://doi.org/10.1074/jbc.M301403200.

[87] He C, Shi Y, Wu R, Sun M, Fang L, Wu W, Liu Z. miR-301a promotes intestinal mucosal inflammation through induction of IL-17A and TNF-alpha in IBD. Gut 2016;65(12):1938. https://doi.org/10.1136/gutjnl-2015-309389.

[88] Fransen F, van Beek AA, Borghuis T, Meijer B, Hugenholtz F, van der Gaast-de Jongh CE, Vos P. The impact of gut microbiota on gender-specific differences in immunity. Front Immunol 2017;8:754. https://doi.org/10.3389/fimmu.2017.00754.

[89] Fish EN. The X-files in immunity: sex-based differences predispose immune responses. Nat Rev Immunol 2008;8(9):737–44. https://doi.org/10.1038/nri2394.

[90] vom Steeg LG, Klein SL. SeXX matters in infectious disease pathogenesis. PLoS Pathog 2016;12(2). https://doi.org/10.1371/journal.ppat.1005374.

[91] Tremellen K. Gut endotoxin leading to a decline IN gonadal function (GELDING) - a novel theory for the development of late onset hypogonadism in obese men. Basic Clin Androl 2016;26(1):7. https://doi.org/10.1186/s12610-016-0034-7.

[92] Tremellen K, McPhee N, Pearce K, Benson S, Schedlowski M, Engler H. Endotoxin-initiated inflammation reduces testosterone production in men of reproductive age. Am J Physiol Endocrinol Metab 2018;314(3):E206–13. https://doi.org/10.1152/ajpendo.00279.2017.

[93] Trumble BC, Blackwell AD, Stieglitz J, Thompson ME, Suarez IM, Kaplan H, Gurven M. Associations between male testosterone and immune function in a pathogenically stressed forager-horticultural population. Am J Phys Anthropol 2016;161(3):494–505. https://doi.org/10.1002/ajpa.23054.

[94] Golombos DM, Ayangbesan A, O'Malley P, Lewicki P, Barlow L, Barbieri CE, Scherr DS. The role of gut microbiome in the pathogenesis of prostate cancer: a prospective, pilot study. Urology 2018;111:122–8. https://doi.org/10.1016/j.urology.2017.08.039.

[95] Porter CM, Shrestha E, Peiffer LB, Sfanos KS. The microbiome in prostate inflammation and prostate cancer. Prostate Cancer Prostatic Dis 2018. https://doi.org/10.1038/s41391-018-0041-1.

[96] Liss MA, White JR, Goros M, Gelfond J, Leach R, Johnson-Pais T, Shah DP. Metabolic biosynthesis pathways identified from fecal microbiome associated with prostate cancer. Eur Urol 2018. https://doi.org/10.1016/j.eururo.2018.06.033.

[97] Slocum C, Kramer C, Genco CA. Immune dysregulation mediated by the oral microbiome: potential link to chronic inflammation and atherosclerosis. J Intern Med 2016;280(1):114–28. https://doi.org/10.1111/joim.12476.

[98] Wang Z, Klipfell E, Bennett BJ, Koeth R, Levison BS, Dugar B, Hazen SL. Gut flora metabolism of phosphatidylcholine promotes cardiovascular disease. Nature 2011;472(7341):57–63. https://doi.org/10.1038/nature09922.

[98a] Koeth R, Levison B, Culley M, Buffa J, Wang Z, Gregory J, et al. gamma-Butyrobetaine is a proatherogenic intermediate in gut microbial metabolism of L-carnitine to TMAO. Cell Metab 2014;20(5):799–812. https://doi.org/10.1016/j.cmet.2014.10.006.

[99] Komaroff AL. The microbiome and risk for atherosclerosis. JAMA 2018;319(23):2381. https://doi.org/10.1001/jama.2018.5240.

[100] Spence JD. Intestinal microbiome and atherosclerosis. EBioMedicine 2016;13:17–8. https://doi.org/10.1016/j.ebiom.2016.10.033.

[101] Bogiatzi C, Gloor G, Allen-Vercoe E, Reid G, Wong RG, Urquhart BL, Spence JD. Metabolic products of the intestinal microbiome and extremes of atherosclerosis. Atherosclerosis 2018;273:91–7. https://doi.org/10.1016/j.atherosclerosis.2018.04.015.

[102] Kasselman LJ, Vernice NA, DeLeon J, Reiss AB. The gut microbiome and elevated cardiovascular risk in obesity and autoimmunity. Atherosclerosis 2018;271:203–13. https://doi.org/10.1016/j.atherosclerosis.2018.02.036.

[103] Suganami T, Tanimoto-Koyama K, Nishida J, Itoh M, Yuan X, Mizuarai S, Ogawa Y. Role of the toll-like receptor 4/NF-kappaB pathway in saturated fatty acid-induced inflammatory changes in the interaction between adipocytes and macrophages. Arterioscler Thromb Vasc Biol 2007;27(1):84.

[104] Williams NT. Probiotics. Am J Health Syst Pharm 2010;67(6):449–58. https://doi.org/10.2146/ajhp090168.

[105] Bell V, Ferrão J, Fernandes T. Nutritional guidelines and fermented food frameworks. Foods 2017;6(8):65. https://doi.org/10.3390/foods6080065.

[106] Zou J, Dong J, Yu X. Meta-analysis: Lactobacillus containing quadruple therapy versus standard triple first-line therapy for helicobacter pylori eradication. Helicobacter 2009;14(5):97. https://doi.org/10.1111/j.1523-5378.2009.00716.x.

[107] Chiu H, Chen Y, Lu Y, Han Y, Shen Y, Venkatakrishnan K, Wang C. Regulatory efficacy of fermented plant extract on the intestinal microflora and lipid profile in mildly hypercholesterolemic individuals. J Food Drug Anal 2017;25(4):819–27. https://doi.org/10.1016/j.jfda.2016.10.008.

[108] Wang C, Nagata S, Asahara T, Yuki N, Matsuda K, Tsuji H, Yamashiro Y. Intestinal microbiota profiles of healthy pre-school and school-age children and effects of probiotic supplementation. Ann Nutr Metab 2015;67(4):257–66. https://doi.org/10.1159/000441066.

[109] Veiga P, Pons N, Agrawal A, Oozeer R, Guyonnet D, Brazeilles R, Kennedy SP. Changes of the human gut microbiome induced by a fermented milk product. Sci Rep 2014/2015;4(1):6328. https://doi.org/10.1038/srep06328.

[110] Elli M, Callegari ML, Ferrari S, Bessi E, Cattivelli D, Soldi S, Antoine J-M. Survival of yogurt bacteria in the human gut. Appl Environ Microbiol 2006;72(7):5113–7. https://doi.org/10.1128/AEM.02950-05.

[111] Bezkorovainy A. Probiotics: determinants of survival and growth in the gut. Am J Clin Nutr 2001;73(2):399s–405s. https://doi.org/10.1093/ajcn/73.2.399s.

[112] Osborn DA, Sinn JKH. Probiotic supplements. Br Med J 2013;347. https://doi.org/10.1136/bmj.f7138.

[113] Gomaa EZ. Effect of prebiotic substances on growth, fatty acid profile and probiotic characteristics of Lactobacillus brevis NM101-1. Microbiology 2017;86(5):618–28. https://doi.org/10.1134/S0026261717050095.

[114] Vandenplas Y, Huys G, Daube G. Probiotics: an update. J Pediatr 2015;91(1):6–21. https://doi.org/10.1016/j.jped.2014.08.005.

[115] Quagliariello A, Aloisio I, Cionci NB, Luiselli D, D'Auria G, Martinez-Priego L, Gioia DD. Effect of Bifidobacterium breve on the intestinal microbiota of coeliac children on a gluten free diet: a pilot study. Nutrients 2016;8(10):660. https://doi.org/10.3390/nu8100660.

[116] Klemenak M, Dolinšek J, Langerholc T, Di Gioia D, Mičetić-Turk D. Administration of Bifidobacterium breve decreases the production of TNF-α in children with celiac disease. Dig Dis Sci 2015;60(11):3386–92. https://doi.org/10.1007/s10620-015-3769-7.

[117] Klarin B, Wullt M, Palmquist I, Molin G, Larsson A, Jeppson B, Institutionen för kirurgiska vetenskaper. Lactobacillus plantarum 299v reduces colonisation of clostridium difficile in critically ill patients treated with antibiotics. Acta Anaesthesiol Scand 2008;52(8):1096–102. https://doi.org/10.1111/j.1399-6576.2008.01748.x.

[118] Castagliuolo I, Riegler MF, Valenick L, LaMont JT, Pothoulakis C. Saccharomyces boulardii protease inhibits the effects of Clostridium difficile toxins A and B in human colonic mucosa. Infect Immun 1999;67(1):302–7.

[119] Akkasheh G, Kashani-Poor Z, Tajabadi-Ebrahimi M, Jafari P, Akbari H, Taghizadeh M, Esmaillzadeh A. Clinical and metabolic response to probiotic administration in patients with major depressive disorder: a randomized, double-blind, placebo-controlled trial. Nutrition 2016;32(3):315–20. https://doi.org/10.1016/j.nut.2015.09.003.

[120] Steenbergen L, Sellaro R, van Hemert S, Bosch JA, Colzato LS. A randomized controlled trial to test the effect of multispecies probiotics on cognitive reactivity to sad mood. Brain Behav Immun 2015;48:258–64. https://doi.org/10.1016/j.bbi.2015.04.003.

[121] Pirbaglou M, Katz J, de Souza RJ, Stearns JC, Motamed M, Ritvo P. Probiotic supplementation can positively affect anxiety and depressive symptoms: a systematic review of randomized controlled trials. Nutr Res 2016;36(9):889–98. https://doi.org/10.1016/j.nutres.2016.06.009.

[122] Gerasimov SV, Vasjuta VV, Myhovych OO, Bondarchuk LI. Probiotic supplement reduces atopic dermatitis in preschool children: a randomized, double-blind, placebo-controlled, clinical trial.

Am J Clin Dermatol 2010;11(5):351–61. https://doi.org/10.2165/11531420-000000000-00000.

[123] Kalliomäki M, Salminen S, Poussa T, Arvilommi H, Isolauri E. Probiotics and prevention of atopic disease: 4-year follow-up of a randomised placebo-controlled trial. Lancet 2003;361 (9372):1869–71. https://doi.org/10.1016/S0140-6736(03)13490-3.

[124] Isolauri E, Arvola T, Sütas Y, Moilanen E, Salminen S. Probiotics in the management of atopic eczema. Clin Exp Allergy 2000;30(11):1604.

[125] Doron S, Gorbach SL. Probiotics: their role in the treatment and prevention of disease. Expert Rev Anti-Infect Ther 2006;4(2):261–75. https://doi.org/10.1586/14787210.4.2.261.

[126] Liu Z, Shen T, Zhang P, Ma Y, Qin H. Lactobacillus plantarum surface layer adhesive protein protects intestinal epithelial cells against tight junction injury induced by enteropathogenic Escherichia coli. Mol Biol Rep 2010;38(5):3471–80. https://doi.org/10.1007/s11033-010-0457-8.

[127] Asemi Z, Zare Z, Shakeri H, Sabihi S, Esmaillzadeh A. Effect of multispecies probiotic supplements on metabolic profiles, hs-CRP, and oxidative stress in patients with type 2 diabetes. Ann Nutr Metab 2013;63(1–2):1–9. https://doi.org/10.1159/000349922.

[128] Collado MC, González R, González A, Hernández M, Ferrús MA, Sanz Y. Antimicrobial peptides are among the antagonistic metabolites produced by bifidobacterium against helicobacter pylori. Int J Antimicrob Agents 2005;25(5):385–91. https://doi.org/10.1016/j.ijantimicag.2005.01.017.

[129] Guslandi M, Mezzi G, Sorghi M, Testoni PA. Saccharomyces boulardii in maintenance treatment of Crohn's disease. Dig Dis Sci 2000;45(7):1462.

[130] Guslandi M, Giollo P, Testoni P. A pilot trial of Saccharomyces boulardii in ulcerative colitis. Eur J Gastroenterol Hepatol 2003;15:697–8. https://doi.org/10.1097/01.meg.0000059138.68845.06.

[130a] Haller D, Antoine J, Bengmark S, Enck P, Rijkers G, Lenoir-Wijnkoop I, Haller D. Guidance for substantiating the evidence for beneficial effects of probiotics: probiotics in chronic inflammatory bowel disease and the functional disorder irritable bowel syndrome. J Nutr 2010;140(3):690S–6907S. https://doi.org/10.3945/jn.109.113746.

[131] Mohammad M, Hussein L, Yamamah G, Rawi S. The impact of probiotic and or honey supplements on gut permeability among egyptian children. J Nutr Environ Med 2007;16(1):10–5. https://doi.org/10.1080/13590840601016387.

[132] Shing CM, Peake JM, Lim CL, Briskey D, Walsh NP, Fortes MB, Vitetta L. Effects of probiotics supplementation on gastrointestinal permeability, inflammation and exercise performance in the heat. Eur J Appl Physiol 2014;114(1):93–103. https://doi.org/10.1007/s00421-013-2748-y.

[133] Guandalini S, Pensabene L, Zikri M, et al. Lactobacillus GG administered in oral rehydration solution to children with acute diarrhea: a multicenter European trial. J Pediatr Gastroenterol Nutr 2000;30:54–60.

[134] Van Niel C, Feudtner C, Garrison M, Christakis D. Lactobacillus therapy for acute infectious diarrhea in children: a meta-analysis. Pediatrics 2002;109:678–84.

[135] Lyseng-Williamson K. Bifidobacterium infantis 35624 as a probiotic dietary supplement: a profile of its use. Drugs Ther Perspect 2017;33(8):368–74. https://doi.org/10.1007/s40267-017-04239.

[136] Björklund M, Ouwehand AC, Forssten SD, Nikkilä J, Tiihonen K, Rautonen N, Lahtinen SJ. Gut microbiota of healthy elderly NSAID users is selectively modified with the administration of Lactobacillus acidophilus NCFM and lactitol. Age 2012;34(4):987–99. https://doi.org/10.1007/s11357-011-9294-5.

References

[137] Andreasen AS, Larsen N, Pedersen-Skovsgaard T, Berg RMG, Møller K, Svendsen KD, Pedersen BK. Effects of Lactobacillus acidophilus NCFM on insulin sensitivity and the systemic inflammatory response in human subjects. Br J Nutr 2010;104(12):1831–8. https://doi.org/10.1017/S0007114510002874.

[138] Hojsak I, Abdovi'c S, Szajewska H, Milosevi'c M, Krznari'c Z, Kolacek S. Lactobacillus G.G. in the prevention of nosocomial gastrointestinal and respiratory tract infections. Pediatrics 2010;125:e1171–7.18.

[139] Zhang H, Zhang L, Yeh C, Jin Z, Ding L, Liu BY, Dannelly HK. Prospective study of probiotic supplementation results in immune stimulation and improvement of upper respiratory infection rate. Synth Syst Biotechnol 2018;3(2):113–20. https://doi.org/10.1016/j.synbio.2018.03.001.

[140] Vaghef-Mehrabany E, Alipour B, Homayouni-Rad A, Sharif S, Asghari-Jafarabadi M, Zavvari S. Probiotic supplementation improves inflammatory status in patients with rheumatoid arthritis. Nutrition 2014;30(4):430–5. https://doi.org/10.1016/j.nut.2013.09.007.

[141] Zamani B, Golkar HR, Farshbaf S, Emadi-Baygi M, Tajabadi-Ebrahimi M, Jafari P, Asemi Z. Clinical and metabolic response to probiotic supplementation in patients with rheumatoid arthritis: a randomized, double-blind, placebo-controlled trial. Int J Rheum Dis 2016;19(9):869–79. https://doi.org/10.1111/1756-185X.12888.

[142] Ozdemir O. Various effects of different probiotic strains in allergic disorders: an update from laboratory and clinical data. Clin Exp Immunol 2010;160(3):295–304. https://doi.org/10.1111/j.1365-2249.2010.04109.x.

[143] Palócz O, Pászti-Gere E, Gálfi P, Farkas O. Chlorogenic acid combined with lactobacillus plantarum 2142 reduced LPS-induced intestinal inflammation and oxidative stress in IPEC-J2 cells. PLoS ONE 2016;11(11). https://doi.org/10.1371/journal.pone.0166642.

[144] Caillard R, Lapointe N. In vitro gastric survival of commercially available probiotic strains and oral dosage forms. Int J Pharm 2017;519(1–2):125–7. https://doi.org/10.1016/j.ijpharm.2017.01.019.

[145] Ellis ML, Shaw KJ, Jackson SB, Daniel SL, Knight J. Analysis of commercial kidney stone probiotic supplements. Urology 2015;85(3):517–21. https://doi.org/10.1016/j.urology.2014.11.013.

[146] Katz J, Pirovano F, Matteuzzi D, et al. Commercially available probiotic preparations: are you getting what you pay for? Gastroenterology 2002;122(Suppl. 1):A-459. Abstract.

[147] Cutting S, Fraser P, Payne K. Probiotic potential spore-forming bacteria. Nutraceutical Bus Technol 2008;4(5):5.

[148] Lakshmi SG, Jayanthi N, Saravanan M, Ratna MS. Safety assessment of Bacillus clausii UBBC07, a spore forming probiotic. Toxicol Rep 2017;4:62–71. https://doi.org/10.1016/j.toxrep.2016.12.004.

[149] Elshaghabee FMF, Rokana N, Gulhane RD, Sharma C, Panwar H. Bacillus as potential probiotics: status, concerns, and future perspectives. Front Microbiol 2017;8:1490. https://doi.org/10.3389/fmicb.2017.01490.

[150] Parnell JA, Reimer RA. Prebiotic fibres dose-dependently increase satiety hormones and alter bacteroidetes and firmicutes in lean and obese JCR: LA-cp rats. Br J Nutr 2012;107(4):601–13. https://doi.org/10.1017/S0007114511003163.

[151] Caetano BFR, de Moura NA, Almeida APS, Dias MC, Sivieri K, Barbisan LF. Yacon (Smallanthus sonchifolius) as a food supplement: health-promoting benefits of fructooligosaccharides. Nutrients 2016;8(7):436. https://doi.org/10.3390/nu8070436.

[152] Shi Y, Zhai Q, Li D, Mao B, Liu X, Zhao J, Chen W. Restoration of cefixime-induced gut microbiota changes by lactobacillus cocktails and fructooligosaccharides in a mouse model. Microbiol Res 2017;200:14–24. https://doi.org/10.1016/j.micres.2017.04.001.

[153] Davis LMG, Martínez I, Walter J, Goin C, Hutkins RW. Barcoded pyrosequencing reveals that consumption of galactooligosaccharides results in a highly specific bifidogenic response in humans. PLoS ONE 2011;6(9). https://doi.org/10.1371/journal.pone.0025200.

[154] Azcarate-Peril MA, Ritter AJ, Savaiano D, Monteagudo-Mera A, Anderson C, Magness ST, Klaenhammer TR. Impact of short-chain galactooligosaccharides on the gut microbiome of lactose-intolerant individuals. Proc Natl Acad Sci U S A 2017;114(3):E367–75.

[155] So D, Whelan K, Rossi M, Morrison M, Holtmann G, Kelly JT, Campbell KL. Dietary fiber intervention on gut microbiota composition in healthy adults: a systematic review and meta-analysis. Am J Clin Nutr 2018;107(6):965–83. https://doi.org/10.1093/ajcn/nqy041.

[156] Carlson JL, Erickson JM, Hess JM, Gould TJ, Slavin JL. Prebiotic dietary fiber and gut health: Comparing the in vitro fermentations of beta-glucan, inulin and xylooligosaccharide. Nutrients 2017;9(12):1361. https://doi.org/10.3390/nu9121361.

[157] Christensen EG, Licht TR, Leser TD, Bahl MI. Dietary xylo-oligosaccharide stimulates intestinal bifidobacteria and lactobacilli but has limited effect on intestinal integrity in rats. BMC Res Notes 2014;7:660. https://doi.org/10.1186/1756-0500-7-660.

[158] Nieto-Domínguez M, de Eugenio LI, York-Durán MJ, Rodríguez-Colinas B, Plou FJ, Chenoll E, ...Jesús Martínez M. Prebiotic effect of xylooligosaccharides produced from birchwood xylan by a novel fungal GH11 xylanase. Food Chem 2017;232:105–13. https://doi.org/10.1016/j.foodchem.2017.03.149.

[159] Van den Abbeele P, Duysburgh C, Jiang TA, Rebaza M, Pinheiro I, Marzorati M. A combination of xylooligosaccharides and a polyphenol blend affect microbial composition and activity in the distal colon exerting immunomodulating properties on human cells. J Funct Foods 2018;47:163–71. https://doi.org/10.1016/j.jff.2018.05.053.

[160] Alfa MJ, Strang D, Tappia PS, Graham M, Van Domselaar G, Forbes JD, Lix LM. A randomized trial to determine the impact of a digestion resistant starch composition on the gut microbiome in older and mid-age adults. Clin Nutr 2017/2018;37(3):797–807. https://doi.org/10.1016/j.clnu.2017.03.025.

[161] Hald S, Schioldan AG, Moore ME, Dige A, Helle NL, Agnholt J, Dahlerup JF. Effects of arabinoxylan and resistant starch on intestinal microbiota and short-chain fatty acids in subjects with metabolic syndrome: a randomised crossover study. PLoS ONE 2016;11(7). https://doi.org/10.1371/journal.pone.0159223.

[162] Maier TV, Lucio M, Lee LH, VerBerkmoes NC, Brislawn CJ, Bernhardt J, Jansson JK. Impact of dietary resistant starch on the human gut microbiome, metaproteome, and metabolome. MBio 2017;8(5). https://doi.org/10.1128/mBio.01343-17.

[163] Venkataraman A, Sieber JR, Schmidt AW, Waldron C, Theis KR, Schmidt TM. Variable responses of human microbiomes to dietary supplementation with resistant starch. Microbiome 2016;4(1):33. https://doi.org/10.1186/s40168-016-0178-x.

[164] Fukunaga T, Sasaki M, Araki Y, Okamoto T, al, e. Effects of the soluble fibre pectin on intestinal cell proliferation, fecal short chain fatty acid production and microbial population. Digestion 2003;67(1):42–9. Retrieved from, https://search-proquest-com.ezproxy2.apus.edu/docview/195194733?accountid=8289.

[165] Bouhnik Y, Alain S, Attar A, Flourié B, Raskine L, Sanson-Le Pors MJ, Rambaud JC. Bacterial populations contaminating the upper gut in patients with small intestinal bacterial overgrowth syndrome. Am J Gastroenterol 1999;94(5):1327–31.

[166] Dukowicz AC, Lacy BE, Levine GM. Small intestinal bacterial overgrowth: a comprehensive review. Gastroenterol Hepatol 2007;3(2):112–22.

[167] Husebye E. The pathogenesis of gastrointestinal bacterial overgrowth. Chemotherapy 2005;51(1):1–22. https://doi.org/10.1159/000081988.

[168] Lewis SJ, Franco S, Young G, O'Keefe SJ. Altered bowel function and duodenal bacterial overgrowth in patients treated with omeprazole. Aliment Pharmacol Ther 1996;10:557–61. https://doi.org/10.1046/j.1365-2036.1996.d01-506.x.

[169] Saltzman JR, Russell R. Nutritional consequences of intestinal bacterial overgrowth. Compr Ther 1994;20(9):523–30.

[170] Crinnion WJ. Organic foods contain higher levels of certain nutrients, lower levels of pesticides, and may provide health benefits for the consumer. Altern Med Rev 2010;15(1):4.

[171] Barbieri R, Coppo E, Marchese A, Daglia M, Sobarzo-Sánchez E, Nabavi SF, Nabavi SM. Phytochemicals for human disease: an update on plant-derived compounds antibacterial activity. Microbiol Res 2017;196:44–68. https://doi.org/10.1016/j.micres.2016.12.003.

[172] He Y, Yuan X, Zhou G, Feng A. Activation of IGF-1/IGFBP-3 signaling by berberine improves intestinal mucosal barrier of rats with acute endotoxemia. Fitoterapia 2018;124:200–5. https://doi.org/10.1016/j.fitote.2017.11.012.

[173] Stermitz FR, Lorenz P, Tawara JN, Zenewicz LA, Lewis K. Synergy in a medicinal plant: antimicrobial action of berberine potentiated by 5′-methoxyhydnocarpin, a multidrug pump inhibitor. Proc Natl Acad Sci U S A 2000;97(4):1433–7. https://doi.org/10.1073/pnas.030540597.

[174] Catinean A, Neag MA, Muntean DM, Bocsan IC, Buzoianu AD. An overview on the interplay between nutraceuticals and gut microbiota. Peerj 2018;6. https://doi.org/10.7717/peerj.4465.

[175] Altintas A, Tabanca N, Tyihák E, Ott PG, Móricz AM, Mincsovics E, Wedge DE. Characterization of volatile constituents from origanum onites and their antifungal and antibacterial activity. J AOAC Int 2013;96(6):1200–8. https://doi.org/10.5740/jaoacint.SGEAltintas.

[176] Fournomiti M, Kimbaris A, Mantzourani I, Plessas S, Theodoridou I, Papaemmanouil V, Alexopoulos A. Antimicrobial activity of essential oils of cultivated oregano (Origanum vulgare), sage (Salvia officinalis), and thyme (Thymus vulgaris) against clinical isolates of Escherichia coli, Klebsiella oxytoca, and Klebsiella pneumoniae. Microb Ecol Health Dis 2015;26:1–7. https://doi.org/10.3402/mehd.v26.23289.

[177] Baskaran SA, Kollanoor-Johny A, Nair MS, Venkitanarayanan K. Efficacy of plant-derived antimicrobials in controlling enterohemorrhagic Escherichia coli virulence in vitro. J Food Protect 2016;79(11):1965–70. https://doi.org/10.4315/0362-028X.JFP-16-104.

[178] Nabavi SF, Nabavi SM, Marchese A, Izadi M, Curti V, Daglia M. Plants belonging to the genus thymus as antibacterial agents: From farm to pharmacy. Food Chem 2015;173:339–47. https://doi.org/10.1016/j.foodchem.2014.10.042.

[179] Utchariyakiat I, Surassmo S, Jaturanpinyo M, Khuntayaporn P, Mullika TC. Efficacy of cinnamon bark oil and cinnamaldehyde on anti-multidrug resistant pseudomonas aeruginosa and the synergistic effects in combination with other antimicrobial agents. BMC Complement Altern Med 2016;16. https://doi.org/10.1186/s12906-016-1134-9.

[180] Zhang X, Jiang A, Qi B, Ma Z, Xiong Y, Dou J, Wang J. Resveratrol protects against helicobacter pylori-associated gastritis by combating oxidative stress. Int J Mol Sci 2015;16(11):27757–69. https://doi.org/10.3390/ijms161126061.

[181] Queipo-Ortuño M, Boto-Ordóñez M, Murri M, Gomez-Zumaquero J, Clemente-Postigo M, Estruch R, Cardona Diaz F, Andrés-Lacueva C, Tinahones F. Influence of red wine polyphenols and ethanol on the gut microbiota ecology and biochemical biomarkers. Am J Clin Nutr 2012;95(6):1323–34. https://doi.org/10.3945/ajcn.111.027847.

[182] Kaulmann A, Bohn T. Bioactivity of polyphenols: Preventive and adjuvant strategies toward reducing inflammatory bowel diseases-promises, perspectives, and pitfalls. Oxid Med Cell Longev 2016. https://doi.org/10.1155/2016/9346470.

[183] Bereswill S, Muñoz M, Fischer A, Plickert R, Haag L, Otto B, Heimesaat MM. Anti-inflammatory effects of resveratrol, curcumin and simvastatin in acute small intestinal inflammation. PLoS ONE 2010;5(12). https://doi.org/10.1371/journal.pone.0015099.

[184] Ohno M, Nishida A, Sugitani Y, Nishino K, Inatomi O, Sugimoto M, Andoh A. Nanoparticle curcumin ameliorates experimental colitis via modulation of gut microbiota and induction of regulatory T cells. PLoS ONE 2017;12(10). https://doi.org/10.1371/journal.pone.0185999.

[185] Shen L, Liu L, Ji H. Regulative effects of curcumin spice administration on gut microbiota and its pharmacological implications. Food Nutr Res 2017;61:1–4. https://doi.org/10.1080/16546628.2017.1361780.

[186] Hassaninasab A, Hashimoto Y, Tomita-Yokotani K, Kobayashi M. Discovery of the curcumin metabolic pathway involving a unique enzyme in an intestinal microorganism. Proc Natl Acad Sci U S A 2011;108(16):6615–20. https://doi.org/10.1073/pnas.1016217108.

[187] Duda-Chodak A, Tarko T, Satora P, Sroka P. Interaction of dietary compounds, especially polyphenols, with the intestinal microbiota: a review. Eur J Nutr 2015;54(3):325–41. https://doi.org/10.1007/s00394-015-0852-y.

[188] Vaquero MJR, Alberto MR, de Nadra MCM. Antibacterial effect of phenolic compounds from different wines. Food Control 2007;18(2):93–101. https://doi.org/10.1016/j.foodcont.2005.08.010.

[189] Williamson G, Clifford MN. Colonic metabolites of berry polyphenols: the missing link to biological activity? Br J Nutr 2010;104:S48–66. https://doi.org/10.1017/S0007114510003946.

[190] Kahle K, Kraus M, Scheppach W, Ackermann M, Ridder F, Richling E. Studies on apple and blueberry fruit constituents: do the polyphenols reach the colon after ingestion? Mol Nutr Food Res 2006;50:418–23. https://doi.org/10.1002/mnfr.200500211.

[191] Sato Y, Itagaki S, Kurokawa T, Ogura J, Kobayashi M, Hirano T, Iseki K. In vitro and in vivo antioxidant properties of chlorogenic acid and caffeic acid. Int J Pharm 2011;403(1–2):136–8. https://doi.org/10.1016/j.ijpharm.2010.09.035.

[192] Jamar G, Estadella D, Pisani LP. Contribution of anthocyanin-rich foods in obesity control through gut microbiota interactions. Biofactors 2017;43(4):507–16. https://doi.org/10.1002/biof.1365.

[193] Zhang P, Zhang M, He S, Cao X, Sun H, Chen X, Ye Y. Extraction and probiotic properties of new anthocyanins from purple sweet potato (Solanum tuberosum), Curr Top Nutraceutical Res 2016;14(2):153–60. Retrieved from, https://search-proquest-com.ezproxy1.apus.edu/docview/1870902562?accountid=8289.

[194] Sahib N, Anwar F, Gilani A, Hamid A, Saari N, Alkharfy K. Coriander (Coriandrum sativum L.): a potential source of high-value components for functional foods and nutraceuticals – a review. Phytother Res 2013;(10):1439.

[195] Xiao S, Li Q, Hu K, He Y, Ai Q, Hu L, Yu J. Vitamin A and retinoic acid exhibit protective effects on necrotizing enterocolitis by regulating intestinal flora and enhancing the intestinal epithelial barrier. Arch Med Res 2018. https://doi.org/10.1016/j.arcmed.2018.04.003.

[196] Holder BS, Grant CR, Liberal R, Ma Y, Heneghan MA, Mieli-Vergani G, Longhi MS. Retinoic acid stabilizes antigen-specific regulatory T-cell function in autoimmune hepatitis type 2. J Autoimmun 2014;53:26–32. https://doi.org/10.1016/j.jaut.2014.02.001.

[197] Clark A, Mach N. Role of vitamin D in the hygiene hypothesis: the interplay between vitamin D, vitamin D receptors, gut microbiota, and immune response. Front Immunol 2016;7:627.

[198] Biesalski HK. Nutrition meets the microbiome: micronutrients and the microbiota. Ann N Y Acad Sci 2016;1372(1):53–64. https://doi.org/10.1111/nyas.13145.

[199] Mielcarz DW, Kasper LH. The gut microbiome in multiple sclerosis. Curr Treat Options Neurol 2015;17. https://doi.org/10.1007/s11940-015-0344-7.

[200] Ghaly S, Kaakoush NO, Lloyd F, McGonigle T, Mok D, Baird A, Hart PH. High dose vitamin D supplementation alters faecal microbiome and predisposes mice to more severe colitis. Sci Rep 2018;8:1–12. https://doi.org/10.1038/s41598-018-29759-y.

[201] Gominak SC. Vitamin D deficiency changes the intestinal microbiome reducing B vitamin production in the gut. The resulting lack of pantothenic acid adversely affects the immune system, producing a "pro-inflammatory" state associated with atherosclerosis and autoimmunity. Med Hypotheses 2016;94:103.

[202] Bashir M, Prietl B, Tauschmann M, Mautner SI, Kump PK, Treiber G, Pieber TR. Effects of high doses of vitamin D3 on mucosa-associated gut microbiome vary between regions of the human gastrointestinal tract. Eur J Nutr 2016;55(4):1479–89. https://doi.org/10.1007/s00394-015-0966-2.

[203] Xiao G, Tang L, Yuan F, Zhu W, Zhang S, Liu Z, Su L. Eicosapentaenoic acid enhances heat stress-impaired intestinal epithelial barrier function in caco-2 cells. PLoS ONE 2013;8(9) https://doi.org/10.1371/journal.pone.0073571.

[204] Menni C, Zierer J, Pallister T, Jackson MA, Long T, Mohney RP, Valdes AM. Omega-3 fatty acids correlate with gut microbiome diversity and production of N-carbamylglutamate in middle aged and elderly women. Sci Rep 2017;7(1):1. https://doi.org/10.1038/s41598-017-10382-2.

[205] Mokkala K, Laitinen K, Röytiö H. Bifidobacterium lactis 420 and fish oil enhance intestinal epithelial integrity in caco-2 cells. Nutr Res 2016;36(3):246–52. https://doi.org/10.1016/j.nutres.2015.11.014.

[206] Costantini L, Molinari R, Farinon B, Merendino N. Impact of omega-3 fatty acids on the gut microbiota. Int J Mol Sci 2017;18(12):2645. https://doi.org/10.3390/ijms18122645.

[207] Liu T, Hougen H, Vollmer AC, Hiebert SM. Gut bacteria profiles of Mus musculus at the phylum and family levels are influenced by saturation of dietary fatty acids. Anaerobe 2012;18 (3):331–7. https://doi.org/10.1016/j.anaerobe.2012.02.004.

[208] Yu H, Zhu J, Pan W, Shen S, Shan W, Das UN. Effects of fish oil with a high content of n-3 polyunsaturated fatty acids on mouse gut microbiota. Arch Med Res 2014;45(3):195–202. https://doi.org/10.1016/j.arcmed.2014.03.008.

[209] Balfego M, Canivell S, Hanzu FA, Sala-Vila A, Martinez-Medina M, Murillo S, Aranda G. Effects of sardine-enriched diet on metabolic control, inflammation and gut microbiota in drug-naive patients with type 2 diabetes: a pilot randomized trial. Lipids Health Dis 2016;15. https://doi.org/10.1186/s12944-016-0245-0.

[210] Beutheu S, Ghouzali I, Galas L, Déchelotte P, Coëffier M. Glutamine and arginine improve permeability and tight junction protein expression in methotrexate-treated caco-2 cells. Clin Nutr 2013;32(5):863–9. https://doi.org/10.1016/j.clnu.2013.01.014.

[211] de Souza AZ, Zambom AZ, Abboud KY, Reis SK, Tannihão F, Guadagnini D, Prada PO. Oral supplementation with l-glutamine alters gut microbiota of obese and overweight adults: a pilot study. Nutrition 2015;31(6):884–9. https://doi.org/10.1016/j.nut.2015.01.004.

[212] Beutheu S, Ouelaa W, Guérin C, Belmonte L, Aziz M, Tennoune N, Coëffier M. Glutamine supplementation, but not combined glutamine and arginine supplementation, improves gut barrier function during chemotherapy-induced intestinal mucositis in rats. Clin Nutr 2013/2014;33(4):694–701. https://doi.org/10.1016/j.clnu.2013.09.003.

[213] Rao R, Samak G. Role of glutamine in protection of intestinal epithelial tight junctions. J Epithel Biol Pharmacol 2012;5(Suppl 1-M7):47–54. https://doi.org/10.2174/1875044301205010047.

[214] Ren W, Wang K, Yin j, Chen S, Liu G, Tan B, Yin Y. Glutamine on intestinal secretory immunoglobulin A secretion: a mechanistic perspective. Front Immunol 2016;7. https://doi.org/10.3389/fimmu.2016.00503.

Further Reading

[215] Andermann TM, Rezvani A, Bhatt AS. Microbiota manipulation with prebiotics and probiotics in patients undergoing stem cell transplantation. Curr Hematol Malig Rep 2016;11(1):19–28. https://doi.org/10.1007/s11899-016-0302-9.

[216] Rothwell JA, Pérez-Jiménez J, Neveu V, Medina-Ramon A, M'Hiri N, Garcia Lobato P, Manach C, Knox K, Eisner R, Wishart D, Scalbert A. Phenol-Explorer 3.0: a major update of the Phenol-Explorer database to incorporate data on the effects of food processing on polyphenol content. Database 2013. https://doi.org/10.1093/database/bat070. [Accessed 2 August 2018].

PART 2

Metabolic Health

CHAPTER 2.1

Metabolic Health: Inflammation and Men's Health

Brent M. Hanson*, James M. Hotaling[†]

*RMA New Jersey, Sidney Kimmel Medical College at Thomas Jefferson University, Basking Ridge, NJ, United States, [†]University of Utah Center for Reconstructive Urology and Men's Health, Department of Surgery—Urology, Salt Lake City, UT, United States

INTRODUCTION

Inflammation within the male genitourinary tract, often resulting from proinflammatory mediators, infectious etiologies, or elevations in reactive oxygen species (ROS), can have sweeping detrimental effects on men's health. The causes, consequences, and various manifestations of inflammation within the male reproductive tract will be discussed throughout this chapter. Traditionally, the genitourinary tract of healthy individuals was thought to be sterile. Recent studies, however, have determined that a complex urinary microbiome is present in both men and women and plays a key role in the overall health of men [1]. The long-standing belief that all bacteria within the urinary tract are pathogens has been proved to be false. Initial research related to the urinary microbiome evaluated the female urinary tract, focusing on conditions such as asymptomatic bacteriuria, interstitial cystitis, and urgency urinary incontinence [2]. A heightened understanding of the presence of bacteria within the urinary tracts of otherwise healthy women led to increased attention on the bacterial composition of the male reproductive tract. There are relatively fewer studies that focus on the male urinary microbiome, but in general, bacteria such as *Lactobacillus, Sneathia, Veillonella, Corynebacterium Prevotella, Streptococcus,* and *Ureaplasma* appear to be present in the healthy male genitourinary tract [3, 4].

Disruptions in the urinary microbiome and elevations in ROS or proinflammatory cytokines have been found to be associated with increased levels of inflammation. This inflammation can lead to various urological disorders, including sexual dysfunction, prostatitis, chronic pelvic pain, male factor infertility, and genitourinary malignancies (Table 1) [5–8]. Several studies have shown associations between elevated levels of ROS and abnormal sperm concentration, motility, morphology, sperm DNA damage, and impaired

Effects of Lifestyle on Men's Health. https://doi.org/10.1016/B978-0-12-816665-9.00005-6
© 2019 Elsevier Inc. All rights reserved.

Table 1 Urogenital Conditions Associated With Inflammation

Condition	Clinical Characteristics	Associated Inflammatory Markers or Mechanism of Action
Chronic prostatitis and pelvic pain syndrome	• Pain • No clear infectious etiology • Often associated with other pain syndromes (IBS, fibromyalgia, etc.)	• Elevations in serum interleukins (IL-1α, IL-1β, IL-4, and IL-13) and TNF-alpha • Mast cell response (tryptase-β and PAR2) • Recruitment of Th1 cells
Bacterial prostatitis	• Pain • *Escherichia coli* and *Enterococcus* species commonly isolated • Represents 5%–10% of prostatitis cases	• Similar to chronic prostatitis and pelvic pain syndrome • Isolation of bacterial species
Erectile dysfunction	• Sexual dysfunction • Associated with diabetes, HTN, obesity, low levels of exercise, and depression • Often associated with prostatic inflammation and BPH	• Increased TNF-alpha, IL-6, and IL-8 • Elevated neutrophil-to-lymphocyte ratio (systemic inflammation)
Hypogonadism	• Low testosterone • Often associated with obesity, sleep apnea, subfertility, and metabolic syndrome	• Systemic inflammation from immunosuppression due to low testosterone • Possible increased systemic inflammation from passage of GI bacteria into circulation
Subfertility/infertility	• Impaired spermatogenesis • Associated with varicocele, cryptorchidism, testicular torsion, GU infections, and environmental exposures or toxins	• Elevated reactive oxygen species (ROS) • Elevated heat-shock proteins, granulysin, glutathione S-transferase, nitric oxide synthase, IL-1, IL-6, IL-8, malondialdehyde, interferon gamma, macrophage migration inhibitory factor, TNF-alpha • Increased BAX, decreased BCL2 genes • Associated with temperature dysregulation of spermatogenesis and mast cell response • Can be associated with infectious etiologies
Prostate cancer	• Inflammation associated with high-grade prostate cancer	• Increased IL-6 correlates with metastatic or castration-resistant prostate cancer • Expression of monocyte chemoattractant protein-1, cyclooxygenase-2, TNF-alpha, and interleukin-1β • Infectious etiologies can result in chronic prostatic inflammation
Bladder cancer	• Known risk factors include genetics, geography, smoking, occupational exposures, arsenic, *Schistosoma haematobium* infection, and certain medications	• Pro-inflammatory cytokines may induce differentiation of cytotoxic cells and tumor proliferation • Cyclooxygenase-2 (COX-2) upregulation • Urinary microbiome dysregulation • Altered binding of PD-L1 to PD-1

sperm-oocyte fusion [6]. There is also increasing evidence demonstrating that chronic inflammation may promote the process of carcinogenesis and could play a role in the development of both bladder and prostate cancer [9]. The specific mechanism by which inflammation impacts the male genitourinary tract is an area of ongoing study, but current literature has clearly demonstrated that inflammation can have significant consequences for overall men's health.

GENITOURINARY CONDITIONS, PAIN SYNDROMES, AND SEXUAL DYSFUNCTION

Prostatitis, Chronic Pelvic Pain, and Inflammation

Chronic prostatitis and chronic pelvic pain syndrome are two overlapping conditions that commonly result in urological morbidity for men younger than 50 years of age. These issues can affect up to 15% of men at some point in their lives [10]. These conditions are characterized by pain and inflammation without a clear infectious etiology [11]. Chronic prostatitis and chronic pelvic pain syndrome must be differentiated from bacterial prostatitis, which only accounts for 5%–10% of prostatitis cases. In cases of bacterial prostatitis, the most commonly isolated microorganisms are *Escherichia coli* and *Enterococcus* species [12]. This section will focus on the inflammatory response seen in the noninfectious subtype of chronic prostatitis.

Numerous inflammatory cytokines and signaling pathways have been evaluated in animal and human models, although a definitive etiology is not well understood. Elevations in serum interleukins (IL-1α, IL-1β, IL-4, and IL-13) and TNF-alpha have been proposed as markers of inflammation in men with chronic prostatitis and chronic pelvic pain [13]. A mast cell-mediated response involving tryptase-β and its cognate receptor protease-activated receptor 2 (PAR2) has also been investigated since levels of tryptase-β and PAR2 have been involved with other pain pathways [14]. It is likely that several factors are involved with a complex process that triggers an inflammatory response within the prostate or pelvic structures of men with this condition. The recruitment of Th1 cells within the prostate and increased activity of mast cells appear to be primary driving forces involved in the development of chronic pelvic pain, although research is ongoing to elucidate a clear mechanistic pathway [11].

The presence of chronic prostatitis or chronic pelvic pain in men is often associated with other chronic pain conditions such as irritable bowel syndrome, fibromyalgia, and chronic fatigue syndrome. From a treatment standpoint, alpha blockers and psychological interventions may alleviate symptoms for some men. Treatment options such as acupuncture and pelvic floor physical therapy may also provide some degree of relief. As the understanding of precise inflammatory pathways causing pain improves, future treatments may focus on therapies that directly target sources of inflammation [15]. Clinical phenotyping of symptom burden using a validated approach such as urinary, psychosocial, organ-specific, infection, neurological/systemic, and tenderness (UPOINT) can guide multi-modal therapy and lead to significant symptom improvement in patients [16].

Erectile Dysfunction and Inflammation

Male sexual function is frequently influenced by inflammation. One of the most commonly reported sexual problems among older men is erectile dysfunction (ED), which has been associated with diabetes, hypertension, obesity, low levels of physical exercise, lower urinary tract symptoms, and depression [17]. The exact prevalence of ED is debated, but this condition is extremely common in high risk groups. For example, a 2018 meta-analysis that included 2525 diabetic men found ED prevalence to be 74.2% (95% CI 59.0–89.4) among men with both diabetes and depression. The overall prevalence of ED among diabetic men without depression was 37.4% (95% CI 16.2–58.6) [18]. A variety of causes of ED have been proposed, including endothelial dysfunction, atherosclerosis, and chronic inflammation [19].

The role of inflammation in the development of ED is supported by research that has found increased levels of pro-inflammatory cytokines such as TNF-alpha, interleukin-6 (IL-6), and interleukin-8 (IL-8) in men with ED [20]. A 2018 study investigated the role of systemic inflammation as a cause of ED by evaluating the neutrophil-to-lymphocyte ratio (NLR) in men with ED. Elevations in NLR were found to be associated with an increased risk of having severe ED, further supporting existing evidence linking systemic inflammation to ED [21]. Prostatic inflammation, benign prostatic hypertrophy (BPH), and elevated levels of IL-8 within prostatic fluid have also been associated with ED [22].

The concern for inflammation as a cause of ED has led to research regarding the use of antiinflammatory medications as a treatment modality for ED. Several studies have claimed that aspirin or nonsteroidal antiinflammatory medications (NSAIDs) are beneficial in the treatment of ED through interactions with nitric oxide synthase. Other studies, however, considered the use of aspirin and NSAIDs to be risk factors for ED since these medications interfere with prostaglandin production [19]. The mechanistic effect of aspirin and NSAIDs on the nitric oxide pathway is still controversial, although inflammation does appear to be a contributing factor in the development of ED.

Hypogonadism and Inflammation

Chronic low-grade inflammation has been proposed as a cause of hypogonadism in men. Low testosterone levels resulting from hypogonadism are commonly seen in men with metabolic syndrome and obesity. Obesity can lead to elevations in adipokines that are produced by adipose tissue. These adipokines have been found to induce inflammation and oxidative stress within the male genitourinary tract, which can directly impair gonadal function [23]. Early publications suggested that elevated systemic inflammation in obese men was also a result of a decline in the immunosuppressive properties of testosterone. Later publications proposed that inflammation originating from the passage of gastrointestinal bacteria into the systemic circulation may provide another explanation for hypogonadism in obese men, the so-called Gut Endotoxin Leading to a Decline IN Gonadal function (GELDING) theory [24, 25].

A 2018 study tested the inflammatory hypothesis of testosterone deficiency among men with metabolic syndrome and found that treatment with an interleukin-1 (IL-1) antagonist, an antiinflammatory drug, was able to increase testosterone levels among obese men with low initial testosterone levels [26]. While the development of male factor infertility is not universal among obese men or men with metabolic syndrome, low levels of circulating testosterone and hypogonadism are certainly risk factors for subfertility and are commonly found during an infertility evaluation. The impact of inflammation on male fertility will be discussed in detail in the following section.

MALE INFERTILITY

In the United States, infertility affects somewhere between 8% and 15% of couples. In approximately 20% of infertile couples, the male partner is the sole cause of infertility, but male factor infertility is believed to be a contributing factor in greater than 50% of infertility cases [27]. Poor sperm quality and male factor infertility are strongly tied to inflammation within the male genitourinary tract. Several decades ago, ROS were proposed as a causative mechanism for poor semen parameters. Compared with fertile men, semen samples from infertile men have shown significantly higher levels of ROS, with up to 80% of infertile men demonstrating elevated seminal ROS [28]. Elevations in ROS or decreased physiological antioxidant capability can result in oxidative stress, which leads to decreased sperm motility, lipid peroxidation, increased sperm DNA damage, and an overall worsening of semen analysis parameters [6, 29]. Oxidative stress can arise from intrinsic abnormalities within the sperm itself, varicocele, cryptorchidism, testicular torsion, localized infections, environmental factors, or inflammation related to aging [6].

The mechanism by which sperm is affected by inflammation and ROS is complex (Fig. 1). Reduced membrane fluidity and decreased axonemal protein phosphorylation from ROS can result in decreased sperm motility [30, 31]. Hydrogen peroxide, a specific ROS, can diffuse across cellular membranes and disrupt the enzymatic activity within cells [32]. Sperm mitochondrial

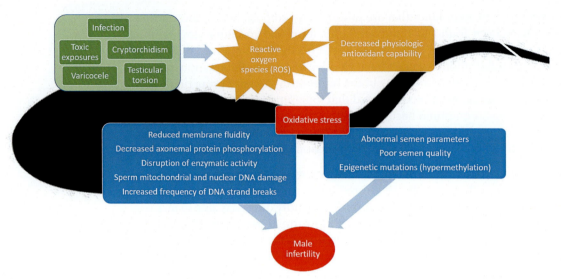

FIG. 1
The effects of inflammation on sperm.

and nuclear DNA can also be damaged as abnormal DNA cross-links, frame shifts, or deletions occur due to ROS exposure [29]. Increased frequency of sperm DNA strand breaks; increased rates of aneuploidy that can result in genetic abnormalities; increased cellular apoptosis; and findings of oligozoospermia, oligoasthenoteratospermia, or oligoasthenospermia have all been associated with elevated levels of ROS [6, 33, 34].

Varicocele, Infertility, and Inflammation

Specific conditions of the male reproductive tract may negatively impact male fertility through inflammatory mechanisms. One such condition is a clinically evident varicocele, caused by dilation of the spermatic veins. The incidence of clinical varicocele among infertile men is estimated at 40% or more, which is significantly higher than the varicocele incidence range of 8%–23% found in the fertile male population [35, 36]. In the presence of a varicocele, spermatogenesis is impaired through heat stress, which leads to excess ROS, ultimately resulting in increased apoptosis. Elevations in scrotal temperatures seen in men with varicocele are the result of increased retrograde blood flow [37]. Men with clinically evident varicoceles have been shown to have decreased expression of heat-shock proteins, elevated polymorphism of glutathione S-transferase and nitric oxide synthase genes, and increased BAX and decreased BCL2 genes [38].

The cytolytic and pro-inflammatory molecule granulysin may also play a role in the development of infertility related to varicocele. A 2018 publication demonstrated that infertile men with varicocele had significantly elevated levels of seminal granulysin when compared with infertile men without varicocele. Seminal granulysin levels appeared to be negatively correlated with sperm motility, concentration, morphology, and testicular volume [39]. Vitamin A (all-trans-retinoic acid) is necessary to maintain male fertility and achieve normal spermatogenesis. In men with normal semen parameters, vitamin A appears to have a lipid-lowering effect and plays a role in the cellular metabolism of triglycerides, lipase, and glucose. The presence of a varicocele appears to disrupt proper functioning of vitamin A, resulting in decreased levels of superoxide dismutase (SOD) and glutathione S-transferase omega 2 (GSTO2). The decreases in SOD and GSTO2 that are seen in men with varicocele can ultimately lead to elevations in malondialdehyde and ROS, markers of inflammation [40].

Fortunately, surgical repair can eliminate varicocele as a source of inflammation within the male genitourinary tract. The exact effect of varicocele repair on spontaneous pregnancy rates and live birth rates is unclear, but varicocelectomy has been found to improve sperm parameters such as count, total motility, and progressive motility. Repair has been found to be a cost-effective way to reduce sperm DNA damage and seminal oxidative stress, improving semen analysis parameters [41, 42].

Cryptorchidism, Infertility, and Inflammation

Similar to varicocele, cryptorchidism has been associated with infertility, increased testicular temperature, and elevated ROS [6]. The prevalence of cryptorchidism, or undescended testes, ranges from between 1% and 3% in full-term infant males to as high as 30% in premature boys [43]. Recommendations suggest that orchidopexy should be performed prior to 1 year of age to minimize the long-term fertility impact for men with a history of cryptorchidism [44, 45]. Despite surgical correction, azoospermia is present in 10%–13% of men with a history of unilateral cryptorchidism [46]. Among men with untreated bilateral cryptorchidism, rates of azoospermia are as high as 89% [47].

In men, spermatogenesis within the scrotum occurs at a temperature approximately 3°C lower than body temperature. Spermatogenesis suffers dramatically at 37°C, which occurs in cryptorchid men due to the presence of the testis within the abdomen [48]. There is also a strong link between inflammation and the development of infertility among men or boys with undescended testes. Undescended testes result in high levels of oxidative stress and a chronic inflammatory state. A 2012 study compared boys with cryptorchidism to boys with normal descent of the testes and found that concentrations of malondialdehyde (MDA) and interleukin-6 (IL-6) were significantly elevated in the group with undescended testes compared with controls. Additionally, mean MDA and IL-6 concentrations were significantly higher in cases of bilateral cryptorchidism compared with unilateral cases (MDA: 4.03±3.68 vs 3.49±5.22, $P=.015$; IL-6: 7.70±6.86 vs 3.48±6.50, $P=.001$) [49].

Mast cells, which are known to be involved with inflammation and fibrosis within a variety of organ systems, have been strongly associated with testicular fibrosis in human studies. This relationship appears to be mediated through the actions of tryptase and chymase. Mast cells have also been implicated in the development of decreased spermatogenesis. Within cryptorchid testes, elevated levels of intratesticular estrogen may promote mast cell migration and proliferation, ultimately leading to increased inflammation and deleterious effects on sperm production [50]. Autoimmune factors may also play a role in the development of infertility among men with a history of cryptorchidism. Carbonic anhydrase (CA) II autoantibody titers in a group of 59 boys with undescended testes were found to be significantly higher than levels of a control group ($P=.048$) [51]. Ultimately, despite surgical correction, the inflammatory effects of cryptorchidism may result in infertility or subfertility that can persist into adulthood.

Inflammation and Testicular Torsion

Testicular torsion is a urological emergency that is most commonly seen in adolescent boys and young men. Torsion results from rotation of the testis around the spermatic cord. Incidence of this condition is approximately 4.5 cases per every 100,000 males less than 25 years of age [52]. The subsequent ischemia can result in a rapid loss of germ cells, often within 4–8h [53, 54]. If misdiagnosed or inappropriately treated, testicular torsion can result in the loss of testicular tissue and the development of male subfertility. Following surgical correction of testicular torsion, reperfusion injury may also occur as circulation is restored to the testicular tissue [55]. As blood supply returns to the tissue, free oxygen radicals are produced, intracellular calcium concentrations rise significantly, and lipid peroxidation occurs [56]. Tissue reperfusion also leads to stimulation of mitogen-activated protein kinases, protein denaturation, DNA damage, and acceleration of apoptosis. The testes are known to be sensitive to free radical damage associated with torsion. Antioxidant mechanisms that protect the testes from oxidative stress at baseline are altered due to elevations in calcium during reperfusion, which can lead to permanent loss of testicular tissue [55, 57].

Surgical correction of torsion may result in orchiectomy in up to one-third of cases, when testicular tissue is considered dead [58]. When testicular conservation is able to be safely achieved during surgery, the reperfusion injury cascade described above is a concerning possibility. While the impact of an acute episode of testicular torsion on long-term fertility is not entirely clear, some publications evaluating semen analyses after testicular torsion have demonstrated abnormal semen parameters compared with controls. Specifically, sperm count, motility, and morphology may be permanently abnormal due to the loss of germ cell tissue [54, 59]. The contribution of testicular torsion to overall rates of male factor infertility, however, appears to be low. Approximately 0.5% of patients with a diagnosis of male factor infertility have a history of testicular torsion as the primary cause [60]. Despite its relative rarity, testicular torsion is a well-documented cause of acute inflammation within the male urogenital tract that can have a significant impact on testicular viability.

Genitourinary Infections and Infertility

Genital tract infections, which include both viruses and bacteria, can result in male infertility. An estimated 6%–10% of male factor infertility has been attributed to urogenital infections [61]. The correlation between infection and inflammation has been heavily studied, and the impact of these two processes on male fertility can be highly problematic. Infections of the male genitourinary tract can involve the urethra, prostate gland, seminal vesicles, vas deferens, epididymis, testes, or any combination of these structures [62]. Inflammation from *Neisseria gonorrhoeae*, *Chlamydia trachomatis*, certain gram-negative bacteria (*E. coli*, *Klebsiella* species, *Proteus*, *Serratia*, or *Pseudomonas* species), or other bacterial pathogens can result in leukocytospermia and increased concentrations of cytokines and ROS [63]. Chronic exposure to the inflammatory effects of infectious agents can also compromise male accessory sex gland function, and the presence of inflammation can have a negative impact on sperm quality and function [62, 64].

Infectious etiologies result in ROS production that may cause sperm abnormalities such as decreased sperm motility, abnormal acrosin activity, and decreased sperm-oocyte fusion capability [65]. Increased oxidative stress that results from genitourinary infections causes seminal plasma hyperviscosity, damage to the sperm chromatin and DNA integrity, elevations in sperm DNA fragmentation rates, and increased sperm DNA protein cross-linking [66]. Infections have also been shown to interfere with proper function of human beta-defensin 1 (DEFB1). Abnormal function of DEFB1 decreases both motility and bactericidal activity in sperm [67]. A number of interleukins (IL-1, IL-6, and IL-8), interferon gamma, macrophage migration inhibitory factor, TNF-alpha, and other pro-inflammatory mediators are associated with detrimental semen parameters in the setting of bacterial infections [62].

Although mechanisms are less clear, the presence of human papilloma virus (HPV) in semen is associated with impaired sperm motility and the presence of antisperm antibodies [68]. Chronic viral infections such as human immunodeficiency virus (HIV), hepatitis B virus (HBV), hepatitis C virus (HCV), herpes simplex virus (HSV), and cytomegalovirus (CMV) may also play a role in the development of male infertility through inflammatory mechanisms [69]. For example, the incidence of infertility has been shown to be 1.59 times higher in patients with HBV infection than in those without HBV infection (2.21 vs 1.39 per 1000 person-years) [70]. Compared with bacterial infections, relatively little emphasis has been placed on the study of inflammation related to viral infections within the male infertility literature.

It is clear that active infection and inflammation have a negative impact on sperm quality. After the treatment of the infectious etiology, however, the likelihood that a man will develop long-term fertility issues is variable. A 2018 study evaluated fertility outcomes following successful treatment of asymptomatic chlamydial infections in men who had abnormal semen parameters at the time of *C. trachomatis* diagnosis. Following treatment with antibiotics, semen parameters improved significantly and uniformly reached normal ranges. Total antioxidant capacity (TAC) improved, and ROS levels also decreased after antibiotic treatment [71]. Similar improvements in semen parameters, TAC, and leukocytes within the seminal fluid have been seen following appropriate treatment of *Mycoplasma genitalium* [72]. If an infection is present within the male genital tract, it should be treated promptly. However, despite treatment, the possibility remains for permanent damage to spermatogenesis if infections trigger permanent immunopathologic changes within the male reproductive system [61].

Environmental Exposures, Inflammation, and Infertility

Since the era of industrialization, exposure to environmental toxins has increased significantly. Several studies have linked environmental exposures to poorer semen quality. Toxins may exert detrimental effects on male fertility through inflammatory pathways and through estrogenic or antiandrogenic effects that can alter the hypothalamic-pituitary-gonadal axis (HPGA) or result in sperm epigenetic changes [73].

Epigenetic mutations (most commonly hypermethylation) can affect several genes. Specific genes involved with epigenetic dysregulation that have been associated with abnormal semen parameters and male factor infertility include MTHFR, PAX8, NTF3, SFN, HRAS, JHM2DA, IGF2, H19, RASGRF1, GTL2, PLAG1, D1RAS3, MEST, KCNQ1, LIT1, and SNRPN. Specific environmental toxins or drugs that can impact fertility through epigenetic changes include the anticancer agent deoxycytidine, the pesticide methoxychlor, and the fungicide vinclozolin [74]. Deoxycytidine causes a decrease in global DNA methylation that leads to altered sperm morphology, decreased sperm motility, decreased fertilization capacity, and decreased embryo survival. Methoxychlor and vinclozolin act in a similar manner to disrupt endocrine functioning and result in spermatogenic defects that can impact subsequent generations [74]. Toxin exposure can also result in inflammation that leads to oxidative stress and elevated ROS production within the testes and epididymis. As described above, elevated ROS may lead to a decrease in sperm viability and motility, ultimately manifesting as male factor infertility. Exposure to radiation, extreme temperature, certain drugs, toxins, heavy metals, smoking, and biological hazards can lead to a pro-inflammatory state with a negative impact on male fertility [75].

The Effects of Inflammation on Assisted Reproductive Technology

While ROS and inflammation clearly result in changes that are detrimental to sperm quality, the clinical implications of elevated ROS levels in the setting of assisted reproductive technology (ART) such as in vitro fertilization (IVF) and intracytoplasmic sperm injection (ICSI) are conflicting. Several studies, including a meta-analysis published in 2005 by Agarwal et al., suggested that high ROS levels negatively impacted fertilization rates during IVF cycles [76]. Other studies, however, have demonstrated no negative effect of elevated ROS on IVF or ICSI rates, and Yeung et al. published a possible positive effect for fertilization rates in couples where the male partner had elevated levels of ROS prior to fertility treatment [77, 78]. To date, the impact of inflammation on fertility outcomes with the use of ART remains uncertain.

UROLOGIC MALIGNANCY

Inflammation and Prostate Cancer

Prostate cancer is the most commonly diagnosed malignancy among men, with an estimated 164,690 new cases to be diagnosed in the United States in 2018 [79]. In recent years, there has been an increased focus on inflammation as a risk factor for the development of prostate cancer. This focus stems from findings within the broader cancer literature, which have estimated that up to 20% of all adult cancers may be associated with chronic inflammatory conditions caused by infectious agents, pro-inflammatory states, or environmental exposures [12, 80]. As a result, multiple studies have proposed that chronic inflammation can lead to or worsen prostate cancer, although this has not been proved definitively (Fig. 2).

A 2014 study found that the presence of inflammation within prostatic biopsies, most of which was chronic, was positively associated with high-grade prostate cancer [80]. A separate study from 2014 reported that chronic inflammation appeared to be involved in the regulation of prostate carcinogenesis. Men with elevated levels of inflammation (elevated IL-6) and prostate cancer were found to have dysregulations in the immune response and tumor microenvironment. Higher levels of IL-6 were correlated with metastatic or castration-resistant prostate cancer. Increased levels of IL-6 were also negatively

Prostate cancer

Most commonly diagnosed malignancy among men.
Estimated 164,690 cases to be diagnosed in 2018 in the United States.

Inflammation positively correlated with high-grade prostate cancer.
Rates of inflammation on biopsy high even in men without cancer.
Malignancy likely results from a combination of genetics, environmental exposures, & inflammation.

IL-6

Elevated IL-6 associated with dysregulations in the immune response and tumor microenvironment.
Higher IL-6 correlates with metastatic disease & castration-resistant cancer.
High IL-6 levels confer impaired chemotherapy response and poor survival.

Mouse model: increased monocyte chemoattractant protein-1, cyclooxygenase-2, TNF-alpha, & interleukin-1β associated with rapid tumor growth, high proliferation indices, & increased rates of invasion

FIG. 2
The relationship between inflammation and prostate cancer.

associated with survival and response to chemotherapy [81]. In a mouse model, significant increases in expression of monocyte chemoattractant protein-1, cyclooxygenase-2, TNF-alpha, and interleukin-1β were seen in prostate tumors that exhibited rapid tumor growth, high proliferation indexes, and increased rates of invasion. This observed inflammatory response appears to be highly correlated with poorer outcomes [82].

In men without prostate cancer, the prevalence of asymptomatic prostatic inflammation is also extremely high. Studies evaluating men who have undergone biopsy and ultimately test negative for cancer have shown levels of prostatic inflammation in as many as 80% of patients [12]. This high rate of inflammation is concerning due to inflammation's role in the subsequent development of malignancy. Levels of inflammation may influence cellular and immune responses that could predispose men to prostate cancer or worsen an existing cancer [83]. Multiple microorganisms (*C. trachomatis, Propionibacterium acnes, Trichomonas vaginalis,* and *Gammaretrovirus*) have also been associated with prostatic inflammation [12]. Prostatic corpora amylacea and calculi, which are laminated bodies and calcified stones, are frequently seen in the adult prostate and can lead to focal acute and chronic inflammation [84]. Genetic polymorphisms within inflammatory pathways may also lead to the production of inflammatory cytokines that could possibly elevate prostate cancer risk or progression [85]. While prostate cancer is likely caused by a combination of genetics and environmental exposures, it also seems evident that chronic inflammation plays a role in the growth of malignant cells.

Inflammation and Bladder Cancer

The risk of urinary bladder cancer, most commonly transitional cell carcinoma in developed countries, increases with increasing age [86]. Certain genetic factors, geography, cigarette smoking, occupational exposures, arsenic exposure, *Schistosoma haematobium* infection, and certain medications are also known risk factors [87]. The role of inflammation in the development of bladder cancer is complex, but many of the known risk factors for bladder cancer act via inflammatory pathways to promote carcinogenesis. Pro-inflammatory cytokines may induce differentiation of cytotoxic cells that can promote bladder cancer cell proliferation. This can result in a highly inflammatory state that can stimulate cancer metastasis via degradation of extracellular matrix proteins. Additionally, the suppression of the cell-mediated immune response by cyclooxygenase-2 (COX-2) upregulation can result in undifferentiated tumor growth. Higher levels of inflammation present in patients with bladder cancer can also be predictive of poor response to treatment [88]. Bladder cancer may also alter the binding of PD-L1 to PD-1, promoting inflammation and decreasing immune activity [89].

The urinary microbiome may also play an important role in the inflammatory pathways that promote bladder cancer. A 2018 study provided an analysis of the urinary microenvironment in patients with bladder cancer compared with controls. Increases in specific bacterial species (*Acinetobacter, Anaerococcus,* and *Sphingobacterium*) and decreases in other bacterial species (*Serratia, Proteus,* and *Roseomonas*) were found in the bladder cancer group when compared with non-cancer group. This study reported that there was a significant difference in the urinary microbiome of cancer patients compared with patients without bladder cancer. Additionally, increased isolation of *Herbaspirillum, Porphyrobacter,* and *Bacteroides* occurred in cancer patients who had high risk of recurrence and progression, implying that these bacterial species could potentially be used as biomarkers for risk stratification in the future [5]. At this time, specific relationships between dysregulation of the urinary microbiome and markers of inflammation have not been established.

While inflammation can promote growth of bladder tumors, current literature has not demonstrated clinical benefit from the use of antiinflammatory drugs such as celecoxib or atorvastatin in the treatment of bladder cancer [88]. Long-term survival for patients with advanced bladder cancer is poor, with a 5-year survival of only 5% when metastases are present [89]. A deeper understanding of specific mechanistic pathways by which inflammation contributes to the development of bladder cancer will likely impact future treatment options for this condition.

IMPLICATIONS

The relationship between inflammation and men's health is an area of ongoing research. Numerous inflammatory pathways and mechanisms involved in proper functioning of the male genitourinary tract have been described. To date, there is strong evidence that an elevated inflammatory state in men contributes to urological conditions, sexual dysfunction, and endocrine disturbances. Inflammation also appears to play a key role in the development of male infertility and certain malignancies within the male reproductive tract. As specific markers of inflammation are identified and inflammatory pathways become clearer, researchers and physicians will hopefully be able to provide targeted antiinflammatory treatments for certain conditions. In the future, the use of inflammatory markers may also be useful to stratify patients into distinct risk categories for specific conditions. Overall, it is clear that the role of inflammation is a complex yet important aspect of male genitourinary health.

References

[1] Thomas-White K, Brady M, Wolfe AJ, Mueller ER. The bladder is not sterile: history and current discoveries on the urinary microbiome. Curr Bladder Dysfunct Rep 2016;11(1):18–24.

[2] Gottshick C, Deng Z, Vital M, Masur C, Abels C, Pieper D, Wagner-Dobler I. The urinary microbiota of men and women and its changes in women during bacterial vaginosis and antibiotic treatment. Microbiome 2017;5:99.

[3] Dong Q, Nelson DE, Toh E, Diao L, Gao X, Fortenberry JD. The microbial communities in male first catch urine are highly similar to those in paired urethral swab specimens. PLoS ONE 2011;6.

[4] Nelson DE, Dong Q, VanderPol B, Toh E, Fan B, Katz BP. Bacterial communities of the coronal sulcus and distal urethral of adolescent males. PLoS ONE 2012;7.

[5] Wu P, Zhang G, Zhao J, Chen J, Chen Y, Huang W, Zhong J, Zeng J. Profiling the urinary microbiota in male patients with bladder cancer in China. Front Cell Infect Microbiol 2018;31(8):167.

[6] Ko EY, Sabanegh ES, Agarwal A. Male infertility testing: reactive oxygen species and antioxidant capacity. Fertil Steril 2014;102(6):1518–27.

[7] Shoskes DA, Altemus J, Polackwich AS, Tucky B, Wang H, Eng C. The urinary microbiome differs significantly between patients with chronic prostatitis/chronic pelvic pain syndrome and controls as well as between patients with different clinical phenotypes. Urology 2016;92:26–32.

[8] Naboka I, Kogan M, Gudima I, Ibishev K, Pasechnik D, Logvinov A, Ilmdarov S. Microbiota of lower urine tract and genital organs of healthy men and in infertility. Zh Mikobiol Epidemiol Immunobiol 2015;1:65–71.

[9] Doat S, Marous M, Rebillard X, Tretarre B, Lamy PJ, Soares P, Delbos O, Thuret R, Segui B, Cenee S, Menegaux F. Prostatitis, other genitourinary infections and prostate cancer risk: influence of non-steroidal anti-inflammatory drugs? Results from the EPICAP study. Int J Cancer 2018;143(7):1644–51.

[10] Ismail M, Mackenzie K, Hashim H. Contemporary treatment options for chronic prostatitis/chronic pelvic pain syndrome. Drugs Today 2013;49(7):457–62.

[11] Breser ML, Salazar FC, Rivero VE, Motrich RD. Immunological mechanisms underlying chronic pelvic pain and prostate inflammation in chronic pelvic pain syndrome. Front Immunol 2017;8(898):1–11.

[12] Sfanos KS, DeMarzo AM. Prostate cancer and inflammation: the evidence. Histopathology 2012;60(1):199–215.

[13] Hu C, Yang H, Zhao Y, Chen X, Dong Y, Li L, Dong Y, Cui J, Zhu T, Zheng P, Lin CS, Dai J. The role of inflammatory cytokines and ERK1/2 signaling in chronic prostatitis/chronic pelvic pain syndrome with related mental health disorders. Sci Rep 2016;6(28608):1–12.

[14] Roman K, Done JD, Schaeffer AJ, Murphy SF, Thumbikat P. Tryptase - PAR2 axis in experimental autoimmune prostatitis, a model for chronic pelvic pain syndrome. Pain 2014;155(7):1328–38.

[15] Pontari M, Giusto L. New developments in the diagnosis and treatment of chronic prostatitis/chronic pelvic pain syndrome. Curr Opin Urol 2013;23(6):565–9.

[16] Polackwich AS, Shoskes DA. Chronic prostatitis/chronic pelvic pain syndrome: a review of evaluation and therapy. Prostate Cancer Prostatic Dis 2016;19(2):132–8.

[17] Shamloul R, Ghanem H. Erectile dysfunction. Lancet 2013;381(9861):153–65.

[18] Wang X, Yang X, Cai Y, Wang S, Weng W. High prevalence of erectile dysfunction in diabetic men with depressive symptoms: a meta-analysis. J Sex Med 2018;15(7):935–41.

[19] Li T, Wu C, Fu F, Qin F, Wei Q, Yuan J. Association between use of aspirin or non-aspirin non-steroidal anti-inflammatory drugs and erectile dysfunction: a systematic review. Medicine 2018;97(28). [epub ahead of print].

[20] Rodrigues FL, Fais RS, Tostes RC, Carneiro FS. There is a link between erectile dysfunction and heart failure: it could be inflammation. Curr Drug Targets 2015;16(5):442–50.

[21] Ventimiglia E, Cazzaniga W, Pederzoli F, Frego N, Chierigo F, Capogrosso P, Boeri L, Deho F, Abbate C, Moretti D, Piemonti L, Montorsi F, Salonia A. The role of neutrophil-to-lymphocyte ratio in men with erectile dysfunction-preliminary findings of a real-life cross-sectional study. Andrology 2018;6(4):559–63.

[22] Agnihotri S, Mittal RD, Kapoor R, Mandhani A. Asymptomatic prostatic inflammation in men with clinical BPH and erectile dysfunction affects the positive predictive value of prostate-specific antigen. Urol Oncol 2014;32(7):946–51.

[23] Liu Y, Ding Z. Obesity, a serious etiologic factor for male subfertility in modern society. Reproduction 2017;154(4):R123–31.

[24] Tremellen K, McPhee N, Pearce K. Metabolic endotoxaemia related inflammation is associated with hypogonadism in overweight men. Basic Clin Androl 2017;27(5):1–9.

[25] Tremellen K. Gut endotoxin leading to a decline IN gonadal function (GELDING) - a novel theory for the development of late onset hypogonadism in obese men. Basic Clin Androl 2016;26(7):1–13.

[26] Ebrahimi F, Urwyler SA, Straumann S, Doerpfeld S, Bernasconi L, Neyer P, Schuetz P, Mueller B, Donath MY, Christ-Crain M. Interleukin-1 antagonism in men with metabolic

syndrome and low testosterone - a randomized clinical trial. J Clin Endocrinol Metab 2018;103(9):3466–76.

[27] ASRM. Diagnostic evaluation of the infertile male: a committee opinion. Fertil Steril 2015; 103(3):18–25.

[28] Agarwal A, Sharma RK, Sharma R, Assidi M, Abuzenadah AM, Alshahrani S. Characterizing semen parameters and their association with reactive oxygen species in infertile men. Reprod Biol Endocrinol 2014;12:33.

[29] Agarwal A, Saleh RA, Bedaiwy MA. Role of reactive oxygen species in the pathophysiology of human reproduction. Fertil Steril 2003;79:829–43.

[30] DeLamirande E, Gagnon C. Reactive oxygen species and human spermatozoa: effects on the motility of intact spermatozoa and on sperm axonemes. J Androl 1992;13:368–78.

[31] DeLamirande E, Gagnon C. Reactive oxygen species and human spermatozoa: depletion of adenosine triphosphate (ATP) plays an important role in the inhibition of sperm motility. J Androl 1992;13:379–86.

[32] Aitken R. Molecular mechanisms regulating human sperm function. Mol Hum Reprod 1997;3:169–73.

[33] Aitken RJ, Krausz C. Oxidative stress, DNA damage and the Y chromosome. Reproduction 2001;122:497–506.

[34] Kodama H, Yamaguchi R, Fukuda J, Kasai H, Tanaka T. Increased oxidative deoxyribonucleic acid damage in the spermatozoa of infertile male patients. Fertil Steril 1997;65:519–24.

[35] Kursh E. What is the incidence of varicocele in a fertile population? Fertil Steril 1987;48(3):510–1.

[36] Muratorio C, Meunier M, Sonigo C, Massart P, Boitrelle F, Hugues JN. Varicocele and infertility: where do we stand in 2013? Gynecol Obstet Fertil 2013;41(11):660–6.

[37] Goldstein M, Eid JF. Elevation of intratesticular and scrotal skin surface temperature in men with varicocele. J Urol 1989;142:743–5.

[38] Hassanin AM, Ahmed HH, Kaddah AN. A global view of the pathophysiology of varicocele. Andrology 2018;6(5):654–61.

[39] Mikhael NW, el-Refaie AM, Sabry JH, Akl EM, Habashy AY, Mostafa T. Assessment of seminal granulysis in infertile men with varicocele. Andrologia 2018;50(8):e13066.

[40] Malivindi R, Rago V, DeRose D, Gervasi MC, Cione E, Russo G, Santoro M, Aquila S. Influence of all-trans retinoic acid on sperm metabolism and oxidative stress: its involvement in the physiopathology of varicocele-associated male infertility. J Cell Physiol 2018;233(12):9526–37.

[41] Sonmez MG, Haliloglu AH. Role of varicocele treatment in assisted reproductive technologies. Arab J Urol 2018;16(1):188–96.

[42] Baazeem A, Belzile E, Ciampi A, Dohle G, Jarvi K, Salonia A, Weidner W, Zini A. Varicocele and male factor infertility treatment: a new meta-analysis and review of the role of varicocele repair. Eur Urol 2011;60(4):796–808.

[43] Kolon F, Patel RP, Huff DS. Cryptorchidism: diagnosis, treatment and long-term prognosis. Urol Clin North Am 2004;31:469–80.

[44] Cortes D, Thorup J, Lindenberg S, Visfeldt J. Infertility despite surgery for cryptorchidism in childhood can be classified by patients with normal or elevated follicle-stimulating hormone and identified at orchidopexy. BJU Int 2003;91(7):670–4.

[45] Trussel JC, Lee PA. The relationship of cryptorchidism to fertility. Curr Urol Rep 2004;5:142–5.

[46] Fawzy F, Hussein A, Mahmoud-Eid M, Mahmoud-El-Kashash A, Khairy-Salem H. Cryptorchidism and fertility. Clin Med Insights Reprod Health 2015;9:39–43.

[47] Hadziselimovic F, Herzog B. The importance of both and early orchidopexy and germ cell maturation. Lancet 2001;358:1156–7.

[48] Yadav SK, Pandey A, Kumar L, Devi A, Kushwaha B, Vishvkarma R, Maikhuri JP, Rajender S, Gupta G. The thermo-sensitive gene expression signatures of spermatogenesis. Reprod Biol Endocrinol 2018;16(1):56.

[49] Imamoglu M, Bulbul SS, Kaklikkaya N, Sarihan H. Oxidative, inflammatory and immunologic status in children with undescended testes. Pediatr Int 2012;54(6):816–9.

[50] Mechlin C, Kogan B. Mast cells, estrogens, and cryptorchidism: a histologic based review. Trans Androl Urol 2012;1(2):97–102.

[51] Alver A, Imamoglu M, Mentese A, Senturk A, Bulbul SS, Kahraman C, Sumer A. Malondialdehyde and CA II autoantibody levels are elevated in children with undescended testes. World J Urol 2014;32(1):209–13.

[52] Mansbach JM, Forbes P, Peters C. Testicular torsion and risk factors for orchiectomy. Arch Pediatr Adolesc Med 2005;159(12):1167–71.

[53] Shimizu S, Tsounapi P, Dimitriadis F, Higashi Y, Shimizu T, Saito M. Testicular torsion-detorsion and potential therapeutic treatments: a possible role for ischemic postconditioning. Int J Urol 2016;23(6):454–63.

[54] DaJusta D, Granberg CF, Villanueva C, Baker LA. Contemporary review of testicular torsion: new concepts, emerging technologies, and potential therapeutics. J Pediatr Urol 2013;9(6). [epub ahead of print].

[55] Arena S, Iacona R, Antonuccio P, Russo T, Salvo V, Gitto E, Impellizzeri P, Romeo C. Medical perspective in testicular ischemia-reperfusion injury. Exp Ther Med 2017;13(5): 2115–22.

[56] Akbas H, Ozden M, Kanko M, Maral H, Bulbul S, Yavuz S, Ozker E, Berki T. Protective antioxidant effects of carvedilol in a rat model of ischaemia-reperfusion injury. J Int Med Res 2005;33:528–36.

[57] Antonuccio P, Minutoli L, Romeo C, Nicotina PA, Bitto A, Arena S, Altavilla D, Zuccarello B, Polito F, Squadrito F. Lipid peroxidation activates mitogen-activated protein kinases in testicular ischemia-reperfusion injury. J Urol 2006;176:1666–72.

[58] Cost N, Bush NC, Barber TD, Huang R, Baker LA. Pediatric testicular torsion: demographics of national orchiopexy versus orchiectomy rates. J Urol 2011;185(6 Suppl):2459–63.

[59] Ferreira U, Netto-Junior NR, Esteves SC, Rivero MA, Schirren C. Comparative study of the fertility potential of men with only one testis. Scand J Urol Nephrol 1991;25(4):255–9.

[60] Sigman M, Jarow JP. In: Alan JW, editor. Campbell-Walsh urology. Philadelphia: Saunders Elsevier; 2007.

[61] Schuppe HC, Pilatz A, Hossain H, Diemer T, Wagenlehner F. Urogenital infection as a risk factor for male infertility. Dtsch Arztebl Int 2017;114:339.

[62] LaVignera S, Condorelli RA, Vicari E, Salmeri M, Morgia G, Favilla V, Cimino S, Calogero AE. Microbiological investigation in male infertility: a practical overview. J Med Microbiol 2014;63:1–14.

[63] LaVignera S, Vicari E, Condorelli RA, D'Agata R, Calogero AE. Male accessory gland infection and sperm parameters (review). Int J Androl 2011;34:330–47.

[64] Vicari E, LaVignera S, Castiglione R, Calogero AE. Sperm parameter abnormalities, low seminal fructose and reactive oxygen species overproduction do not discriminate patients with unilateral or bilateral post-infectious inflammatory prostato-vesiculo-epididymitis. J Endocrinol Investig 2006;29:18–25.

[65] Lanzafame FM, LaVignera S, Vicari E, Calogero AE. Oxidative stress and medical antioxidant treatment in male infertility. Reprod BioMed Online 2009;19:638–59.

[66] Twigg JP, Irvine DS, Aitken RJ. Oxidative damage to DNA in human spermatozoa does not preclude pronucleus formation at intracytoplasmic sperm injection. Hum Reprod 1998;13:1864–71.

[67] Diao R, Fok KL, Chen H, Yu MK, Duan Y, Chung CM, Li Z, Wu H, Li Z, Zhang H, Ji Z, Zhen W, Ng CF, Gui Y, Cai Z, Chan HC. Deficient human beta-defensin 1 underlies male infertility associated with poor sperm motility and genital tract infection. Sci Transl Med 2014; 6(249):249.

[68] Foresta C, Noventa M, DeToni L, Gizzo S, Garolla A. HPV-DNA sperm infection and infertility: from a systematic literature review to a possible clinical management proposal. Andrology 2015;3(2):163–73.

[69] Garolla A, Pizzol D, Bertoldo A, Menegazzo M, Barzon L, Foresta C. Sperm viral infection and male infertility: focus on HBV, HCV, HIV, HPV, HSV, HCMV, and AAV. J Reprod Immunol 2013;100(1):20–9.

[70] Su F, Chang S, Sung F, Su C, Shieh Y, Lin C, Yeh C. Hepatitis B virus infection and the risk of male infertility: a population-based analysis. Fertil Steril 2014;102(6):1677–84.

[71] Ahmadi MH, Mirsalehian A, Sadighi-Gilani MA, Bahador A, Afraz K. Association of asymptomatic Chlamydia trachomatis infection with male infertility and the effect of antibiotic therapy in improvement of semen quality in infected infertile men. Andrologia 2018;50(4): e12944.

[72] Hossein-Ahmadi M, Mirsalehian A, Sadighi-Gilani M, Bahador A, Talebi M. Improvement of semen parameters after antibiotic therapy in asymptomatic infertile men infected with Mycoplasma genitalium. Infection 2018;46(1):31–8.

[73] Mima M, Greenwald D, Ohlander S. Environmental toxins and male fertility. Curr Urol Rep 2018;19(7):50.

[74] Rajender S, Avery K, Agarwal A. Epigenetics, spermatogenesis and male infertility. Mutat Res 2011;727(3):62–71.

[75] Lavranos G, Balla M, Tzortzopoulou A, Syriou V, Angelopoulou R. Investigating ROS sources in male infertility: a common end for numerous pathways. Reprod Toxicol 2012;34 (3):298–307.

[76] Agarwal A, Allamaneni SS, Nallella KP, George AT, Mascha E. Correlation of reactive oxygen species levels with the fertilization rate after in vitro fertilization: a qualified meta-analysis. Fertil Steril 2005;84:223–31.

[77] Yeung CH, DeGeyter C, DeGeyter M, Nieschlag E. Production of reactive oxygen species by and hydrogen peroxide scavenging activity of spermatozoa in an IVF program. J Assist Reprod Genet 1996;13:495–500.

[78] Hammadeh ME, AlHasani S, Rosenbaum P, Schmidt W, Fisher-Hammadeh C. Reactive oxygen species, total antioxidant concentration of seminal plasma and their effect on sperm parameters and outcome of IVF/ICSI patients. Arch Gynecol Obstet 2008;277:515–26.

[79] SEER. Cancer stat facts: prostate cancer, https://seer.cancer.gov/statfacts/html/prost.html; 2018. Accessed 14 July 2018.

[80] Gurel B, Lucia MS, Thompson IM, Goodman PJ, Tangen CM, Kristal AR, Parnes HL, Hoque A, Lippman SM, Sutcliffe S, Peskoe SB, Drake CG, Nelson WG, DeMarzo AM, Platz EA. Chronic inflammation in benign prostate tissue is associated with high-grade prostate cancer in the placebo arm of the prostate Cancer prevention trial. Cancer Epidemiol Biomark Prev 2014;23(5):847–56.

[81] Nguyen DP, Li J, Tewari AK. Inflammation and prostate cancer: the role of interleukin 6 (IL-6). BJU Int 2014;113(6):986–92.

[82] Adissu HA, McKerlie C, DiGrappa M, Waterhouse P, Xu Q, Fang H, Khokha R, Wood GA. Timp3 loss accelerates tumour invasion and increases prostate inflammation in a mouse model of prostate cancer. Prostate 2015;75(16):1831–43.

[83] Sfanos KS, Hempel HA, DeMarzo AM. The role of inflammation in prostate cancer. Adv Exp Med Biol 2014;816:153–81.

[84] Klimas R, Bennett B, Gardner WA. Prostatic calculi: a review. Prostate 1985;7:91–6.

[85] Palapattu GS, Sutcliffe S, Bastian PJ. Prostate carcinogenesis and inflammation: emerging insights. Carcinogenesis 2005;26:1170–81.

[86] Zhang X, Zhang Y. Bladder cancer and genetic mutations. Cell Biochem Biophys 2015;73(1):65–9.

[87] Malats N, Real FX. Epidemiology of bladder cancer. Hematol Oncol Clin North Am 2015;29(2):177–89.

[88] Gakis G. The role of inflammation in bladder cancer. Adv Exp Med Biol 2014;816:183–96.

[89] Barclay J, Creswell J, Leon J. Cancer immunotherapy and the PD-1/PD-L1 checkpoint pathway. Arch Esp Urol 2018;71(4):393–9.

CHAPTER 2.2

Diabetes and Men's Health

Carolyn A. Salter, John P. Mulhall
Department of Urology, Memorial Sloan-Kettering Cancer Center, New York, NY, United States

FUNCTIONAL ANATOMY

Erectile Tissue

The penis is composed of three chambers: the paired corpora cavernosa and the ventral corpus spongiosum that contains the urethra and widens distally to form the glans penis. All three structures are covered with a dense, collagenous layer called the tunica albuginea. The tunica albuginea has an outer longitudinal and inner circular layer [1]. The tunica is thickest over the corpora cavernosa, thinner over the corpus spongiosum, and absent over the glans penis [2].

The erectile tissue has a spongy appearance on gross examination and is composed of bundles of smooth muscle fibers that form the endothelial-lined sinuses (lacunar spaces) [1]. The corpora cavernosa share a perforated septum that allows for communication between the two chambers [3]. This is how unilateral intracavernosal injection leads to bilateral cavernosal engorgement. Superficial to the tunica albuginea, the erectile bodies are covered with Buck's (deep) fascia, dartos (superficial) fascia, and the penile skin [1].

Nerves

The dorsal penile nerves run superficially above Buck's fascia along the dorsal aspect of the penis starting at the 11 and 1 o'clock positions proximally and moving laterally. These afferent nerves provide penile skin sensation [3] and help form the pudendal nerve that runs through Alcock's canal and enters the spinal cord at the S2–S4 level [2]. The location of these nerves is clinically significant as the 12 o'clock position is devoid of nerve fibers and thus is the ideal location for a dorsal plication. The mixed perineal nerves are also a branch of the pudendal nerve and supply the bulbospongiosus muscle, the ventral shaft skin, and the frenulum [3].

The cavernous nerves branch from the pelvic plexus [3] and innervate the penis, running between the endopelvic fascia and the prostate capsule toward the periprostatic venous plexus and to the base of the penis [2, 4]. These nerves contain a mixture of parasympathetic nervous system (PNS) and sympathetic nervous system (SNS) fibers [2]. Sympathetic fibers originate from the T11 to L2 level and enter the superior hypogastric and inferior mesenteric plexuses and then send fibers in the hypogastric nerve to the pelvic plexus. Parasympathetic and somatic fibers emanate from the S2 to S4 level. The parasympathetic nerves are nonadrenergic, noncholinergic (NANC) in type and use nitric oxide (NO) as the primary neurotransmitter. Onuf's nucleus is the somatic spinal center for the penis and innervates the bulbocavernous and ischiocavernosus muscles [2].

Blood Vessels

The venous drainage of the penis is separated into superficial, intermediate, and deep venous systems. Superficial venous drainage involves the superficial dorsal vein that is formed from small veins running between Buck's and Colles' fascia. The superficial dorsal vein drains into the saphenous vein [4]. The intermediate venous system involves the circumflex and deep dorsal veins. The circumflex veins start on the ventral aspect of the corpus spongiosum and receive venous drainage from the emissary veins that run obliquely through the tunica. The circumflex veins also receive drainage from the corpus spongiosum and then drain into the deep dorsal veins that empty into the periprostatic plexus [3, 4]. Emissary veins from the proximal penis and veins from the urethral bulb form the cavernosal veins that function as the primary venous drainage of the penis [4]. This vein merges with the spongiosal (bulbourethral) vein that drains the proximal spongiosum, and together, they drain into the internal pudendal vein [3]. Crural veins originate from the crus of the penis and also drain into the internal pudendal veins [4]. Thus, only two of these venous drainage systems exist on the shaft of the penis, and thus, constriction bands can only compress the dorsal and spongiosal venous systems, as the cavernosal and crural veins exist proximal to the base of the penis.

The arterial supply to the penis has more variation than the venous system. The common penile artery arises from the internal pudendal artery and gives off small crural branches that enter the crus of the penis before splitting into three terminal branches: the dorsal, cavernosal, and bulbourethral arteries. The dorsal arteries run along the dorsum of the penis sitting between the dorsal vein and nerves. The bulbourethral artery travels to the bulb of the penis. Lastly, the cavernosal arteries, which lie along the septum, run centrally within each corpus cavernosum to give off helicine artery branches. Note that some men may also have an accessory pudendal artery (APA), an artery emanating from above the urogenital diaphragm, which travels along the anterolateral prostate to the base of the penis [4]. These APAs may be branches of the obturator, internal iliac, and vesical arteries. Given that an APA may be a major source of arterial inflow into the corpora cavernosa, it is therefore recommended to make an effort to preserve such arteries during a radical prostatectomy.

Ejaculatory Apparatus

Sperm from the testes travel through the epididymis, vas deferens, and vasal ampulla before emptying into the ejaculatory ducts. The epididymis is 3–4 m long and is composed of the head (caput), body (corpus), and tail (cauda). The epididymis leads to the vas deferens that is 25–45 cm long and travels from the scrotum through the inguinal canal to the abdomen. The seminal vesicles are lateral to the vas deferens and are 2 cm wide by 4 cm long. The ejaculatory ducts are formed from the convergence of the vasal ampullae and the seminal vesicles and empty into the urethra via the verumontanum. In terms of musculature, the bulbocavernosus muscle originates from the central perineal tendon and surrounds the corpus spongiosum. Its rhythmic contraction results in the expulsion of semen in an antegrade fashion. The ischiocavernosus muscle has its origin from the ischial tuberosity [5]. The latter muscle functions to increase intracavernosal pressure during the terminal phase of erection prior to and during orgasm.

FUNCTIONAL PHYSIOLOGY

Erectile Physiology

The mechanisms regulating the flaccid and erect states of the penis involve a complex interplay between the neural and vascular systems and the cavernosal smooth muscle and endothelium within the penis. Any disruption in these processes can lead to erectile dysfunction (ED).

At baseline, the penis is in the flaccid state resulting from the tonic contraction of the cavernosal smooth muscle within the trabeculae resulting in low resting arterial inflow into the penis [2]. The parasympathetic nervous system governs the erect state. The PNS nerve fibers arise from S2 to S4, and these fibers form the pelvic plexus along with fibers from the sympathetic ganglia [6].

Sexual stimulation leads to a neurotransmitter release that triggers cavernosal smooth muscle relaxation and the subsequent erectile response. Nitric oxide (NO) is the predominant neurotransmitter for the erect state. NO is synthesized via the enzyme nitric oxide synthase (NOS) that is released from endothelial cells (eNOS) and nerve terminals (nNOS) [6, 7]. NO leads to smooth muscle relaxation by activating guanylate cyclase that leads to the production of cyclic guanosine monophosphate (cGMP) within the sarcoplasm of the cavernosal smooth muscle cells [6]. This lowers intracytoplasmic calcium through a combination of sequestration within the sarcoplasmic reticulum and the closing of calcium channels preventing influx of calcium into the smooth muscle cells. The decrease in intracytoplasmic calcium leads to smooth muscle relaxation and an increase in blood flow to the penis. The sinusoids within the cavernosal bodies dilate and fill with blood, thus expanding. The emissary veins traveling between the layers of the tunica albuginea are occluded by the expanding cavernosa. Thus, there is a reduction in venous outflow, and the penis becomes rigid. Erectile tissue requires good elasticity to effect this expansion and venous occlusion [7].

The pressure during an erection is lower in both the corpus spongiosum, which has thinner tunica, and the glans that lacks tunica albuginea altogether. Thus, there is little venous occlusion, and these structures do not become rigid. The glans does engorge passively from venous occlusion of the circumflex and deep dorsal veins between Buck's fascia and the corpora cavernosa [2].

Note that phosphodiesterase type 5 (PDE5) released by the cavernosal smooth muscle cells degrades cGMP resulting in detumescence [7]. The oral erectogenic drugs, phosphodiesterase 5 inhibitors (PDE5i), prevent cGMP degradation, maintaining high intrasarcoplasmic calcium levels, thus preventing detumescence [6].

The sympathetic nervous system controls the flaccid state. SNS nerves arise from the T11 to L2 level of the spinal cord. The sympathetic fibers travel in the hypogastric nerve to join the pelvic plexus [6]. Detumescence is the process through which the erect penis returns to a baseline flaccid state. Norepinephrine and endothelin are the predominant neurotransmitters leading to detumescence [6, 7]. Norepinephrine is released from the cavernosal nerve terminal, and endothelin is released from the endothelial cells of cavernosal sinusoids. This leads to an increase in intracellular calcium that then activates myosin light chain kinase and results in smooth muscle contraction. Thus, these neurotransmitters result in vasoconstriction. The RhoA/Rho kinase pathway also contributes to this smooth muscle contraction by making these cells more sensitive to calcium [7]. The resultant opening of the venous drainage system leads the penis to return to its baseline flaccid state [6].

Physiology of Ejaculation

Ejaculation has two phases: emission, when the ejaculate enters the urethra, and expulsion, when rhythmic muscular contractions expel ejaculate from the urethral meatus [8]. The epididymis and bladder neck are under pure sympathetic innervation. The vas deferens and seminal vesicles have parasympathetic innervation as well, although sympathetic innervation predominates.

The external urinary sphincter and perineal periurethral muscles are under somatic control [5]. The SNS is responsible for emission and bladder neck closure, whereas the somatic nervous system is responsible for contraction of the pelvic floor musculature and the bulbourethral and ischiocavernosal muscles [9].

During normal ejaculation, the sperm stored in the vas deferens travels through the ejaculatory ducts and merges with fluid from the seminal vesicles. The bladder neck (with internal urethral sphincter) closes, the external sphincter opens, and the periurethral muscles contract in order to propel the fluid antegrade in the process of ejaculation [5, 8, 9]. This complex process is coordinated by the spinal ejaculation generator, also known as the ejaculatory reflex center, which is in the lateral ventral horn of the spinal cord at the T12–L2 levels. This center receives afferent input on cognitive arousal from higher cortical centers such as the medial preoptic area and paraventricular nucleus as well as genital sensory and autonomic input from the pudendal nerve [5, 10]. The ejaculatory reflex center then sends efferent output via sympathetics to the bladder neck, seminal vesicles, vas deferens, and prostate, causing contractions that result in emission. There is also motor output to Onuf's nucleus at S2–S4 that then sends output via the perineal branch of the pudendal nerve to the bulbocavernosus and ischiocavernosus muscles causing them to contract, resulting in antegrade ejaculation [5].

The hormonal control of ejaculation is less well understood. Oxytocin is a hormone produced in the hypothalamus but released from the posterior pituitary. Studies have demonstrated a surge of oxytocin in males during sexual stimulation, and the levels peak around orgasm. Oxytocin has been shown to cause contraction of the epididymis [11]. Estrogens are also implicated in regulating the epididymal contractility. Prolactin is involved with ejaculation, but the exact role is not known. Androgens are involved in emission and ejaculation, and low testosterone levels can be associated with semen volume changes. Thyroid hormone dysregulation is thought to result in ejaculatory dysfunction as well [11].

Physiology of Testosterone Production

Testosterone is the predominant male sex hormone and is responsible for the male phenotype and virilization [12]. It is a steroid hormone and is a cholesterol derivative. The vast majority (95%) of testosterone is produced in the testes with the remainder coming from the adrenal glands [13]. The hormone pathway leading to testicular testosterone production starts with gonadotropin-releasing hormone (GnRH) that is released from the mediobasal hypothalamus in a periodic fashion every 60–90 min. GnRH activates the anterior pituitary to release follicle-stimulating hormone (FSH) and luteinizing hormone (LH) in a pulsatile manner [12]. LH then stimulates the Leydig cells within the testes to produce testosterone while FSH stimulates the Sertoli cells leading to spermatogenesis. This system has a negative feedback on the hypothalamus and pituitary from the testosterone metabolites estradiol and dihydrotestosterone (DHT). Testosterone is converted to estradiol via aromatase and converted to DHT via 5α-reductase (which is predominantly in the skin and prostate), and these hormones then act on the hypothalamus and pituitary to inhibit further testosterone release [12, 13]. Note that testosterone levels are highest in the morning and lowest in the evening. However, in older men, these diurnal changes are less marked [13].

In the body, most testosterone is bound to proteins, and only the minority is free. About 60% binds strongly to sex hormone-binding globulin (SHBG), and 38% is bound weakly to albumin. Only 2% of testosterone is free. Bioavailable testosterone is composed of free and albumin-bound testosterone. Altered levels of SHBG can thus change the amount of bioavailable testosterone. Causes of elevated SHBG include chronic liver disease, HIV-AIDS, and antiepileptics, while obesity, DM, or hypothyroidism can lead to decreased SHBG [12, 13].

ERECTILE DYSFUNCTION

Overview

ED is a common problem in adult males and increases in incidence with age. According to the NIH, ED is defined as "consistent or recurrent inability to attain and/or maintain penile erection sufficient for sexual performance" [17]. The rate of ED in the general population varies in the literature, likely due in part to the definitions of ED used and whether validated questionnaires were employed. A review of the literature shows that despite differences in individual studies, the trend confirms that older men have a higher prevalence of ED. Men <40 years old have a 1%–10% chance of ED, men 40–49 years have a 2%–15% rate, men 50–59 years have the greatest variation in prevalence rates but show higher ED rates compared with younger men, men 60–69 years had 20%–40% risk of ED, and men in their 70s and 80s have a 50%–100% prevalence of ED [18] (Table 1).

Table 1 Prevalence of Sexual Dysfunctions in Diabetic Men

Disorder	Prevalence in Diabetics
Hypogonadism [7, 14]	33%–50%
Erectile dysfunction [7, 15]	32%–90%
Ejaculatory and orgasmic dysfunction [7, 16]	19%–35%

Not only is ED more common in diabetic men, but also their ED is more severe and less likely to respond to treatments [7, 19]. The percentage of diabetics with ED varies in the literature based on the method used to diagnose this condition. When men were simply asked whether or not they had ED, 52% were categorized as having the disease. However, if they were given the International Index of Erectile Function (IIEF), then 88% of the men were diagnosed with ED [7]. Approximately half of male diabetics will develop ED within 10 years of their diabetes diagnosis. The increase in ED as diabetics age is thought to be attributed to longer duration of DM and more diabetes-related complications [7]. Data from the Health Professionals Follow-up Study confirmed that overall, diabetics had a higher rate of ED with a relative risk (RR) of 1.32 compared with men without DM. However, when looking at men who had been diagnosed with DM for 20 years or more, the RR rose to 1.7 compared with nondiabetics, which depicts that there is a higher risk for ED with increasing duration of DM [20].

Pathophysiology of ED in Diabetics

Diabetes mellitus (DM) is a common disease worldwide, and its prevalence is rising. DM type 2 (DM2) is also referred to as adult-onset diabetes or noninsulin-dependent diabetes. These patients still make insulin, but their cells develop a resistance to the hormone, and they are unable to respond normally. The main causes of DM2 are being overweight or obese and having a sedentary lifestyle [21, 22]. About 90% of the cases of DM in adults are DM2. This disease has equal prevalence in males and females [21].

The global burden of DM is staggering, and it has risen sharply over the past 2–3 decades. In 1980, 108 million adults had DM compared with 422 million in 2014. This corresponds to an increase from 4.7% to 8.5% of the total adult population worldwide [22].

The causes of ED in diabetics are numerous and include testosterone deficiency, smooth muscle degeneration, endothelial dysfunction, abnormal collagenization, and diabetic neuropathy. ED in diabetics can be classified as a late complication of the disease [7]. A study on diabetic men <60 years old revealed that poor glycemic control and duration of DM were both independent risk factors for ED [23]. Thus, there appear to be cumulative effects over time and with increasing severity of the disease (Fig. 1).

Arteriosclerosis
Chronic hyperglycemia leads to advanced glycation end products (AGEs) that form from reactions between glucose and other cellular components like proteins, lipids, and nucleic acids [15]. AGEs have many ways in which they contribute to ED, but one of these mechanisms is by causing atherosclerosis

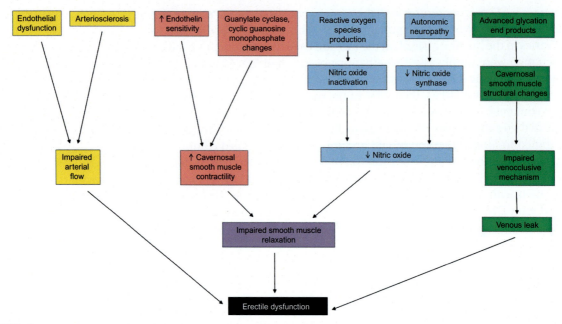

FIG. 1
Pathophysiology of erectile dysfunction in diabetics.

[15, 19]. This can impair the influx of arterial blood into the penis and thus contribute to vasculogenic ED.

Endothelial Dysfunction
Diabetes impairs the relaxation of arterial and cavernosal smooth muscle within the penis through numerous mechanisms. DM impairs the endothelium-based relaxation of penile arteries because diabetics produce less eNOS, and thus, their NO levels can be decreased. This relaxation is further hindered in diabetics as cGMP in the cavernosal tissue is decreased in all men with ED but has been found to be lower in diabetics with ED. Additionally, endothelin and its vasoconstrictive effects are increased in diabetic men [7, 15, 19], and endothelin receptors are also upregulated in these patients [15]. Advanced glycation end products (AGEs) are formed when glucose reacts with cellular components such as proteins or lipids and can also directly lead to endothelial dysfunction and reduced elasticity that further contributes to ED [15].

Diabetic Neuropathy
Diabetic neuropathy is the most common complication of DM and affects as many as 50% of diabetics [24]. The pathogenesis of diabetic neuropathy is secondary to hyperglycemia that can cause axonal degeneration, impair nerve

conduction velocity, and reduce the amplitude of conducted potentials. Additionally, excess glucose converts to sorbitol that can build up in nerves and lead to water accumulation. Hyperglycemia can also lead to ischemia and hypoxia within the nerves that further damages this tissue [24].

Compared with diabetics without ED, diabetics with ED are more likely to have abnormal vibratory sensation and nerve conduction velocities [7]. This has obvious implications for sexual function as diminished sensation can cause a decrease in pleasure from sexual stimulation, and this can lead to sexual dysfunction such as delayed orgasm or anorgasmia. Additionally, diabetics have less nNOS, and this is another contributing factor to lower NO and impaired erectile function in these patients [19].

Smooth Muscle Alterations

Over time, DM can result in an increased collagenization of the cavernosal smooth muscle resulting in fibrosis. These changes result in a loss of elasticity and muscle relaxation resulting in reduced cavernosal expansion. Additionally, the low testosterone levels that are often concomitant in diabetics can lead to apoptosis of the smooth muscle cells in the corpora cavernosa [7]. Furthermore, AGEs lead to cross-linking of smooth muscle, and reactive oxygen species (ROS) generation leads to further impairment of smooth muscle relaxation [19]. When the corporal bodies do not fully expand, the emissary veins are not compressed, and excessive venous outflow (venous leak) results.

Cellular Events

Due to chronic hyperglycemia, there is an increase in ROS in diabetics. Not only do these ROS lead to inflammation and oxidative damage to tissues within the penis, but also they interact with NO and thus decrease its bioavailability [7]. Chronic hyperglycemia also results in upregulation of the RhoA/Rho kinase pathway that results in less NO because this process suppresses eNOS [15].

Patient Evaluation

The evaluation of diabetic men with ED is similar to a man in the general population. A routine history and physical exam should be obtained with a focus on sexual history. It is important to ascertain if the ED is situational (only with a partner or certain partners) or whether it is generalized (occurs in all sexual situations, including masturbation). Providers should also ask about nocturnal erections. These questions help differentiate between a psychogenic and organic etiology. Routine laboratory tests such as metabolic panel and testosterone should be obtained. HbA1c is an important measure of DM severity and should be obtained in diabetics with ED. Specialized tests are generally not needed for most patients [19].

Management

Diabetic men are treated for their ED using the same clinical care pathway as nondiabetic men. However, treatments for ED are generally less effective in diabetic men [25]. These patients are 50% more likely to need a higher level of treatment compared with nondiabetics. Diabetics are twice as likely to undergo penile implant surgery than nondiabetics with ED [7]. In general, the International Consultation on Sexual Medicine (ICSM) recommends PDE5i as the first line for ED treatment unless there are specific contraindications. Intracavernosal injections (ICI) or intraurethral alprostadil represents second-line therapy [26]. Note that improved glycemic control may lead to improved erectile function in diabetic men [23].

PDE5 Inhibitors (PDE5i)

These drugs are the only FDA-approved oral medications for ED. They inhibit the PDE5-mediated breakdown of cGMP and thus prevent detumescence. The drugs are competitive inhibitors that bind to cGMP and block its natural substrate [26]. Unlike other treatments, these drugs only augment the natural erectile response (facilitators of erection) and thus do not work well in patients with poor NO production such as men with DM-associated autonomic nerve damage [6]. Sexual stimulation is required to achieve an erection when using these medications to maximize NO generation. There are four PDE5i approved in the United States: sildenafil, vardenafil, tadalafil, and avanafil (Table 2). Other drugs, such as mirodenafil, lodenafil, and udenafil [26], are only available in other countries and will not be discussed in detail.

These drugs are generally effective and well tolerated and should be first-line agents unless contraindications exist. There is no difference in efficacy between the various drugs [26]. Note that because diabetics typically respond suboptimally to ED treatment, it is recommended that men start on the highest doses of PDE5i [7]. The ICSM recommends using the maximum tolerated dose due to higher efficacies [26].

Table 2 Comparison of Pharmacokinetics of Phosphodiesterase 5 Inhibitors

Drug	T_{max} (h)	$T_{1/2}$ (h)	Food Effects
Sildenafil [27, 31]	1	4	Increases T_{max} by 60 min. Decreases concentration by 29%
Vardenafil [28, 32, 33]	0.7–0.9	4	Increases T_{max} by 60 min. Decreases concentration by 18%
Tadalafil [30, 34]	2	17.5	None
Avanafil [29, 35]	0.5–0.75	5	Increases T_{max} by 75 min. Decreases concentration by 24%–39%

Most PDE5i need to be taken on an empty stomach for maximal effect [27–29]. The exception is tadalafil that can be taken with food [30]. Alcohol should be avoided with these medications as well because it can slow absorption. The food effect on PDE5i is amplified in diabetics due to the associated gastroparesis and inefficient stomach emptying. Thus, we dose all PDE5i in DM patients in a preprandial fashion. In practical terms, during the trial phase, we suggest patients take the medication 2 hours before a meal with an 8-hour window of opportunity, focusing on the application of sexual stimulation, mental and physical.

Sildenafil was the first PDE5i and was approved in 1998 [26]. Sildenafil comes in 25, 50, and 100 mg tablets. Labeling recommends administration 1 hour prior to desired sexual activity, but it can be taken 30 minutes to 4 hours before [31]. Sildenafil has a time to maximum concentration (T_{max}) of 1 hour and a $T½$ (half-life, time for drug concentration to be reduced by half) of 4 hours. This drug does have a significant food effect, and fatty food has been shown to delay T_{max} by 60 minutes and decrease the drug concentration by 29% [27].

Vardenafil has been available in the United States since 2003 [26]. This drug is available in 2.5, 5, 10, and 20 mg tablets. Labeling recommends patients take this medication 1 hour prior to planned sexual intercourse [32]. The T_{max} for this drug is 0.7–0.9 hours depending on the dose, and the $T½$ is 4.2 hours [33]. Like sildenafil, vardenafil also has food effects on its pharmacokinetics. Consumption of a high-fat meal delays vardenafil T_{max} by 1 hour and reduces its maximum concentration from 17.14 to 14.0 μm/L [28].

Tadalafil has been available in the United States since 2003 [26]. It is unique in that it has the longest T_{max} in this drug class at 17.5 hours. Unlike other PDE5i, tadalafil has been shown to experience no food effect, and concentrations are the same while fasting or after a fatty meal [30]. Additionally, tadalafil is available for low-dose daily use in 2.5 or 5 mg tablets or on demand dosing using 5, 10, or 20 mg tablets [34]. It has also been approved for the treatment of lower urinary tract symptoms secondary to prostate enlargement [26, 34]. The package insert does not specify when to take tadalafil with respect to sexual activity. However, it does note that the medication can improve erectile function for up to 36 hours [34]. The T_{max} for this drug has been noted to be 2 hours [30].

Avanafil is the newest PDE5i available in the United States and has been available since 2013. It has the fastest T_{max} at 30–45 minutes [29] and is designed to be taken only 30 minutes prior to intercourse [26, 35]. The $T½$ of this medication is 5 hours [29]. Avanafil is available in 50, 100, or 200 mg tablets [35]. Note that food does effect absorption and reduces the concentration by 24%–39% depending on the medication dose and delays T_{max} by 75 minutes [29].

PDE5i are less effective in men with DM (56% response rate) compared with men without DM (87% response rate). Even when diabetic men respond initially, these drugs do not typically work long term [7]. ICI and topical or intraurethral alprostadil should be second-line therapy [26].

These drugs are generally well tolerated. Class side effects are due to smooth muscle relaxation in other tissues of the body secondary to cross-reactivity. Thus, patients can get rhinorrhea, facial flushing, or headaches. Sildenafil and vardenafil can lead to visual disturbances such as reactivity blurred vision, diplopia, and chromatopsia due to PDE6 cross-reactivity. Tadalafil has higher rate of myalgias than the other drugs [26].

Vacuum Erection Device

A vacuum erection device (VED) is composed of a cylinder that is placed over the penis and creates a seal at the base. The cylinder has either a battery-operated or a hand-operated pump that creates negative pressure that engorges the penis with blood. A VED is typically used with a penile constriction band to trap this blood within the penis [15]. It's important to educate patients that they cannot wear the constriction band for over 30 minutes at a time due to ischemia concerns [15, 36]. The first VED was marketed in the United States in 1974, and the first marketed VED with FDA approval was in 1982 [36].

Patient satisfaction with the VED can range from 70% to 82% [15, 36] depending on the study. Partner satisfaction has been reported to be up to 87% [36]. The most common side effects from this device are that it can lead to a decrease in the temperature of the penis (as much as 1°C) and penile bruising [15]. Patients may also report discomfort from the constriction band [23], penile pain, and numbness [36].

Intraurethral Alprostadil

Alprostadil (synthetic prostaglandin E1) increases cyclic adenosine monophosphate (cAMP) by binding to smooth muscle cells and activating intracytoplasmic adenylate cyclase [26, 37]. Thus, like the PDE5i, it leads to smooth muscle relaxation within the penis. This drug effect starts within 5–10 minutes of administration, and 80% of the drug is absorbed within 10 minutes. Effects typically last between 30 and 60 minutes [38].

Patients are instructed to urinate prior to insertion to moisten the urethra [26]. First-time administration should occur in the office under supervision due to small risk of hypotension and syncope (2% at maximum dose of 1000 μg). The PGE1 is transferred from the corpus spongiosum to the corpora cavernosa through communicating venules between the chambers that are maximally dilated upon standing. While standing, the applicator is inserted into the penis and the plunger depressed. The applicator should remain in place for 5 seconds before withdrawing while rocking the applicator back and forth to ensure the suppository has been dislodged. After the applicator is removed, the penis should be rolled between the fingers or hands to ensure even distribution of the pellet [38]. Additionally, men should be informed of the need to use a condom when having intercourse with a pregnant woman [38]. This drug is available in 4 dosages: 125, 250, 500, and 1000 µg. Each suppository is individually wrapped with a single-use applicator [38].

Intraurethral alprostadil has limited success and only leads to erections in about 50% of men, with issues with consistency of response [6]. In men who get a response, there is about a 70% improvement in erectile function parameters. This medication is typically less effective than PDE5i [37]. Intraurethral alprostadil can lead to urethral burning or mild bleeding per urethra [26]. Over one-third of patients report penile pain. The most serious adverse event noted was hypotension (3%) and syncope (0.4%) [38]. This explains the recommendation that the drug should be administered in the office for the first time.

Intracavernosal Injections

Intracavernosal injections (ICI) are medications injected directly into the corpora cavernosa by the patient and/or his partner. Injections are delivered with a fine-gauge diabetic needle (we utilize a 29-gauge needle) inserted midshaft at the 2 o'clock or 10 o'clock positions. Men need to be trained on proper injection techniques in the office prior to home use and dose titration occurring in the office to prevent priapism occurrence during at-home use.

Alprostadil is the only drug with worldwide approval for ICI use. It can lead to localized penile pain. The vast majority (90%) of men report increased rigidity and 70% endorse sexual satisfaction [26].

Papaverine is a nonspecific PDE inhibitor and thus increases both cAMP and cGMP. Papaverine is a weak agent and is typically used in combination with other drugs. A study evaluating papaverine monotherapy and papaverine plus an alpha blocker found that overall there was an 84.8% patient satisfaction rate and 91% of men achieved normal erectile function with the injections [26].

Phentolamine is an alpha blocker and promotes vascular relaxation and increased blood inflow to the penis [6]. This drug is used in combinations such as bimix or trimix and is not used as a monotherapy.

Bimix is a combination drug typically formulated as papaverine + phentolamine [26]. We use a 30 mg/mL + 1 mg/mL mixture. Because it lacks prostaglandin, bimix is a good option for patients with autonomic neuropathy with resultant PGE1 hypersensitivity who have the potential to experience penile pain associated with trimix or alprostadil injections.

Trimix is a more potent and efficacious drug as it combines three medications: papaverine + prostaglandin + phentolamine [26]. We use a 30 mg/mL + 1 mg/mL + 10 µg/mL mixture. Given the prostaglandin component, it can lead to penile pain.

Diffuse penile pain associated with ICI is only seen in men receiving alprostadil monotherapy or trimix. All men using ICI are at risk for priapism. They need to be cautioned that erections lasting over 4 hours need medical attention usually requiring intracavernosal phenylephrine injection with or without aspiration.

Penile Implants

Surgery is considered a third-line treatment modality. Prior to surgery, all diabetics should be counseled about the risk for implant infection. However, the ICSM does not recommend a definitive HbA1c level for preop optimization [25].

There are two basic types of penile implants: malleable/semirigid and inflatable. The inflatable penile implant is more commonly used in the United States (not the case in the remainder of the world) and has been available since 1973. It is available in a two-piece or three-piece model. The two-piece devices have paired cylinders and a pump, whereas the three-piece devices also have a separate reservoir. The two-piece model is ideal for men who have had pelvic or inguinal surgery as it negates the need for reservoir placement. The three-piece inflatable penile prosthesis (IPP) is more commonly used. The reservoir can be placed by entering the external inguinal ring and piercing the transversalis fascia and placing the reservoir in the space of Retzius. More recently, "ectopic" locations have been advocated by some authorities. The most common ectopic location is placement behind the rectus superficial to the transversalis fascia [25].

The two main companies producing three-piece implants in the United States are Coloplast and Boston Scientific. The Coloplast devices are made of Bioflex and are coated with polyvinylpyrrolidone that absorbs antibiotics, and thus, the surgeon can choose the antibiotic coverage desired for each individual patient. These implants contain a lockout valve in the reservoir. The reservoirs are cloverleaf shaped and available in 75 or 125 mL sizes [25]. The Boston Scientific devices include the 700CX™, 700CXR™, and LGX™ (Minneapolis, MN). These are three-layered devices (silicone, Dacron, and silicone) impregnated with parylene to increase the longevity of the device. Unlike the Coloplast devices, these are coated with Inhibizone, which is a combination of rifampin and minocycline. The Boston Scientific devices contain a lockout valve in the pump. There are 65 and 100 mL spherical reservoirs available and a 100 mL Conceal™ (Minneapolis, MN) reservoir designed for submuscular placement [25].

A disadvantage of the two-piece IPP is that they do not deflate completely and are more difficult to conceal in the flaccid state [25]. In terms of the risks of surgery, these include standard operative risks, mechanical device failure, implant infection, device migration, erosion, crural perforation, and urethral perforation. Mechanical failure rates increase over time. For the three-piece IPP, this changes from 85% mechanical survival at 5 years to only 57% at 15 years [25].

The most feared complication of a penile implant is an infection. This requires device removal. Many experienced surgeons will attempt a salvage procedure with a device explant, antibiotic irrigation, and new implant placed in the same operation. Most implant infections are thought to be from bacterial seeding from the patient's skin [25]. Most infections (75%) are milder and include local symptoms such as pain, erythema, pump fixation to the skin, and pus expression. These are usually caused by coagulase-negative *Staphylococcus*. The remaining infections are caused by more virulent bacteria such as *Staphylococcus aureus*, *Escherichia coli*, or *Enterococcus*. These patients manifest more significant local signs such as crepitus and systemic symptoms such as a fever, and these patients may be ill-appearing [39].

Historically, the role of diabetes and penile prosthesis infections is unclear. While DM in general is a known risk factor for wound infections, a previous review of the penile prosthesis literature had insufficient evidence to suggest that DM was a risk factor for device infection [39]. However, a more recent multicenter, 6-year prospective trial of over 900 implants has demonstrated the role of poor glycemic control in penile prosthesis infection. This study had an overall 8.9% infection rate in a patient population with 74.8% being diabetic. Poorly controlled DM, as defined by an HbA1c of 7.5% or higher, was present in 61.3% of the study population. The data show that the infection rate was higher with more severe DM. The infection rate was 1.3% in men with an HbA1c of 6.5 or less and rose to 22.4% in men with an HbA1c of >9.5. The receiver operating characteristic (ROC) curve suggested a sharp rise in infections with an HbA1c of 8.5% or greater. The recommended cutoff for implant surgery was 8.5% as anything higher than this resulted in an unacceptable infection rate [40]. These data clearly demonstrate the role of poorly controlled diabetes and implant infection.

EJACULATORY DYSFUNCTION

Overview

To begin, some definitions are given: Retrograde ejaculation (RE) is described as the expulsion of seminal fluid into the bladder because of bladder neck dysfunction in the presence of otherwise normal emission; failure of emission

(FOE) is the failure for any ejaculate to enter the prostatic urethra via the ejaculatory ducts; anejaculation is the absence of normal antegrade ejaculation during orgasm due to either retrograde ejaculation or failure of emission.

Ejaculatory dysfunction is common in diabetics. Delayed ejaculation has been reported in 33% and retrograde ejaculation in 19% [7]. Another study demonstrated even higher rates of RE in diabetics. Compared with matching controls who had no RE, 35% of diabetic men suffered from this disorder. The diabetics with RE had a longer duration of DM compared with those without ejaculatory dysfunction, but this was not statistically significant [16].

Anejaculation Evaluation

Anejaculation is the result of three major etiologies: pharmacological, neurological, and iatrogenic. The classic medication groups causing anejaculation are alpha adrenergic blockers used for benign prostatic hypertrophy and lower urinary tract symptoms. Iatrogenic etiologies include bladder neck operations such as transurethral incision or resection of the prostate or pediatric bladder neck reconstructive procedures. Neurological causes include any cause of autonomic neuropathy such as DM, any retroperitoneal surgery (such as a retroperitoneal lymph node dissection), and chemotherapy.

The key step is differentiating between RE and FOE. These conditions can be distinguished by a postorgasm urinalysis. This entails inserting a Foley catheter and draining the bladder before filling the bladder in a retrograde fashion with an alkalinizing fluid to protect potential sperm. The catheter is then removed; the patient masturbates and then submits a urine sample that is evaluated for sperm. If the urine contains sperm, then this confirms retrograde ejaculation. The absence of sperm (and semen) in the urine leads to a diagnosis of FOE. All patients with anejaculation without an overt etiology should be screened for DM.

Treatment of Anejaculation

Retrograde ejaculation is potentially responsive to alpha adrenergic agonists, such as pseudoephedrine. This medication aims to close the bladder neck and prevent the ejaculate from entering the bladder. However, a review of this literature found that most studies have small sample sizes and anejaculation of various etiologies [9]. A novel idea recently studied was endourethral collagen injection in diabetic men with RE who had failed medical therapy. The men who received a collagen injection had an increase in antegrade ejaculate volume (0.71 mL, $P < .05$) compared with their preinjection volumes [41]. While this study only included men with type 1 DM, the results are potentially generalizable to patients with DM2.

ORGASMIC DYSFUNCTION

Pathophysiology

The ICSM defines delayed orgasm (DO) as a distressing lengthening of ejaculatory latency that occurs in most (>50%) coital experiences after a period of normal ejaculatory function and/or clinically meaningful change that results in distress [17]. This occurs in about 1%–10% of men of all ages although it is thought to increase with age [18]. Another more objective definition used in the literature involves intravaginal ejaculatory latency time (IELT) that is the time from vaginal penetration to male orgasm. The median IELT is 5–8 minutes, and two standard deviations above this are 21–23 minutes. Delayed orgasm can thus also be defined as an IELT of 25–30 minutes or a time when the man stops due to exhaustion [8].

Anorgasmia by definition is the inability to reach orgasm despite adequate and prolonged sexual stimulation leading to adequate sexual arousal [17]. It is difficult to ascertain prevalence as many men do not distinguish correctly between orgasm and ejaculation; however, a literature review showed that most studies have a range of 11%–19% of men [18].

Delayed orgasm or anorgasmia may result from selective serotonin reuptake inhibitor (SSRI)-type medications, low testosterone, penile sensory loss, hyperstimulation, and psychogenic etiologies [42]. SSRIs are known to contribute to sexual dysfunction. Corona et al. evaluated 2040 men presenting with sexual dysfunction, and 139 of these men had DO as defined as a self-reported "slowness to ejaculate." Men on SSRIs were seven times more at risk for DO. Even non-SSRI antidepressants led to an increased risk of DO ($P < .005$) [43]. Hyperstimulation results from penile desensitization—for example, as a result of frequent vigorous masturbation and/or using an idiosyncratic techniques for masturbation [42]. A study conducted by Perelman involved 80 men with DO. He found that many of these men were "high-frequency masturbators" with over one-third of the men masturbating at least once daily. Over one-third of the population also had an "idiosyncratic style" of masturbation that was defined as stimulation not easily replicated with a partner using their vagina, mouth, or hand [44]. DO increases in prevalence with age, and this is thought to be from a combination of decreasing testosterone levels in older men and a physiological decrease in penile sensation [42]. DM is a common neurogenic cause of DO related to the loss of penile sensation [8].

Patient Evaluation

Patients should undergo a history and physical exam, to include a detailed sexual history and genital exam. Aspects of the history that must be ascertained include potential stressors, masturbation frequency and idiosyncrasies, relationship status, and penile sensation changes. Providers should also determine if the orgasmic delay is situational or generalized and acquired versus lifelong [42]. An early morning total testosterone level should be checked [8]. If the history and physical are concerning for penile sensation loss, providers can consider additional testing. This includes biothesiometry to evaluate the threshold of vibratory sensation. If abnormal, consideration may be given to doing the more sophisticated pudendal somatosensory evoked potentials (SSEP), which evaluates afferent conduction velocities of the dorsal penile nerve and can define the location of the neural problem and its magnitude [42]. The process of a pudendal SSEP basically involves anal and penile stimulation with a bipolar electrode placed at the anal orifice and the base of the penis. The responses are recorded via electrodes on the scalp that are placed approximately at the site of the anal and genital regions of the somatosensory cortex. This enables cortical latencies to be recorded, and these data provide information on nerve conduction [45].

Treatment

Treatment should focus on treating or managing the underlying conditions. Even if the nerve damage from DM is irreversible, further damage to the nerves should be prevented by achieving tight glycemic control. Specific treatments vary depending on the underlying etiology. For example, if the orgasmic dysfunction is drug-related, stop the offending drug if possible. If a patient is on an SSRI, a trial of cessation should be considered by the SSRI prescribing clinician, with substitution by medications such as bupropion or buspirone [42]. If psychological factors are contributing, then the patients may benefit from counseling.

Low testosterone, if present, should also be addressed [42]. A randomized control trial of 190 men with low testosterone and DM demonstrated that compared with placebo, testosterone injections improved intercourse satisfaction as measured by improvement in the IIEF score (5.95 ± 5.12 vs 4.72 ± 5.11, $P < .015$) and the orgasm domain in the IIEF (5.68 ± 4.04 vs 3.99 ± 3.92, $P = .04$) [46].

There are numerous drugs that have been evaluated for the treatment of DO. A study of 19 men with lifelong DO, as defined by a persistent or recurrent delay in orgasm after a period of normal sexual activity, evaluated the effects of bupropion. While on this medication, the percentage of men reporting "fair to good" control over ejaculatory latency went from 0% to 21.1%. There was also a 0.74-fold decrease in the mean IELT that demonstrates that this drug has some potential for treating DO [47]. However, these results are modest at best as this is a small decrease in IELT. Another potential treatment is intranasal oxytocin, although this has been studied less extensively.

A case study of one patient with secondary anorgasmia showed significant improvement with oxytocin, and the man was able to orgasm regularly with this treatment [48].

A review on DO showed that amantadine increased sexual behavior in some rat studies. In a small human study of men with DO or anorgasmia secondary to SSRIs, 8/19 men noted improvement with amantadine [42]. A study on 131 men with DO or anorgasmia revealed subjective improvement in orgasmic function in 66.4% of men with twice weekly cabergoline [49]. Not only is yohimbine an herbal supplement traditionally used to treat ED, but also it has been used for DO and anorgasmia. A review revealed only small studies with subjective improvement in DO and a study of 29 men with anorgasmia where 19 of the men were able to orgasm after using yohimbine although some men used this in conjunction with vibratory therapy [42].

Vibratory therapy for DO warrants a special focus as diabetic neuropathy, and reduced penile sensation is a common etiology of this condition in diabetics. Penile vibratory stimulation (PVS) may be useful in men with penile sensory loss. This therapy entails a vibrator being applied to the frenulum to induce orgasm [42]. A study on men with secondary anorgasmia evaluated the efficacy of PVS in men who had not had any orgasms during sexual activity for at least 3 months. Men were told to apply a vibrator to their penile frenulum for 1 minute on, 1 minute off, for three cycles. Orgasms occurred in 72% of the 36 men in the study. Of those who were responders, they achieved orgasms on an average of 62% of their sexual encounters. This therapy thought to work by stimulating afferent nerves and triggering a normal ejaculatory reflex [50]. PVS is a noninvasive and cost-effective management strategy for men with orgasmic dysfunction.

TESTOSTERONE DEFICIENCY

Overview

Low testosterone (T) is common in older men as T levels decline with age. In order to have the diagnosis of testosterone deficiency, men need to have low testosterone levels and clinical symptoms or signs. It is estimated that 2%–6% of adult men meet this definition, that is an important distinction as 39% of adult men have low testosterone based on labs alone [13]. The Massachusetts Male Aging Study followed a general population of men aged 40–70 years old in the Boston area at baseline and then at a follow-up that averaged 8.8 years but ranged 7.0–10.4 years later. At baseline, 6% of men had T deficiency as defined by a testosterone <400 ng/dL and at least three signs and symptoms. At follow-up, this had increased to 12% of men. The risk of

T deficiency increased with age, and at follow-up, low T was in 23% of men 70–79 years old compared with only 7% in men 48–59 years old [51].

T deficiency is particularly common in diabetics. Fifty percent of diabetic males will experience low T [7]. These two disorders are so often concomitant that the International Society for Sexual Medicine [13] and the American Association of Clinical Endocrinologists recommend screening diabetic men for testosterone deficiency. A study of 1717 men demonstrated that hypogonadism was almost three times higher in diabetics (24.3%) than nondiabetics (8.3%). The total testosterone in the diabetic group was 378 ng/dL compared with 492 ng/dL in the nondiabetic men ($P < .005$) [52].

Pathophysiology

T deficiency (TD) is considered a clinical and biochemical syndrome. In order to be diagnosed, men need a low testosterone level (<300 ng/dL or 11 nmol/L) and associated clinical symptoms or signs. This is because a review of the literature reveals that less than one-third of men with low testosterone levels actually have multiple symptoms that can be attributed to TD [53]. Symptoms of TD fall into the general categories of sexual function, sleep, cognitive function, affect, and physical function. Specific symptoms include low libido, decreased nocturnal erections, impaired erectile, ejaculatory or orgasmic function, reduced muscle strength, decreased endurance, impaired concentration, afternoon fatigue, and irritability [13].

There are four categories of T deficiency. The first is hypergonadotropic hypogonadism that involves impaired testicular synthesis of testosterone [13]. These patients have low testosterone and elevated gonadotropins (LH and FSH). Causes include Klinefelter's disease, chemotherapy, radiation to the testes, mumps orchitis, or testicular trauma [12]. The second category is hypogonadotropic hypogonadism where the Leydig cells are inadequately stimulated to produce testosterone. Testosterone is low, but the gonadotropins are also low or low-normal. This is often caused by pituitary pathology such as a tumor, trauma, surgery, or hyperprolactinemia [12]. Diabetes can result in an acquired form of hypogonadotropic hypogonadism [13]. The third group is mixed, and this is the most common etiology of testosterone deficiency in aging males [13]. Hypogonadism in older men is from a combination of lower LH and FSH due to chronic medical conditions, sedentary lifestyle, and increasing BMI and from a decline in the number and function of Leydig cells within the testes [54]. Lastly, there is compensated hypogonadism that refers to testosterone within the normal range but only because of abnormally elevated gonadotropins that results from a compensatory response that is needed to maintain normal testosterone levels. There are also certain classes of drugs such as opioids and chronic corticosteroid use that can contribute to testosterone deficiency [13].

Treatment of Testosterone Deficiency and the Impact on DM

Treating diabetic men who have TD may improve their glycemic control [13, 53]. A study of men with idiopathic hypogonadotropic hypogonadism demonstrated that they experienced an increase in insulin resistance after cessation of testosterone therapy [14]. Another study on diabetics with TD showed improvement in insulin sensitivity with testosterone therapy but only a trend toward significance in terms of improving HbA1c [14]. The most compelling data are from a meta-analysis by Zhang et al. They evaluated a total of 596 men from 8 randomized control trials. Results demonstrated that testosterone supplementation in hypogonadal diabetic man improved glycemic control as evidenced by lower fasting glucose (mean difference −0.98 and confidence interval from −1.13 to −0.54) and HbA1c (mean difference −0.45 and confidence interval from −0.73 to −0.16) [55]. However, while treating symptomatic hypogonadal diabetic men with testosterone is recommended for overall health, it is not recommended for testosterone supplementation to be the only treatment for hypogonadal men with DM [12].

Patient Evaluation

Men who are suspected of having TD should have two early morning testosterone levels drawn. The lab gold standard is liquid chromatography mass spectrometry (LCMS), and this is the preferred method for evaluating testosterone due its very low coefficient of variation (CV). If the first result shows low testosterone, then the second lab draw should include total T, free T (using an accurate assay such as equilibrium dialysis), SHBG, and LH [13]. Men should also be asked a thorough review of systems to ascertain if they have any symptoms related to TD. Examination should focus on testicular presence, volume, and consistency as well as an examination to define if varicoceles are present.

Treatment

The goal of treating TD is to supplement testosterone so that men fall into the midnormal range, approximately 450–600 ng/dL. There are multiple formulations available based on patient preference and insurance coverage (Table 3). Testosterone therapy can have numerous benefits in TD men. This includes decreased body fat, improved energy and mood, increased bone density, and improved nocturnal erections [12]. According to the American Urological Association (AUA) guidelines on testosterone deficiency, testosterone replacement can also result in improved libido and erectile function [56]. However, the role of testosterone in improving erectile dysfunction has mixed results in the data.

Table 3 Comparisons of Medication Routes for Testosterone Replacement Therapy

Route	Advantages	Disadvantages
Oral [57]	Convenient administration	Hepatotoxicity
Buccal patch [13]	Noninvasive	Twice daily dosing Gum irritation
Transdermal gels [13]	Noninvasive Daily dosing	Transference risk
Transdermal patch [13]	Noninvasive Daily dosing No transference risk	Skin irritation
Short-acting IM injection [13, 53]	Self-administered Minimally invasive	Widest range in T levels Higher risk of polycythemia
Long-acting IM injection [13, 53]	Infrequent dosing Minimally invasive	Pulmonary oil embolism Administered in office in the United States
Subcutaneous pellets [53]	Infrequent insertion	Invasive

It is thought that in men over 60 years of age, erectile function does not correlate with testosterone levels, and thus, ED in this population is thought to be from other causes [12]. This could explain the lack of correlation between testosterone therapy and improved erectile function in the literature.

Oral
Oral testosterone replacement is problematic, and currently, there is only one oral testosterone drug available in the United States. Methyltestosterone is marketed as Android or Testred and does have FDA approval. However, this drug is hepatically cleared and can cause jaundice or cholestatic hepatitis even at low doses, which results in its rare usage [57]. Oral testosterone preparations are difficult to develop because most formulations are substantially inactivated by first-pass hepatic metabolism, and drug modifications to prevent this are associated with hepatotoxicity [13, 53].

Buccal
Buccal application of testosterone is not commonly used in the United States. This involves twice a day application of a drug patch onto the gums. Side effects include altered taste and gum irritation [13].

Intramuscular Injections
Short-acting agents are the most commonly used intramuscular (IM) injections of testosterone including testosterone cypionate or enanthate. These are typically injected intramuscularly into the thigh/gluteals every 1–2 weeks [53].

Injections are associated with the widest range of testosterone levels between injections. These medications also carry the highest risk of polycythemia [13].

Although used in Europe for years, the long-acting testosterone undecanoate was only recently approved by the FDA. This intramuscular injection needs a loading dose 4–6 weeks after the initial injection and then only every 10–12 weeks thereafter [53]. This drug can cause an irritative cough thought to be associated with pulmonary oil microembolism (POME) [13].

Topical

Topical therapies provide 5–10 mg of testosterone each day. Each mode has different benefits and risks. Choice of therapy depends on patient preference and whether they cohabitate with young children that should make patient/clinicians concerned about the risk of transference.

Testosterone gels are applied every 24 hours and are available in 1%–2% concentrations. Location of gel placement varies based on the specific agent but includes the axilla, arms/chest, or thigh. There is a risk of transference with this medication [13], so men are advised not to have skin-to-skin contact with children or females for at least 6 hours after application.

Patches are another topical formulation that minimizes risk of transference compared with the gels. These are placed on the skin and replaced every 24 hours. These have a higher rate of skin irritation compared with gels [13].

Subcutaneous Pellets

Testosterone pellets can be inserted under the skin, typically in the upper buttocks or flank. These pellets can provide therapeutic testosterone levels for up to 3–4 months after insertion. They are implanted in the office using local anesthetic. A typical dose is 8–10 pellets. However, men who are overweight or obese may need a higher dose for therapeutic levels [53].

Nontestosterone Treatment Methods

These alternative therapies are less commonly used but are ideal for men who fall under certain categories such as men wanting to preserve their fertility. Clomiphene is a selective estrogen receptor modulator (SERM) that blocks the negative feedback loop that estradiol has on the hypothalamus and pituitary and thus results in more GnRH and LH/FSH and ultimately more testosterone release. Aromatase inhibitors are another class of drug that can be used to treat TD [13]. These drugs raise testosterone levels by blocking its conversion to estrogen. Given that testosterone therapy inhibits spermatogenesis [12], these alternate therapies are useful in men who want to preserve fertility.

Complications

Patients on testosterone therapy are at risk for adverse effects specific to the modality of treatment as discussed previously. However, as a class, there are numerous shared complications with all types of testosterone. Patients are at risk for elevated estradiol levels due to peripheral conversion of testosterone. This can lead to gynecomastia or mastodynia. Polycythemia is another adverse event. A hematocrit of over 54% needs to be addressed, either with therapeutic phlebotomy or cessation of testosterone therapy [12]. The aforementioned two potential adverse events are why T therapy patients need to have laboratory testing every 6–12 months.

CONCLUSION

In summation, sexual health problems are common in aging men and even more so in diabetic men. Key issues facing the aging diabetic from a men's health perspective include ED, hypogonadism, ejaculatory dysfunction, and orgasmic dysfunction. There are significant data to support that the duration and severity of DM increase the risk for developing these sexual health disorders. Thus, strict glycemic control could lessen the chances of these health conditions. Not only are diabetics more prone to sexual health disorders, but also, with regard to ED, they have more severe symptoms and respond suboptimally to treatment compared with nondiabetics. These patients are more prone to complications from penile implant surgery, with higher infection rates in men with poorly controlled DM. Diabetics need to be appropriately counseled on these increased risks prior to any surgery. Given the high prevalence of DM in men with sexual health concerns, men presenting with these conditions should be screened for diabetes.

References

[1] Wein A, Kavoussi L, Novick A, Partin A, Peters C, editors. Campbell-Walsh urology. 10th ed. Saunders; 2012.

[2] Dean RC, Lue TF. Physiology of penile erection and pathophysiology of erectile dysfunction. Urol Clin North Am 2005;32(4):379–95.

[3] Yiee JH, Baskin LS. Penile embryology and anatomy. Sci World J 2010;10:1174–9.

[4] Breza J, Aboseif SR, Orvis BR, Lue TF, Tanagho EA. Detailed anatomy of penile neurovascular structures: surgical significance. J Urol 1989;141(2):437–43.

[5] Phillips E, Carpenter C, Oates RD. Ejaculatory dysfunction. Urol Clin North Am 2014;41(1):115–28.

[6] Hawksworth DJ, Burnett AL. Pharmacotherapeutic management of erectile dysfunction. Clin Pharmacol Ther 2015;98(6):602–10.

[7] Kamenov ZA. A comprehensive review of erectile dysfunction in men with diabetes. Exp Clin Endocrinol Diabetes 2015;123(3):141–58.

[8] Shin DH, Spitz A. The evaluation and treatment of delayed ejaculation. Sex Med Rev 2014;2(3–4):121–33.

[9] Parnham A, Serefoglu EC. Retrograde ejaculation, painful ejaculation and hematospermia. Transl Androl Urol 2016;5(4):592–601.

[10] Gaunay G, Nagler HM, Stember DS. Reproductive sequelae of diabetes in male patients. Endocrinol Metab Clin N Am 2013;42(4):899–914.

[11] Corona G, Jannini EA, Vignozzi L, Rastrelli G, Maggi M. The hormonal control of ejaculation. Nat Rev Urol 2012;9(9):508–19.

[12] Basaria S. Male hypogonadism. Lancet 2014;383(9924):1250–63.

[13] Dean JD, McMahon CG, Guay AT, et al. The International Society for Sexual Medicine's process of care for the assessment and management of testosterone deficiency in adult men. J Sex Med 2015;12(8):1660–86.

[14] Gibb FW, Strachan MW. Androgen deficiency and type 2 diabetes mellitus. Clin Biochem 2014;47(10–11):940–9.

[15] Thorve VS, Kshirsagar AD, Vyawahare NS, Joshi VS, Ingale KG, Mohite RJ. Diabetes-induced erectile dysfunction: epidemiology, pathophysiology and management. J Diabetes Complicat 2011;25(2):129–36.

[16] Fedder J, Kaspersen MD, Brandslund I, Hojgaard A. Retrograde ejaculation and sexual dysfunction in men with diabetes mellitus: a prospective, controlled study. Andrology 2013;1(4):602–6.

[17] McCabe MP, Sharlip ID, Atalla E, et al. Definitions of sexual dysfunctions in women and men: a consensus statement from the Fourth International Consultation on Sexual Medicine 2015. J Sex Med 2016;13(2):135–43.

[18] McCabe MP, Sharlip ID, Lewis R, et al. Incidence and prevalence of sexual dysfunction in women and men: a consensus statement from the Fourth International Consultation on Sexual Medicine 2015. J Sex Med 2016;13(2):144–52.

[19] Hatzimouratidis K, Hatzichristou D. How to treat erectile dysfunction in men with diabetes: from pathophysiology to treatment. Curr Diab Rep 2014;14(11):545.

[20] Bacon CG, Hu FB, Giovannucci E, Glasser DB, Mittleman MA, Rimm EB. Association of type and duration of diabetes with erectile dysfunction in a large cohort of men. Diabetes Care 2002;25(8):1458–63.

[21] Tao Z, Shi A, Zhao J. Epidemiological perspectives of diabetes. Cell Biochem Biophys 2015;73(1):181–5.

[22] WHO. Global report on diabetes. World Health Organization; 2016.

[23] Redrow GP, Thompson CM, Wang R. Treatment strategies for diabetic patients suffering from erectile dysfunction: an update. Expert Opin Pharmacother 2014;15(13):1827–36.

[24] Pasnoor M, Dimachkie MM, Kluding P, Barohn RJ. Diabetic neuropathy part 1: overview and symmetric phenotypes. Neurol Clin 2013;31(2):425–45.

[25] Levine LA, Becher E, Bella A, et al. Penile prosthesis surgery: current recommendations from the International Consultation on Sexual Medicine. J Sex Med 2016;13(4):489–518.

[26] Hatzimouratidis K, Salonia A, Adaikan G, et al. Pharmacotherapy for erectile dysfunction: recommendations from the Fourth International Consultation for Sexual Medicine (ICSM 2015). J Sex Med 2016;13(4):465–88.

[27] Nichols DJ, Muirhead GJ, Harness JA. Pharmacokinetics of sildenafil after single oral doses in healthy male subjects: absolute bioavailability, food effects and dose proportionality. Br J Clin Pharmacol 2002;53(Suppl 1):5S–12S.

[28] Rajagopalan P, Mazzu A, Xia C, Dawkins R, Sundaresan P. Effect of high-fat breakfast and moderate-fat evening meal on the pharmacokinetics of vardenafil, an oral

phosphodiesterase-5 inhibitor for the treatment of erectile dysfunction. J Clin Pharmacol 2003;43(3):260–7.
[29] Kyle JA, Brown DA, Hill JK. Avanafil for erectile dysfunction. Ann Pharmacother 2013;47(10):1312–20.
[30] Forgue ST, Patterson BE, Bedding AW, et al. Tadalafil pharmacokinetics in healthy subjects. Br J Clin Pharmacol 2006;61(3):280–8.
[31] Viagra (sildenafil citrate) [package insert]. New York, NY: Pfizer Labs; 2014.
[32] Levitra (vardenafil HCl) [package insert]. West Haven, CT: Bayer Pharmaceuticals Corporation; 2007.
[33] Klotz T, Sachse R, Heidrich A, et al. Vardenafil increases penile rigidity and tumescence in erectile dysfunction patients: a RigiScan and pharmacokinetic study. World J Urol 2001;19(1):32–9.
[34] Cialis (tadalafil) [package insert]. Indianapolis, IN: Eli Lilly and Company; 2011.
[35] Stendra (avanafil) [package insert]. Mountain View, CA: Vivus, Inc.; 2012.
[36] Brison D, Seftel A, Sadeghi-Nejad H. The resurgence of the vacuum erection device (VED) for treatment of erectile dysfunction. J Sex Med 2013;10(4):1124–35.
[37] Hanchanale V, Eardley I. Alprostadil for the treatment of impotence. Expert Opin Pharmacother 2014;15(3):421–8.
[38] Muse (alprostadil suppository) [package insert]. Somerset, NJ: Meda Pharmaceuticals Inc.; 2015.
[39] Pineda M, Burnett AL. Penile prosthesis infections-a review of risk factors, prevention, and treatment. Sex Med Rev 2016;4(4):389–98.
[40] Habous M, Tal R, Tealab A, et al. Defining a glycated haemoglobin (HbA1c) level that predicts increased risk of penile implant infection. BJU Int 2018;121(2):293–300.
[41] Kurbatov D, Russo GI, Galstyan GR, et al. Correction of retrograde ejaculation in patients with diabetes mellitus using endourethral collagen injection: preliminary results. J Sex Med 2015;12(11):2126–9.
[42] Jenkins LC, Mulhall JP. Delayed orgasm and anorgasmia. Fertil Steril 2015;104(5):1082–8.
[43] Corona G, Ricca V, Bandini E, et al. Selective serotonin reuptake inhibitor-induced sexual dysfunction. J Sex Med 2009;6(5):1259–69.
[44] Perelman MA. Idiosyncratic masturbation patterns: a key unexplored variable in the treatment of retarded ejaculation by the practicing urologist. J Urol 2005;173(4, Suppl):340.
[45] Pelliccioni G, Piloni V, Sabbatini D, Fioravanti P, Scarpino O. Sex differences in pudendal somatosensory evoked potentials. Tech Coloproctol 2014;18(6):565–9.
[46] Hackett G, Cole N, Bhartia M, Kennedy D, Raju J, Wilkinson P. Testosterone replacement therapy with long-acting testosterone undecanoate improves sexual function and quality-of-life parameters vs. placebo in a population of men with type 2 diabetes. J Sex Med 2013;10(6):1612–27.
[47] Abdel-Hamid IA, Saleh eS. Primary lifelong delayed ejaculation: characteristics and response to bupropion. J Sex Med 2011;8(6):1772–9.
[48] IsHak WW, Berman DS, Peters A. Male anorgasmia treated with oxytocin. J Sex Med 2008;5(4):1022–4.
[49] Hollander AB, Pastuszak AW, Hsieh TC, et al. Cabergoline in the treatment of male orgasmic disorder-a retrospective pilot analysis. Sex Med 2016;4(1):e28–33.
[50] Nelson CJ, Ahmed A, Valenzuela R, Parker M, Mulhall JP. Assessment of penile vibratory stimulation as a management strategy in men with secondary retarded orgasm. Urology 2007;69(3):552–5. discussion 555–556.

[51] Araujo AB, O'Donnell AB, Brambilla DJ, et al. Prevalence and incidence of androgen deficiency in middle-aged and older men: estimates from the Massachusetts Male Aging Study. J Clin Endocrinol Metab 2004;89(12):5920–6.

[52] Al Hayek AA, Khawaja NM, Khader YS, Jaffal SK, Ajlouni KM. The prevalence of hypogonadism among diabetic and non-diabetic men in Jordan. J Diabetes Complicat 2014;28(2):135–40.

[53] Khera M, Adaikan G, Buvat J, et al. Diagnosis and treatment of testosterone deficiency: recommendations from the Fourth International Consultation for Sexual Medicine (ICSM 2015). J Sex Med 2016;13(12):1787–804.

[54] Decaroli MC, Rochira V. Aging and sex hormones in males. Virulence 2017;8(5):545–70.

[55] Zhang J, Yang B, Xiao W, Li X, Li H. Effects of testosterone supplement treatment in hypogonadal adult males with T2DM: a meta-analysis and systematic review. World J Urol 2018;36(8):1315–26.

[56] Mulhall JP, Trost LW, Brannigan RE, et al. Evaluation and management of testosterone deficiency: AUA guideline. J Urol 2018;200(2):423–32.

[57] Android, Testred (methyltestosterone) [package insert]. Bridgewater, NJ: Valeant Pharmaceuticals North America; 2015.

CHAPTER 2.3

Obesity and Men's Health

Ahmet Tevfik Albayrak*, Ege Can Serefoglu[†]
*Health Science University, Sisli Etfal Training & Research Hospital, Istanbul, Turkey,
[†]Bahceci Health Group, Istanbul, Turkey

OVERVIEW

Obesity has emerged as a global public health concern over the past several decades. The World Health Organization (WHO) defines obesity as a body mass index (BMI) of 30 or higher [1], and more than one-third of adults (and 17% of adolescents) in the United States were found to be obese according to that definition [2]. Numerous factors such as genetics, dietary intake, physical activity, environmental and socioeconomic parameters may be responsible for this relatively high prevalence [3].

Obesity is associated with a chronic inflammatory state in the body [4], and each $5\,kg/m^2$ increase in the BMI was found to be 30% higher overall mortality, mainly due to increased risk of type 2 diabetes mellitus (T2DM) and cardiovascular diseases (CVD) [5]. Obesity also plays a role in the pathophysiology of several men's health conditions, such as benign prostate hyperplasia (BPH)/lower urinary tract symptoms (LUTS) [6, 7], erectile dysfunction (ED) [8–10], hypogonadism [11–13], infertility [14–17], and prostate cancer [18, 19] (Fig. 1).

This chapter will be summarizing the impact of obesity on men's health with an emphasis on the pathophysiological mechanisms and treatment options.

OBESITY AND BPH/LUTS

Introduction

Obesity is a common health problem among men who suffer from BPH/LUTS [20–23]. The results of a recent meta-analysis revealed the positive association between the BMI and the risk of BPH/LUTS [24]. Several epidemiological studies such as the Prostate Cancer Prevention Trial [25], the Southern Community

Cohort Study [26], and the Osteoporotic Fractures in Men Study [27] also confirmed that higher BMI is directly associated with an increased risk of BPH/LUTS. Data from several well-designed epidemiological data provided further evidence regarding the association between obesity and BPH/LUTS [25–30]. Especially, the analysis of the data from the Health Professionals Follow-Up Study clearly showed that obesity may be an important target for LUTS prevention [31]. This large, population-based, prospective study, enrolling 51,529 US men aged 40–75 years, with a follow-up duration of >16 years, demonstrated that increased adiposity (measured by using BMI, waist circumference (WC), and hip circumference) and weight gain since young ages were highly correlated with the incidence of BPH/LUTS [31]. More importantly, the risk of LUTS progression also increased with the increase in BMI, WC, and weight gain from the age of 21 years [31].

BPH/LUTS is more common in the aging male population; therefore, the impact of obesity on prostate enlargement is commonly studied among the elderly patients [27]. However, a sonographic study also demonstrated that enlargement of the prostate could also be detected in young obese men (mean age 36.4 years) seeking medical care for infertility [32]. Semen samples of young obese men also had elevated IL-8 levels, indicating that prostate inflammation starts at an earlier age among obese patients [32].

The findings of the aforementioned epidemiological and case-control studies support the hypothesis that treatment of obesity might be crucial in the prevention and treatment of BPH/LUTS. Treatment of obesity may also improve the outcomes of the surgical treatment of BPH/LUTS, as it has been found to be the main determinant of persistent storage symptoms after open prostatectomy or transurethral resection of the prostate [33].

Underlying Mechanisms

In spite of the clear association between obesity and BPH/LUTS, the exact pathophysiological mechanisms that explain this relation are not clear. Central obesity usually leads to dyslipidemia characterized by high triglyceride and low HDL cholesterol levels [34]. Dyslipidemia can induce and sustain inflammation in the prostate [35]. A recent study has provided more evidence confirming that obesity-induced dyslipidemia may be associated with chronic inflammation and BPH [36]. After examining the prostate specimens of 132 BPH/LUTS patients who underwent transurethral prostate resection, the authors found significantly lower heme oxygenase levels in the resected tissues of men with dyslipidemia, indicating higher levels of oxidative stress [36]. These findings reveal that visceral and subcutaneous fat tissue may lead to increased oxidative stress

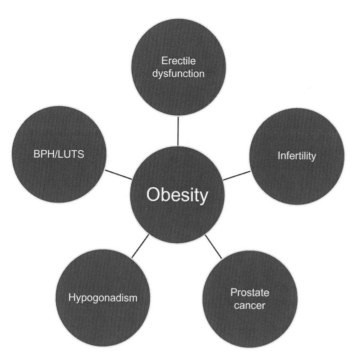

FIG. 1
Obesity-related men's health issues.

and simultaneously decrease the activity of cytoprotective enzymes, such as the heme oxygenase system.

Behavioral Modifications

It has already been established that lifestyle changes, such as weight loss and increased intake of fruits and vegetables in everyday diets, may prevent pathologies caused by obesity including coronary artery disease [37], stroke [38], hypogonadism [39], and type 2 diabetes (T2DM) [40]. However, these lifestyle changes may also ameliorate the BPH/LUTS [6]. Apart from diet, physical activity has also been shown to prevent BPH/LUTS development [41]. In the last few years, some experimental studies have suggested a new motion of range for phosphodiesterase-5 inhibitors (PDE5is) in BPH/LUTS, acting as an antiinflammatory drug and reducing metabolic syndrome (MetS)-associated prostate inflammation [42]. However, these findings must be confirmed in well-designed clinical studies before PDE5is can be recommended for the obese

BPH/LUTS patients. Meanwhile, physical activity and a healthy diet should be advised to all obese men for their prostate health.

OBESITY AND ERECTILE DYSFUNCTION

Introduction

Erectile dysfunction (ED) is defined as the inability to reach and/or maintain an erection to successfully perform sexual intercourse [43]. The condition is present in up to 30 million men in the United States and approximately 100 million men worldwide [44]. Obesity is one of the main risk factors for ED as it has a detrimental effect on penile endothelium and smooth muscle tissue [45]. According to the cross-sectional Massachusetts Male Aging Study and Health Professionals Follow-up Study, obesity doubles the risk of having ED, even after adjusting for lifestyle confounders [46, 47]. Similar results were obtained from the European epidemiological studies [48–52]. The findings of European Male Aging Study survey [49], which included a sample of 3369 community-dwelling men aged 40–79 years old, confirmed the significant association between severity of ED and obesity.

Underlying Mechanisms

Although it has been well established that central obesity is correlated with ED, it is not clear how visceral fat tissue can impair penile hemodynamics [53]. The association between decreased testosterone and increased visceral fat storage has been extensively described in the previous studies [54–57]. Since testosterone levels are crucial for the function of every component involved in erection (e.g., pelvic ganglions, smooth muscle, and endothelial cells), decreased testosterone levels among obese patients may be one of the factors that deteriorate erectile function [58]. Besides, the aromatase enzyme activity is also enhanced among obese men, which results in conversion of the already lowered testosterone to estradiol [59].

In addition to those hormonal changes, obesity is also associated with a low-inflammatory state because of the increased production of the cytokines (adipokines) in the adipose tissues [60]. In fact, adipose tissue regulates the pathways, which is responsible for energy balance via hormonal, neural, and chemical signals, such as tumor necrosis factor-alpha (TNF-α), angiotensinogen, and leptin complex [61, 62]. All these products may contribute to the enhanced inflammation and result in endothelial dysfunction.

Increased adiposity is also associated with many metabolic pathologies related to ED, including insulin resistance, dyslipidemia, increased oxidative stress, lipid peroxidation, and hypertension [63, 64]. All these metabolic changes may be involved in the erectile function impairment in obese men.

Obesity, ED and CVD Risk

ED and CVD are considered different manifestations of a common underlying vascular pathology [65]. Three independent meta-analyses have documented that ED should be regarded as a predictor of coronary heart diseases [66–68] and future cardiac events [69], emphasizing the importance of early diagnosis and correct management of ED-associated morbidities.

The hypothesis behind this concern is the smaller penile arteries might reach critical narrowing, with insufficient blood flow far more early than larger vessels like cardiac arteries (the artery size hypothesis) [70, 71]. Therefore, clinicians should be concerned about possible CVDs when they detect ED in relatively younger adults (<55 years of age) [68]. Considering the pathways linking obesity to ED and CVD (Fig. 2), treatment of obesity becomes essential in the management of these two conditions.

Treatment of Obesity-Related ED

The first study analyzing the effect of lifestyle modifications on erectile function was performed by Esposito et al. [72]. This study showed that weight loss and dietary changes could improve ED, without any further need of medical treatment. The beneficial effect of lifestyle modifications on erectile function has been confirmed by other studies [73–75] and is recognized as the first step in the management of ED [76]. Therefore, ED may be considered as an opportunity for the management of obesity, which will decrease the risk of future CVD.

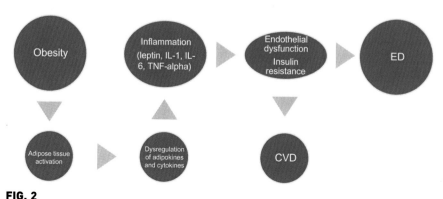

FIG. 2
The pathological mechanism of obesity-related CVD and ED.

OBESITY AND HYPOGONADISM

Introduction

Hypogonadism is defined as a clinical syndrome caused by decreased testosterone levels ($T < 10$ nmol) and concomitant symptoms suggesting androgen deficiency [77]. Compared with healthy men, hypogonadism is a more common problem among obese patients [39, 78–82]. Several epidemiological studies have demonstrated a negative correlation between BMI and testosterone levels [83–87]. Moreover, weight loss can significantly increase testosterone levels in obese men and reverts obesity-associated hypogonadism [39]. Considering this close relation, some authors define this clinical entity as male obesity-associated secondary hypogonadism (MOSH) [88].

Underlying Mechanism

Adipocytes have high expression of aromatase enzyme that converts testosterone to estradiol and thus lowers the circulating androgens. Concurrently, estrogens affect the hypothalamic-pituitary (HP) axis with a negative feedback mechanism and suppress gonadotropin-releasing hormone (GnRH). Consequently, luteinizing hormone (LH) levels start to decrease, and this situation leads to a reduction in gonadal testosterone release [11] (Fig. 3).

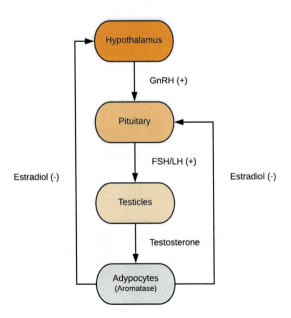

FIG. 3
Hypothalamo-pituitary-testicle axis and the negative feedback mechanism.

Extending this theory, adipose tissue may be regarded as an endocrine organ secreting many factors that cause the pathological changes related to both obesity and hypogonadism. In addition to estrogens suppressing the HP axis, the inflammatory adipocytokines (e.g., TNF-α and IL-6) have similar inhibiting effects via the suppression of hypothalamic GnRH secretion. Conversely, leptin, an adipose-derived hormone, can stimulate GnRH neurons in the hypothalamus to induce LH release and enhance testosterone levels in the healthy subjects [89, 90]. However, exaggerated leptin production among obese men may desensitize hypothalamus and deteriorate the testosterone production [91]. Furthermore, elevated leptin levels can directly inhibit the testosterone production of the Leydig cells and decrease androgen levels even more [91].

Treatment of Obesity-Related Hypogonadism

Several studies have shown that testosterone replacement therapy (TRT) reduces weight, BMI, WC, and fat mass in obese men [92–97]. The change in these anthropometric measures can be observed between 9 and 12 months of the TRT [98–102]. Moreover, men who are treated with varying concentrations of testosterone supplementation gain muscle mass, in addition to the loss of visceral adipose tissue [103, 104]. Several meta-analyses have demonstrated that TRT not only decrease WC but also improve the components of metabolic syndrome [105–107]. Therefore, treatment of MOSH with TRT must be considered as the standard of care, and clinicians dealing with obesity must check the testosterone levels of their patients routinely before prescribing testosterone containing gels, creams, pills, injections, or implants.

However, standard TRT may be associated with fertility problems among obese men, whose sperm cell production may already be impaired because of obesity. Therefore, alternative therapies must be adopted for obese men who want to father a child. In a single-center randomized controlled trial, Soares et al. evaluated the efficacy of clomiphene citrate treatment among 78 men with MOSH [108]. Considering the significant improvements in the hormonal profile and body composition, the authors recommended clomiphene citrate as an alternative to TRT for the treatment of MOSH. On the other hand, a recently performed meta-analysis revealed that bariatric surgery may also be considered for the treatment of MOSH [88]. In fact, MOSH started to be recognized as an additional indication for the bariatric surgery [109].

OBESITY AND INFERTILITY

Introduction

Infertility is defined as the absence of conception after 1 year of unprotected intercourse [110]. According to epidemiological studies, infertility affects

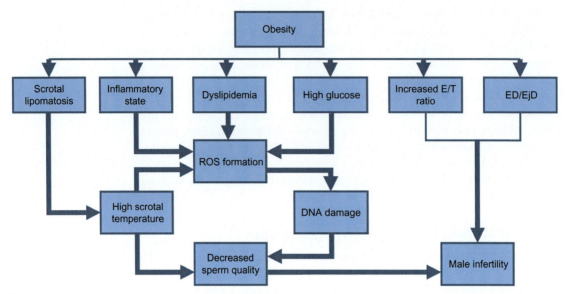

FIG. 4
Possible mechanisms of obesity-related male infertility. *E/T*, estrogen/testosterone; *ED*, erectile dysfunction; *EjD*, ejaculatory disorders; and *ROS*, reactive oxygen species.

an estimated 15% of couples globally, and male-related factors are responsible for 20%–30% of infertility cases (contributing to 50% of cases overall) [110]. In the recent years, obesity rates and male infertility prevalence have been increasing worldwide [14, 111, 112]. Reduced semen quality and quantity due to the dramatic changes in the lifestyle of civilized communities have been considered as the main reason behind the increased prevalence of infertility [14].

Not all obese men are infertile, but those who do can experience this problem due to reduced semen quality/quantity, impaired erectile function, hypogonadism, and increased scrotal temperatures [85, 113–116]. Obese men may commonly have oligozoospermia and asthenozoospermia [111, 117–122]. The mechanisms of obesity-related male infertility could be these alterations in semen parameters, especially sperm concentration [123], total motile sperm count [123, 124], sperm morphology, and DNA fragmentation [124] (Fig. 4).

Underlying Mechanisms
Endocrinopathy
Suppression of sex hormone-binding globulin (SHBG) and increased estrogen production may lead to reduced gonadotropin secretion in the obese men [114, 124–126]. In fact, obese men have decreased total and bioavailable testosterone levels [59, 124, 125, 127–130] and reduced inhibin-B concentrations

[124, 126, 131] and diminished LH pulse amplitude [59]. This hormonal profile may be the result of the estrogen inhibition on hypothalamo-pituitary-gonadal axis (HPG axis) [132]. Consequently, obesity in men is accompanied by decreased LH secretion, which leads to impaired Leydig cell function and low testosterone levels [130, 133]. Inhibition of HPG axis may also result in the decrease in follicle-stimulating hormone (FSH) levels and impair sperm production.

Insulin Resistance
Peripheral and central insulin resistance and the effect of proinflammatory cytokines (TNF-α and IL-6) on the HPG axis can also cause secondary hypogonadism in obese men with T2DM [134]. Low SHBG levels and high insulin levels have been shown previously in obese men, suggesting that insulin resistance has inhibitory effects on testosterone synthesis [135].

Thermal Effect of Obesity on Spermatogenesis
Increased scrotal adiposity can elevate the testicular heat, which contributes to altered sperm parameters. Spermatogenesis is a heat-sensitive process and optimal temperature for spermatogenesis in humans is 34–35°C [136]. The increased scrotal heat can lead to reduced sperm motility, increased sperm DNA fragmentation, and increased sperm oxidative stress [137]. Increased sperm DNA damage and oxidative stress are identified in obese patients, and the surgical removal of scrotal fat tissue has been reported to improve sperm parameters [138].

Structural Changes in Sperm Cells
In addition to the aforementioned pathophysiological mechanisms, alterations in sperm function [125], increase in sperm DNA damage [139–144], decrease in sperm mitochondrial activity [142, 143], and enhanced seminal oxidative stress [128] have been commonly observed in obese men, probably because of the adipokine-associated inflammatory response of the body.

Treatment of Obesity-Related Infertility
Although the beneficial effect of weight loss on fertility has been demonstrated in obese women, there is a relative paucity of studies in obese men with infertility. However, low-calorie diets and exercise interventions are associated with improvements in sperm parameters [145]. Weight loss may also improve the sperm DNA integrity in obese men, which may be linked to an improved live birth rate [146].

OBESITY AND PROSTATE CANCER

Introduction

In the last decade, epidemiologic studies have suggested that obesity is associated with increased risk of numerous cancer types including PCa [147–149]. Three meta-analyses [149–151] reported a positive association between obesity and PCa incidence. Several large-scale clinical studies have suggested a positive dose-response relationship between increasing BMI and fatal PCa [147, 152, 153]. Increase in the BMI has also been found to be associated with PCa-specific mortality and biochemical recurrence after primary PCa treatment [154].

Underlying Mechanisms

Obesity is associated with reduced PSA levels due to plasma hemodilution [155]. Therefore, in countries with PSA screening programs where biopsies are primarily driven by PSA levels, obese men may have lesser chance of undergoing prostate biopsy compared with normal-weight men. This may lead to the detection of fewer early-stage cancers in obese individuals. This phenomenon may explain the association between obesity and increased PCa-specific mortality after the introduction of PSA screening in the United States [153, 154]. However, obesity-associated detection bias cannot explain the impact of obesity on PCa alone as PCa-specific mortality was higher among obese men before the launch of PSA screening program as well [153, 154].

The insulin/insulin growth factor-1 (IGF-1) pathway has been highly relevant to obesity-related oncogenesis [156]. Survival of a cancer cell from apoptosis can be enhanced by increased IGF signaling. Moreover, IGF can induce angiogenesis by stimulating the secretion of vascular endothelial growth factor (VEGF), which is one of the fundamental steps in cancer development [157]. A recent meta-analysis confirmed the moderate positive association between circulating concentrations of IGF-1 and PCa risk [158]. Experimental studies also support the mitotic and antiapoptotic effects of IGF in the development of PCa [157].

Insulin molecules can also bind to IGF receptors, and hyperinsulinemia has also been shown to accelerate tumor growth in PCa xenograft models [159, 160]. Moreover, primary human PCa cells commonly express insulin receptors [161]; therefore, insulin may stimulate human PCa growth as well. A study reported that higher serum C-peptide concentrations were associated with increased PCa-specific mortality [162]. The antidiabetic drug metformin has been shown to reduce PCa risk in diabetic patients [163, 164]. According to these studies, insulin may be the key element to explain the relevance between obesity and PCa.

Adipokines may also explain the increased PCa risk among obese patients [165]. Leptin is an adipokine that is associated with a protumor effect in

PCa cell lines causing increased proliferation, decreased apoptosis [166], and increased cell migration [167]. However, another adipokine, adiponectin, has a potent antitumor effect. Serum leptin levels are generally elevated in obese individuals, whereas adiponectin levels are reduced significantly in metastatic PCa patients compared with those with organ-confined disease [168]. Other adipokines such as TNF-α and IL-6 may also play a role in the development of PCa among obese men [169–171].

Possible Interventions

Several studies evaluated the impact of weight loss on PCa [172, 173], which reported mixed effects on tumor markers, gene expressions, and proliferation predictors. A recent phase II prospective randomized trial of weight loss prior to radical prostatectomy could not reveal any changes in malignant epithelium apoptosis or proliferation [174]. Therefore, more prospective controlled studies are warranted before weight loss can be recommended to obese men with PCa.

Since obesity is a state of chronic inflammation and increased oxidative stress [165], antioxidant micronutrients may be protective against the development of PCa among obese men [175]. There are several studies about dietary modifications and their effects on PCa progression and development [176, 177]. Increased intake of lycopene, which is a carotenoid antioxidant found in tomato, pink grapefruit, and watermelon, has been shown to have a protective effect from PCa [175, 178–181]. In a meta-analysis, high intakes of tomato or tomato-based products were associated with a 10%–20% reduction in PCa risk [182]. However, further studies are needed to demonstrate if lycopene or any other nutrient has the ability to decrease the elevated PCa risk among men with obesity.

Key Points: Obesity and Men's Health

- Obesity is a constellation of clinical factors associated with an increased risk of incident CVD and T2DM.
- Obesity is associated with increased risks of the following urological conditions:
  - BPH/LUTS
  - ED
  - Hypogonadism
  - Infertility
  - PCa
- Weight loss improves erectile function, BPH/LUTS, hypogonadism, and infertility in obese men.
- Further studies are required before weight loss can be recommended for obese men with PCa.

References

[1] National Institutes of Health. Clinical guidelines on the identification, evaluation, and treatment of overweight and obesity in adults–the evidence report. Obes Res 1998;6(Suppl. 2):51S–209S.

[2] Ogden CL, et al. Prevalence of obesity among adults and youth: United States, 2011-2014. NCHS Data Brief 2015;(219):1–8.

[3] Zhang ZY, Wang MW. Obesity, a health burden of a global nature. Acta Pharmacol Sin 2012;33(2):145–7.

[4] Fernandez-Sanchez A, et al. Inflammation, oxidative stress, and obesity. Int J Mol Sci 2011;12(5):3117–32.

[5] Prospective Studies Collaboration, et al. Body-mass index and cause-specific mortality in 900 000 adults: collaborative analyses of 57 prospective studies. Lancet 2009;373(9669):1083–96.

[6] Raheem OA, Parsons JK. Associations of obesity, physical activity and diet with benign prostatic hyperplasia and lower urinary tract symptoms. Curr Opin Urol 2014;24(1):10–4.

[7] Vignozzi L, Gacci M, Maggi M. Lower urinary tract symptoms, benign prostatic hyperplasia and metabolic syndrome. Nat Rev Urol 2016;13(2):108–19.

[8] Kaya E, Sikka SC, Gur S. A comprehensive review of metabolic syndrome affecting erectile dysfunction. J Sex Med 2015;12(4):856–75.

[9] Shamloul R, Ghanem H. Erectile dysfunction. Lancet 2013;381(9861):153–65.

[10] Li R, et al. Metabolic syndrome in rats is associated with erectile dysfunction by impairing PI3K/Akt/eNOS activity. Sci Rep 2017;7(1).

[11] Cohen PG. The hypogonadal-obesity cycle: role of aromatase in modulating the testosterone-estradiol shunt–a major factor in the genesis of morbid obesity. Med Hypotheses 1999;52(1):49–51.

[12] Huhtaniemi I. Late-onset hypogonadism: current concepts and controversies of pathogenesis, diagnosis and treatment. Asian J Androl 2014;16(2):192–202.

[13] Kelly DM, Jones TH. Testosterone and obesity. Obes Rev 2015;16(7):581–606.

[14] Katib A. Mechanisms linking obesity to male infertility. Cent European J Urol 2015;68(1):79–85.

[15] Ehala-Aleksejev K, Punab M. The effect of metabolic syndrome on male reproductive health: A cross-sectional study in a group of fertile men and male partners of infertile couples. PLoS ONE 2018;13(3).

[16] Brewer CJ, Balen AH. The adverse effects of obesity on conception and implantation. Reproduction 2010;140(3):347–64.

[17] Practice Committee of the American Society for Reproductive Medicine. Obesity and reproduction: a committee opinion. Fertil Steril 2015;104(5):1116–26.

[18] Allott EH, Masko EM, Freedland SJ. Obesity and prostate cancer: weighing the evidence. Eur Urol 2013;63(5):800–9.

[19] Rundle A, et al. Obesity and future prostate cancer risk among men after an initial benign biopsy of the prostate. Cancer Epidemiol Biomark Prev 2013;22(5):898–904.

[20] Lee SH, et al. Effects of obesity on lower urinary tract symptoms in Korean BPH patients. Asian J Androl 2009;11(6):663–8.

[21] Parsons JK. Benign prostatic hyperplasia and male lower urinary tract symptoms: epidemiology and risk factors. Curr Bladder Dysfunct Rep 2010;5(4):212–8.

[22] Kim JM, et al. Effect of obesity on prostate-specific antigen, prostate volume, and international prostate symptom score in patients with benign prostatic hyperplasia. Korean J Urol 2011;52(6):401–5.

[23] Dahle SE, et al. Body size and serum levels of insulin and leptin in relation to the risk of benign prostatic hyperplasia. J Urol 2002;168(2):599–604.

[24] Wang S, et al. Body mass index and risk of BPH: a meta-analysis. Prostate Cancer Prostatic Dis 2012;15(3):265–72.

[25] Kristal AR, et al. Race/ethnicity, obesity, health related behaviors and the risk of symptomatic benign prostatic hyperplasia: results from the prostate cancer prevention trial. J Urol 2007;177(4):1395–400. quiz 1591.

[26] Penson DF, et al. Obesity, physical activity and lower urinary tract symptoms: results from the Southern Community Cohort Study. J Urol 2011;186(6):2316–22.

[27] Parsons JK, et al. Obesity increases and physical activity decreases lower urinary tract symptom risk in older men: the Osteoporotic Fractures in Men study. Eur Urol 2011;60(6):1173–80.

[28] Burke JP, et al. Association of anthropometric measures with the presence and progression of benign prostatic hyperplasia. Am J Epidemiol 2006;164(1):41–6.

[29] Maserejian NN, et al. Treatment status and progression or regression of lower urinary tract symptoms in a general adult population sample. J Urol 2014;191(1):107–13.

[30] Kok ET, et al. Risk factors for lower urinary tract symptoms suggestive of benign prostatic hyperplasia in a community based population of healthy aging men: the Krimpen Study. J Urol 2009;181(2):710–6.

[31] Mondul AM, Giovannucci E, Platz EA. A prospective study of obesity, and the incidence and progression of lower urinary tract symptoms. J Urol 2014;191(3):715–21.

[32] Lotti F, et al. Elevated body mass index correlates with higher seminal plasma interleukin 8 levels and ultrasonographic abnormalities of the prostate in men attending an andrology clinic for infertility. J Endocrinol Investig 2011;34(10):e336–42.

[33] Gacci M, et al. Central obesity is predictive of persistent storage lower urinary tract symptoms (LUTS) after surgery for benign prostatic enlargement: results of a multicentre prospective study. BJU Int 2015;116(2):271–7.

[34] Russo GI, et al. Increase of Framingham cardiovascular disease risk score is associated with severity of lower urinary tract symptoms. BJU Int 2015;116(5):791–6.

[35] Nandeesha H, et al. Hyperinsulinemia and dyslipidemia in non-diabetic benign prostatic hyperplasia. Clin Chim Acta 2006;370(1–2):89–93.

[36] Vanella L, et al. Correlation between lipid profile and heme oxygenase system in patients with benign prostatic hyperplasia. Urology 2014;83(6):1444.e7–1444.e13.

[37] Dauchet L, et al. Fruit and vegetable consumption and risk of coronary heart disease: a meta-analysis of cohort studies. J Nutr 2006;136(10):2588–93.

[38] He FJ, Nowson CA, MacGregor GA. Fruit and vegetable consumption and stroke: meta-analysis of cohort studies. Lancet 2006;367(9507):320–6.

[39] Corona G, et al. Body weight loss reverts obesity-associated hypogonadotropic hypogonadism: a systematic review and meta-analysis. Eur J Endocrinol 2013;168(6):829–43.

[40] Hamer M, Chida Y. Intake of fruit, vegetables, and antioxidants and risk of type 2 diabetes: systematic review and meta-analysis. J Hypertens 2007;25(12):2361–9.

[41] Platz EA, et al. Physical activity and benign prostatic hyperplasia. Arch Intern Med 1998;158(21):2349–56.

[42] Morelli A, et al. Mechanism of action of phosphodiesterase type 5 inhibition in metabolic syndrome-associated prostate alterations: an experimental study in the rabbit. Prostate 2013;73(4):428–41.

[43] Lue TF. Erectile dysfunction. N Engl J Med 2000;342(24):1802–13.

[44] Chew KK. Prevalence of erectile dysfunction in community-based studies. Int J Impot Res 2004;16(2):201–2.

[45] Muller A, Mulhall JP. Cardiovascular disease, metabolic syndrome and erectile dysfunction. Curr Opin Urol 2006;16(6):435–43.

[46] Derby CA, et al. Modifiable risk factors and erectile dysfunction: can lifestyle changes modify risk? Urology 2000;56(2):302–6.

[47] Bacon CG, et al. A prospective study of risk factors for erectile dysfunction. J Urol 2006;176(1):217–21.

[48] Andersen I, Heitmann BL, Wagner G. Obesity and sexual dysfunction in younger Danish men. J Sex Med 2008;5(9):2053–60.

[49] Corona G, et al. Age-related changes in general and sexual health in middle-aged and older men: results from the European Male Ageing Study (EMAS). J Sex Med 2010;7(4 Pt 1):1362–80.

[50] Blanker MH, et al. Correlates for erectile and ejaculatory dysfunction in older Dutch men: a community-based study. J Am Geriatr Soc 2001;49(4):436–42.

[51] Han TS, et al. Impaired quality of life and sexual function in overweight and obese men: the European Male Ageing Study. Eur J Endocrinol 2011;164(6):1003–11.

[52] Corona G, et al. Low levels of androgens in men with erectile dysfunction and obesity. J Sex Med 2008;5(10):2454–63.

[53] Corona G, et al. Erectile dysfunction and central obesity: an Italian perspective. Asian J Androl 2014;16(4):581–91.

[54] Corona G, et al. Hypogonadism, ED, metabolic syndrome and obesity: a pathological link supporting cardiovascular diseases. Int J Androl 2009;32(6):587–98.

[55] Corona G, Forti G, Maggi M. Why can patients with erectile dysfunction be considered lucky? The association with testosterone deficiency and metabolic syndrome. Aging Male 2008;11(4):193–9.

[56] Corona G, et al. How to recognize late-onset hypogonadism in men with sexual dysfunction. Asian J Androl 2012;14(2):251–9.

[57] Saad F, et al. Testosterone as potential effective therapy in treatment of obesity in men with testosterone deficiency: a review. Curr Diabetes Rev 2012;8(2):131–43.

[58] Isidori AM, et al. A critical analysis of the role of testosterone in erectile function: from pathophysiology to treatment-a systematic review. Eur Urol 2014;65(1):99–112.

[59] Vermeulen A, et al. Attenuated luteinizing hormone (LH) pulse amplitude but normal LH pulse frequency, and its relation to plasma androgens in hypogonadism of obese men. J Clin Endocrinol Metab 1993;76(5):1140–6.

[60] Ye J, Keller JN. Regulation of energy metabolism by inflammation: a feedback response in obesity and calorie restriction. Aging (Albany, NY) 2010;2(6):361–8.

[61] Rondinone CM. Adipocyte-derived hormones, cytokines, and mediators. Endocrine 2006;29(1):81–90.

[62] Rahmouni K. Obesity-associated hypertension: recent progress in deciphering the pathogenesis. Hypertension 2014;64(2):215–21.

[63] Gimeno RE, Klaman LD. Adipose tissue as an active endocrine organ: recent advances. Curr Opin Pharmacol 2005;5(2):122–8.

[64] Vincent HK, Innes KE, Vincent KR. Oxidative stress and potential interventions to reduce oxidative stress in overweight and obesity. Diabetes Obes Metab 2007;9(6):813–39.

[65] Nehra A, et al. The Princeton III consensus recommendations for the management of erectile dysfunction and cardiovascular disease. Mayo Clin Proc 2012;87(8):766–78.

[66] Guo W, et al. Erectile dysfunction and risk of clinical cardiovascular events: a meta-analysis of seven cohort studies. J Sex Med 2010;7(8):2805–16.

[67] Dong JY, Zhang YH, Qin LQ. Erectile dysfunction and risk of cardiovascular disease: meta-analysis of prospective cohort studies. J Am Coll Cardiol 2011;58(13):1378–85.

[68] Vlachopoulos CV, et al. Prediction of cardiovascular events and all-cause mortality with erectile dysfunction: a systematic review and meta-analysis of cohort studies. Circ Cardiovasc Qual Outcomes 2013;6(1):99–109.

[69] Yamada T, et al. Erectile dysfunction and cardiovascular events in diabetic men: a meta-analysis of observational studies. PLoS ONE 2012;7(9).

[70] Montorsi P, Montorsi F, Schulman CC. Is erectile dysfunction the "tip of the iceberg" of a systemic vascular disorder? Eur Urol 2003;44(3):352–4.

[71] Montorsi P, et al. The artery size hypothesis: a macrovascular link between erectile dysfunction and coronary artery disease. Am J Cardiol 2005;96(12B):19M–23M.

[72] Esposito K, et al. Effect of lifestyle changes on erectile dysfunction in obese men: a randomized controlled trial. JAMA 2004;291(24):2978–84.

[73] Esposito K, et al. Effects of intensive lifestyle changes on erectile dysfunction in men. J Sex Med 2009;6(1):243–50.

[74] Esposito K, et al. Mediterranean diet improves erectile function in subjects with the metabolic syndrome. Int J Impot Res 2006;18(4):405–10.

[75] Wing RR, et al. Effects of weight loss intervention on erectile function in older men with type 2 diabetes in the Look AHEAD trial. J Sex Med 2010;7(1 Pt 1):156–65.

[76] Hatzimouratidis K (Chair), Giuliano F, Moncada I, Muneer A, Salonia A (Vice-chair), Verze P, Guideline Associates: Parnham A, Serefoglu EC. EAU guidelines on erectile dysfunction, premature ejaculation, penile curvature and priapism. Arnhem, The Netherlands: EAU Guidelines Office; 2018.

[77] Khera M, et al. Diagnosis and treatment of testosterone deficiency: recommendations from the Fourth International Consultation for Sexual Medicine (ICSM 2015). J Sex Med 2016;13(12):1787–804.

[78] Haffner SM, et al. Relationship of sex hormones to lipids and lipoproteins in nondiabetic men. J Clin Endocrinol Metab 1993;77(6):1610–5.

[79] Phillips GB. Relationship between serum sex hormones and the glucose-insulin-lipid defect in men with obesity. Metabolism 1993;42(1):116–20.

[80] Couillard C, et al. Contribution of body fatness and adipose tissue distribution to the age variation in plasma steroid hormone concentrations in men: the HERITAGE Family Study. J Clin Endocrinol Metab 2000;85(3):1026–31.

[81] Hofstra J, et al. High prevalence of hypogonadotropic hypogonadism in men referred for obesity treatment. Neth J Med 2008;66(3):103–9.

[82] Dhindsa S, et al. Testosterone concentrations in diabetic and nondiabetic obese men. Diabetes Care 2010;33(6):1186–92.

[83] Allen NE, et al. Lifestyle and nutritional determinants of bioavailable androgens and related hormones in British men. Cancer Causes Control 2002;13(4):353–63.

[84] Gapstur SM, et al. Serum androgen concentrations in young men: a longitudinal analysis of associations with age, obesity, and race. The CARDIA male hormone study. Cancer Epidemiol Biomark Prev 2002;11(10 Pt 1):1041–7.

[85] Jensen TK, et al. Body mass index in relation to semen quality and reproductive hormones among 1,558 Danish men. Fertil Steril 2004;82(4):863–70.

[86] Svartberg J, et al. Association of endogenous testosterone with blood pressure and left ventricular mass in men. The Tromso study. Eur J Endocrinol 2004;150(1):65–71.

[87] Svartberg J, et al. Waist circumference and testosterone levels in community dwelling men. The Tromso study. Eur J Epidemiol 2004;19(7):657–63.

[88] Escobar-Morreale HF, et al. Prevalence of 'obesity-associated gonadal dysfunction' in severely obese men and women and its resolution after bariatric surgery: a systematic review and meta-analysis. Hum Reprod Update 2017;23(4):390–408.

[89] Finn PD, et al. The stimulatory effect of leptin on the neuroendocrine reproductive axis of the monkey. Endocrinology 1998;139(11):4652–62.

[90] Roseweir AK, Millar RP. The role of kisspeptin in the control of gonadotrophin secretion. Hum Reprod Update 2009;15(2):203–12.

[91] Isidori AM, et al. Leptin and androgens in male obesity: evidence for leptin contribution to reduced androgen levels. J Clin Endocrinol Metab 1999;84(10):3673–80.

[92] Rebuffe-Scrive M, Marin P, Bjorntorp P. Effect of testosterone on abdominal adipose tissue in men. Int J Obes 1991;15(11):791–5.

[93] Marin P, Oden B, Bjorntorp P. Assimilation and mobilization of triglycerides in subcutaneous abdominal and femoral adipose tissue in vivo in men: effects of androgens. J Clin Endocrinol Metab 1995;80(1):239–43.

[94] Saad F, et al. An exploratory study of the effects of 12 month administration of the novel long-acting testosterone undecanoate on measures of sexual function and the metabolic syndrome. Arch Androl 2007;53(6):353–7.

[95] Marin P, Krotkiewski M, Bjorntorp P. Androgen treatment of middle-aged, obese men: effects on metabolism, muscle and adipose tissues. Eur J Med 1992;1(6):329–36.

[96] Marin P, et al. The effects of testosterone treatment on body composition and metabolism in middle-aged obese men. Int J Obes Relat Metab Disord 1992;16(12):991–7.

[97] Agledahl I, Hansen JB, Svartberg J. Impact of testosterone treatment on postprandial triglyceride metabolism in elderly men with subnormal testosterone levels. Scand J Clin Lab Invest 2008;68(7):641–8.

[98] Saad F, et al. A dose-response study of testosterone on sexual dysfunction and features of the metabolic syndrome using testosterone gel and parenteral testosterone undecanoate. J Androl 2008;29(1):102–5.

[99] Jones TH, et al. Testosterone replacement in hypogonadal men with type 2 diabetes and/or metabolic syndrome (the TIMES2 study). Diabetes Care 2011;34(4):828–37.

[100] Kapoor D, et al. Testosterone replacement therapy improves insulin resistance, glycaemic control, visceral adiposity and hypercholesterolaemia in hypogonadal men with type 2 diabetes. Eur J Endocrinol 2006;154(6):899–906.

[101] Boyanov MA, Boneva Z, Christov VG. Testosterone supplementation in men with type 2 diabetes, visceral obesity and partial androgen deficiency. Aging Male 2003;6(1):1–7.

[102] Svartberg J, et al. Testosterone treatment in elderly men with subnormal testosterone levels improves body composition and BMD in the hip. Int J Impot Res 2008;20(4):378–87.

[103] Woodhouse LJ, et al. Dose-dependent effects of testosterone on regional adipose tissue distribution in healthy young men. J Clin Endocrinol Metab 2004;89(2):718–26.

[104] Allan CA, et al. Testosterone therapy prevents gain in visceral adipose tissue and loss of skeletal muscle in nonobese aging men. J Clin Endocrinol Metab 2008;93(1):139–46.

[105] Corona G, Rastrelli G, Maggi M. Diagnosis and treatment of late-onset hypogonadism: systematic review and meta-analysis of TRT outcomes. Best Pract Res Clin Endocrinol Metab 2013;27(4):557–79.

[106] Corona G, et al. Testosterone and metabolic syndrome: a meta-analysis study. J Sex Med 2011;8(1):272–83.

[107] Cai X, et al. Metabolic effects of testosterone replacement therapy on hypogonadal men with type 2 diabetes mellitus: a systematic review and meta-analysis of randomized controlled trials. Asian J Androl 2014;16(1):146–52.

[108] Soares AH, et al. Effects of clomiphene citrate on male obesity-associated hypogonadism: a randomized, double-blind, placebo-controlled study. Int J Obes 2018;42(5):953–63.

[109] Samavat J, et al. Hypogonadism as an additional indication for bariatric surgery in male morbid obesity? Eur J Endocrinol 2014;171(5):555–60.

[110] Agarwal A, et al. A unique view on male infertility around the globe. Reprod Biol Endocrinol 2015;13:37.

[111] Hammoud AO, et al. Male obesity and alteration in sperm parameters. Fertil Steril 2008;90(6):2222–5.

[112] Ring JD, Lwin AA, Kohler TS. Current medical management of endocrine-related male infertility. Asian J Androl 2016;18(3):357–63.

[113] Sallmen M, et al. Reduced fertility among overweight and obese men. Epidemiology 2006;17(5):520–3.

[114] Pasquali R. Obesity and androgens: facts and perspectives. Fertil Steril 2006;85(5):1319–40.

[115] Ramlau-Hansen CH, et al. Subfecundity in overweight and obese couples. Hum Reprod 2007;22(6):1634–7.

[116] Cabler S, et al. Obesity: modern man's fertility nemesis. Asian J Androl 2010;12(4):480–9.

[117] Bakos HW, et al. Paternal body mass index is associated with decreased blastocyst development and reduced live birth rates following assisted reproductive technology. Fertil Steril 2011;95(5):1700–4.

[118] Relwani R, et al. Semen parameters are unrelated to BMI but vary with SSRI use and prior urological surgery. Reprod Sci 2011;18(4):391–7.

[119] Umul M, et al. Effect of increasing paternal body mass index on pregnancy and live birth rates in couples undergoing intracytoplasmic sperm injection. Andrologia 2015;47(3):360–4.

[120] Sermondade N, et al. BMI in relation to sperm count: an updated systematic review and collaborative meta-analysis. Hum Reprod Update 2013;19(3):221–31.

[121] Braga DP, et al. Food intake and social habits in male patients and its relationship to intracytoplasmic sperm injection outcomes. Fertil Steril 2012;97(1):53–9.

[122] Hammiche F, et al. Body mass index and central adiposity are associated with sperm quality in men of subfertile couples. Hum Reprod 2012;27(8):2365–72.

[123] Kort HI, et al. Impact of body mass index values on sperm quantity and quality. J Androl 2006;27(3):450–2.

[124] MacDonald AA, et al. The impact of body mass index on semen parameters and reproductive hormones in human males: a systematic review with meta-analysis. Hum Reprod Update 2010;16(3):293–311.

[125] Palmer NO, et al. Impact of obesity on male fertility, sperm function and molecular composition. Spermatogenesis 2012;2(4):253–63.

[126] Teerds KJ, de Rooij DG, Keijer J. Functional relationship between obesity and male reproduction: from humans to animal models. Hum Reprod Update 2011;17(5):667–83.

[127] Al-Ali BM, et al. Body mass index has no impact on sperm quality but on reproductive hormones levels. Andrologia 2014;46(2):106–11.

[128] Tunc O, Bakos HW, Tremellen K. Impact of body mass index on seminal oxidative stress. Andrologia 2011;43(2):121–8.

[129] Zumoff B, et al. Plasma free and non-sex-hormone-binding-globulin-bound testosterone are decreased in obese men in proportion to their degree of obesity. J Clin Endocrinol Metab 1990;71(4):929–31.

[130] Pitteloud N, et al. Increasing insulin resistance is associated with a decrease in Leydig cell testosterone secretion in men. J Clin Endocrinol Metab 2005;90(5):2636–41.

[131] Stewart TM, et al. Associations between andrological measures, hormones and semen quality in fertile Australian men: inverse relationship between obesity and sperm output. Hum Reprod 2009;24(7):1561–8.

[132] Jarow JP, et al. Effect of obesity and fertility status on sex steroid levels in men. Urology 1993;42(2):171–4.

[133] Hofny ER, et al. Semen parameters and hormonal profile in obese fertile and infertile males. Fertil Steril 2010;94(2):581–4.

[134] Bhasin S, et al. Testosterone therapy in men with androgen deficiency syndromes: an Endocrine Society clinical practice guideline. J Clin Endocrinol Metab 2010;95(6):2536–59.

[135] Tsai EC, et al. Association of bioavailable, free, and total testosterone with insulin resistance: influence of sex hormone-binding globulin and body fat. Diabetes Care 2004;27(4):861–8.

[136] Robinson D, Rock J, Menkin MF. Control of human spermatogenesis by induced changes of intrascrotal temperature. JAMA 1968;204(4):290–7.

[137] Sheynkin Y, et al. Increase in scrotal temperature in laptop computer users. Hum Reprod 2005;20(2):452–5.

[138] Shafik A, Olfat S. Lipectomy in the treatment of scrotal lipomatosis. Br J Urol 1981;53(1):55–61.

[139] Thomsen L, et al. The impact of male overweight on semen quality and outcome of assisted reproduction. Asian J Androl 2014;16(5):749–54.

[140] Chavarro JE, et al. Body mass index in relation to semen quality, sperm DNA integrity, and serum reproductive hormone levels among men attending an infertility clinic. Fertil Steril 2010;93(7):2222–31.

[141] Dupont C, et al. Obesity leads to higher risk of sperm DNA damage in infertile patients. Asian J Androl 2013;15(5):622–5.

[142] Fariello RM, et al. Association between obesity and alteration of sperm DNA integrity and mitochondrial activity. BJU Int 2012;110(6):863–7.

[143] La Vignera S, et al. Negative effect of increased body weight on sperm conventional and nonconventional flow cytometric sperm parameters. J Androl 2012;33(1):53–8.

[144] Rybar R, et al. Male obesity and age in relationship to semen parameters and sperm chromatin integrity. Andrologia 2011;43(4):286–91.

[145] Best D, Avenell A, Bhattacharya S. How effective are weight-loss interventions for improving fertility in women and men who are overweight or obese? A systematic review and meta-analysis of the evidence. Hum Reprod Update 2017;23(6):681–705.

[146] Faure C, et al. In subfertile couple, abdominal fat loss in men is associated with improvement of sperm quality and pregnancy: a case-series. PLoS ONE 2014;9(2).

[147] Calle EE, et al. Overweight, obesity, and mortality from cancer in a prospectively studied cohort of U.S. adults. N Engl J Med 2003;348(17):1625–38.

[148] Reeves GK, et al. Cancer incidence and mortality in relation to body mass index in the Million Women Study: cohort study. BMJ 2007;335(7630):1134.

[149] Renehan AG, et al. Body-mass index and incidence of cancer: a systematic review and meta-analysis of prospective observational studies. Lancet 2008;371(9612):569–78.

[150] MacInnis RJ, English DR. Body size and composition and prostate cancer risk: systematic review and meta-regression analysis. Cancer Causes Control 2006;17(8):989–1003.

[151] Bergstrom A, et al. Overweight as an avoidable cause of cancer in Europe. Int J Cancer 2001;91(3):421–30.

[152] Andersson SO, et al. Body size and prostate cancer: a 20-year follow-up study among 135006 Swedish construction workers. J Natl Cancer Inst 1997;89(5):385–9.

[153] Wright ME, et al. Prospective study of adiposity and weight change in relation to prostate cancer incidence and mortality. Cancer 2007;109(4):675–84.

[154] Cao Y, Ma J. Body mass index, prostate cancer-specific mortality, and biochemical recurrence: a systematic review and meta-analysis. Cancer Prev Res (Phila) 2011;4(4):486–501.

[155] Banez LL, et al. Obesity-related plasma hemodilution and PSA concentration among men with prostate cancer. JAMA 2007;298(19):2275–80.

[156] Roberts DL, Dive C, Renehan AG. Biological mechanisms linking obesity and cancer risk: new perspectives. Annu Rev Med 2010;61:301–16.

[157] Grimberg A. Mechanisms by which IGF-I may promote cancer. Cancer Biol Ther 2003;2(6):630–5.

[158] Travis RC, et al. A meta-analysis of individual participant data reveals an association between circulating levels of IGF-I and prostate cancer risk. Cancer Res 2016;76(8):2288–300.

[159] Venkateswaran V, et al. Association of diet-induced hyperinsulinemia with accelerated growth of prostate cancer (LNCaP) xenografts. J Natl Cancer Inst 2007;99(23):1793–800.

[160] Freedland SJ, et al. Carbohydrate restriction, prostate cancer growth, and the insulin-like growth factor axis. Prostate 2008;68(1):11–9.

[161] Cox ME, et al. Insulin receptor expression by human prostate cancers. Prostate 2009;69(1):33–40.

[162] Ma J, et al. Prediagnostic body-mass index, plasma C-peptide concentration, and prostate cancer-specific mortality in men with prostate cancer: a long-term survival analysis. Lancet Oncol 2008;9(11):1039–47.

[163] Wright JL, Stanford JL. Metformin use and prostate cancer in Caucasian men: results from a population-based case-control study. Cancer Causes Control 2009;20(9):1617–22.

[164] Decensi A, et al. Metformin and cancer risk in diabetic patients: a systematic review and meta-analysis. Cancer Prev Res (Phila) 2010;3(11):1451–61.

[165] Ferro M, et al. The emerging role of obesity, diet and lipid metabolism in prostate cancer. Future Oncol 2017;13(3):285–93.

[166] Hoda MR, Popken G. Mitogenic and anti-apoptotic actions of adipocyte-derived hormone leptin in prostate cancer cells. BJU Int 2008;102(3):383–8.

[167] Huang WC, et al. Activation of androgen receptor, lipogenesis, and oxidative stress converged by SREBP-1 is responsible for regulating growth and progression of prostate cancer cells. Mol Cancer Res 2012;10(1):133–42.

[168] Goktas S, et al. Prostate cancer and adiponectin. Urology 2005;65(6):1168–72.

[169] Shariat SF, et al. Plasma levels of interleukin-6 and its soluble receptor are associated with prostate cancer progression and metastasis. Urology 2001;58(6):1008–15.

[170] Shariat SF, et al. Association of preoperative plasma levels of vascular endothelial growth factor and soluble vascular cell adhesion molecule-1 with lymph node status and biochemical progression after radical prostatectomy. J Clin Oncol 2004;22(9):1655–63.

[171] Michalaki V, et al. Serum levels of IL-6 and TNF-alpha correlate with clinicopathological features and patient survival in patients with prostate cancer. Br J Cancer 2004;90(12):2312–6.

[172] Demark-Wahnefried W, et al. Presurgical weight loss affects tumour traits and circulating biomarkers in men with prostate cancer. Br J Cancer 2017;117(9):1303–13.

[173] Wright JL, et al. A study of caloric restriction versus standard diet in overweight men with newly diagnosed prostate cancer: a randomized controlled trial. Prostate 2013;73(12):1345–51.

[174] Henning SM, et al. Phase II prospective randomized trial of weight loss prior to radical prostatectomy. Prostate Cancer Prostatic Dis 2018;21(2):212–20.

[175] Zu K, et al. Dietary lycopene, angiogenesis, and prostate cancer: a prospective study in the prostate-specific antigen era. J Natl Cancer Inst 2014;106(2):djt430.

[176] Parsons JK, et al. A randomized trial of diet in men with early stage prostate cancer on active surveillance: rationale and design of the Men's Eating and Living (MEAL) Study (CALGB 70807 [Alliance]). Contemp Clin Trials 2014;38(2):198–203.

[177] Parsons JK, et al. Men's Eating and Living (MEAL) study (CALGB 70807 [Alliance]): recruitment feasibility and baseline demographics of a randomized trial of diet in men on active surveillance for prostate cancer. BJU Int 2018;121(4):534–9.

[178] Giovannucci E, et al. Intake of carotenoids and retinol in relation to risk of prostate cancer. J Natl Cancer Inst 1995;87(23):1767–76.

[179] Giovannucci E, et al. A prospective study of tomato products, lycopene, and prostate cancer risk. J Natl Cancer Inst 2002;94(5):391–8.

[180] Bosetti C, et al. Retinol, carotenoids and the risk of prostate cancer: a case-control study from Italy. Int J Cancer 2004;112(4):689–92.

[181] Jian L, Lee AH, Binns CW. Tea and lycopene protect against prostate cancer. Asia Pac J Clin Nutr 2007;16(Suppl. 1):453–7.

[182] Etminan M, Takkouche B, Caamano-Isorna F. The role of tomato products and lycopene in the prevention of prostate cancer: a meta-analysis of observational studies. Cancer Epidemiol Biomark Prev 2004;13(3):340–5.

CHAPTER 2.4

Cardiovascular Disease and Men's Health

İyimser Üre*, John M. Masterson[†], Ranjith Ramasamy[†]
*Eskişehir Osmangazi University, Faculty of Medicine, Department of Urology, Eskişehir, Turkey,
[†]University of Miami, Miller School of Medicine, Department of Urology, Miami, FL, United States

CARDIOVASCULAR DISEASE AND ERECTILE DYSFUNCTION

Erectile dysfunction (ED) is defined as inability to achieve adequate penile erection or to maintain an erection for satisfactory sexual performance. The incidence of ED increases with age with more than half of the over 70 population affected [1]. Many chronic diseases associated with advanced age are considered risk factors for ED. Hypertension, diabetes, and dyslipidemia have been found to be highly linked with ED [2]. Additionally, a strong correlation has been demonstrated between modifiable risk factors such as obesity and smoking and ED. In light of all these data, ED is thought to be a precursor and possibly predictor of systemic disease. Other chronic diseases may arise after some time in men who present with ED [3].

Pathophysiology

Normal erection physiology depends on the interactions between vascular, hormonal, neurological, and psychological factors. For an erection to be suitable for sexual intercourse, there must be adequate arterial blood flow. This blood must then be trapped in penile structures and must not be able to leak back [4]. Thus, any problem affecting penile blood flow will disrupt erectile function. The mechanism of blood flow disruption in cardiovascular disease (CVD) is endothelial dysfunction. It is endothelial dysfunction that establishes the link between CVD and ED.

According to results from the Massachusetts Male Aging Study, erectile dysfunction is more prevalent in men with CVD, regardless of age, than in men without CVD [5]. The first step in the formation of atherosclerotic plaques in CVD is endothelial dysfunction [6]. Endothelial dysfunction leads to disruption of nitric oxide (NO) production and increased permeability to cellular components such as low-density lipoproteins (LDL). As a result, free oxygen radicals and oxidative stress cause endothelial damage, and the clinical manifestations of CVD begin to emerge [7, 8].

Erectile dysfunction is associated with increased inflammatory and endothelial prothrombotic activation in the penile vasculature. Interestingly, this activation closely resembles the processes occurring in the coronary arteries of patients with CAD and no ED. Atherosclerotic plaque burden is greatest in patients with both CAD and ED [9]. The arterial structure of the penis is narrower than the coronary vascular structures. Additionally, penile endothelium and smooth muscle tissue are very sensitive to functional and structural changes. For this reason, plaques occurring in the penile arteries may result in earlier clinical findings than the same amount of plaque in the coronary arteries. Therefore, erectile dysfunction is more likely to occur before clinical entities like CAD, transient ischemic attack, or stroke [10]. This allows ED to be used as a marker of cardiovascular disease [11]. Vasculogenic ED can be considered an early manifestation of CVD. However, the presence of different factors in the etiology of ED may limit this relationship [12].

ED as a Predictor of CVD

It has long been debated whether or not ED can predict CVD. The prevalence of ED in CVD patients was previously evaluated by many studies. As a result, ED was found to be more prevalent patients with CVD [13, 14]. In one of the first comprehensive studies, it was determined that 50% of patients with ED had comorbid CVD as proved by coronary angiography [15]. Of the CVD patients in this study, 70% had ED for more than 3 years. This study is one of the first to suggest that ED can be used as a marker for CVD that will occur later in life.

In the Prostate Cancer Prevention Trial study of men aged 55 years and older, compelling data have been demonstrated regarding the association of ED and subsequent cardiovascular events [16]. This study primarily investigates whether finasteride can reduce the prevalence of prostate cancer. In order to assess the possible sexual side effects of finasteride, the prevalence of ED was measured in the subjects. In order to assess one of the possible side effects of the drug, prevalence of CVD was measured at baseline, and incidence of CVD was followed throughout the study period. The authors have detected that in men with ED, the risk of a cardiovascular incident is 45% greater than those without ED. Prostate Cancer Prevention Trial concluded that newly diagnosed ED is a risk factor for CVD on par with risk factors such as family history of myocardial infarction (MI) and medical history of smoking or hyperlipidemia. This study is considered a landmark study in establishing that ED is a marker for CVD and cardiovascular evaluation should be performed in patients presenting with ED.

Further studies have shown that ED is a precursor not only to CVD but also to many cardiovascular atherosclerotic diseases. ED has also been associated with coronary artery calcification, which is an important marker of coronary artery disease [17]. Additionally, the occurrence of ED at a younger age, smoking, the presence of other comorbidities, and poor socioeconomic status pose risks for possible atherosclerotic cardiovascular events [18].

In the Cost and Outcome of Behavioural Activation (COBRA) study investigating the relationship between ED and coronary artery disease, it is stated that the prevalence of ED differs according to the clinical presentation of coronary artery disease and the number of vessels involved [19]. In this study, the clinical presentation of coronary artery disease was classified as acute or chronic coronary artery syndrome. Overall prevalence of ED was found to be 47% in patients with coronary artery disease of any kind, as opposed to 24% for patients with normal coronary angiography. The rate of ED in patients with single-vessel acute coronary artery syndrome is 22%. Prevalence of ED increases to 55% in patients with multivessel acute coronary syndrome. This rate was found to be 65% in patients with chronic coronary syndrome. Men with severe coronary artery involvement as assessed by coronary angiography were more likely to have severe ED (International Index for Erectile Dysfunction (IEF) <10) or ED lasting greater than 24 months. The findings of this study show that systemic atherosclerosis significantly affects penile arterial structures and leads to higher ED rates.

Not only are patients with ED at risk for coronary artery disease, but also there is evidence to suggest that ED is also a risk factor for all-cause death [9]. However, there are studies that do not accept ED as an independent factor in deaths due to coronary heart disease. The Vitamins and Lifestyle (VITAL) study included over 30,000 men over 50 years of age [20]. The authors reported that, when adjusting for age, marital status, and education, men with ED had a 23% increased risk of cardiovascular death (hazard ratio (HR) 1.23 and 95% confidence interval (CI) 1.01 and 1.49). When adjusted for known risk factors (diabetes, treatment for hypertension or hyperlipidemia, family history of MI, stroke, elevated body mass index, and exercise), ED no longer predicted cardiovascular death (HR 0.93 and 95% CI 0.76 and 1.15). These data serve to further obviate the fact that ED can be used as a reliable marker for systemic vascular diseases, considering the relationship with different vascular diseases such as cerebrovascular events and peripheral artery disease. Therefore, patients with ED should be investigated for possible underlying CVD [21].

SEXUAL HEALTH IN CVD PATIENTS

During sexual intercourse, many physiological and physical changes occur in the human body. Most of these changes are cardiovascular and endocrine processes.

Cardiovascular changes and the risks of these changes have been attractive research topics for some time. During foreplay, the blood pressure rises slightly, and the pulse rate increases. The pulse rate is usually below 130 bpm, and systolic blood pressure remains below 170 mmHg [22]. During orgasm, these values reach the highest point and then descend to within normal range values after orgasm. These changes occur similarly in both men and women [23, 24].

Experiments have demonstrated that the amount of oxygen required for myocardium during sexual intercourse is equivalent to the amount of oxygen needed to climb the stairs for two flights of stairs [25]. However, since the studies that determine these data are generally done on young and healthy population, it is difficult to obtain similar findings in elderly and unhealthy populations. People with erectile problems or people who have a sexual desire problem for any reason may need more effort to reach orgasm, which means an increase in cardiovascular burden. The same can be said for older individuals, regardless of their health status [26].

Increased cardiovascular burden during sexual activity is the result of metabolic response to sexual stimulation rather than physical effort itself. It has been determined that the energy spent during different sexual activities did not differ according to the type of sexual activity [27]. Cardiovascular events that occur during sexual activity may be due to overactivation of the sympathetic system, and those who live sedentary lives at baseline are at increased risk. For individuals with atherosclerotic coronary disease, sexual activity should be considered heavy exercise, carrying with it some risk of acute MI [28]. According to a meta-analysis on this point, sexual activity was found to increase the risk of MI (RR = 2.70; 95% CI 1.48–4.91) [29].

The risk of acute MI during sexual activity is reduced by having intercourse in a familiar environment, monogamy, no excessive eating before intercourse, and no prior alcohol intake. [30]. Sudden cardiac death rates due to sexual intercourse are low as reported in the literature. In one of the most extensive studies on this subject, only 68 deaths were detected in Berlin between 1972 and 2004 [31]. Of these 68 deaths, 92.6% were male. The most common cause of death was acute MI; 20 of these patients had coronary artery disease with no history of MI. In a similar study conducted in Korea, it was found that only 14 sudden cardiac deaths occurred during sexual intercourse in over a 4-year period [32]. Nine of these 14 people were men.

Another issue related to sexual activity in CVD patients is ED after acute MI. Studies have shown that erectile function is adversely affected in patients following acute MI. It should be noted; however, there is no definitive information about the mechanism of this pathological change [33]. Interestingly, the majority of patients with ED after MI do not have any sexual dysfunction prior to their MI.

The likely mechanism responsible for ED after acute cardiac event is the progression of existing atherosclerotic disease in the cavernosal artery [34]. According to the Thompson study, 55% of 152 patients who experienced a vascular incident (acute MI, angina, stroke, TIA, chronic heart failure, and arrhythmia) with no prior ED developed ED after their vascular incident [16]. There are insufficient data to suggest whether the cause of ED in these patients is psychological, physiological, or medication side effect. If the development of ED is truly due to advancing atherosclerotic changes already present in the cavernosal arteries, it should be kept in mind that coronary artery disease may progress and new vascular events may occur.

CVD Drugs and ED

While ED is common in men with CVD, a potentially confounding factor is that many CVD medications can cause ED as a side effect.

One of the most frequently mentioned drug classes on this subject are beta-blockers. Studies have shown that beta-blockers can lead to reduced sexual desire and ED [35]. In the most comprehensive study on this topic, 35,000 patients were evaluated, and it was reported that ED may develop with beta-blocker use [36]. Within this group of drugs, nebivolol stands out as an exception. Some studies have shown that nebivolol may preserve erectile function due to its effect on nitric oxide, thereby promoting vasodilation [37]. There are also data to suggest that erectile function may be improved by switching to nebivolol from atenolol, metoprolol, or bisoprolol [38].

Unfortunately, information on the mechanism by which beta-blockers cause ED is limited. It is speculated that beta-blockade leads to low perfusion pressure on penis and smooth muscle cells secondary to the alpha receptor stimulatory effect of the drug [39]. An associated decrease in testosterone level due to the suppression of Leydig cell activity may be the source of beta-blocker use leading to reduced sexual desire [40].

Another group of drugs that may have a negative impact on erectile function is thiazide diuretics. In a study investigating the treatment of mild hypertension, different patient groups were administered placebo, beta-blocker, calcium channel blocker, thiazide diuretic, alpha-adrenergic blocker, and angiotensin-converting enzyme (ACE) inhibitor [41]. Although all treatment agents were used at low doses, ED was more common in the diuretic group. No difference was observed between the other agents and placebo in terms of causing ED. It has not yet been elucidated how thiazide diuretics affect erectile function. It is possible that these drugs lead to contraction of penile vascular smooth muscle cells or that thiazides impact the catecholamine synthesis pathway [42].

Other agents used in the treatment of hypertension are calcium channel blockers, aldosterone receptor antagonists, angiotensin-converting enzyme inhibitors, and angiotensin receptor blockers (ARB). There are insufficient data in the current literature to show that these drug groups have negative effects on sexual function. In fact, there are studies to suggest that calcium channel blockers, ACE inhibitors, and ARBs do not adversely impact erectile function and may even show some beneficial effects [43–45].

HYPOGONADISM AND CVD

Historically, androgens were thought to increase the risk of cardiovascular disease. Studies investigating the risk of cardiovascular events of athletes and body builders with excessive exogenous testosterone use supported this view [46]. However, currently, testosterone is considered to have a protective role against cardiometabolic diseases.

Low testosterone can lead to increased body fat content and increase the risk of obesity [47]. In fact, the relationship between testosterone and fat can be considered bidirectional in that increased aromatization activity in people with a larger waist circumference decreases testosterone levels [48]. Low testosterone is also associated with the development of impaired glucose tolerance and metabolic syndrome in men. Considering the absence of a protective effect of estrogen, it is reasonable to conclude that the risk of CVD will increase in such patients [47].

In most men, serum testosterone levels decrease with age. In a meta-analysis of 70 studies, it was found that testosterone levels were lower in men with a history of CVD than in the other subjects [49]. This study showed that baseline testosterone levels were lower in patients who died from cardiovascular causes. It has also been reported that testosterone replacement therapy (TRT) improves cardiac stress tests and prolongs the time to 1 mm ST segment depression. There are, however, conflicting results in the literature about the effect of TRT on this issue. Prior meta-analyses have reported that TRT has no effect on the risk of cardiovascular events [50].

In men with high serum testosterone and free testosterone levels, the risk of abdominal aortic atherosclerosis is found to be lower than in men with lower levels of testosterone [51]. In men with heart failure, total and free testosterone levels may be lower compared with healthy individuals. There are also data to suggest that the prevalence of hypogonadism in these men is more than 40% [50].

While testosterone has historically been thought to increase the risk of CVD, the amount of evidence to suggest that testosterone, like estrogen, has a protective role against CVD is rapidly growing. Another important question requiring investigation is, does low testosterone increase the risk of CVD or does the development of CVD lead to low testosterone?

References

[1] Gandaglia G, Briganti A, Jackson G, Kloner RA, Montorsi F, Montorsi P, et al. A systematic review of the association between erectile dysfunction and cardiovascular disease. Eur Urol 2014;65(5):968–78.

[2] Fung MM, Bettencourt R, Barrett-Connor E. Heart disease risk factors predict erectile dysfunction 25 years later: the Rancho Bernardo Study. J Am Coll Cardiol 2004;43(8):1405–11.

[3] Montorsi P, Ravagnani PM, Galli S, Salonia A, Briganti A, Werba JP, et al. Association between erectile dysfunction and coronary artery disease: matching the right target with the right test in the right patient. Eur Urol 2006;50(4):721–31.

[4] Shin D, Pregenzer G, Gardin JM. Erectile dysfunction: a disease marker for cardiovascular disease. Cardiol Rev 2011;19(1):5–11.

[5] Feldman HA, Johannes CB, Derby CA, Kleinman KP, Mohr BA, Araujo AB, et al. Erectile dysfunction and coronary risk factors: prospective results from the Massachusetts male aging study. Prev Med 2000;30(4):328–38.

[6] Cooke JP. The endothelium: a new target for therapy. Vasc Med Lond Engl 2000;5(1):49–53.

[7] Jeremy JY, Angelini GD, Khan M, Mikhailidis DP, Morgan RJ, Thompson CS, et al. Platelets, oxidant stress and erectile dysfunction: an hypothesis. Cardiovasc Res 2000;46(1):50–4.

[8] Guay AT. ED2: erectile dysfunction = endothelial dysfunction. Endocrinol Metab Clin N Am 2007;36(2):453–63.

[9] Vlachopoulos C, Jackson G, Stefanadis C, Montorsi P. Erectile dysfunction in the cardiovascular patient. Eur Heart J 2013;34(27):2034–46.

[10] Montorsi P, Montorsi F, Schulman CC. Is erectile dysfunction the "tip of the iceberg" of a systemic vascular disorder? Eur Urol 2003;44(3):352–4.

[11] Katsiki N, Wierzbicki AS, Mikhailidis DP. Erectile dysfunction and coronary heart disease. Curr Opin Cardiol 2015;30(4):416–21.

[12] Jain A, Harvey D, Robertson L, Mikhailidis DP, Nair DR. Gender-based cardiometabolic risk evaluation in minority and non-minority men grading the evidence of non-traditional determinants of cardiovascular risk. Int J Clin Pract 2011;65(6):715–6.

[13] Jackson G. Erectile dysfunction and cardiovascular disease. Int J Clin Pract 1999;53(5):363–8.

[14] O'Kane PD, Jackson G. Erectile dysfunction: is there silent obstructive coronary artery disease? Int J Clin Pract 2001;55(3):219–20.

[15] Montorsi F, Briganti A, Salonia A, Rigatti P, Margonato A, Macchi A, et al. Erectile dysfunction prevalence, time of onset and association with risk factors in 300 consecutive patients with acute chest pain and angiographically documented coronary artery disease. Eur Urol 2003;44(3):360–5.

[16] Thompson IM, Tangen CM, Goodman PJ, Probstfield JL, Moinpour CM, Coltman CA. Erectile dysfunction and subsequent cardiovascular disease. JAMA 2005;294(23):2996–3002.

[17] Lee JH, Ngengwe R, Jones P, Tang F, O'Keefe JH. Erectile dysfunction as a coronary artery disease risk equivalent. J Nucl Cardiol Off Publ Am Soc Nucl Cardiol 2008;15(6):800–3.

[18] Chew K-K, Finn J, Stuckey B, Gibson N, Sanfilippo F, Bremner A, et al. Erectile dysfunction as a predictor for subsequent atherosclerotic cardiovascular events: findings from a linked-data study. J Sex Med 2010;7(1 Pt 1):192–202.

[19] Montorsi P, Ravagnani PM, Galli S, Rotatori F, Veglia F, Briganti A, et al. Association between erectile dysfunction and coronary artery disease. Role of coronary clinical presentation and extent of coronary vessels involvement: the COBRA trial. Eur Heart J 2006;27(22):2632–9.

[20] Hotaling JM, Walsh TJ, Macleod LC, Heckbert S, Pocobelli G, Wessells H, et al. Erectile dysfunction is not independently associated with cardiovascular death: data from the vitamins and lifestyle (VITAL) study. J Sex Med 2012;9(8):2104–10.

[21] Polonsky TS, Taillon LA, Sheth H, Min JK, Archer SL, Ward RP. The association between erectile dysfunction and peripheral arterial disease as determined by screening ankle-brachial index testing. Atherosclerosis 2009;207(2):440–4.

[22] Levine GN, Steinke EE, Bakaeen FG, Bozkurt B, Cheitlin MD, Conti JB, et al. Sexual activity and cardiovascular disease: a scientific statement from the American Heart Association. Circulation 2012;125(8):1058–72.

[23] Exton NG, Truong TC, Exton MS, Wingenfeld SA, Leygraf N, Saller B, et al. Neuroendocrine response to film-induced sexual arousal in men and women. Psychoneuroendocrinology 2000;25(2):187–99.

[24] Holloway IW. Substance use homophily among geosocial networking application using gay, bisexual, and other men who have sex with men. Arch Sex Behav 2015;44(7):1799–811.

[25] Bohlen JG, Held JP, Sanderson MO, Patterson RP. Heart rate, rate-pressure product, and oxygen uptake during four sexual activities. Arch Intern Med 1984;144(9):1745–8.

[26] Lindau ST, Schumm LP, Laumann EO, Levinson W, O'Muircheartaigh CA, Waite LJ. A study of sexuality and health among older adults in the United States. N Engl J Med 2007;357(8):762–74.

[27] Bispo GS, de Lima Lopes J, de Barros ALBL. Cardiovascular changes resulting from sexual activity and sexual dysfunction after myocardial infarction: integrative review. J Clin Nurs 2013;22(23–24):3522–31.

[28] Stein RA. Cardiovascular response to sexual activity. Am J Cardiol 2000;86(2A):27F–9F.

[29] Dahabreh IJ, Paulus JK. Association of episodic physical and sexual activity with triggering of acute cardiac events: systematic review and meta-analysis. JAMA 2011;305(12):1225–33.

[30] DeBusk R, Drory Y, Goldstein I, Jackson G, Kaul S, Kimmel SE, et al. Management of sexual dysfunction in patients with cardiovascular disease: recommendations of The Princeton Consensus Panel. Am J Cardiol 2000;86(2):175–81.

[31] Parzeller M, Bux R, Raschka C, Bratzke H. Sudden cardiovascular death associated with sexual activity: a forensic autopsy study (1972-2004). Forensic Sci Med Pathol 2006;2(2):109–14.

[32] Lee S, Chae J, Cho Y. Causes of sudden death related to sexual activity: results of a medicolegal postmortem study from 2001 to 2005. J Korean Med Sci 2006;21(6):995–9.

[33] Hardin SR. Cardiac disease and sexuality: implications for research and practice. Nurs Clin North Am 2007;42(4):593–603. vii.

[34] Montorsi P, Ravagnani PM, Vlachopoulos C. Clinical significance of erectile dysfunction developing after acute coronary event: exception to the rule or confirmation of the artery size hypothesis? Asian J Androl 2015;17(1):21–5.

[35] Baumhäkel M, Schlimmer N, Kratz M, Hackett G, Hacket G, Jackson G, et al. Cardiovascular risk, drugs and erectile function–a systematic analysis. Int J Clin Pract 2011;65(3):289–98.

[36] Ko DT, Hebert PR, Coffey CS, Sedrakyan A, Curtis JP, Krumholz HM. Beta-blocker therapy and symptoms of depression, fatigue, and sexual dysfunction. JAMA 2002;288(3):351–7.

[37] Brixius K, Middeke M, Lichtenthal A, Jahn E, Schwinger RHG. Nitric oxide, erectile dysfunction and beta-blocker treatment (MR NOED study): benefit of nebivolol versus metoprolol in hypertensive men. Clin Exp Pharmacol Physiol 2007;34(4):327–31.

[38] Doumas M, Tsakiris A, Douma S, Grigorakis A, Papadopoulos A, Hounta A, et al. Beneficial effects of switching from beta-blockers to nebivolol on the erectile function of hypertensive patients. Asian J Androl 2006;8(2):177–82.

[39] Shiri R, Koskimäki J, Häkkinen J, Auvinen A, Tammela TLJ, Hakama M. Cardiovascular drug use and the incidence of erectile dysfunction. Int J Impot Res 2007;19(2):208–12.

[40] Boydak B, Nalbantgil S, Fici F, Nalbantgil I, Zoghi M, Ozerkan F, et al. A randomised comparison of the effects of nebivolol and atenolol with and without chlorthalidone on the sexual function of hypertensive men. Clin Drug Investig 2005;25(6):409–16.

[41] Grimm RH, Grandits GA, Prineas RJ, McDonald RH, Lewis CE, Flack JM, et al. Long-term effects on sexual function of five antihypertensive drugs and nutritional hygienic treatment in hypertensive men and women Treatment of Mild Hypertension Study (TOMHS). Hypertension 1997;29(1 Pt 1):8–14.

[42] Chrysant SG. Antihypertensive therapy causes erectile dysfunction. Curr Opin Cardiol 2015;30(4):383–90.

[43] Omvik P, Thaulow E, Herland OB, Eide I, Midha R, Turner RR. Double-blind, parallel, comparative study on quality of life during treatment with amlodipine or enalapril in mild or moderate hypertensive patients: a multicentre study. J Hypertens 1993;11(1):103–13.

[44] Speel TGW, Kiemeney LA, Thien T, Smits P, Meuleman EJ. Long-term effect of inhibition of the angiotensin-converting enzyme (ACE) on cavernosal perfusion in men with atherosclerotic erectile dysfunction: a pilot study. J Sex Med 2005;2(2):207–12.

[45] Böhm M, Baumhäkel M, Teo K, Sleight P, Probstfield J, Gao P, et al. Erectile dysfunction predicts cardiovascular events in high-risk patients receiving telmisartan, ramipril, or both: the ONgoing Telmisartan Alone and in combination with Ramipril Global Endpoint Trial/Telmisartan Randomized AssessmeNt Study in ACE iNtolerant subjects with cardiovascular Disease (ONTARGET/TRANSCEND) Trials. Circulation 2010;121(12):1439–46.

[46] Cheever K, House MA. Cardiovascular implications of anabolic steroid abuse. J Cardiovasc Nurs 1992;6(2):19–30.

[47] Cattabiani C, Basaria S, Ceda GP, Luci M, Vignali A, Lauretani F, et al. Relationship between testosterone deficiency and cardiovascular risk and mortality in adult men. J Endocrinol Investig 2012;35(1):104–20.

[48] Snyder PJ. Decreasing testosterone with increasing age: more factors, more questions. J Clin Endocrinol Metab 2008;93(7):2477–8.

[49] Corona G, Rastrelli G, Monami M, Guay A, Buvat J, Sforza A, et al. Hypogonadism as a risk factor for cardiovascular mortality in men: a meta-analytic study. Eur J Endocrinol 2011;165(5):687–701.

[50] Corona G, Rastrelli G, Vignozzi L, Mannucci E, Maggi M. Testosterone, cardiovascular disease and the metabolic syndrome. Best Pract Res Clin Endocrinol Metab 2011;25(2):337–53.

[51] Hak AE, Witteman JCM, de Jong FH, Geerlings MI, Hofman A, Pols HAP. Low levels of endogenous androgens increase the risk of atherosclerosis in elderly men: the Rotterdam study. J Clin Endocrinol Metab 2002;87(8):3632–9.

PART 3

Mental Health

CHAPTER 3.1

Sleep, Shift Work, and Men's Health

Jorge Rivera Mirabal*, Mohit Khera*, Alexander W. Pastuszak[†]

*Scott Department of Urology, Baylor College of Medicine, Houston, TX, United States, [†]Division of Urology, Department of Surgery, University of Utah School of Medicine, Salt Lake City, UT, United States

INTRODUCTION

Approximately 50–70 million Americans suffer from a sleep disorder [1]. Sleep disorders have been linked to increased risk for chronic illnesses including diabetes mellitus (DM), hypertension (HTN), cardiovascular disease (CVD), and obesity [1–5]. There is a link between sleepiness/fatigue and worse job outcomes, including loss of productivity and more accidents [6, 7]. Annually, an estimated loss of about 8 workdays in individuals with insomnia has been identified, even when these individuals showed up to work [6]. In addition, employees sleeping <5 h per night have a 2.65-fold increased chance of work-related injury compared with peers who slept between 7 and 8 h per night [7].

More than 21 million members of the US labor force work nonstandard shifts outside of a 6 a.m.–6 p.m. work period [8]. Male nonstandard shift workers account for 19% of all employed males. A nonstandard shift work schedule is associated with decreased sleep time and predisposes workers to excessive sleepiness, impaired function, and disturbed sleep, which has profound health-care and financial ramifications. A report by the National Commission on Sleep Disorders Research noted indirect losses of upward of $70 billion annually due to accidents related to insomnia [9]. Up to 20% of male shift workers may suffer from an entity called shift work sleep disorder (SWSD) [10]. The International Classification of Sleep Disorders defines SWSD as "insomnia or excessive sleepiness due to decreased total sleep time as a result of work schedule for a minimum of 3 months, causing significant distress or impairment in mental, physical, or social functioning" [11].

SWSD has been associated with an increased risk of numerous medical conditions, including gastric ulcers, HTN, and major depressive disorder

(MDD) [12]. Various studies have also identified links between urological conditions, men's health, and sleep disorders [12–17]. Shift workers appear to have a notable incidence of lower urinary tract symptoms (LUTS) [14, 16], hypogonadal symptoms/sexual dysfunction [15], and elevated prostate-specific antigen (PSA) [17]. Semen parameter alterations are seen in work overexertion [18]. Gonadotropin dysregulation is appreciated in nonstandard shift work [19]. In addition, there have been studies evaluating the relationship between shift work and incidence of prostate cancer [20, 21].

This chapter aims to discuss the relationship between sleep disorders, shift work, and urological conditions. We will start by discussing shift work and its effect on overall health and chronic medical conditions, followed by a review of urological conditions associated with sleep disorders and shift work.

SLEEP, SHIFT WORK, INSOMNIA AND OCCUPATIONAL HEALTH

Sleep disorders are closely linked to shift work, as seen in Fig. 1 [22]. Yong et al. found that night shift workers who worked between 5 p.m. and 8 a.m. had a 1.5-fold increase in reported poor sleep compared with daytime workers who worked between 6 a.m. and 6 p.m. [23]. The authors also found that night shift workers had a 1.4-fold increase in impaired sleep-related activities of daily living such as being able to concentrate or remember events, eating, driving, paying bills, maintaining a telephone conversation, or working on a hobby. In addition to its relationship with poor sleep, shift work has also been increasingly related to workplace accidents. Kecklund performed a literature review on the effect of shift work on occupational hazards, finding a 1.2- to 1.3-fold increased risk of workplace accidents among night shift workers and rotating shift workers compared with daytime workers [24].

A similar trend is also present in health care. An article from the Joint Commission Journal on Quality and Patient Safety reported a threefold increased risk of medical errors and a higher needle stick injury rate in nurses that work ≥ 12.5 h shifts compared with nurses who did not work extended shifts. This report also analyzed resident physician errors and work-related injuries as a function of work schedule. Residents who worked 24 h shifts or longer committed 36% more serious medical errors had a higher needle stick injury rate and had 2.3 times more automobile accidents than residents who worked 16 h shifts [25]. These findings underscore the negative impact of shift work in occupational health.

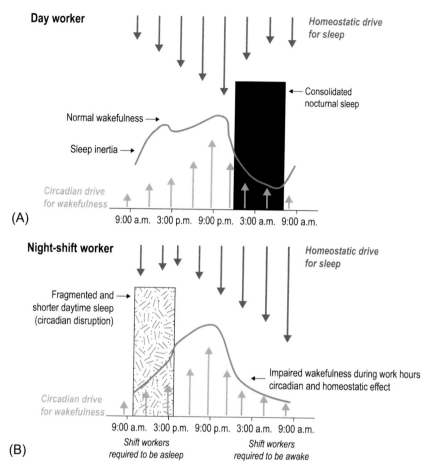

FIG. 1
Interplay between arousal and drive for sleep in (A) day workers and (B) night shift workers. *Reproduced with permission from Wright KP, Bogan RK, Wyatt JK. Shift work and the assessment and management of shift work disorder (SWD). Sleep Med Rev 2013;17:41–54. https://doi.org/10.1016/j.smrv.2012.02.002.*

SLEEP, SHIFT WORK AND CHRONIC ILLNESS

Currently, there is no established explanation as to why shift workers have an increased incidence of chronic illness. As a possible explanation, Puttonen and colleagues suggested a combination of psychosocial, behavioral, and physiological mechanisms to explain the link between shift work and chronic diseases [26]. From a psychosocial standpoint, shift workers can suffer from increased stress, problems with work-life balance and more difficult recovery from work, particularly in those working nonstandard shifts. Shift workers are more prone to smoking, physical inactivity, and poor nutrition than nonshift workers.

When considering physiological elements, shift work may result in altered inflammation, coagulation, autonomic function, and changes in the hypothalamic-pituitary-adrenal axis and blood pressure. None of these have been consistently linked to shift work, and a causal relationship has been hard to define.

From a population standpoint, some of the most common medical conditions that affect men, such as HTN, obesity, DM, and CVD, have been increasingly linked to sleep disorders and shift work [24]. Gan et al. performed a meta-analysis of 226,652 shift workers and the association with DM [2]. Men and women that participated in rotating shift work had an increased risk of having DM compared with nonshift workers. Subgroup analysis by sex showed that male shift workers had a 1.37-fold increased risk of DM compared with male nonshift workers, adjusted for family history and occupation. In addition to shift work, other studies have evaluated the association between sleep duration and DM; this may be mediated by effects of sleep deprivation on insulin resistance, as seen in Fig. 2 [27]. A meta-analysis performed by Shan et al. assessed the dose-response relationship between sleep duration and risk of type 2 DM in 482,502 men and women [28]. The meta-analysis comprised 10 studies and a median follow-up time of 7.5 years. Sleep duration of <5 h was associated with a 1.37-fold increased risk of DM compared with those sleeping 7 h. Interestingly, the authors also observed a 1.40-fold increased risk of DM in individuals who slept 9 h or more compared with those sleeping 7 h, suggesting that both long and short sleep duration may be risk factors for DM as well.

In addition to DM, there is research evaluating the link between HTN and shift work. A cross-sectional study examined the association between HTN and shift work in 25,343 German auto workers [29]. Hypertension was more prevalent among rotating shift and night shift workers (11.5% and 11.0%, respectively) compared with daytime workers (7.8%). After adjusting for confounders such as age, gender, socioeconomic status, noise exposure, and heat exposure at work, the authors found that rotating shift work resulted in a higher risk of HTN (OR 1.15, 95% CI 1.02–1.30) when compared with daytime work. Another study included the roles of race and HTN among shift workers. A 2010 National Health Interview Survey reported that black shift workers had an increased likelihood of reporting a HTN diagnosis compared with black nonshift workers (OR 1.35, 95% CI 1.06–1.72) [3]. This association was stronger in black shift workers who slept <6 h (OR 1.81, 95% CI 1.29–2.54). While various studies have proposed a link between HTN and shift work, a causal relationship between these variables has not been determined.

The link between CVD and shift work has been examined in prospective and case–control studies. Wang et al. performed a cohort study that described a threefold increased incidence of acute myocardial infarction (MI) among

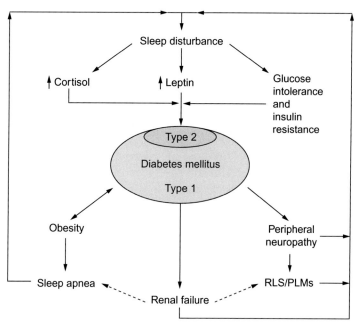

FIG. 2

Sleep disorders and diabetes mellitus. *Solid arrows* indicate strong relationships. *Dashed arrows* demonstrate relationships that are still under investigation. *PLMs*, periodic limb movements; *RLS*, restless legs syndrome. *Reproduced with permission from George CFP, Avidan AY. Diabetes mellitus. In: Kryger M, editor. Atlas of clinical sleep medicine. 2nd ed. Philadelphia: Elsevier Sanders; 2013. p. 346–9. https://doi.org/10.1016/B978-0-323-18727-5.00015-X.*

men with preexisting ischemic heart disease who spent the night working away from home more than three times per week [30]. A case-control study from Sweden evaluated the interaction between shift work, parental history of CVD, and occurrence of MI [31]. It was interesting to note that the authors evaluated the combination of shift work and other risk factors for CVD. Shift work, as defined by fixed work between 6 p.m. and 6 a.m. or rotating shift work 5 years before the study, compared with daytime work resulted in an increased risk of CVD (OR 1.39, 95% CI 1.10–1.74). Paternal history of MI before age 65 versus no history in daytime workers had a comparably increased risk (OR 1.55, 95% CI 1.14–2.09). An even higher risk for both exposures was observed (OR 2.04, 95% CI 1.04–4.00) that was greater than the effect from the individual components. Based on this, the authors concluded that 40% of MI events in the study resulted from this interaction. Other studies have attempted to assess the role of sleep duration in men and women with a recent cardiovascular event. A study by Barger et al. evaluated risk factors for major coronary events (coronary heart disease death, MI, or urgent coronary revascularization for myocardial ischemia) within 30 days of an acute coronary syndrome diagnosis [5]. The authors found that after multivariable adjustment, patients who reported sleeping fewer than 6 h per night had a 29% higher risk of a major coronary event than those who slept >6 h per night. In addition, those patients who worked overnight shifts for at least 1 year had a 15% increased risk of a major coronary event compared with those who did not work overnight shifts. These studies highlight a strong relationship between CVD and sleep disorders/shift work and demonstrate the possible synergy between established cardiac risk factors and sleep disorders/shift work in contributing to cardiovascular events.

Obesity is one of the most prevalent diseases in the United States, affecting approximately 35% of men [32]. As obesity was recognized as a disease by the American Medical Association in 2013 [33], it is imperative to identify new risk factors that could aid in decreasing the incidence and impact of this condition, which carries significant morbidity. Kawabe et al. examined four components (blood pressure >130/85 mmHg, triglycerides >150 mg/dL or high-density lipoprotein (HDL) <40 mg/dL, blood glucose >110 mg/dL, and BMI >25 kg/m^2) of metabolic syndrome derived from data obtained from 4227 participants, of which 81% were male [34]. The authors found that fixed night time work and shift work contributed to an increase in the number of metabolic syndrome components compared with daytime work. A cross-sectional study analyzed the prevalence of obesity in 1206 poultry processing plant employees 18–50 years old in Brazil and compared it among night and day shift workers [4]. Night shift workers had a higher prevalence of overweight (42.2% vs 34.3%; $P=.020$) or obesity (24.9% vs 19.5%; $P=.037$) compared with day shift workers. After controlling for sociodemographic factors, parental overweight status, behavioral characteristics, and sleep characteristics, night shift workers had a higher likelihood of being obese (OR 1.45, 95% CI 1.10–1.92). Grundy et al. published a case-control study that evaluated the relationship between shift work and obesity among 1622 men [35]. The authors concluded that rotating shift work was associated with both being overweight (OR 1.34, 95% CI 1.05–1.73) and obese (OR 1.57, 95% CI 1.12–2.21). Of note, evening shift work was not associated with an increased risk of being overweight or obese compared with day work.

SLEEP DISORDERS, SHIFT WORK AND ERECTILE DYSFUNCTION

Erectile dysfunction (ED) affects 18.4% of the adult US male population [36]. There is interest in evaluating the role of obstructive sleep apnea (OSA), sleep impairment, and nonapnea sleep disorders as risk factors for ED. A clinical review by Pepin et al. reviewed the pathogenic mechanisms of OSA and its impact on ED [37]. The authors concluded that there is blunted endothelial cell-dependent vasodilation of large and small blood vessels in OSA patients [38], that there are clinical trials demonstrating that alleviation of OSA can facilitate repair of the endothelium with subsequent vasodilation, and that endothelium-independent dysfunction is not commonly observed in men with OSA. It also seems that vascular damage because of OSA could take years to occur. This is clinically relevant as men with both OSA and ED have a higher risk of CVD due to endothelial dysfunction. Hoyos et al. evaluated the causal relationship of OSA on ED and how treatment options for OSA can improve erectile function [39]. The authors reported that there is sufficient clinical trial evidence to conclude that OSA causes endothelial impairment that can be reversed by continuous positive airway pressure (CPAP) [40]. CPAP use alone improves ED in men with OSA, albeit to a lesser degree than phosphodiesterase type 5 (PDE5) inhibitors [41]. These studies highlight the importance of understanding the pathophysiology of OSA and ED, as OSA is a common condition among patients presenting for urological care. Mcbride et al. evaluated the risk of OSA in 312 men presenting to a high-volume andrology clinic [42]; 27% of men were deemed high risk for OSA and referred for polysomnography (PSG). Of those men who underwent PSG, 80% were diagnosed with OSA. Due to the physiological relationship and prevalence of OSA in men with ED, evaluation for OSA in this patient population is reasonable.

In addition to the role of sleep apnea in ED, there has been interest in evaluating the risk of ED in patients with nonsleep apnea sleep disorders. Chen et al. published a 13-year longitudinal population-based cohort study to investigate whether sleep disorders increase the risk of ED [43]. The examined cohorts consisted of 603 men with sleep apnea, 17,182 with nonapnea sleep disorders, and 35,570 matched comparisons as controls. The authors found that men with sleep apnea and those with nonapnea sleep disorder had 9.44-fold and 3.72-fold increased risks of developing ED compared with controls when adjusting for age and comorbidities. Currently, the pathophysiological mechanisms resulting in ED in nonsleep apnea sleep disorders are poorly defined. However, multiple mechanisms have been proposed, including lower levels of luteinizing hormone and testosterone, increased sympathetic activity, peripheral nerve dysfunction, decreased nitric oxide levels, and increased production of endothelin [44].

The clinical impact of poor sleep on ED has also been studied. Kohn et al. analyzed the impact of poor sleep quality on ED using the validated International Index of Erectile Function (IIEF), Patient Health Questionnaire 2 (PHQ-2), Pittsburgh Sleep Quality Index (PSQI), Sleep Hygiene Index (SHI), and STOP-BANG in a group of 377 men [45]. The authors found that men had worse erectile function as global sleep quality decreased. A score of 6 on the PSQI, which is considered poor sleep, was associated with a 2.22-point decrease in IIEF-EF score. Interestingly, improved SHI was associated with improved erectile function ($P<.02$). This last point is intriguing as improving sleep hygiene adds a possible behavioral therapy to improve ED in men who suffer from sleep problems. Pastuszak et al. performed a cross-sectional study evaluating the relationship between poor sleep quality and ED among nonstandard shift workers [15]. The authors found that worse sleep quality was associated with decreased sexual function as measured using the IIEF. This association was also present among all IIEF domains (erectile function, orgasmic function, sexual desire, intercourse satisfaction, and overall satisfaction). Rodriguez et al. explored a specific relationship between men with SWSD and ED, by examining a subset of 2517 shift workers, 196 of whom had SWSD [46]. The authors found that men who worked night shifts had IIEF scores that were 7.58 points lower on average than men who worked during the day or evening. In addition, men who suffered from SWSD had IIEF scores that were 2.8 points lower than those from men without SWSD ($P<.01$). These results controlled for testosterone use, testosterone levels, PDE5 inhibitor use, age, and comorbidities. Of note, the authors also found that testosterone use improved IIEF scores for men with SWSD by 2.90 points.

SLEEP DISORDERS, SHIFT WORK AND HYPOGONADISM

Hypogonadism affects about 4 million men, with only about 5% of those receiving treatment [47]. There have been efforts to understand the physiological impact of low testosterone on sleep, as sleep restriction may be related to lower testosterone levels, as seen in Fig. 3 [48]. Miller et al. performed a literature review of the possible mechanisms between low testosterone and sleep dysfunction [49]. The authors reported that testosterone inhibits and is inversely correlated with cortisol, which promotes wakefulness, increases with age, and impairs sleep in the elderly. Jensen et al. evaluated the diurnal rhythms of melatonin, cortisol, and testosterone and how these respond to different numbers of consecutive night shifts [50]. A cohort of 73 police officers participated in this study. They were asked to perform 2, 4, and 7 consecutive night shifts with that same amount of recovery days (defined as days off and/or day shifts). Lab results were obtained at the end of the night shift period and recovery day period. The authors found that the lowest cortisol value was delayed by 33 min per night shift and melatonin concentration decreased further with more consecutive night shifts. Testosterone diurnal rhythm did not vary between the night shift groups. When comparing the last recovery day in each intervention, the authors found no significant difference in the rhythms of melatonin, cortisol, and testosterone between the interventions. Leibenluft et al. evaluated leuprolide administration and testosterone replacement to study the effects of hypogonadism and gonadal steroid replacement in men [51]. The authors gauged timing, duration, and microarchitecture of sleep

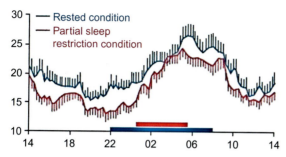

FIG. 3

Mean (± standard error mean) testosterone levels after resting condition (bedtimes from 22:00 to 8:00 for three nights, in *blue*) and after sleep restriction (bedtimes from 00:30 to 05:30 for seven nights, in *red*). Reproduced with permission from Leproult R, van Cauter E. Endocrine physiology. In: Kryger M, editor. Atlas of clinical sleep medicine. 2nd ed. Philadelphia: Elsevier Sanders; 2013. p. 58–64. https://doi.org/10.1016/B978-0-323-18727-5.00003-3. Data obtained from Leproult R, Van Cauter E. Effect of 1 week of sleep restriction on testosterone levels in young healthy men. JAMA 2011;305(21):2173–4.

(recorded using electroencephalography) and 24h secretory profiles of melatonin and prolactin (PRL). After 1 month of leuprolide alone, testosterone (testosterone enanthate 200mg IM every 2weeks) and placebo injections were administered for 1 month each in a double-blind, randomized, crossover design. The authors found that pharmacologically induced hypogonadism in men caused significant decreases in the amount of stage 4 sleep (mean: 25.27min vs 13.17min; $P=.05$) and in PRL levels (mean: 13.4 ± 9.9 ng/mL vs 7.8 ± 5.6 ng/mL, $P=.02$) compared with measures taken during testosterone replacement.

As changes in sleep affect hormone secretion patterns, appreciated in Fig. 4 [48], it is important to further understand the relationship between sleep disorders and hypogonadism. Shigehara et al. performed a subgroup analysis of the *effects of long-term androgen replacement therapy on the physical and mental statuses of aging males with late-onset hypogonadism (EARTH) trial* with regard to the efficacy of testosterone therapy (TTh) on sleep disturbance and quality of life among hypogonadal men without OSA [52]. Subjects received 250 mg testosterone enanthate every 4 weeks. Results were recorded as change from baseline questionnaires (aging male symptom questions 4 and 7 for sleep symptom assessment, Sexual Health Inventory for Men (SHIM), International Prostate Symptom Score (I-PSS), and Short Form 36). The authors used a serum testosterone threshold of <8.5 pg/mL as the biochemical definition of hypogonadism and threshold for treatment initiation. A total of 48 patients with sleep disturbances were divided into two groups—24 who received treatment for a year and 24 who did not. No significant differences in baseline characteristics,

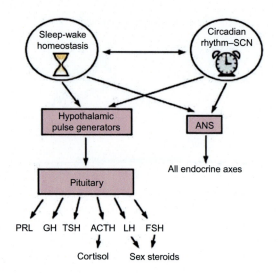

FIG. 4

Relationship between the sleep wake cycle and pituitary hormone secretions over a 24-h cycle. *ACTH*, adrenocorticotropic hormone; *ANS*, autonomic nervous system; *FSH*, follicle-stimulating hormone; *GH*, growth hormone; *LH*, luteinizing hormone; *PRL*, prolactin; *SCN*, suprachiasmatic nuclei; *TSH*, thyroid-stimulating hormone. *Reproduced with permission from Leproult R, van Cauter E. Endocrine physiology. In: Kryger M, editor. Atlas of clinical sleep medicine. 2nd ed. Philadelphia: Elsevier Sanders; 2013. p. 58–64. https://doi.org/10.1016/B978-0-323-18727-5.00003-3.*

other than higher physical functioning in the TTh group, were observed between the two groups. After 1 year of treatment, significant improvements in sleep, erectile function, and sexual desire in the treatment versus no treatment group were observed. Ismailogullari and colleagues compared sleep stages and sleep-breathing parameters during PSG in male patients with idiopathic hypogonadotropic hypogonadism (IHH) to age-matched controls with normal pubertal development [53]. They also investigated the impact of 12-month treatment with gonadotropins using human chorionic gonadotropic (hCG) for 6 months followed by a combination of HCG and follicle-stimulating hormone (FSH) for 6 months on sleep parameters. The proportion of slow-wave sleep (SWS), which is considered restorative sleep, was higher in the IHH group compared with controls (22.3 vs 14.5%; $P=.009$). After treatment, the IHH group had a significant increase in sleep efficiency ($82.6 \pm 11.6\%$ at baseline vs $87.8 \pm 8.9\%$ at follow-up; $P=.013$) and decrease in SWS proportion, closer to that of controls ($22.3 \pm 6.3\%$ at baseline vs $18.6 \pm 6.4\%$ at follow-up; $P=.05$). A recent literature review evaluated the relationship between testosterone deficiency and nocturia [54]. The authors found that there is some evidence supporting a feedback cycle in hypogonadal men with nocturia, in which testosterone deficiency leads to the development of nocturia, and nocturia contributes to a decline in testosterone levels.

Based on this idea, TTh may also improve nocturia in hypogonadal men. However, there are limited clinical data that evaluate the effects of TTh on nocturia in hypogonadal men.

Pastuszak and colleagues have also sought to understand the relationship between sleep quality and hypogonadism among nonstandard shift workers [15]. The authors used the validated androgen deficiency in the aging male (ADAM) questionnaire when interviewing men and obtained serum testosterone (T), free testosterone (FT), estradiol (E), dehydroepiandrosterone (DHEA), luteinizing hormone (LH), and follicle-stimulating hormone (FSH) levels concurrently. The authors performed multivariate regression analysis that demonstrated significant linear associations between sleep quality and hypogonadal symptoms. This was in spite of the fact that overall serum hormone levels were comparable between men working standard and nonstandard shifts. According to the study, those who reported being "somewhat dissatisfied" with their sleep quality had lower ADAM scores than those who were "very satisfied" with their sleep quality.

New studies have tried to shed light on the role of OSA in hypogonadism. Kim and Cho reviewed recent literature on the relationship between OSA and low testosterone [55]. The authors found articles showing that the apnea-hypopnea index (AHI) is associated with decreased testosterone levels in men with OSA [56–58]. Hammoud and colleagues analyzed the effect of sleep apnea on reproductive hormone levels and sexual function in obese men recruited in the Utah obesity study [59]. After controlling for BMI and age, a significant inverse relationship between severity of sleep apnea and free testosterone levels was observed (mild sleep apnea 74.4 ± 3.8 pg/mL, moderate sleep apnea 68.6 ± 4.2 pg/mL, and severe sleep apnea 60.2 ± 2.92 pg/mL, $P=.014$). Since then, other authors have evaluated the effect of TTh on OSA and CPAP treatment on hypogonadism. Cole et al. compared various clinical end points in a population of 3422 men, including risk for OSA, between a cohort of middle-aged men using TTh for low testosterone levels and a cohort of age- and comorbidity-matched controls [60]. Results showed that the risk of developing OSA was higher among men on TTh, with a 2-year absolute risk of 16.5% (95% CI 15.1%–18.1%) compared with 12.7% (95% CI 11.4%–14.2%) among controls. Even though there may be a possible relationship between TTh and OSA, CPAP treatment does not seem to impact hypogonadism. Vloka and colleagues compared sex hormone levels in saliva before and after a night without and with CPAP in 42 patients (28 men) with OSA [61]. No significant differences in sex hormones were observed when comparing both groups. Knapp et al. performed a similar study among men with type 2 DM and OSA being treated with CPAP for 3 months, with no significant changes in serum testosterone or free testosterone observed after treatment [62].

SLEEP DISORDERS, SHIFT WORK AND LOWER URINARY TRACT SYMPTOMS

Lower urinary tract symptoms (LUTS) is a term coined by Abrams in 1994 to describe a collection of symptoms related to the lower urinary tract without implying the direct cause [63]. In the United States, approximately 97% of men older than 65 years of age suffer from LUTS [64]. Due to the prevalence of this constellation of symptoms, there has been interest in establishing the relationship between LUTS and sleep disorders. Fantus and colleagues evaluated the National Health and Nutrition Examination Survey data set to verify the relationship between LUTS and sleep disorders in 3071 community-dwelling men [65]. When controlling for age, BMI, C-reactive protein, and medical comorbidities, men with sleep disorders had increased odds of having two or more urinary symptoms (OR 1.128, 95% CI 1.125–1.131), nocturia (OR 1.278, 95% CI 1.275–1.281), and daytime LUTS (OR 1.269, 95% CI 1.267–1.272, all $P<.0001$) when compared with men with no documented history of sleep disorders.

Branche and colleagues performed a post hoc analysis of *reduction* by dutasteride of prostate cancer events (REDUCE), a prospective randomized trial of dutasteride compared with placebo for prostate cancer prevention in men with elevated PSA and negative biopsy to further investigate the relationship between sleep and LUTS [66]. Men were administered the 6-item Medical Outcomes Study Sleep Scale survey (MOS-Sleep; a higher score means greater sleep disturbance) at baseline and were followed every 6 months for 4 years to monitor LUTS. Men with I-PSS scores <8 were followed to determine LUTS development, and men with I-PSS scores of 8 or greater were followed to determine LUTS progression. The authors defined LUTS development as 2 I-PSS total scores >14 or medical or surgical treatment for benign prostatic hyperplasia (BPH). LUTS progression was defined as an I-PSS score increase of 4 points, any surgical procedure for BPH, or the start of a new BPH drug. LUTS developed in 209 of 1452 initially asymptomatic men (14%) and 580 of 1136 (51%) with LUTS initially demonstrated progression. When the authors performed multivariable analysis, higher sleep scores were correlated with and increased likelihood of LUTS development in asymptomatic men (HR 1.41, 95% CI 0.92–2.17, $P=.12$), although this did not achieve statistical significance. The only MOS-Sleep questions that were correlated with increased likelihood of LUTS were trouble falling asleep (HR 1.07, 95% CI 1.01–1.13, $P=.020$) and awakening during sleep time and having trouble falling back asleep (HR 1.08, 95% CI 1.02–1.14, $P=.010$). The authors did find that a higher sleep score was associated with and increased risk of LUTS progression (HR 1.06, 95% CI 1.01–1.12, $P=.029$). In order to investigate if LUTS have an impact on sleep quality, Helfand and colleagues analyzed bother and severity of LUTS secondary to BPH and the severity of sleep disturbance in cohorts of the Complementary and Alternative Medicine for Urological Symptoms (CAMUS) trial [67].

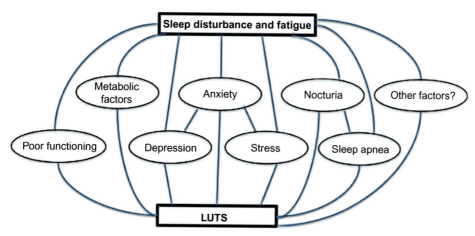

FIG. 5
Interplay between lower urinary tract symptoms (LUTS) and sleep disorders. *Modified from Ge TJ, Vetter J, Lai HH. Sleep disturbance and fatigue are associated with more severe urinary incontinence and overactive bladder symptoms. Urology 2017;109:67–73. https://doi.org/10.1016/j.urology.2017.07.039.*

The baseline characteristics of the 339 men were similar, and these were followed for 72 weeks. Multivariate analyses demonstrated that changes in total I-PSS score and nocturia score were highly correlated with changes in the Jenkins sleep scale and its component questions on sleep abilities. The study also found that improvements in I-PSS questions 1–6 were associated with more improvements in the Jenkins sleep score than improvements in nocturia alone (question 7), as a 3-point improvement in I-PSS was associated with a 0.73-point improvement in the Jenkins sleep scale.

In order to objectively assess the impact of sleep dysfunction on LUTS, Matsushita and colleagues verified the relationship between sleep quality and LUTS using electroencephalography (EEG) in 47 men [68]. Results indicated that better sleep quality was associated with higher LUTS quality of life and longer slow-wave sleep. Multivariate analysis demonstrated that lower nocturnal urinary volume was an independent variable predictive of higher sleep efficiency. With regard to the EEG, multivariate analyses revealed a relationship between good sleep quality, increased maximum urinary flow (Q_{max}), and longer SWS time. With regard to the component of nocturia, Shao and colleagues performed a cross-sectional study that observed a higher number of nocturia events that were significantly correlated with higher sleepiness score [69].

As we better understand the relationship between LUTS and sleep, studies have tried to assess if LUTS affects sleep, if sleep affects LUTS, or if both affect each other, as seen in Fig. 5 [70]. Araujo et al. evaluated data from a prospective cohort of 1610 men and 2535 women over a 5-year period [71]. The authors assessed the bidirectional association between LUTS and sleep-related variables. After controlling for possible confounders, subjects with self-reported poor sleep had increased odds of LUTS (OR 1.73, 95% CI 1.20–2.51), including irritative (OR 1.55, 95% CI 1.06–2.27) and obstructive (OR 2.15, 95% CI 1.39–3.32) LUTS. These subjects also had an increased likelihood of nocturia (OR 1.42, 95% CI 1.02–1.97). Subjects with sleep restriction (<5 h/night) were more likely to report LUTS (OR 1.98, 95% CI 1.03–3.78), including increased irritative symptoms (OR 2.58, 95% CI 1.41–4.73) but not obstructive symptoms. When evaluating the effect of LUTS on sleep, subjects with obstructive symptoms had a lower likelihood of sleep medication use (OR 0.59, 95% CI 0.35–0.99) but a higher likelihood of having suffered from sleep restriction (OR 2.20, 95% CI 1.11–4.33). Subjects with nocturia were more likely to have poor sleep quality (OR 1.98, 95% CI 1.38–2.85). This study supports the argument that LUTS and sleep are interrelated, with each variable affecting the other.

Even though there are multiple studies evaluating the relationship between sleep quality and LUTS, there are few publications that evaluate the formal effect of shift work on LUTS. Scovell et al. evaluated the relationship between sleep quality and severity of LUTS in male nonstandard shift workers [16]. The authors examined 228 nonstandard shift workers using an unvalidated survey that asked about the subjects' work patterns. Subjects were also asked to complete the I-PSS questionnaire and answer questions about quality of life and cognitive function. The authors found a direct correlation between worse LUTS severity and difficulty staying asleep or falling back asleep ($P<.001$). In addition, men who reported a decreased sense of well-being or decreased physical and cognitive function had more severe LUTS than men who did not report such changes ($P<.001$ and $P=.016$, respectively).

LUTS have also been linked to obstructive sleep apnea (OSA). Martin et al. examined the association between nocturia (with and without other LUTS) and sleep disorders as assessed using overnight PSG testing in a large cohort of community-dwelling middle-aged and elderly men [72]. One thousand four hundred forty-five men without a previous diagnosis of OSA were invited to participate in an in-home PSG testing session; 837 agreed and successfully recorded the PSG. Moderate to severe OSA was identified in 32.2 and 21.5% of men with and without nocturia, respectively ($P=.003$). Nocturia was associated with increased daytime sleepiness and decreased sleep quality when controlling for potential confounders. To better understand the pathophysiology behind this association, Abler and Vezina used mice to study the effect of intermittent hypoxia on bladder tissue [73]. Hematoxylin and eosin staining of bladder sections revealed microanatomical changes in the bladder urothelium following intermittent hypoxia, including edema and the presence of enlarged urothelial cells. In addition, the total number of voids and bladder weight increased in mice exposed to intermittent hypoxia compared with mice exposed only to room air. Yilmaz et al. evaluated bladder wall thickness in men and women with OSA and controls using ultrasonography [74]. While there were no significant differences in bladder wall thickness between OSA patients and controls, a significant increase in bladder wall thickness of OSA patients who had worse overactive bladder symptom scores (OABSS; $P=.002$) was observed.

As there appears to be a strong relationship between OSA and LUTS, particularly in the setting of nocturia, some studies have evaluated whether treatment of OSA using CPAP improves nocturia. Miyauchi et al. performed a prospective study measuring the effect of CPAP treatment on nocturnal urine volume, night time urine frequency, and quality of life (QOL) in patients with OSA [75]. The authors excluded patients with active urinary tract infection; urolithiasis; interstitial cystitis; and treated with alpha-blockers, 5α-reductase inhibitors, or anticholinergic agents within 4 weeks prior to study initiation. They also excluded subjects who were treated with diuretics for congestive heart failure or renal failure and patients with an elevated hemoglobin A1c. Night time urinary frequency and urine volume were significantly decreased after 1 month of CPAP treatment, resulting in significant improvement in total I-PSS and QOL score. The remaining I-PSS questions were not significantly changed after treatment. Park et al. evaluated changes in nocturia episodes after surgical therapy (uvulopalatopharyngoplasty) for OSA [76]. Intervention success was defined as a 50% decrease in snoring based on the patient's subjective judgment. Patients, of whom 76% were men, were divided into success and failure groups. Questionnaires were administered 3 months after surgery. Nocturia episodes were significantly decreased in the success group (from 1.9 to 0.7, $P<.001$), with 41% of patients reporting no nocturia episodes. Total I-PSS score (from 8.8 to 5.1, $P<.001$), OABSS score (from 2.6 to 1.7, $P=.015$), and QOL score (from 1.7 to 1.1, $P=.017$) were all significantly decreased as well. Of note, the patients who failed to achieve more than a 50% decrease in snoring did not have significant changes in any parameters.

SLEEP DISORDERS, SHIFT WORK, OCCUPATIONAL EXPOSURES AND INFERTILITY

Infertility affects approximately 15% of couples worldwide, with males being the sole contributors in 20%–30% of cases and cocontributors with their female partners in 50% of cases [77]. Recent work has elucidated several risk factors for male infertility related to work and sleep, as appreciated in Fig. 6 [78]. Kohn et al. evaluated the impact of shift work on semen parameters and reproductive hormones in infertile men [79]. The study comprised 198 men, of which 75 were infertile shift workers, 96 infertile nonshift workers,

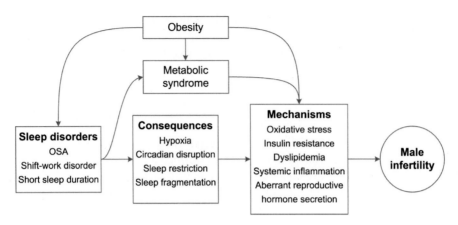

FIG. 6

The role of sleep disorders in male infertility. In addition, we demonstrate the impact of obesity and metabolic syndrome in male infertility and sleep. *Reproduced with permission from Palnitkar G, Phillips CL, Hoyos CM, Marren AJ, Bowman AMC, Brendon AJ. Linking sleep disturbance to idiopathic male infertility. Sleep Med Rev 2018. https://doi.org/10.1016/j.smrv.2018.07.006.*

and 27 fertile controls. The authors found that sperm density, total motile count (TMC), and testosterone levels were lower in infertile shift workers compared with infertile nonshift workers ($P=.012$, .019, .026, respectively). No differences were observed in semen volume, sperm motility, LH, or FSH levels. Eisenberg and colleagues assessed the relationship between occupation, health, and semen quality in a cohort of men attempting to conceive using data from the Longitudinal Investigation of Fertility and the Environment (LIFE) Study, which comprised a prospective cohort of 501 couples attempting to conceive in Texas and Michigan between 2005 and 2009 [80]. Work-related heavy exertion was associated with lower sperm concentration and total sperm count. Oligospermia was identified in 13% of men who reported heavy exertion, compared with 6% of those who did not. Further, average sperm counts were 16% lower in men who reported workplace exertion. Importantly, these associations persisted after adjustments were made for age, BMI, race, and smoking. Of note, more medications were associated with a higher risk of low sperm count (P-value $<.05$). In addition to shift work, toxic exposure and stress can contribute to male infertility. Sheiner et al. investigated the influence of working conditions, occupational exposures to potential reproductive toxins, and psychological stress on male fertility [81]. The case group was composed of men who presented to clinic for male infertility, and the control group was composed of men who accompanied their female partners for female infertility. Couples with both a male and female factor were excluded from the study. No significant baseline differences were observed between the groups besides a higher rate of couple primary infertility among cases as compared with the controls (61.3% vs 40.9%; $P=.009$). The study results indicated that infertility was associated with working in industry and construction as compared with other occupations (78.6% vs 58.3%, $P=.044$). In contrast, Tuntiserane et al. performed a cross-sectional study in 1201 pregnant couples from Thailand. There was no association in men working long hours (>71h per week) or performing shift work and increased time to pregnancy [82].

Apart from the possible hormonal changes that may occur because of disruption of normal circadian rhythms in shift workers, there are occupational exposures that could affect male fertility. Gracia and colleagues examined 1348 cases and controls for links between occupational exposures and male infertility [83]. In this cohort, infertile men were more likely to be Caucasian, employed in blue-collar jobs, and less educated than fertile men. The only significant finding was that infertile men were less likely to report radiation exposure and video display terminals exposure than fertile men. No significant associations were observed between infertility and exposure to shift work, metal fumes, electromagnetic fields, solvents, lead, paint, pesticides, work-related stress, or vibration. In contrast to this study, El-Helaly and colleagues performed a case-control study of 255 infertile men and 267 fertile men, finding that infertile men had increased odds of being exposed to solvents and painting materials (OR 3.88, 95% CI 1.50–10.03), lead (OR 5.43, 95% CI 1.28–23.13), computers (OR 8.01, 95% CI 4.03–15.87), shift work (OR 3.60, 95% CI 1.12–11.57), and work-related stress (fairly present: OR 3.11, 95% CI 1.85–5.24; often present: OR 3.76, 95% CI 1.96–7.52) [84]. Based on these results, there is conflicting evidence as to whether shift work, occupational stress, and workplace exposures are strong risk factors for male infertility.

There is interest in elucidating the pathophysiological relationship between OSA and infertility. Torres et al. performed an elegant animal model study that evaluated this relationship, with the goal being to determine if episodes of intermittent hypoxia, as occurs in men with OSA, induce changes in testicular tissue oxygenation that could lead to increases in testicular oxidative stress with reduction of sperm motility [85]. The authors compared mice that were exposed to intermittent hypoxia with control mice exposed to constant room air and evaluated testicular oxidative stress and pregnancy rates. Increased testicular oxidative stress was observed via lower expression of the antioxidant enzymes Gpx1 and Sod1 (37% and 57% reduction) in experimental mice when compared with controls. Sperm motility was decreased from $31.5 \pm 3.5\%$ in the control group to $22.9 \pm 1.8\%$ in the experimental group, and the number of pregnant females per mating was significantly higher in the control group (0.72 ± 0.16) than in the experimental group (0.33 ± 0.10) ($P=.04$). In addition, the number of fetuses per mating was reduced in the intermittent hypoxia group (2.45 ± 0.73) when compared with room air controls (5.80 ± 1.24) ($P=.02$). In order to further assess the link between reproduction and OSA in humans, Macrea et al. investigated the effect of CPAP therapy on serum

PRL in patients with OSA [86] as there have been data linking hyperprolactinemia to secondary hypogonadism [87]. The authors found a significant decrease in PRL after initiation of CPAP therapy for a period of 11–36 months (10.7 ± 1.6 ng/dL to 6.7 ± 0.5 ng/dL; $P = .037$). No changes in FSH, LH, testosterone, and TSH were observed. It should be noted, however, that PRL levels decreased but remained within the normal laboratory range at both the beginning and end of the study.

SLEEP DISORDERS, SHIFT WORK AND PROSTATE CANCER

Prostate cancer is the most commonly diagnosed nonskin cancer and the second leading cause of cancer death in males in the United States, with 1 of every 6 men being diagnosed and 1 in 33 dying of the disease [88]. Currently, age, family history, and race/ethnicity are the only widely agreed-upon risk factors for prostate cancer. However, shift work has recently been explored as a possible risk factor for the development of prostate cancer in cross-sectional case–control and cohort studies. Flynn-Evans et al. analyzed data obtained from the National Health and Nutrition Examination Survey (NHANES) in order to determine whether total PSA levels were higher in male shift workers than in those working during the day [17]. On baseline analysis, nonshift workers had a significantly longer average sleep duration than shift workers ($P < .001$). When adjusting for age, BMI, race/ethnicity, health insurance, average hours of sleep per night, and months on the current job, shift workers had a higher likelihood of having a PSA >4.0 ng/mL (OR 2.62, 95% CI 1.16–5.95, $P = .02$) compared with nonshift workers. Rao and colleagues performed a meta-analysis of eight case-control or cohort studies that encompassed 2,459,845 men in order to assess the relationship between night shift work and prostate cancer risk [20]. The authors observed an increased risk of prostate cancer for men working night shifts (RR 1.24, 95% CI 1.05–1.46, $P = .011$). In contrast, Du et al. found no increased risk of prostate cancer in male night shift workers (RR 1.05, 95% CI 1.00–1.11, $P = .06$) in a meta-analysis of nine cohort studies [89]. The authors also found no difference when subgrouping men into night shift work (RR 1.04, CI 0.95–1.14, $P = .22$) or shift work in general (RR 0.85, CI 0.43–1.69, $P = .24$).

Several studies have evaluated the biochemical implications of sleep disorders on prostate cancer. Sigurdardottir et al. examined 928 Icelandic men without prostate cancer to investigate the prospective association between urinary melatonin metabolite levels and the subsequent risk for prostate cancer [90]. Men with lower melatonin metabolite levels had a fourfold increased risk for advanced cancer (HR 4.04, 95% CI 1.26–12.98, $P = .02$) and lethal disease

(HR 4.83, CI 1.26–18.45, $P=.02$). However, no increased risk was observed for the development of prostate cancer (HR 1.47, CI 0.94–2.30, $P=.09$). The median follow-up time from urine collection to prostate cancer diagnosis was 2.3 years. Kakizaki and colleagues analyzed sleep duration and risk of prostate cancer in a prospective cohort of 22,230 men [91]. Long sleep, defined as >9h per night, appears to be protective against prostate cancer when compared with 7–8h of sleep (HR 0.48, 95% CI 0.29–0.79, $P=.02$). Short sleep (<6h.) was not associated with an increased risk of prostate cancer risk (HR 1.34, 95% CI 0.83–2.17). Wendeu-Foyet and colleagues used a case-control study to assess the role of night shift work, either permanent or rotating, in the development of prostate cancer [92]. The authors considered prostate cancer aggressiveness in their analysis, and while no increased risk of prostate cancer was observed among men who ever performed shift work (OR 0.97, CI 0.79–1.19), an increased risk of prostate cancer was observed in men who performed >30 years of permanent night work in addition to shifts longer than 10h (OR 3.07, CI 1.07–8.80). An increased risk of prostate cancer was also observed in men who worked >1314 total nights with shifts longer than 10h (OR 2.36, CI 1.21–4.56). In addition, 20 years of permanent night work was associated with an increased risk of aggressive prostate cancer (OR 1.76, 95% CI 1.13–2.75), the risk of which was higher when 20 years of permanent shift work was combined to shifts longer than 10h (OR 4.64, 95% CI 1.78–12.13). Another study by Sigurdardottir et al. found a strong association between severe sleep problems and aggressive prostate cancer (T3 or lethal disease) (HR 3.2, 95% CI 1.1–9.7) [93].

In contrast, several large cohort studies have found no association between prostate cancer risk and sleep. Markt and colleagues performed a 23-year prospective study of 51,529 US males that investigated the association between altered sleep patterns or reduced sleep duration and prostate cancer [94]. The authors did not find increased prostate cancer risk in men who reported short (<6h) or long (>9h) sleep duration compared with men who reported consistently normal sleep duration. Markt also published a 13-year follow-up prospective cohort study involving 14,041 Swedish men [95]. The authors evaluated the association between sleep duration and sleep disruption and risk of prostate cancer. They found no association between sleep duration and prostate cancer risk (HR 0.93, 95% CI 0.76–1.14 for ≤6h; HR 0.89, 95% CI 0.54–1.49 for ≥9h).

OSA appears to have a systemic response by primarily affecting oxygenation and vascular remodeling. However, we do not know the full extent of the impact of OSA on malignancy risk, particularly prostate cancer. Fang and colleagues performed a case-control study involving cancer patients as cases compared with age-matched controls [96]. The authors looked for sleep disorders, including OSA, as possible risk factors for various cancers, and found a significantly increased risk of prostate cancer in men with OSA (OR 3.69, 95% CI

1.98–6.89) after adjusting for age, sex, income, and rural/urban region. Chung and colleagues performed a cross-sectional study of 6180 men and evaluated the link between OSA with a broad spectrum of urological comorbidities in large Taiwanese data set, finding that men with OSA had an increased risk of prostate cancer (OR 2.45, 95% CI 1.19–5.05, $P=.021$) compared with men without OSA [97]. In contrast, Gozal et al. found a lower risk of prostate cancer in men with OSA (HR 0.93, 95% CI 0.90–0.96) when evaluating an existing longitudinal nationwide-based health insurance database [98].

CONCLUSIONS

In this chapter, we have explored the relationship between sleep disorders, shift work, and men's health. We have discussed the known physiological and behavioral mechanisms that explain these relationships, though many mechanisms remain incompletely elucidated. Sleep disorders and shift work have deleterious consequences in chronic illnesses that commonly affect men, including cardiovascular disease. With regard to urological disease, a bidirectional relationship between lower urinary tract symptoms and sleep disorders exists. Obstructive sleep apnea worsens and increases the risk for many urological diseases, to the point that it may be reasonable to consider it as a contributing factor in patients presenting with lower urinary tract symptoms or erectile dysfunction. Currently available studies demonstrate a link between shift work and urological diseases, though additional work is needed to validate these findings and to better define the mechanisms that result in this relationship.

References

[1] Colten HR, Altevogt BM. Sleep disorders and sleep deprivation: an unmet public health problem. Washington: The National Academies Press; 2006. https://doi.org/10.17226/11617.

[2] Gan Y, Yang C, Tong X, Sun H, Cong Y, Yin X, et al. Shift work and diabetes mellitus: a meta-analysis of observational studies. Occup Environ Med 2015;72:72–8. https://doi.org/10.1136/oemed-2014-102150.

[3] Ceide M, Pandey A, Ravenell J, Donat M, Ogedegbe G, Jean-louis G. Associations of short sleep and shift work status with hypertension among black and white Americans. Int J Hypertens 2015;2015:1–7. https://doi.org/10.1155/2015/697275.

[4] Macagnan J, Pattussi MP, Canuto R, Henn RL, Fassa AG, Olinto MTA, et al. Impact of nightshift work on overweight and abdominal obesity among workers of a poultry processing plant in southern Brazil impact of nightshift work on overweight and abdominal obesity among workers of a poultry processing plant in Southern Brazil. Chronobiol Int 2012;29:336–43. https://doi.org/10.3109/07420528.2011.653851.

[5] Barger LK, Rajaratnam SMW, Cannon CP, Lukas MA, Im K, Goodrich EL, et al. Short sleep duration, obstructive sleep apnea, shiftwork, and the risk of adverse cardiovascular events in patients after an acute coronary syndrome. J Am Heart Assoc 2017;(10):1–9. https://doi.org/10.1161/JAHA.117.006959.

[6] Kessler RC, Berglund PA, Coulouvrat C, Hajak G, Roth T, Shahly V, et al. Insomnia and the performance of US workers: results from the America Insomnia Survey. Sleep 2011;34:1161–71. https://doi.org/10.5665/sleep.1230.

[7] Lombardi DA, Folkard S, Willetts JL, Smith GS, Lombardi DA, Folkard S, et al. Daily sleep, weekly working hours, and risk of work-related injury: US National Health Interview Survey (2004–2008). Chronobiol Int 2010;27:1013–30. https://doi.org/10.3109/07420528.2010.489466.

[8] McMenamin TM. A time to work: recent trends in shift work and flexible schedules. Mon Labor Rev 2007;3–15.

[9] Leger D. The cost of sleep-related accidents: a report for the National Commission on Sleep Disorders Research. Sleep 1994;(1):84–93.

[10] Gumenyuk V, Howard R, Roth T, Korzyukov O, Drake CL. Sleep loss, circadian mismatch, and abnormalities in reorienting of attention in night workers with shift work disorder. Sleep 2014;37:545–56.

[11] Sateia MJ. International Classification of Sleep Disorders-third edition highlights and modifications. Chest 2014;146:1387–94. https://doi.org/10.1378/chest.14-0970.

[12] Deng N, Kohn TP, Lipshultz LI, Pastuszak AW. The relationship between shift work and men's health. Sex Med Rev 2018;1–11. https://doi.org/10.1016/j.sxmr.2017.11.009.

[13] Deng N, Haney NM, Kohn TP, Pastuszak AW, Lipshultz LI. The effect of shift work on urogenital disease: a systematic review. Curr Urol Rep 2018;19:1–15.

[14] Kim JW. Effect of shift work on nocturia. Urology 2016;87:153–60. https://doi.org/10.1016/j.urology.2015.07.047.

[15] Pastuszak AW, Moon YM, Scovell J, Badal J, Lamb D, Link RE, et al. Poor sleep quality predicts hypogonadal symptoms and sexual dysfunction in male non-standard shift workers. Urology 2017;121–5. https://doi.org/10.1016/j.urology.2016.11.033.Poor.

[16] Scovell J, Pastuszak AW, Slawin J, Badal J, Link RE, Lipshultz LI. Impaired sleep quality is associated with more significant lower urinary tract symptoms in male shift workers. Urology 2017;99:197–202. https://doi.org/10.1016/j.urology.2016.05.076.Impaired.

[17] Flynn-evans EE, Mucci L, Stevens RG, Lockley SW. Shiftwork and prostate-specific antigen in the National Health and Nutrition Examination Survey. J Natl Cancer Inst 2013;105:1291–7. https://doi.org/10.1093/jnci/djt169.

[18] Jensen TK, Andersson A, Skakkebæk NE, Joensen UN, Jensen MB, Lassen TH, et al. Original contribution association of sleep disturbances with reduced semen quality: a cross-sectional study among 953 healthy young Danish men. Am J Epidemiol 2013;177:1027–37. https://doi.org/10.1093/aje/kws420.

[19] Gamble KL, Resuehr D, Johnson CH. Shift work and circadian dysregulation of reproduction. Front Endocrinol (Lausanne) 2013;4:1–9. https://doi.org/10.3389/fendo.2013.00092.

[20] Rao D, Yu H, Yu Z, Xiangyi B, Xie L. Does night-shift work increase the risk of prostate cancer? a systematic review and meta-analysis. Onco Targets Ther 2015;8:2817–26.

[21] Parent M-É, El-zein M, Rousseau M, Pintos J, Siemiatycki J. Night work and the risk of cancer among men. Am J Epidemiol 2015;176:751–9. https://doi.org/10.1093/aje/kws318.

[22] Wright KP, Bogan RK, Wyatt JK. Shift work and the assessment and management of shift work disorder (SWD). Sleep Med Rev 2013;17:41–54. https://doi.org/10.1016/j.smrv.2012.02.002.

[23] Yong LC, Li J, Calvert GM. Sleep-related problems in the US working population: prevalence and association with shiftwork status. Occup Environ Med 2017;74:93–104. https://doi.org/10.1136/oemed-2016-103638.

[24] Kecklund G, Axelsson J. Health consequences of shift work and insufficient sleep. BMJ 2016;355:1–13. https://doi.org/10.1136/bmj.i5210.

[25] Lockley SW, Barger LK, Ayas NT, Rothschild JM, Czeisler CA, Landrigan CP. Effects of health care provider work hours and sleep deprivation on safety and performance. Jt Comm J Qual Patient Saf 2007;33:7–18.

[26] Puttonen S, Harma M, Hublin C. Shift work and cardiovascular disease – pathways from circadian stress to morbidity. Scand J Work Environ Health 2010;36:96–108. https://doi.org/10.5271/sjweh.2894.

[27] George CFP, Avidan AY. Diabetes mellitus. In: Kryger M, editor. Atlas of clinical sleep medicine. 2nd ed. Philadelphia: Elsevier Sanders; 2013. p. 346–9. https://doi.org/10.1016/B978-0-323-18727-5.00015-X.

[28] Shan Z, Ma H, Xie M, Yan P, Guo Y, Bao W, et al. Sleep duration and risk of type 2 diabetes: a meta-analysis of prospective studies. Diabetes Care 2015;38:529–37. https://doi.org/10.2337/dc14-2073.

[29] Ohlander J, Keskin ĀM, Stork J, Radon K. Shift work and hypertension: prevalence and analysis of disease pathways in a German Car manufacturing company. Am J Ind Med 2015;58:549–60. https://doi.org/10.1002/ajim.22437.

[30] Wang A, Arah OA, Kauhanen J, Krause N. Shift work and 20-year incidence of acute myocardial infarction: results from the Kuopio Ischemic Heart Disease Risk Factor Study. Occup Environ Med 2016;73:588–94. https://doi.org/10.1136/oemed-2015-103245.

[31] Hermansson J, Hallqvist J, Karlsson B, Knutsson A, Gådin KG. Shift work, parental cardiovascular disease and myocardial infarction in males. Occup Med (Chicago, IL) 2018;68:120–5. https://doi.org/10.1093/occmed/kqy008.

[32] Fryar CD, Margaret DC, Ogden CL. Prevalence of overweight, obesity, and extreme obesity among adults aged 20 and over: United States, 1960–1962 through 2013–2014, Cent Dis Control Prev 2016;1–6. https://www.cdc.gov/nchs/data/hestat/obesity_adult_15_16/obesity_adult_15_16.htm (accessed August 31, 2018).

[33] Kyle TK, Dhurandhar EJ, Allison DB. Regarding obesity as a disease: evolving policies and their implications. Endocrinol Metab Clin N Am 2016;45:511–20. https://doi.org/10.1016/j.ecl.2016.04.004.Regarding.

[34] Kawabe Y, Nakamura Y, Kikuchi S, Murakami Y, Tanaka T. Relationship between shift work and clustering of the metabolic syndrome diagnostic components. J Atheroscler Thromb 2014;21:703–11.

[35] Grundy A, Cotterchio M, Kirsh VA, Nadalin V. Rotating shift work associated with obesity in men from northeastern Ontario. Health Promot Chronic Dis Prev Can Res Policy Pract 2017;37:238–47.

[36] Selvin E, Burnett AL, Platz EA. Prevalence and risk factors for erectile dysfunction in the US. Am J Med 2007;120:151–7. https://doi.org/10.1016/j.amjmed.2006.06.010.

[37] Pepin J-L, Tamisier R, Godin-Ribuot D, Levy PA. Erectile dysfunction and obstructive sleep apnea: from mechanisms to a distinct phenotype and combined therapeutic strategies. Sleep Med Rev 2015;20:1–4. https://doi.org/10.1016/j.smrv.2014.12.004.

[38] Hoyos CM, Melehan KL, Liu PY, Grunstein RR, Phillips CL. Does obstructive sleep apnea cause endothelial dysfunction? A critical review of the literature. Sleep Med Rev 2015;20:15–26. https://doi.org/10.1016/j.smrv.2014.06.003.

[39] Hoyos CM, Melehan KL, Phillips CL, Grunstein RR, Liu PY. To ED or not to ED—is erectile dysfunction in obstructive sleep apnea related to endothelial dysfunction? Sleep Med Rev 2015;20:5–14. https://doi.org/10.1016/j.smrv.2014.03.004.

[40] Kohler M, Stoewhas A, Ayers L, Senn O, Bloch KE, Russi EW, et al. Effects of continuous positive airway pressure therapy withdrawal in patients with obstructive sleep apnea a randomized controlled trial. Am J Respir Crit Care Med 2011;184:1192–9. https://doi.org/10.1164/rccm.201106-0964OC.

[41] Perimenis P, Karkoulias K, Markou S, Gyftopoulos K, Athanasopoulos A, Barbalias G. Erectile dysfunction in men with obstructive sleep apnea syndrome: a randomized study of the efficacy of sildenafil and continuous positive airway pressure. Int J Impot Res 2004;256–60. https://doi.org/10.1038/sj.ijir.3901219.

[42] McBride JA, Kohn TP, Rodriguez KM, Thirumavalavan N, Mazur DJ, Pastuszak AW, et al. Incidence and characteristics of men at high risk for sleep apnea in a high volume andrology clinic. J Urol 2018;199:e559. https://doi.org/10.1016/j.juro.2018.02.1357.

[43] Chen K, Liang S, Lin C, Liao W. Sleep disorders increase risk of subsequent erectile dysfunction in individuals without sleep apnea: a nationwide population-base cohort study. Sleep Med 2016;17:64–8. https://doi.org/10.1016/j.sleep.2015.05.018.

[44] Jankowski JT, Seftel AD, Strohl KP. Erectile dysfunction and sleep related disorders. J Urol 2008;179:837–41. https://doi.org/10.1016/j.juro.2007.10.024.

[45] Kohn TP, Rodriguez KM, Sigalos JT, Hasan A, Pastuszak AW, Lipshultz LI. Poor sleep quality is associated with clinically significant erectile dysfunction. J Urol 2018;199. https://doi.org/10.1016/j.juro.2018.02.1359.

[46] Rodriguez KM, Kohn TP, Kohn JR, Kirby EW, Pickett SM, Pastuszak AW, et al. Shift work sleep disorder and night shift work significantly impair erectile function. J Urol 2018;199:e559.

[47] Seftel A. Male hypogonadism. Part I: epidemiology of hypogonadism. Int J Impot Res 2006;18:115–20. https://doi.org/10.1038/sj.ijir.3901397.

[48] Leproult R, van Cauter E. Endocrine physiology. In: Kryger M, editor. Atlas of clinical sleep medicine. 2nd ed. Philadelphia: Elsevier Sanders; 2013. p. 58–64. https://doi.org/10.1016/B978-0-323-18727-5.00003-3.

[49] Miller CM, Rindflesch TC, Fiszman M, Hristovski D, Shin D. A closed literature-based discovery technique finds a mechanistic link between hypogonadism and diminished sleep quality in aging men. Sleep 2012;35:279–85.

[50] Jensen MA, Hansen ÅM, Kristiansen J, Nabe K, Garde AH, Aarrebo M, et al. Changes in the diurnal rhythms of cortisol, melatonin, and testosterone after 2, 4, and 7 consecutive night shifts in male police officers. Chronobiol Int 2016;33:1280–92. https://doi.org/10.1080/07420528.2016.1212869.

[51] Leibenluft E, Schmidt PJ, Turner EH, Danaceau MA, Ashman SB, Wehr TA, et al. Effects of leuprolide-induced hypogonadism and testosterone replacement on sleep, melatonin, and prolactin secretion in men. J Clin Endocrinol Metab 1997;82:3203–7.

[52] Shigehara K, Konaka H, Sugimoto K, Izumi K, Kadono Y, Namiki M, et al. Sleep disturbance as a clinical sign for severe hypogonadism: efficacy of testosterone replacement therapy on sleep disturbance among hypogonadal men without obstructive sleep apnea. Aging Male 2018;21:99–105. https://doi.org/10.1080/13685538.2017.1378320.

[53] Ismailogullari S, Korkmaz C, Peker Y, Bayram F. Impact of long-term gonadotropin replacement treatment on sleep in men with idiopathic hypogonadotropic hypogonadism. J Sex Med 2011;8:2090–7. https://doi.org/10.1111/j.1743-6109.2010.02143.x.

[54] Shigehara K, Izumi K, Mizokami A, Namiki M. Testosterone deficiency and nocturia: a review. World J Mens Heal 2017;35:14–21.

[55] Kim S, Cho K. Obstructive sleep apnea and testosterone deficiency. World J Mens Health 2018;36:1–7.

[56] Molina FD, Suman M, de Carvalho TBO, Piatto VB. Evaluation of testosterone serum levels in patients with obstructive sleep apnea syndrome. Braz J Otorhinolaryngol 2011;77:88–95. https://doi.org/10.1590/S1808-86942011000100015.

[57] Luboshitzky R, Lavie L, Shen-orr Z, Herer P, Lavie L, Shen Z, et al. Risk factors and chronic disease altered luteinizing hormone and testosterone secretion in middle-aged obese men with obstructive sleep apnea. Obes Res 2005;13:780–6.

[58] Gambineri A, Pelusi C, Pasquali R. Testosterone levels in obese male patients with obstructive sleep apnea syndrome: relation to oxygen desaturation, body weight, fat distribution and the metabolic parameters. J Endocrinol Investig 2003;26:493–8.

[59] Hammoud AO, Walker J, Gibson M, Cloward TV, Hunt SC, et al. Sleep apnea, reproductive hormones and quality of sexual life in severely obese men. Obesity 2011;19:1118–23. https://doi.org/10.1038/oby.2010.344.Sleep.

[60] Cole AP, Hanske J, Jiang W, Kwon NK, Lipsitz SR, Kathrins M, et al. Impact of testosterone replacement therapy on thromboembolism, heart disease and obstructive sleep apnoea in men. BJU Int 2018;121:811–8. https://doi.org/10.1111/bju.14149.

[61] Vlkova B, Mucska I, Hodosy J, Celec P. Short-term effects of continuous positive airway pressure on sex hormones in men and women with sleep apnoea syndrome. Andrologia 2014;5:386–90. https://doi.org/10.1111/and.12092.

[62] Knapp A, Myhill PC, Davis WA, Peters KE, Hillman D, Hamilton EJ, et al. Effect of continuous positive airway pressure therapy on sexual function and serum testosterone in males with type 2 diabetes and obstructive sleep apnoea. Clin Endocrinol 2014;81:254–8. https://doi.org/10.1111/cen.12401.

[63] Abrams P. New words for old: lower urinary tract symptoms for "prostatism" Br Med J 1994;929–30.

[64] Taylor BC, Wilt TJ, Fink HA, Lambert LC, Marshall LM, Hoffman AR, et al. Prevalence, severity, and health correlates of lower urinary tract symptoms among older men: the MrOS study. Urology 2006;68:804–9. https://doi.org/10.1016/j.urology.2006.04.019.

[65] Fantus RJ, Packiam VT, Wang CH, Erickson BA, Helfand BT. Voiding dysfunction the relationship between sleep disorders and lower urinary tract symptoms: results from the NHANES. J Urol 2018;200:161–6. https://doi.org/10.1016/j.juro.2018.01.083.

[66] Branche BL, Howard LE, Moreira DM, Roehrborn C, Castro-santamaria R, Andriole GL, et al. Sleep problems are associated with development and progression of lower urinary tract symptoms: results from REDUCE. J Urol 2018;199:536–42. https://doi.org/10.1016/j.juro.2017.08.108.

[67] Helfand BT, Lee JY, Sharp V, Foster H, Naslund M, Williams OD, et al. Associations between improvements in lower urinary tract symptoms and sleep disturbance over time in the CAMUS trial. J Urol 2012;188:2288–93. https://doi.org/10.1016/j.juro.2012.07.104.

[68] Matsushita C, Torimoto K, Goto D, Morizawa Y, Kiba K, Shinohara M, et al. Linkage of lower urinary tract symptoms to sleep quality in elderly men with nocturia: a community based study using home measured electroencephalogram data. J Urol 2017;197:204–9. https://doi.org/10.1016/j.juro.2016.07.088.

[69] Shao I-H, Wu C-C, Hsu H-S, Chang S-C, Wang H-H, Chuang H-C, et al. The effect of nocturia on sleep quality and daytime function in patients with lower urinary tract symptoms: a cross-sectional study. Clin Interv Aging 2016;11:879–85.

[70] Ge TJ, Vetter J, Lai HH. Sleep disturbance and fatigue are associated with more severe urinary incontinence and overactive bladder symptoms. Urology 2017;109:67–73. https://doi.org/10.1016/j.urology.2017.07.039.

[71] Araujo AB, Yaggi HK, Yang M, McVary KT, Fang S, Bliwise DL. Sleep-related problems and urologic symptoms: testing the hypothesis of bi-directionality in a longitudinal, population- based study. J Urol 2014;191:1–12. https://doi.org/10.1016/j.juro.2013.07.011. Sleep-related.

[72] Martin SA, Appleton SL, Adams RJ, Taylor AW, Catcheside PG, Vakulin A, et al. Nocturia, other lower urinary tract symptoms and sleep dysfunction in a community-dwelling cohort of men. Urology 2016;97:219–26. https://doi.org/10.1016/j.urology.2016.06.022.

[73] Abler LL, Vezina CM. Respiratory physiology & neurobiology links between lower urinary tract symptoms, intermittent hypoxia and diabetes: causes or cures? Respir Physiol Neurobiol 2017;1–10. https://doi.org/10.1016/j.resp.2017.09.009.

[74] Yilmaz Z, Voyvoda B. Overactive bladder syndrome and bladder wall thickness in patients with obstructive sleep apnea syndrome. Int Braz J Urol 2017;44:330–7. https://doi.org/10.1590/S1677-5538.IBJU.2017.0253.

[75] Miyauchi Y, Okazoe H, Okujyo M, Inada F, Kakehi T, Kikuchi H, et al. Effect of the continuous positive airway pressure on the nocturnal urine volume. Urology 2015;85:333–6. https://doi.org/10.1016/j.urology.2014.11.002.

[76] Park HK, Paick SH, Kim HG, Park D, Cho JH, Hong S, et al. Nocturia improvement with surgical correction of sleep apnea. Int Neurourol J 2016;20:329–34.

[77] Agarwal A, Mulgund A, Hamada A, Chyatte MR. A unique view on male infertility around the globe. Reprod Biol Endocrinol 2015;13:1–9. https://doi.org/10.1186/s12958-015-0032-1.

[78] Palnitkar G, Phillips CL, Hoyos CM, Marren AJ, Bowman AMC, Brendon AJ. Linking sleep disturbance to idiopathic male infertility. Sleep Med Rev 2018. https://doi.org/10.1016/j.smrv.2018.07.006.

[79] Kohn TP, Alexander W, Pastuszak SMP, Kohn JR, Lipshultz LI. Shift work is associated with altered semen parameters in infertile men. J Urol 2017;197:e273–4. https://doi.org/10.1016/j.juro.2017.02.696.

[80] Eisenberg ML, Chen Z, Ye A, Germaine M, Louis B. The relationship between physical occupational exposures and health on semen quality: data from the LIFE study. Fertil Steril 2015;103:1271–7. https://doi.org/10.1016/j.fertnstert.2015.02.010.The.

[81] Sheiner EK, Carel R. Potential association between male infertility and occupational psychological stress. J Occup Environ Med 2002;44:1093–9. https://doi.org/10.1097/01.jom.0000044116.59147.64.

[82] Tuntiseranee P, Olsen J, Geater A, Kor-anantakul O. Are long working hours and shiftwork risk factors for subfecundity? A study among couples from southern Thailand. Occup Environ Med 1998;55:99–105.

[83] Gracia CR, Sammel MD, Coutifaris C, Guzick DS, Barnhart KT. Exposures and male infertility. Am J Epidemiol 2005;162:729–33. https://doi.org/10.1093/aje/kwi269.

[84] El-helaly M, Awadalla N, Mansour M, El-biomy Y. Workplace exposures and male infertility — a case-control study. Int J Occup Med Environ Health 2010;23:331–8. https://doi.org/10.2478/v10001-010-0039-y.

[85] Torres M, Laguna-barraza R, Dalmases M, Calle A, Pericuesta E. Male fertility is reduced by chronic intermittent hypoxia mimicking sleep apnea in mice. Sleep 2014;37:1757–65.

[86] Macrea MM, Martin TJ, Zagrean L. Infertility and obstructive sleep apnea: the effect of continuous positive airway pressure therapy on serum prolactin levels. Sleep Breath 2010;14:253–7. https://doi.org/10.1007/s11325-010-0373-0.

[87] De Rosa M, Zarrilli S, Di SA, Milano N, Gaccione M, Boggia B, et al. Hyperprolactinemia in men clinical and biochemical features and response to treatment. Endocrine 2003;20:75–82.

[88] Brawley OW. Prostate cancer epidemiology in the United States. World J Urol 2012;30:195–200. https://doi.org/10.1007/s00345-012-0824-2.

[89] Du H, Bin K, Liu W, Yang F. Shift work, night work, and the risk of prostate cancer. Medicine (Baltimore) 2017;96:1–6.

[90] Sigurdardottir LG, Markt SC, Rider JR, Haneuse S, Fall K, Schernhammer ES, et al. Urinary melatonin levels, sleep disruption, and risk of prostate cancer in elderly men. Eur Urol 2015;67:191–4. https://doi.org/10.1016/j.eururo.2014.07.008.Urinary.

[91] Kakizaki M, Inoue K, Kuriyama S, Sone T, Nakaya N, Fukudo S, et al. Sleep duration and the risk of prostate cancer: the Ohsaki Cohort Study. Br J Cancer 2008;99:176–8. https://doi.org/10.1038/sj.bjc.6604425.

[92] Wendeu-foyet MG, Bayon V, Cénée S, Trétarre B, Rébillard X, Cancel-tassin G, et al. Night work and prostate cancer risk: results from the EPICAP study. Occup Environ Med 2018;75:573–81. https://doi.org/10.1136/oemed-2018-105009.

[93] Sigurdardottir LG, Valdimarsdottir UA, Mucci LA, Fall K, Rider JR, Schernhammer E, et al. Sleep disruption among older men and risk of prostate cancer. Cancer Epidemiol Biomark Prev 2013;22:872–80. https://doi.org/10.1158/1055-9965.EPI-12-1227-T.

[94] Markt SC, Flynn-evans EE, Valdimarsdottir UA, Sigurdardottir LG, Tamimi RM, Batista JL, et al. Sleep duration and disruption and prostate cancer risk: a 23-year prospective study. Cancer Epidemiol Biomark Prev 2016;25:302–8. https://doi.org/10.1158/1055-9965.EPI-14-1274.

[95] Markt SC, Grotta A, Nyren O, Adami H, Mucci LA. Insufficient sleep and risk of prostate cancer in a large Swedish cohort. Sleep 2015;38:1405–10.

[96] Fang H, Miao N, Chen C, Sithole T, Chung M. Risk of cancer in patients with insomnia, parasomnia, and obstructive sleep apnea: a nationwide nested case-control study. J Cancer 2015;6:1140–7. https://doi.org/10.7150/jca.12490.

[97] Chung S, Hung S, Lin H, Tsai M. Obstructive sleep apnea and urological comorbidities in males: a population-based study. Sleep Breath 2016;20:1203–8. https://doi.org/10.1007/s11325-016-1336-x.

[98] Gozal D, Ham SA, Mokhlesi B. Sleep apnea and cancer: analysis of a nationwide population sample. Sleep 2016;39:1493–500.

CHAPTER 3.2

Stress, Depression, Mental Illness, and Men's Health

Nnenaya Q. Agochukwu, Daniela Wittmann
Department of Urology, University of Michigan, Ann Arbor, MI, United States

INTRODUCTION

It is impossible to address the topic of this chapter without recognizing the interaction between men's concept of themselves as men and their experience of health-related issues of stress, depression, and mental illness. Bosson and colleagues discuss "masculinity" as a socioculturally constructed concept that is, in many cultures, including ours, characterized by physical strength, self-sufficiency, emotional toughness, risk taking, and dominance; men earn their status on an ongoing basis through competition and demonstration of these qualities [1]. A number of authors who have written on masculinity suggest that internalizing this concept can be detrimental to men's health because it limits their ability to fully experience their emotions. The male role has been called "hazardous and lethal" [2–5].

Included in this conceptualization of masculinity is men's difficulty in seeking help. Moller-Leimkuhler explains that from a young age, boys are taught to be stoic: "boys don't cry." A boy learns early that becoming a man means learning "I can do it alone. I don't need help." Asking for help then becomes uncomfortable and makes him vulnerable and anxious [2]. This difficulty with accepting feelings of vulnerability and reluctance to seek help is associated with adverse psychosomatic effects [6].

An additional challenge for the boy who is learning to become a man is the fact that he is typically raised with a strong feminine imprint from his mother. He thus develops a feminine psychological component of emotional expression and receptivity, which he must learn to suppress [2]. The suppression of this deep identification alienates him from a crucial component of himself and can result in emotional inflexibility. Research has shown that the difficulty in expressing a range of emotions and seeking help can result in psychological and physical problems and in higher rates of accidents, disease, suicide, and substance use [7, 8].

STRESS AND THE CONTEMPORARY MALE ROLE

The concept of "stress" and its relationship to health, mental health, and coping was developed by Lazarus in the 1980s [9]. It employs cognitive psychology, in which an experience is processed for its relevance to oneself and one's ability to cope with it. When the appraisal signals that one does not have sufficient emotional, cognitive, and social resources with which to cope, stress can be experienced as anxiety, depression, panic, and withdrawal [10]. When men are expected to be emotionally tough, their tendency to suffer in silence may add to the stress of an already difficult situation.

The contemporary male role is complex, and it is evolving. It retains traditional elements with new expectations added. Men are traditionally expected to compete in order to advance in the workplace and also provide for the family [11]. Men are also expected to be emotionally available, and since many families now have both parents working, men are expected to be involved in the day-to-day running of the family. Rather than benefiting from support for their expanded role, men continue to be stigmatized when they display a traditionally assigned "feminine behavior," such as being emotionally expressive, or when they embrace traditionally feminine roles, such as being a stay-at-home parent [12–15]. de Visser et al. tested this concept in an online survey, with 731 university students, which evaluated gender-role stereotypes and beliefs about the association of gender with various health behaviors. Results demonstrated that both men and women gave higher masculinity ratings to men who participated in a greater number of stereotypically masculine behaviors, even if some of those behaviors were detrimental to men's health [16]. Tensions between aspects of this expanded role and resultant stress can lead men to experience depression, anxiety, and other mental health conditions. It is important to note that while balancing the multiple facets of the male role is challenging, success in doing so can be beneficial. Working and being a partner is associated with better mental health [17–20].

In the work environment, imbalances in the masculine role can be detrimental due to family–work conflict [21, 22]. Men are tasked with being the breadwinners in families; work achievements are also part of identity formation in men [23]. There are additional psychosocial factors that can have strong effects on the mental health of men, such as workplace relationships and job insecurity, which can lead to stress, anxiety, major depression, and mental illness [24–26]

CONTRIBUTION OF CLASS, RACE/ETHNICITY, AND SEXUAL MINORITY STATUS TO MEN'S EXPERIENCE OF STRESS

Social factors, other than masculinity, are also implicated in the experience of stress, and can significantly impact many men's health outcomes [27].

Social status and social roles determine the types and amount of stress. Baum et al. describes socioeconomic status as a predictor of health and illness outcomes [28]. Dowd and colleagues evaluated the effect of socioeconomic status on stress and discovered that low socioeconomic status was highly associated with higher levels of perceived stress and stressful life events [29]. Low socioeconomic status is also associated with adverse psychosocial situations that can result in high levels of stress [30].

Men's other identities, including ethnicity and class, can shape the male role and result in added stress [31]. Minority men have higher age-adjusted morbidity, mortality, and death rates. This is a pattern that has been documented for over 150 years [32]. Compared with White men, African American men have higher odds of stressful life events, including racism and social exclusion with resulting higher levels of psychological distress [29, 33]. Income inequality is also a contributor given large racial differences in socioeconomic status [34]. In studies of racial differences in health, adjustment for socioeconomic status markedly reduces and, in instances, eliminates racial disparities in health [32, 35]. Men who cannot reach traditional norms of social success and status due to poverty, minority status, and/or marginalization may compensate by demonstrating their masculinity and mastery by engaging in risky behaviors that can be detrimental to their health, such as high-risk sexual behavior [36].

Gay men experience a different form of minority stress, resulting from social stigma and rejection. For some gay men, repeated social discrimination may result in internalized homophobia, which adds an inner conflict to the experience of hostility from the social environment [37, 38]. Research shows that gay men suffer from more mental health problems than their heterosexual counterparts as a result of stigma, prejudice, and discrimination. They also suffer from inadequate response to their health-care needs by health-care providers through the lack of providers' understanding of their needs and, in some cases, through rejection by providers. Lesbian, gay, bisexual, and transgender populations have been declared as suffering from health disparities by the Institute of Mental Health, and research to improve health-care provision has been encouraged by the National Institutes of Health [39, 40].

It is important to recognize that some men experience all the factors that tend to be risk factors for poor health: struggle with their masculine role, low economic status, non-White race, and gay sexual orientation. This combination of factors, also dubbed as "intersectionality," multiplies the risk beyond the contribution of any individual one and must be taken into consideration when a health assessment is conducted [41].

STRESS AND PHYSICAL HEALTH

Men in the Unites States lead less healthy lifestyles [5], suffer from more severe chronic conditions, have higher death rates, and die nearly 7 years younger than women [42, 43]. Coronary heart disease, cancers, pulmonary diseases, and cerebrovascular disease are among the conditions attributed to stress and unhealthy lifestyles [44–51].

Adherence to the masculine role leads men to avoid health care and engage in risk behaviors rather than in health-promoting behaviors [42] even though the adoption of health-promoting behaviors is the most effective method of sustaining good health and avoiding disease [52–55]. As demonstrated by Pinkhasov and colleagues, men make fewer health-care visits than women, including visiting a doctor's office, emergency department, or utilizing home visits. Pinkhasov and colleagues also reported that men accessed preventive care at a rate of 44.8 per 100 as compared with women's rate of 74.4 per 100 [56]. It is possible that the high value placed on masculine self-reliance, control, and physical and emotional strength inhibits men's willingness to engage in their health and impinges on their motivation to seek health care [57]. Avoidance of positive health behaviors and health care leads to unintended progression of diseases and subsequent poorer health outcomes, including greater morbidity, higher rate of mortality, and diminished longevity [58, 59].

DEPRESSION, MENTAL ILLNESS, AND MEN'S HEALTH

While the specific etiology of mental health conditions based on gender is not well understood, there are trends that invite hypotheses about the relationship between mental illness diagnosis and gender. An association has been made about the impact of the masculine gender role and stress on men's psychological outcomes [12]. In this chapter, we specifically address the prevalence of depression and other mental illnesses, such as schizophrenia and bipolar disorder.

Depression has been described as the leading cause of worldwide disability [60, 61]; depression is diagnosed twice as commonly in women than in men [62]. In a study by Corney and colleagues of 80 men and 204 women seen in a general practice, men reported fewer somatic, emotional, and depressive symptoms than women and men did not report psychosocial distress as a reason for seeking help [63]. The investigators concluded that the different rates of help seeking were due in part to behaviors associated with masculinity, including emotional strength, fortitude, and toughness. This finding is corroborated in a study by Horwitz and colleagues, which evaluated 120 patients (80 women and 40 men) [64]. In this study, patients entering a mental health center were interviewed. Interviews revealed that men were slower than women to pursue psychiatric help. In addition, men were less likely to feel that they had a psychiatric problem and entry into psychiatric treatment was more often coercive. The investigators suggested that men demonstrated culturally learned ways of under recognizing and minimizing emotional problems, which made them less likely to discuss them and seek treatment [64, 65].

Although men and women with depressed mood have similar levels of subjective, social, and occupational impairment, men report fewer depressive symptoms than women, are less likely than women to express symptoms consistent with depression including loss of appetite and somatic symptoms, and men adopt emotional suppression as a coping strategy [62, 66, 67]. Addis et al. suggest that men may turn to alcohol and other drugs for coping and as a method of hiding depression and other psychological problems [68].

It may be that men's conformity to traditional masculine norms including stoicism, emotional restriction, and self-reliance contributes to lower rates of depression diagnoses in men. Flat affect in men can be perceived by providers as emotional restriction and lead to the lack of further evaluation of a man's depression [57, 69]. In Western societies, men are twice as likely to attempt suicide. In the United States, men are six times more likely to attempt suicide and have higher rates of completed suicide than women [70, 71]. The likelihood of underdiagnosis combined with men's aversion to help-seeking may be implicated in the higher rates of suicide in men [72]. Men's estrangement from mental health services and social support, which results in untreated mental health issues, may possibly be the reasons for both the high rates of suicide attempts and completions. It appears from these findings that the cultural construction of masculinity has contributed to a health disparity for men who are coping with depression.

When considering other mental illnesses such as schizophrenia, the mean ratio of male to female in schizophrenia incidence was recently estimated to be 1.42 [73]. In addition, most psychiatrists generally believe that women have a later onset of schizophrenia and a better course of illness. This may be due to the fact that women seek help earlier in the disease course than men. Hambrecht et al. report that there are also differences in symptom expression and disease course. Men more often have negative symptoms [74] (i.e., apathy, absent, blunted responses, and social withdrawal) and substance abuse, whereas women have higher levels of positive (i.e., hallucinations, paranoia, and delusions) and depressive symptoms [75]. For bipolar disorder, most studies report an almost equal gender ratio in the 12-month and lifetime prevalence of bipolar disorder [76, 77]. This, however, is not consistent in all studies as other studies show higher rates of bipolar disorder in men [78, 79]. The lifetime prevalence of manic episodes is significantly higher in men, and men are more likely to have a higher history of unipolar mania [77]. Understanding gender differences in mental illness is key in diagnosis and treatment.

Although our understanding of the effect of the male gender on schizophrenia and bipolar disorder is limited to the knowledge of specific prevalent symptoms by gender, it would be safe to assume that just like with depression, there may be a tendency to underreport. In the case of bipolar disorder, for example, a manic episode could take longer to diagnose in a cultural context in which men are less quickly censured for acting out behavior, as in "boys will be boys."

In general, more research is needed to understand some of the ways in which men may remain under the radar and not receive the care they need as their mental health conditions develop. Research clearly demonstrates that the social and cultural environment in which men are expected to be self-sufficient and emotionally strong does not provide an adequately safe environment for men to seek help and access care that would mitigate the mental health conditions that currently lead to poorer outcomes.

MASCULINITY AND SEXUAL HEALTH

Sexual health, an aspect of overall health, can be seen as both a source of fulfillment and sense of well-being and a potential source of distress. Being able to function well sexually and be a competent partner in a sexual relationship is a core value for most men and, for many, an integral part of their sense as a man. As a result, sexual dysfunction can be a source of distress when a man feels that he is unable to perform and cannot fulfill his sexual role with a partner effectively. This may lead to escape-avoidance where one seeks less support in encounters that threaten self-esteem and may subsequently result in distancing and disengagement [9]. His masculine body image, identity, and self-esteem suffer.

Many health conditions and their treatments affect sexual function [80]. These include heart disease, diabetes, neurogenic conditions, cancer, and depression. Men are distressed by their sexual problems, which negatively affects their relationships [81]. For example, men with depression on treatment are more likely to be dissatisfied with their sex lives [80]. Conversations about sex have been shown to cause discomfort to health-care providers; thus, sexual health solutions are not easy for men to access if their sexual problems are not easily assisted by phosphodiesterase-5 inhibitors. Research has shown that sexual health counseling can help men maintain intimate relationships but such expertise is rarely available in usual care where sexual health counseling typically focuses on the prevention of sexually transmitted sexual diseases, including the human immunodeficiency virus (HIV) [6, 63, 82–86].

MEN AND COPING

Help-Seeking, Athleticism, and Risk Taking Behaviors

Help-seeking, as discussed above, is an important coping behavior for individuals who are faced with stress but feel unable to cope effectively. Men are described as engaging in help-seeking behavior less than women. Help-seeking implies the loss of control and can be seen by some men as damaging to their identity. As a result, they may not receive care until they are in a health crisis. Conformity to traditional masculine norms results not only in less help-seeking but also in negative attitudes to seeking help, especially for psychological concerns [87]. The stigma attached to help-seeking has been reported across men of all ethnicities and throughout the life span [57, 88–90].

Hegemonic masculinity can lead men to engage in extreme demonstration of physical and emotional strength and risk taking, such as extreme demonstrations of muscularity and sporting, excessive alcohol use, and unprotected sex [4, 57, 91–94]. Pleck and colleagues demonstrated that men who held traditional attitudes toward masculinity had more sexual partners, less consistent use of condoms with negative attitudes toward condoms despite knowledge that condomless sex leads to sexually transmitted diseases (STDs), and a belief that pregnancy validates masculinity [95]. The masculine ideology states that women should be responsible for condom use [96] and that men should have sex with multiple women. This has serious consequences not only for transmission of STDs such as chlamydia and gonorrhea but also for transmission of HIV, which has grave health consequences [97].

Drug and alcohol abuse are risk behaviors driven, in part, by the masculine fear of acknowledging vulnerability [2, 42]. Men are also more likely than women to engage in heavy alcohol use, smoking, and illicit drug use [56]. Alcohol abuse can cause acute and chronic health conditions that are injurious to health including motor vehicle and bicycle accidents, cancers, cardiovascular disease, and mortality. Smoking, a risk behavior in which men engage more frequently than women, can result in pulmonary disease, cancer, cardiovascular and cerebrovascular disease [45–51, 98, 99]. The risk of myocardial infarction from smoking is four times that of nonsmokers. Of the 400,000 deaths from smoking among the Danish adult population in 1997–2001, over half were among men [100]. A higher rate of illicit drug use also contributes to higher morbidity and mortality in men.

While health-promoting behavior, such as help-seeking for stress, is considered incompatible with masculinity, there is one health-promoting behavior that men consider compatible with masculinity and that has been shown to be protective: exercise [101]. Physical exercise is the preferred form of health-promoting practice for men. The strong association of masculinity and sports allow men to position themselves as strong, muscular, and competitive [102]. Given men's drive for muscularity, exercise can be leveraged as a health-promoting intervention strategy for men. Men may be encouraged to pursue positive health outcomes, if health-promoting behavior such as exercise is framed by masculine attributes (e.g., sporting targets and goals and autonomy) [103].

HEALTHCARE PROVIDERS AND MEN'S HEALTH

While we have demonstrated, in this chapter, that men are culturally discouraged to seek health care, it is important to realize that physicians and other health-care providers may contribute to this bias. Male and female providers may have the same gender bias and thereby not recognize emotional or health problems in a male patient. Research has shown that physicians are more likely to detect psychiatric illness in women than men [104, 105]. Studies also show that health-care providers do not spend much time on men's health issues during the usual clinic visit [106]. As a result of this, men get less information about their health status and less advice on how behavior may impact it negatively and how they might engage in behaviors that would improve their health. Given these gaps in health-care discussions, men are likely to obtain fewer services that would promote their physical and mental health [42].

Studies have reported that men receive briefer explanations in medical encounters and less physician time in their health visits when compared with women [107–109]. While there are plausible explanations for why men present differently than women [110–112], failure to recognize gender differences in expression leads to further underdiagnosis.

It is important to recognize and understand the masculinity construct and how behaviors associated with masculinity impact men's experience and help seeking, with regard to stress, physical, and mental health. Fig. 1 demonstrates the complex nature of masculinity and the diverse elements that impact each man's masculinity [113]. An open discussion between patients and providers is imperative and would enhance understanding, thereby allowing for appropriate interventions.

SUMMARY

In this chapter, we argue that being male carries with it unique stress that has to be recognized: the concept of masculinity is a significant aspect of men's health and mental health. As a cultural mandate, it asks that men be strong physically and mentally while being self-sufficient when it comes to coping. This prescription may guide how men process and cope with stress. We have also described the fact that the newer trends for an expanded male role that includes a greater

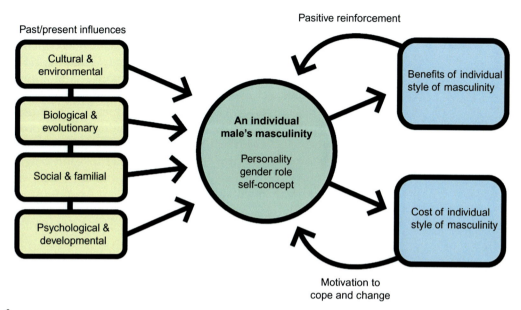

FIG. 1
Integrative model of masculinity—Meek (2011). This model demonstrates that each man has a different way of being identified as male—masculinity is complex and is made up of diverse elements.

emotional repertoire, a willingness to be vulnerable, and an assumption of some traditionally feminine roles, which have not yet commanded sufficient social support to offset the stress of that expansion. The masculine role deprives men's ability to develop and use internal resources of coping and seek external resources in the form of social and professional support. Instead, men may turn to confrontive coping, as explained by Lazarus, which are problem-focused coping strategies associated with unsatisfactory outcomes that ultimately worsen rather than improve the situation [9]. In this context, it is easy for men to feel uncertain about how to cope which increases stress and can lead to reluctance to seek help and turn to alcohol and substance abuse and risk behaviors, all which result in detrimental effects on both physical and mental health. A positive sign of men's efforts to engage in behaviors that can reduce stress and promote both physical and mental health is their engagement in athleticism with all its benefits. It is fair to say, however, that there continues to be a health disparity for men because, to date, they do not access care to help manage stress, depression, mental illness, and other health conditions.

Empowering men as they cope with stress, depression, mental illness, and other health conditions can only be effective if the social construction of masculinity can be expanded to include acceptance of emotional vulnerability, physical infirmity, and the positive nature of help seeking. It is not only men who critically need to embrace this expansion of the concept of masculinity. Women who are often men's partners and health-care providers must also recognize their own biases and overcome them. When health-care providers can normalize the experience of stress, depression, and other mental and physical symptoms by inquiring about them, men may become more communicative or less isolative when they feel overwhelmed and have difficulty coping. It is also important to engage partners and other family members in health and mental health assessments [70]. In the long term, reframing the concept of masculinity to support men's strengths while supporting their wider emotional and help-seeking repertoire is a worthy goal that will lead to addressing stress, depression, and other mental health and health conditions with neutral problem-solving, rather than with a gendered lens.

References

[1] Bosson JK, Vandello JA. Precarious manhood and its links to action and aggression. Curr Dir Psychol Sci 2011;20(2):82–6.

[2] Goldberg H. The hazards of being male: surviving the myth of masculine privilege. New York: New American Library; 1976.

[3] Jourard SM. The transparent self. New York: Van Nostrand Reinhold; 1971.

[4] Kimmel MS, Messner M. Men's lives. Needham Heights, MA: Allyn&Bacon; 1995.

[5] White A, De Sousa B, De Visser R, et al. Men's health in Europe. J Men's Health 2011;8(3):192–201.

[6] Moller-Leimkuhler AM. Barriers to help-seeking by men: a review of sociocultural and clinical literature with particular reference to depression. J Affect Disord 2002. ISSN: 0165-0327.

[7] Public Health Service. Healthy people 2000: national health promotion and disease prevention objectives. Washington, DC: U.S. Department of Health and Human Services; 1990.

[8] Lemle R, Mishkind ME. Alcohol and masculinity. J Subst Abuse Treat 1989. ISSN: 0740-5472.

[9] Folkman S, Lazarus RS, Dunkel-Schetter C, DeLongis A, DeLongis A, Gruen RJ. Dynamics of a stressful encounter: cognitive appraisal, coping, and encounter outcomes. J Pers Soc Psychol 1986. ISSN: 0022-3514.

[10] Goodnite PM. Stress: a concept analysis. Nurs Forum 2014;49(1):71–4.

[11] Harrison J. Warning: the male sex role may be dangerous to your health. J Soc Issues 1978;34(1):65–86.

[12] McCreary DR. The male role and avoiding femininity. Sex Roles 1994;31(9–10):517–31.

[13] Archer J. Childhood gender roles: social context and organisation. In: Childhood social development: contemporary perspectives. Hillsdale, NJ: Lawrence Erlbaum Associates, Inc; 1992. p. 31–61.

[14] Fagot BI. Consequence of moderate cross-gender behaviors in pre-school children. Child Dev 1977;48(3):902–7.

[15] Martin CL. Attitudes and expectations about children with nontraditional and traditional gender roles. Sex Roles 1990;22(3–4):151–65.

[16] de Visser RO, McDonnell EJ. "Man points": masculine capital and young men's health. Health Psychol 2013. ISSN: 1930-7810.

[17] Plaisier I, de Bruijn JG, Smit JH, de Graaf R, et al. Work and family roles and the association with depressive and anxiety disorders: differences between men and women. J Affect Disord 2008. ISSN: 0165-0327.

[18] Alonso J, Angermeyer MC, Bernert S, Bruffaerts R, et al. Prevalence of mental disorders in Europe: results from the European Study of the Epidemiology of Mental Disorders (ESEMeD) project. Acta Psychiatr Scand Suppl 2004. ISSN: 0065-1591.

[19] Horwitz AV, White HR. Becoming married, depression, and alcohol problems among young adults. J Health Soc Behav 1991. ISSN: 0022-1465.

[20] Helbig-Lang S, Lampert T, Klose M, Jacobi F. Is parenthood associated with mental health? Findings from an epidemiological community survey. Soc Psychiatry Psychiatr Epidemiol 2006;41.

[21] Allen TD, Herst DE, Bruck CS, Sutton M. Consequences associated with work-to-family conflict: a review and agenda for future research. J Occup Health Psychol 2000. ISSN: 1076-8998.

[22] Cairney J, Boyle M, Offord DR, Racine Y. Stress, social support and depression in single and married mothers. Soc Psychiatry Psychiatr Epidemiol 2003. ISSN: 0933-7954.

[23] Griffin JM, Fuhrer R, Stansfeld SA, Marmot M. The importance of low control at work and home on depression and anxiety: do these effects vary by gender and social class? Soc Sci Med 2002. ISSN: 0277-9536.

[24] Ferrie JE, Shipley MJ, Stansfeld SA, Marmot MG. Effects of chronic job insecurity and change in job security on self reported health, minor psychiatric morbidity, physiological measures, and health related behaviours in British civil servants: the Whitehall II study. J Epidemiol Community Health 2002. ISSN: 0143-005X.

[25] Wang J, Patten SB. Perceived work stress and major depression in the Canadian employed population, 20–49 years old. J Occup Health Psychol 2001. ISSN: 1076-8998.

[26] Wang JL, Lesage A, Schmitz N, Drapeau A. The relationship between work stress and mental disorders in men and women: findings from a population-based study. J Epidemiol Community Health 2008. ISSN: 0143-005X.

[27] Seeman T, Epel E, Gruenewald T, Karlamangla A, McEwen BS. Socio-economic differentials in peripheral biology: cumulative allostatic load. Ann N Y Acad Sci 2010;1186:223–39.

[28] Baum A, Garofalo JP, Yali AM. Socioeconomic status and chronic stress. Does stress account for SES effects on health? Ann N Y Acad Sci 1999;896:131–44.

[29] Dowd JB, Palermo T, Chyu L, Adam E, McDade TW. Race/ethnic and socioeconomic differences in stress and immune function in The National Longitudinal Study of Adolescent Health. Soc Sci Med (1982) 2014;115:49–55.

[30] Tamashiro KL. Metabolic syndrome: links to social stress and socioeconomic status. Ann N Y Acad Sci 2011;1231:46–55.

[31] Griffith DM, Ellis KR, Allen JO. An intersectional approach to social determinants of stress for African American men: men's and women's perspectives. Am J Mens Health 2013. ISSN: 1557-9891.

[32] Krieger N, Rowley DL, Herman AA, Avery B, Phillips MT. Racism, sexism, and social class: implications for studies of health, disease, and well-being. Am J Prev Med 1993;9(6 Suppl):82–122.

[33] Vega WA, Rumbaut RG. Ethnic minorities and mental health. Annu Rev Sociol 1991;17:351–83.

[34] Williams DR, Yan Y, Jackson JS, Anderson NB. Racial differences in physical and mental health: socio-economic status, stress and discrimination. J Health Psychol 1997;2(3):335–51.

[35] Lillie-Blanton M, Parsons PE, Gayle H, Dievler A. Racial differences in health: not just black and white, but shades of gray. Annu Rev Public Health 1996;17:411–48.

[36] Whitehead TL. Urban low-income African American men, HIV/AIDS, and gender identity. Med Anthropol Q 1997. ISSN: 0745-5194.

[37] Cook SH, Calebs BJ. The integrated attachment and sexual minority stress model: understanding the role of adult attachment in the health and well-being of sexual minority men. Behav Med 2016;42(3):164–73.

[38] Meyer IH. Prejudice, social stress, and mental health in lesbian, gay, and bisexual populations: conceptual issues and research evidence. Psychol Bull 2003;129(5):674–97.

[39] Alexander R, Parker K, Schwetz T. Sexual and gender minority Health Research at the National Institutes of Health. LGBT Health 2015;3(1):7–10.

[40] Institute of Medicine (US) Committee on Lesbian, Gay, Bisexual, and Transgender Health Issues and Research Gaps and Opportunities. The health of lesbian, gay, bisexual, and transgender people: building a foundation for better understanding. Washington, DC: National Academies Press; 2011.

[41] Hankivsky O. Women's health, men's health, and gender and health: implications of intersectionality. Soc Sci Med (1982) 2012;74(11):1712–20.

[42] Courtenay WH. Constructions of masculinity and their influence on men's well-being: a theory of gender and health. Soc Sci Med 2000. ISSN: 0277-9536.

[43] Xu J, Kochanek KD, Murphy SL, Arias E. Mortality in the United States. NCHS Data Brief 2012. ISSN: 1941-4927.

[44] Bunker SJ, Colquhoun DM, Esler MD, et al. "Stress" and coronary heart disease: psychosocial risk factors. Med J Aust 2003;178(6):272–6.

[45] Cai S, Li Y, Ding Y, Chen K, Jin M. Alcohol drinking and the risk of colorectal cancer death: a meta-analysis. Eur J Cancer Prev 2014. ISSN: 1473-5709.

[46] Klarich DS, Brasser SM, Hong MY. Moderate alcohol consumption and colorectal cancer risk. Alcohol Clin Exp Res 2015. ISSN: 1530-0277.

[47] Rehm J, Shield KD. Global alcohol-attributable deaths from cancer, liver cirrhosis, and injury in 2010. Alcohol Res 2013. ISSN: 2168-3492.

[48] Fernandez-Sola J. Cardiovascular risks and benefits of moderate and heavy alcohol consumption. Nat Rev Cardiol 2015. ISSN: 1759-5010.

[49] Matsumoto C, Miedema MD, Ofman P, Gaziano JM, et al. An expanding knowledge of the mechanisms and effects of alcohol consumption on cardiovascular disease. J Cardiopulm Rehabil Prev 2014. ISSN: 1932-751X.

[50] O'Keefe JH, Bhatti SK, Bajwa A, et al. Alcohol and cardiovascular health: the dose makes the poison...or the remedy. Mayo Clin Proc 2014. ISSN: 1942-5546.

[51] Arranz S, Chiva-Blanch G, Valderas-Martinez P, Medina-Remon A, Lamuela-Raventos RM, Estruch R. Wine, beer, alcohol and polyphenols on cardiovascular disease and cancer. Nutrients 2012. ISSN: 2072-6643.

[52] Valencia WM, Stoutenberg M, Florez H. Weight loss and physical activity for disease prevention in obese older adults: an important role for lifestyle management. Curr Diab Rep 2014. ISSN: 1539-0829.

[53] Bauer UE, Briss PA, Goodman RA, Bowman BA. Prevention of chronic disease in the 21st century: elimination of the leading preventable causes of premature death and disability in the USA. Lancet 2014. ISSN: 1474-547X.

[54] Ratner PA, Bottorff JL, Johnson JL, Hayduk LA. The interaction effects of gender within the health promotion model. Res Nurs Health 1994. ISSN: 0160-6891.

[55] Mechanic D, Cleary PD. Factors associated with the maintenance of positive health behavior. Prev Med 1980. ISSN: 0091-7435.

[56] Pinkhasov RM, Wong J, Kashanian J, Lee M, et al. Are men shortchanged on health? Perspective on health care utilization and health risk behavior in men and women in the United States. Int J Clin Pract 2010. ISSN: 1742-1241.

[57] Addis ME, Men Mahalik JR. masculinity, and the contexts of help seeking. Am Psychol 2003. ISSN: 0003-066X.

[58] Doyal L. Sex, gender, and health: the need for a new approach. BMJ 2001. ISSN: 0959-8138.

[59] Banks I. No man's land: men, illness, and the NHS. BMJ 2001. ISSN: 0959-8138.

[60] Kessler RC, Berglund P, Demler O, Jin R, Merikangas KR, Walters EE. Lifetime prevalence and age-of-onset distributions of DSM-IV disorders in the National Comorbidity Survey Replication. Arch Gen Psychiatry 2005. ISSN: 0003-990X.

[61] World Health Organization. The global burden of diease 2004 update; 2008.

[62] Piccinelli M, Wilkinson G. Gender differences in depression. Crit Rev 2000. ISSN: 0007-1250.

[63] Corney RH. Sex differences in general practice attendance and help seeking for minor illness. J Psychosom Res 1990. ISSN: 0022-3999.

[64] Horwitz A. The pathways into psychiatric treatment: some differences between men and women. J Health Soc Behav 1977. ISSN: 0022-1465.

[65] Gove WR, Tudor J. Sex differences in mental illness: a comment on Dohrenwend and Dohrenwend. Am J Sociol 1977;82(6):1327–49.

[66] Angst J, Dobler-Mikola A. Do the diagnostic criteria determine the sex ratio in depression? J Affect Disord 1984. ISSN: 0165-0327.

[67] Kendler KS, Kessler RC, Neale MC, Heath AC, Eaves LJ. The prediction of major depression in women: toward an integrated etiologic model. Am J Psychiatry 1993. ISSN: 0002-953X.

[68] Addis ME. Gender and depression in men. Clin Psychol Sci Pract 2008;15(3):153–68.

[69] Mahalik JR, Rochlen AB. Men's likely responses to clinical depression: what are they and do masculinity norms predict them? Sex Roles 2006;55(9–10):659–67.

[70] Murphy GE. Why women are less likely than men to commit suicide. Compr Psychiatry 1998. ISSN: 0010-440X.

[71] Poorolajal J, Rostami M, Mahjub H, Esmailnasab N. Completed suicide and associated risk factors: a six-year population based survey. Arch Iran Med 2015. ISSN: 1735-3947.

[72] Oliffe J, Phillips MJ. Men, depression and masculinities: a review and recommendations. J Men Health 2008;5.

[73] Aleman A, Kahn RS, Selten J-P. Sex differences in the risk of schizophrenia: evidence from meta-analysis. Arch Gen Psychiatry 2003. ISSN: 0003-990X.

[74] Morgan VA, Castle DJ, Jablensky AV. Do women express and experience psychosis differently from men? Epidemiological evidence from the Australian National Study of Low Prevalence (Psychotic) Disorders. Aust N Z J Psychiatry 2008. ISSN: 0004-8674.

[75] Hambrecht M, Maurer K, Hafner H, Sartorius N. Transnational stability of gender differences in schizophrenia? An analysis based on the WHO study on determinants of outcome of severe mental disorders. Eur Arch Psychiatry Clin Neurosci 1992. ISSN: 0940-1334.

[76] Wells JE, Browne MA, Scott KM, McGee MA, Baxter J, Kokaua J. Prevalence, interference with life and severity of 12 month DSM-IV disorders in Te Rau Hinengaro: the New Zealand Mental Health Survey. Aust N Z J Psychiatry 2006. ISSN: 0004-8674.

[77] Grant BF, Stinson FS, Hasin DS, Dawson DA, et al. Prevalence, correlates, and comorbidity of bipolar I disorder and axis I and II disorders: results from the National Epidemiologic Survey on Alcohol and Related Conditions. J Clin Psychiatry 2005. ISSN: 0160-6689.

[78] Negash A, Alem A, Kebede D, Deyessa N, Shibre T, Kullgren G. Prevalence and clinical characteristics of bipolar I disorder in Butajira, Ethiopia: a community-based study. J Affect Disord 2005. ISSN: 0165-0327.

[79] Almeida-Filho N, Mari Jde J, Coutinho E, Franca JF, et al. Brazilian multicentric study of psychiatric morbidity. Methodological features and prevalence estimates. Br J Psychiatry 1997. ISSN: 0007-1250.

[80] Field N, Prah P, Mercer CH, et al. Are depression and poor sexual health neglected comorbidities? Evidence from a population sample. BMJ Open 2016. ISSN: 2044-6055.

[81] Montejo AL, Montejo L, Navarro-Cremades F. Sexual side-effects of antidepressant and antipsychotic drugs. Curr Opin Psychiatry 2015. ISSN: 1473-6578.

[82] World Health Organization DoRHaR. Developing sexual health programmes: a framework for action; 2010.

[83] Fleming PJ, DiClemente RJ, Barrington C. Masculinity and HIV: dimensions of masculine norms that contribute to Men's HIV-related sexual behaviors. AIDS Behav 2016;20(4):788–98.

[84] Marwick C. Survey says patients expect little physician help on sex. JAMA 1999;281(23):2173–4.

[85] Satcher D, Hook 3rd EW, Coleman E. Sexual health in America: improving patient care and public health. JAMA 2015;314(8):765–6.

[86] Swartzendruber A, Zenilman JM. A national strategy to improve sexual health. JAMA 2010;304(9):1005–6.

[87] Levant RF, Stefanov DG, Rankin TJ, Halter MJ, Mellinger C, Williams CM. Moderated path analysis of the relationships between masculinity and men's attitudes toward seeking psychological help. J Couns Psychol 2013. ISSN: 0022-0167.

[88] Englar-Carlson M. Masculine norms and the therapy process. In: In the room with men: a casebook of therapeutic change. Washington, DC: American Psychological Association; 2006. p. 13–47.

[89] Richards M, Bedi RP. Gaining perspective: How men describe incidents damaging the therapeutic alliance. Psychol Men Masculinity 2015;16(2):170–82.

[90] Spendelow JS. Cognitive-behavioral treatment of depression in men: tailoring treatment and directions for future research. Am J Mens Health 2015. ISSN: 1557-9891.

[91] Deborah Sarah David RB. The forty-nine percent majority: the male sex role. Reading, MA: Addison-Wesley Publishing Company; 1976.

[92] White PG, Young K, McTeer WG. Sport, masculinity, and the injured body. In: Men's health and illness: gender, power, and the body. Thousand Oaks, CA: Sage Publications, Inc; 1995. p. 158–82.

[93] de Visser RO, Smith JA. Alcohol consumption and masculine identity among young men. Psychol Health 2007;22(5):595–614.

[94] Santana MC, Raj A, Decker MR, La Marche A, Silverman JG. Masculine gender roles associated with increased sexual risk and intimate partner violence perpetration among young adult men. J Urban Health 2006. n.d. ISSN: 1099-3460.

[95] Pleck JH, Sonenstein FL, Ku LC. Masculinity ideology: Its impact on adolescent males' heterosexual relationships. J Soc Issues 1993;49(3):11–29.

[96] Bowleg L, Teti M, Massie JS, Patel A, Malebranche DJ, Tschann JM. 'What does it take to be a man? What is a real man?': ideologies of masculinity and HIV sexual risk among black heterosexual men. Cult Health Sex 2011;13(5):545–59.

[97] Schwarz D, Feyler N, Cella J, Brady K. HIV/AIDS in the city of Philadelphia–2008 report. Philadelphia, PA: AIDS Activities Coordinating Office, Philadelphia Department of Public Health; 2008.

[98] Arora P, Chanana A, Tejpal HR. Estimation of blood alcohol concentration in deaths due to roadside accidents. J Forensic Leg Med 2013. ISSN: 1878-7487.

[99] Veldstra JL, Brookhuis KA, de Waard D, Molmans BHW, et al. Effects of alcohol (BAC 0.5 per thousand) and ecstasy (MDMA 100 mg) on simulated driving performance and traffic safety. Psychopharmacology (Berl) 2012. ISSN: 1432-2072.

[100] Bronnum-Hansen H, Juel K. Abstention from smoking extends life and compresses morbidity: a population based study of health expectancy among smokers and never smokers in Denmark. Tob Control 2001. ISSN: 0964-4563.

[101] Gordon DM, Hawes SW, Reid AE, Callands TA, et al. The many faces of manhood: examining masculine norms and health behaviors of young fathers across race. Am J Mens Health 2013. ISSN: 1557-9891.

[102] Messner MA. Messner MA, editor. Power at play: sports and the problem of masculinity. Boston: Beacon Press; 1992.

[103] Sloan C, Gough B, Conner M. Healthy masculinities? How ostensibly healthy men talk about lifestyle, health and gender. Psychol Health 2010;25(7):783–803.

[104] Marks JN, Goldberg DP, Hillier VF. Determinants of the ability of general practitioners to detect psychiatric illness. Psychol Med 1979. ISSN: 0033-2917.

[105] Stoppe G, Sandholzer H, Huppertz C, Duwe H, Staedt J. Gender differences in the recognition of depression in old age. Maturitas 1999. ISSN: 0378-5122.

[106] Bartlett EE. NCQA gender-specific standards: is there a place for men's health? Manag Care Q 2000. ISSN: 1064-5454.

[107] Waitzkin H. Doctor-patient communication. Clinical implications of social scientific research. JAMA 1984. ISSN: 0098-7484.

[108] Hall JA, Roter DL, Katz NR. Meta-analysis of correlates of provider behavior in medical encounters. Med Care 1988. ISSN: 0025-7079.

[109] Blanchard CG, Ruckdeschel JC, Blanchard EB, Arena JG, Saunders NL, Malloy ED. Interactions between oncologists and patients during rounds. Ann Intern Med 1983. ISSN: 0003-4819.

[110] Grigoriadis S, Robinson GE. Gender issues in depression. Ann Clin Psychiatry 2007. ISSN: 1547-3325.

[111] Checkley S. The neuroendocrinology of depression and chronic stress. Br Med Bull 1996. ISSN: 0007-1420.

[112] Weiss EL, Longhurst JG, Mazure CM. Childhood sexual abuse as a risk factor for depression in women: psychosocial and neurobiological correlates. Am J Psychiatry 1999. ISSN: 0002-953X.

[113] Meek W. Psychology of men, http://www.psychologyofmen.org; 2011.

CHAPTER 3.3

Meditation, Yoga, and Men's Health

Claire Postl, Lawrence C. Jenkins

Men's Health Program, Department of Urology, The James Comprehensive Cancer Center, The Ohio State University Wexner Medical Center, Columbus, OH, United States

INTRODUCTION

Stress is ubiquitous. *Unmanaged* stress has been linked to some health issues. Increasingly, providers have turned their attention toward helping patients manage stress more effectively. Research has demonstrated that a number of meditative practices can help patients lessen the toxic effects of stress [1, 2]. While stress management can mitigate the effects of adverse health outcomes, social and psychological barriers often inhibit men from seeking stress management interventions [3]. Societal masculinity standards limit the number of tools available for coping with stress. Men experience increasing rates of heart disease, diabetes, anxiety, depression, and sexual dysfunction, all of which may be impacted by stress [4–8]. Providers can recommend meditative practices as effective, research-based tools for managing stress and improving overall health.

STRESS

Stress is an autonomic response to threatening or unpleasant life situations, causing the endocrine system to release pituitary and adrenal hormones. Decades of research has demonstrated the deleterious effects of stress on mental and physical health [9]. The various pituitary and adrenal hormones all play roles in cardiovascular disease, sexual dysfunction, infertility, and depression [9]. Men are socialized for stress due to stigmata associated with the use of stress reduction mechanisms, such as counseling and emotional expression [3, 10]. Multiple variables *contribute* to stress; however, the ability to *manage* stress is often seeded in childhood and grown through life by socialization [10, 11].

Socialized for Stress

Parental, cultural, and social influences shape constructs of masculinity [10]. Often, boys are discouraged from expressing negative emotions such as sadness, fear, anger, depression, shame, or embarrassment. Additionally, boys are often characterized as less "manly" for having emotions or asking for support [10].

For many young men, being a man entails hiding feelings [12]. Boys will regulate their male peers by teasing and discrediting one another's emotions, perpetuating a connotation of masculinity, and creating feelings of shame and emasculation [12]. These emotional restrictions are associated with increased anger, anxiety, negative attitudes toward help-seeking, and decreased interpersonal functioning and well-being [13, 14]. What's more, men experience high emotional pain but due to culturally constructed masculinity restraints have difficulty disclosing distress or using coping mechanisms [15]. Shaming men for expressing emotions or seeking help limits their ability to cope with stress, which can have catastrophic implications; for example, men have higher rates of suicide than women [15]. Meditative practices are a good option for stress reduction in men because meditation practices do not require outward emotional expression, falling within the standards of socialized masculinity.

MEDITATIVE PRACTICES

Meditation is a mind-body practice focused on the awareness of the interaction between the mind, body, brain, and behavior [16]. Among the types of meditation, mindfulness has experienced recent popularity. Mindfulness is the "awareness that arises through paying attention to the present moment, nonjudgmentally." Mindfulness is an effective way to reduce stress and improve physical and psychological health [1, 2, 16]. Yoga is another mind-body practice, which commonly includes meditation, controlled breathing, and body poses [16].

The use of meditative practices, such as meditation, mindfulness, and yoga, has increased within the general population over the last 50 years [17]. The Centers for Disease Control and Prevention reported that, in the United States from 2002 to 2007, the percent of the general population practicing a form of meditation increased from 8.0% to 9.9% and yoga as a meditative practice increased from 6.0% to 11.0% between 2002 and 2012 [18]. Though meditative practice has increased in popularity in the United States, it is predominantly practiced by women, with 71%–74% of practitioners of meditation, mindfulness, or yoga being women [18–20]. Men could benefit from the use of the various meditative practices as complementary and alternative medicine in the treatment of the most common men's health issues.

HEALTH BENEFITS OF MEDITATIVE PRACTICES

Sexual Health

The stress of experiencing sexual dysfunction can impact confidence and self-esteem, causing increased anxiety and depression [21]. Similarly, depression and anxiety are often causes of sexual dysfunction [22]. Meditative practices have been a primary treatment approach for sexual dysfunction since Masters and Johnson's "sensate focus" [6, 23]. Sensate focus entails being mindful and present of sensations experienced during sexual touch and removing the pressure of *forced* pleasure and arousal [23]. Staying present, mindful, and nonjudgmental has a positive impact on sexual function; such practices alleviate stress and the pressure to perform [6, 24].

Sex positivity can also be used as a meditative practice for the treatment of sexual dysfunction and sexual insecurity [6]. Sex positivity, "the belief that all consensual expressions of sexuality are valid," entails "deliberate focus on the [patient's] personal meaning of their sexuality and its relationship to their well-being" [6]. Deliberate positive focus helps people remain mindful during a sexual encounter, decreasing the risk of negative self-judgment. Higher levels of mindfulness are associated with *fewer* sexual insecurities and *greater* sexual satisfaction in men and women, regardless of their relationship status or length [25, 26].

Men are less likely to engage in help-seeking behaviors. However, the likelihood of participation in help-seeking activities is increased when men hope to improve sexual functioning or sexual esteem [25, 27]. Specifically, men are more likely to seek help with sexual functioning or sexually related concerns to maintain, preserve, or restore feelings of masculinity [28]. Medical providers working with men's sexual health should promote more meditative practice as a complementary and alternative medicine for stress-related sexual function. Additionally, medical providers could use this help-seeking opportunity to educate male patients further on the other harmful effects stress has on their health.

Fertility

When struggling with fertility, men and women experience increased stress. Increased stress can also *cause* infertility in both men and women [29, 30]. The cycle of fertility and stress makes the management of stress important for men struggling with fertility.

Research shows that the pressure for men to perform and conceive creates significant psychological stress and leads to less sexual satisfaction [29]. A study looking at sexual satisfaction of infertile couples found that "helpful interventions" such as timed sexual intercourse and assisted reproductive technology add pressure and stress on couples [31]. Alternatively, men and women who are experiencing fertility issues benefit from stress reduction treatment options [31]. Meditative practices help couples maintain healthy levels of relationship and sexual satisfaction and provide stress management for couples struggling with fertility [31, 32].

Correspondingly, meditative practices can increase male fertility by improving men's overall health [33, 34]. Mindfulness-based stress reduction (MBSR) has been proved to improve sperm health by reducing oxidative stress. After a 4-week MBSR training, participants practicing MBSR had significant improvement in sperm DNA integrity and increased regulation of oxidative stress [35]. Integration of mind-body interventions into medical practice can improve patients' physical well-being and overall quality of life [35].

Weight and Diabetes Management

Obesity is a growing epidemic in our society. Stress has been linked to weight gain and obesity, while obesity often precedes a type 2 diabetes diagnosis [36]. Meditative practices have been shown to be an effective complementary and alternative medicine for weight and diabetes management [37, 38].

Psychosocial stressors are positively associated with weight gain [36]. "Comfort food" is a phrase used to describe foods that often cause abdominal obesity and that are consumed in an attempt to reduce stress and anxiety [7]. In these cases, eating is a maladaptive coping skill providing instant gratification, which is why people tend to turn toward food in times of high stress. Meditative practices provide people with alternative ways to manage emotions and stress, without the need to turn toward food to cope. Additionally, being mindful of hunger and stress can help challenge maladaptive coping behaviors, such as overeating [39]. Mindful eating is a meditative practice focused on "awareness of physical hunger, satiety cues, environmental and emotional triggers to eat, and making healthier food choices" [37]. Mindful eating is shown to be an effective intervention for weight loss and diabetes management [37, 38].

Cardiovascular Disease

Cardiovascular disease is the leading cause of death worldwide, and men have a higher risk of heart attack than women [40]. Health conditions, including obesity, diabetes, depression, and stress, compound the risk of cardiovascular disease in men [8, 40, 41].

Meditative practices have proved to be an effective intervention in decreasing cardiovascular disease-related stress [42, 43]. An eight-session patient education program for patients with congestive heart failure focused on general wellness and "mind-body skills," improved patient depression, fatigue, and reported life satisfaction [44]. Depression and anxiety are common comorbid concerns for cardiovascular patients [43]. The collaboration between medical and mental health professions to manage these comorbidities is essential in providing holistic care to men with cardiovascular disease [43].

Anxiety and Depression

Anxiety and depression are mental health diagnoses that often occur simultaneously. Depression and anxiety commonly coexist due to similar overlapping symptoms and biological mechanisms [45]. Meditative practices are used in psychotherapy for the treatment of depression and anxiety [1]. Meditation has an impact on neurochemical changes in the brain, eliciting the production of the neurotransmitters dopamine, serotonin, and norepinephrine [4]. These neurotransmitters are known to increase mood and decrease depression and anxiety. In an 8-week stress management mindfulness training program for patients on active surveillance for prostate cancer, meditation-based training improved effective coping with stressors associated with their health concerns [46].

Men who experience anxiety and depression are often limited in their ability to cope, due to shaming surrounding emotional expression [5]. Meditative practices allow for coping that does not involve outward emotional expression, thus still falling within the standards of socialized masculinity.

RECOMMENDATIONS

Social constructs of masculinity and socialization play a key role in men's ability to seek help. Men struggle with emotional expression and the identification of coping mechanisms due to constructs of masculinity. Meditative practices, which are female dominated in our society, are not as frequently sought out by men [19, 20, 27]. Medical and mental health providers can help men utilize meditative practices as complementary and alternative medicine.

Firstly, providers should incorporate a social worker or mental health provider into their referral network. Social workers can provide men with resources to help manage psychosocial stress, including meditative practices. Awareness of psychosocial adjustment, support systems, and coping skills are needed and should be assessed regularly [47]. Having an integrated social worker or mental health provider can help to normalize the stress that cycles alongside health issues and provide easier linking to services.

Secondly, providers should talk about masculinity and the barriers men face when seeking help. Men are more likely to seek help if barriers are actively addressed by medical providers [3]. Discussing masculinity and social influences that shape emotional expression can help to normalize the stress that the patient may be experiencing.

Lastly, patients should be educated on the different services and interventions available for the management of stress [3]. Providers should educate men on meditative practices and potential benefits related to patients' stress-related issues. Similarly, providers should encourage participation in meditative practices pre- and postmedical diagnoses. In order for providers to gain comfort in talking with men about meditative practices, it is encouraged that providers practice meditation, mindfulness, or yoga themselves [1].

CONCLUSION

Meditative practices have proved to be beneficial in reducing stress-related health issues, including sexual dysfunction, obesity and diabetes, infertility, cardiovascular disease, depression, and anxiety. Society standards of masculinity often inhibit men from emotional expression, and men are often discouraged and shamed from seeking help for stress management. Meditative practices, such as meditation, mindfulness, and yoga, are possible ways for men to manage stress that does not require emotional expression, thus adhering to our societies' masculinity constructs. Incorporating meditative practices as complementary and alternative medicine into medical treatment will provide men with effective, research-based interventions for stress reduction.

References

[1] Kabat-Zinn J. Mindfulness-based interventions in context: past, present, and future. Clin Psychol Sci Pract 2003;10(2):144–56. https://doi.org/10.1093/clipsy/bpg016.

[2] Davidson RJ, Kabat-Zinn J, Schumacher J, et al. Alterations in brain and immune function produced by mindfulness meditation. Psychosom Med 2003;65(4):564–70. https://doi.org/10.1097/01.PSY.0000077505.67574.E3.

[3] Lynch L, Hons BA, Long M, Hons BA, Moorhead A. Young men, help-seeking, and mental health services: exploring barriers and solutions. Am J Mens Health 2018. https://doi.org/10.1177/1557988315619469.

[4] Ursin H, Baade E, Levine S. Psychobiology of stress: a study of coping men. Elsevier Inc.; 1978.

[5] Ptacek JT, Smith RE, Dodge KL. Gender differences in coping with stress: when stressor and appraisals do not differ. Pers Soc Psychol Bull 1994;20(4):421–30. https://doi.org/10.1177/0146167294204009.

[6] Kimmes JG, Mallory AB, Cameron C, Köse Ö. A treatment model for anxiety-related sexual dysfunctions using mindfulness meditation within a sex-positive framework. Sex Relation Ther 2015;30(2):286–96. https://doi.org/10.1080/14681994.2015.1013023.

[7] Dallman MF, Pecoraro N, Akana SF, et al. Chronic stress and obesity: a new view of "comfort food" Proc Natl Acad Sci U S A 2003;100(20):11696–701. https://doi.org/10.1073/pnas.1934666100.

[8] Pickering TG. Mental stress as a causal factor in the development of hypertension and cardiovascular disease. Proc Curr Hypertens Rep 2001;3(3):249–54. https://doi.org/10.1007/s11906-001-0047-1.

[9] Pressman A, Hernandez A, Sikka SC. Lifestyle stress and its impact on male reproductive health. Elsevier; 2017. https://doi.org/10.1016/B978-0-12-801299-4.00005-0.

[10] Fischer A. Gender and emotion: a psychological perspective. New York, NY: Cambridge University Press; 2000.

[11] Consedine NS, Magai C. Attachment and emotion experience in later life: the view from emotions theory. Attach Hum Dev 2003;5(2):165–87. https://doi.org/10.1080/1461673031000108496.

[12] Oransky M, Marecek J. "I'm Not Going to Be a Girl" in Boys' friendships and peer groups. J Adolesc Res 2009;24(2):218–41. https://doi.org/10.1177/0743558408329951.

[13] Blazina C, Watkins CE. Masculine gender role conflict: effects on college men's psychological well-being, chemical substance usage, and attitudes towards help-seeking. J Couns Psychol 1996;43(4):461–5.

[14] Gross JJ, John OP. Individual differences in two emotion regulation processes: implications for affect, relationships, and well-being. J Pers Soc Psychol 2003;85(2):348–62. https://doi.org/10.1037/0022-3514.85.2.348.

[15] Cleary A. Suicidal action, emotional expression, and the performance of masculinities. Soc Sci Med 2012;74(4):498–505. https://doi.org/10.1016/j.socscimed.2011.08.002.

[16] National Center for Complementary and Integrative Health. Meditation: in depth, https://nccih.nih.gov/health/meditation/overview.htm; 2016. Accessed 7 August 2018.

[17] Kessler RC, Davis RB, Foster DF, et al. Long term trends in the use of complementary and alternative medical therapies in the United States. Ann Intern Med 2001;135(4):262–8. https://doi.org/10.7326/0003-4819-135-4-200108210-00011.

[18] Kachan D, Olano H, Tannenbaum SL, et al. Prevalence of mindfulness practices in the US workforce: National Health Interview Survey. Prev Chronic Dis 2017;14. https://doi.org/10.5888/pcd14.160034.

[19] Cramer H, Hall H, Leach M, et al. Prevalence, patterns, and predictors of meditation use among US adults: a nationally representative survey. Sci Rep 2016;6:1–9. https://doi.org/10.1038/srep36760.

[20] Cramer H, Ward L, Steel A, Lauche R, Dobos G, Zhang Y. Prevalence, patterns, and predictors of yoga use. Am J Prev Med 2016;50(2):230–5. https://doi.org/10.1016/j.amepre.2015.07.037.

[21] McCabe MP. The role of performance anxiety in the development and maintenance of sexual dysfunction in men and women. Int J Stress Manag 2005;12(4):379–88. https://doi.org/10.1037/1072-5245.12.4.379.

[22] Kennedy SH, Rizvi S. Sexual dysfunction, depression, and the impact of antidepressants. J Clin Psychopharmacol 2009;29(2):157–64. https://doi.org/10.1097/JCP.0b013e31819c76e9.

[23] Weiner L, Avery-Clark C. Sensate focus: clarifying the Masters and Johnson's model. Sex Relation Ther 2014;29(3):307–19. https://doi.org/10.1080/14681994.2014.892920.

[24] Stephenson KR. Mindfulness-based therapies for sexual dysfunction: a review of potential theory-based mechanisms of change. Mindfulness (N Y) 2017;8(3):527–43. https://doi.org/10.1007/s12671-016-0652-3.

[25] Dunkley CR, Goldsmith KM, Gorzalka BB. The potential role of mindfulness in protecting against sexual insecurities. Can J Hum Sex 2015;24(2):92–103. https://doi.org/10.3138/cjhs.242-A7.

[26] Pepping CA, Cronin TJ, Lyons A, Caldwell JG. The effects of mindfulness on sexual outcomes: the role of emotion regulation. Arch Sex Behav 2018;1–12. https://doi.org/10.1007/s10508-017-1127-x.

[27] Addis ME, Mahalik JR. Men, masculinity, and the contexts of help seeking. Am Psychol 2003;58(1):5–14. https://doi.org/10.1037/0003-066X.58.1.5.

[28] O'Brien R, Hunt K, Hart G. "It's caveman stuff, but that is to a certain extent how guys still operate": men's accounts of masculinity and help seeking. Soc Sci Med 2005;61(3):503–16. https://doi.org/10.1016/j.socscimed.2004.12.008.

[29] Monga M, Alexandrescu B, Katz SE, Stein M, Ganiats T. Impact of infertility on quality of life, marital adjustment, and sexual function. Urology 2004;63(1):126–30. https://doi.org/10.1016/j.urology.2003.09.015.

[30] Bowlby J. Attachment and loss: retrospect and prospect. Am J Orthop 1982;52(4):664–78. https://doi.org/10.1111/j.1939-0025.1982.tb01456.x.

[31] Shoji M, Hamatani T, Ishikawa S, et al. Sexual satisfaction of infertile couples assessed using the Golombok-Rust Inventory of Sexual Satisfaction (GRISS). Sci Rep 2014;4 [Figure 1]:1–5. https://doi.org/10.1038/srep05203.

[32] Darbandi S, Darbandi M, Khorram Khorshid HR, Sadeghi MR. Yoga can improve assisted reproduction technology outcomes in couples with infertility, Altern Ther Health Med 2017; (c). http://www.ncbi.nlm.nih.gov/pubmed/29112941.

[33] Sengupta P, Chaudhuri P, Bhattacharya K. Male reproductive health and yoga. Int J Yoga 2014;6(6):85–137. https://doi.org/10.4103/0973-6131.113391.

[34] Dada R, Tolahunase M. Yoga meditation lifestyle intervention: impact on male reproductive health. Elsevier Inc.; 2017. https://doi.org/10.1016/B978-0-12-801299-4.00009-8

[35] Gautam S, Chawla B, Bisht S, Tolahunase M, Dada R. Impact of mindfulness based stress reduction on sperm DNA damage. J Anat Soc India 2018. https://doi.org/10.1016/j.jasi.2018.07.003.

[36] Block JP, He Y, Zaslavsky AM, Ding L, Ayanian JZ. Psychosocial stress and change in weight among US adults. Am J Epidemiol 2009;170(2):181–92. https://doi.org/10.1093/aje/kwp104.

[37] Miller CK, Kristeller JL, Headings A, Nagaraja H. Comparison of a mindful eating intervention to a diabetes self-management intervention among adults with type 2 diabetes: a randomized controlled trial. Health Educ Behav 2014;41(2):145–54. https://doi.org/10.1177/1090198113493092.

[38] Dalen J, Smith BW, Shelley BM, Sloan AL, Leahigh L, Begay D. Pilot study: Mindful Eating and Living (MEAL): weight, eating behavior, and psychological outcomes associated with a mindfulness-based intervention for people with obesity. Complement Ther Med 2010;18(6):260–4. https://doi.org/10.1016/j.ctim.2010.09.008.

[39] Godsey J. The role of mindfulness based interventions in the treatment of obesity and eating disorders: an integrative review. Complement Ther Med 2013;21(4):430–9. https://doi.org/10.1016/j.ctim.2013.06.003.

[40] American Heart Association. Understand your risk to prevent a heart attack. American Heart Association, Inc; 2016.http://www.heart.org/en/health-topics/heart-attack/understand-your-risk-to-prevent-a-heart-attack.

[41] Shively CA, Register TC, Clarkson TB. Social stress, visceral obesity, and coronary artery atherosclerosis: product of a primate adaptation. Am J Primatol 2009;71(9):742–51. https://doi.org/10.1002/ajp.20706.

[42] Keyworth C, Knopp J, Roughley K, Dickens C, Bold S, Coventry P. A mixed-methods pilot study of the acceptability and effectiveness of a brief meditation and mindfulness intervention for people with diabetes and coronary heart disease. Behav Med 2014;40(2):53–64. https://doi.org/10.1080/08964289.2013.834865.

[43] Ai AL, Rollman BL, Berger CS. Comorbid mental health symptoms and heart diseases: can health care and mental health care professionals collaboratively improve the assessment and management? Health Soc Work 2006;27–39.

[44] Kemper KJ, Carmin C, Mehta B, Binkley P. Integrative medical care plus mindfulness training for patients with congestive heart failure: proof of concept. J Evid Based Complementary Altern Med 2016;21(4):282–90. https://doi.org/10.1177/2156587215599470.

[45] Beth Salcedo M. The comorbidity of anxiety and depression, National Alliance of Mental Illness; 2018. https://www.nami.org/Blogs/NAMI-Blog/January-2018/The-Comorbidity-of-Anxiety-and-Depression.

[46] Victorson D, Hankin V, Burns J, et al. Feasibility, acceptability and preliminary psychological benefits of mindfulness meditation training in a sample of men diagnosed with prostate cancer. Psychooncology 2017;1155–63.

[47] Andrea B, Schulze T, Maercker A, Horn AB. Mental health and multimorbidity: psychosocial adjustment as an important process for quality of life. Gerontology 2014;60:249–54. https://doi.org/10.1159/000358559.

Further Reading

[48] Luu K, Hall PA. Examining the acute effects of hatha yoga and mindfulness meditation on executive function and mood. Mindfulness (NY) 2017;8(4):873–80. https://doi.org/10.1007/s12671-016-0661-2.

PART 4

Hormones

CHAPTER 4.1

Testosterone and Men's Health

James Anaissie*, Alexander W. Pastuszak[†], Mohit Khera*

*Scott Department of Urology, Baylor College of Medicine, Houston, TX, United States, [†]Division of Urology, Department of Surgery, University of Utah School of Medicine, Salt Lake City, UT, United States

INTRODUCTION

Driven by a rapidly aging male population, the discussion surrounding testosterone deficiency (TD) has become a quickly evolving focus of interest in the field of sexual medicine. Although the literature is often inconsistent in defining TD, for the purposes of this chapter, we will distinguish between "hypogonadism" (HG), as defined by the presence of both hypogonadal symptoms and low serum testosterone levels, and "androgen deficiency" (AD), which is defined solely by the presence of low serum testosterone levels. The diagnosis of HG remains controversial, leading to a wide range in estimated prevalence, from 2% to 77% [1–5]. The prevalence of HG increases as patients age, particularly between 45 and 50 years old, ultimately affecting almost half of all men over 60 [1–4]. While aging alone is not a significant risk factor for HG, aging increases the likelihood of developing the condition through the development of other comorbidities. HG is a disease that can significantly affect quality of life, morbidity, and mortality, making the diagnosis of HG, its treatment with testosterone therapy (TTh), and the safety and lasting effects of TTh a topic of central importance to both patients and their providers.

The symptoms of HG can be very bothersome to patients and are summarized in Table 1. Symptoms of HG are often nonspecific, though delayed sexual development, the loss of body hair, and very small testis (<6 mL) are strongly suggestive of the diagnosis. Suggestive symptoms include diminished libido, often with associated erectile dysfunction (ED), gynecomastia, eunuchoid body proportions, infertility, height loss and decreased bone density, and hot flushes and sweats. Nonspecific signs and symptoms may include decreased energy, motivation, and self-confidence; poor concentration and memory; sleep disturbances; depressive symptoms; and weight gain with the loss of muscle mass and strength [6]. Studies suggest that HG is associated with an increased incidence

Table 1 Signs and Symptoms of Hypogonadism

Physical	Mental	Sexual
Reduced energy	Depressive symptoms	Reduced erectile function
Diminished physical performance	Decreased cognition	Reduced libido
Diminished work performance	Loss of memory	
Loss of body hair	Poor concentration	
Reduced lean muscle mass	Decreased motivation	
Increased fat mass	Irritability	

of major cardiac events including myocardial infarction, hypertension (HTN), dyslipidemia, and obesity, suggesting its treatment may reduce cardiovascular morbidity and mortality [7].

There are also important long-term effects of chronically low testosterone, including an increased incidence of osteoporosis, decreased muscle mass, anemia, reduced quality of life, infertility, and cardiovascular morbidity and mortality [8–13]. Conversely, several risk factors have been identified for the development of TD, namely, not only age but also several conditions with rapidly increasing prevalence in the United States including diabetes mellitus (DM), HTN, and the metabolic syndrome (MetS) [1–4, 14–18]. An increased morbidity and mortality from HG, paired with a higher frequency of metabolic comorbidities, have caused the discussion around HG diagnosis and treatment to gain significant traction in both professional and lay communities.

DIAGNOSIS OF HYPOGONADISM

The diagnosis of HG is a somewhat controversial topic, leading to a variety of practice patterns for diagnosis and treatment. It is estimated that up to 27% of men who receive testosterone do not have serum testosterone levels checked prior to treatment initiation and nearly half do not receive any follow-up testing after initiating treatment [19, 20], and conversely, concerns about the potential side effects have led many men who may need therapy to be left without it [19–21]. One of the main points of contention in the diagnosis of HG stems from a wide range in what is defined as a low serum testosterone. In the literature, this ranges from 200 to 400 ng/dL, which can be partly attributed to the lack of agreed-upon testosterone laboratory reference values and to a lack of consensus on what the lower limit of normal for men across the adult lifespan is [6, 22–25]. In each laboratory, a reference range for serum testosterone is determined for that specific laboratory based on a set of controls chosen to represent a sample of the population that presents to that lab; however, all literature on the subject uses absolute thresholds, challenging the clinician to utilize his/her own laboratory's metrics or the absolute values published in the literature. As an example, a study comparing 12 academic laboratories, 12 community laboratories, and a national laboratory showed a 350% difference in the reference ranges suggested for subnormal serum testosterone [26]. Additionally, the weight practitioners place on the symptoms of HG versus the serum testosterone level when diagnosing HG varies widely, with most settling on a combination of the two, the latter of which is recommended by most-society guidelines [6, 7, 27].

Earlier this year, in an attempt to provide clarity on this issue, the American Urological Association (AUA) released new guidelines on the diagnosis and treatment of HG [7]. In the newly published guidelines, the AUA recommends that patients diagnosed with HG have both a low serum testosterone value *and* the presence of the signs or symptoms of androgen deficiency. In the guidelines, a total testosterone threshold of 300 ng/dL is recommended as the lower limit of normal. Additionally, the AUA guideline states that the value must be subnormal on two separate occasions and measured in the early morning for proper diagnosis. While the updated AUA guidelines are an important step in clarifying the diagnosis of TD, much work is needed to decrease the variability in global practice patterns.

TREATMENT OF HYPOGONADISM

TTh is the mainstay of treatment for HG and can dramatically improve the symptoms, quality of life, and arguably mortality for patients. Proper treatment of HG consists of thorough counseling, optimizing TTh, and close follow-up.

The first step after a patient is diagnosed with HG is thorough counseling. Discussing the potential benefits of TTh, including improvements in erectile function, libido, anemia, bone density, lean body mass, and mood [7, 28–34], is important. Some evidences, although currently inconclusive, suggest that TTh may even improve cognitive function, diabetes, lipid levels, energy, and several quality-of-life measures [29, 35–40]. It is equally important that patients are aware that TD can lead to an increased risk of cardiovascular disease, although at this time it is unclear whether placing patients on TTh increases or decreases cardiovascular risk [7, 41].

TTh dosing should be carefully adjusted such that men achieve a serum total testosterone level in the middle tercile of normal (450–600 ng/dL), which should theoretically ameliorate any symptoms [6, 7, 27]. Patients should be reevaluated 6–12 weeks after initiating treatment to determine if dosing adjustments are needed, although follow-up can vary based on TTh formulation utilized. If patients' serum testosterone is within the physiologically normal range after initiating TTh, but they do not feel improvement in symptoms, therapy should be discontinued [7]. TTh may be initiated in several formulations including a variety of topical formulations, oral agents (only available in Europe), an intranasal gel, intramuscular injections, and subcutaneous pellets (Table 2). Of note, apart from warning against cardiovascular risks, topical formulations carry an additional FDA warning for a small risk of transference of the medication to others in contact, which can lead to precocious puberty in

Table 2 Formulations of Testosterone Replacement Therapy With Associated Efficacy and Adverse Events

Agent/Route	Dosing	Efficacy	Adverse Events
Transdermal			
Gels/solutions	Applied to the skin	74%–87% [42–44]	Variable absorption, site reactions, transference
Patches	Applied to the skin	77%–100% [45]	Application site reactions (60%) [46]
Buccal	Pellet applied to top of gums twice daily	73% [47]	Gum pain and edema, displacement of pellet during exertion
Intranasal	Two intranasal pumps twice daily	90% [48]	Nasopharyngitis, rhinorrhea, epistaxis (all 7%–10%) [48]
Injectable			
Short acting	Weekly vs biweekly	100% [7]	More often supra- vs subtherapeutic, local reactions (7%–33%), polycythemia (19%–44%) [7]
Long acting	Given at time week 0, 4, and every 10 weeks thereafter	94% [49]	Injection site acne, pain, fatigue, pulmonary microemboli (2%)[a], polycythemia [7]
Subcutaneous pellets	2–6 pellets every 3–6 months[b]	100% [50]	Polycythemia (50%), ecchymoses (35%), pain (30%), and swelling [51][c]
Oral			
Alkylated[d]			Not recommended by AUA due to difficulty achieving therapeutic levels and high rates of liver toxicity [52]
Undecanoate[e]	Pill three times daily	77% [53]	Erythrocytosis [53]

[a]Coughing episodes lasted 1–10 min in duration and were managed conservatively in the clinic without need for supplemental oxygen [54].
[b]Most providers administered more pellets than recommended by the FDA, with the majority of providers studied electing to use ≥10 pellets (63%), 27% using 8–9 pellets, and only 10% of cases using 6–7 pellets (all of which are higher than FDA-recommended 2–6 pellets) [50].
[c]All mentioned symptoms resolved within 4 months post insertion.
[d]17-alpha-alkylated androgens not recommended by the AUA due to difficulty in achieving therapeutic levels and liver toxicity, with up to one-third of patients having abnormal liver function tests.
[e]Not yet approved in the United States. Only approved for use in certain European countries.

children and virilization in women [55, 56]. One exception for topical formulations is the Natesto, intranasal gel, which does not carry the FDA warning on transference. Lastly, but importantly, patients should be encouraged to pursue lifestyle modifications including weight loss and exercise [57–59]. Follow-up appointments and testing are crucial for ensuring a therapeutic serum testosterone level [7].

EFFICACY OF TESTOSTERONE THERAPY

Testosterone therapy can lead to significant improvements in several domains of quality of life in men with HG. Many of the most prevalent men's health disorders such as ED and lower urinary tract symptoms (LUTS) can be influenced by TTh. Similarly, recent literature has found that TTh can improve glucose control in diabetes and can positively impact components of the metabolic syndrome.

Erectile Dysfunction

Erectile dysfunction is one of the most common reasons that men seek treatment with testosterone [7]. While seen in just 2%–10% of men <50 years old, the prevalence of ED increases dramatically with age, climbing to 30%–40% in men between 60 and 70 years old and reaching over 50% in men older than 70 [60–62]. While on TTh, patients with both ED and HG often see an improvement in nocturnal erections, ease in attaining erections, and improved penile rigidity [7].

In two well-powered studies, men with HG treated with TTh compared with placebo showed significant improvement in the erectile function domain of the validated International Index of Erectile Function (IIEF) [28–30]. Similarly, a pooled analysis of six trials showed a mean improvement of 1.32 points in the same score, which is consistent with data from meta-analyses [39, 63–68]. Additionally, TTh and normalization of testosterone levels result in improved response to phosphodiesterase type-5 inhibitor (PDE5i) treatment [68]. While the AUA review panel was unable to determine what subset of or what percentage of men with HG have improvements in erectile function from TTh, it is clear that some of these patients will have improvement [7]. Although these studies do show an improvement in erectile function, study design limitations such as failure to record baseline ED, variable inclusion criteria, and short follow-up periods have led to the conclusion that the true benefits of TTh may actually be underestimated. Long-term prospective data suggest that there may be slow but progressive improvements in erectile function for up to 3 years after beginning therapy [69].

TTh may also lead to a meaningful improvement in libido, classically measured using the sexual desire section of the IIEF, though the data remain inconclusive. In prospective studies, nonsignificant improvements in the IIEF were observed, although a large randomized control trial using the Derogatis Interview for Sexual Functioning in Men-II (DISF-M-II) score demonstrated a statistically significant improvement in sexual desire [28–30, 64]. A meta-analysis showed a pooled improvement of a 31% increase in libido among men treated with TTh, noting greater improvement in men with HG [70]. These mixed results suggest that libido represents a multifaceted metric that is difficult to measure using standardized questionnaires, though there appears to be some improvement in libido in patients on TTh.

Urinary Function

Several studies have suggested a relationship between HG and LUTS in aging men [71]. Although TTh can lead to significant prostate growth in young men with HG, these effects may be lost in older men [72–77]. LUTS are often the clinical expression of benign prostatic hypertrophy (BPH), yet TTh does not appear to worsen LUTS and may in fact improve them. In a cross-sectional study of more than 2000 eugonadal men, higher serum testosterone levels positively correlated with improved flow rate (Q_{max}) and negatively with postvoid residual (PVR), although these results were not significant when adjusting for age and/or metabolic syndrome [78]. These results correlate with a prospective observational study of 261 hypogonadal men, in which TTh led to significant improvement in LUTS during a median follow-up of 42.3 months [79]. Two other studies with a smaller subset of men similarly showed small but significant improvements in LUTS, suggesting that TTh may play a beneficial role in the treatment of LUTS in older men [80, 81]. While testosterone products contain warnings against possible prostate enlargement and worsening of LUTS, there is no conclusive evidence supporting either claim; in fact, emerging evidence suggests that TTh may result in improvement in LUTS.

Metabolic Syndrome

The metabolic syndrome (MetS) consists of obesity, dyslipidemia, HTN, and insulin resistance and has a rapidly rising prevalence in the United States, affecting approximately one of every five Americans and increasing 56% over the past 8 years [82–84]. Interestingly, in patients with these comorbidities, the prevalence of HG climbs dramatically, reaching up to 80% [14–18]. Several studies have shown that over half of all obese men also have low serum testosterone values, and conversely, over 70% of men with HG are also obese [4, 85, 86]. In the setting of DM, a large systemic review and meta-analysis found that serum testosterone levels were much lower in men with DM, with a mean difference of

76.6 ng/dL [87]. The authors predicted that people without TD had a 42% lower risk of developing DM than those with TD. Many studies have gone on to suggest that TD may be a significant predictor for the onset of DM [88–91]. Similarly, serum testosterone is inversely related to systolic blood pressure and left ventricular hypertrophy [92] as well as serum cholesterol levels [93, 94]. Although several studies demonstrate strong positive associations between HG and MetS, it is important to consider the inherent difficulty in linking a clinical syndrome with several components to TD.

TTh may positively influence many parameters of the MetS [95]. In a large meta-analysis, TTh in patients with MetS led to significant reductions in fasting glucose, triglycerides, and waist circumference [96]. In another prospective study, TTh improved all components of the MetS including obesity, cholesterol, DM, and blood pressure, with improvements in erectile function and overall quality of life [97]. These benefits were again seen in a cumulative registry study, in which 5 years of TTh in men with MetS led to reductions in cholesterol, blood pressure, blood glucose, and hemoglobin A1c levels [98]. Several double-blind randomized controlled trials have confirmed these results, leading to a promising future for TTh in the treatment of an increasingly prevalent disease with significant morbidity and mortality [99–101].

SAFETY OF TESTOSTERONE THERAPY

Equally as important as describing the expected benefits from TTh is educating patients on the potential yet controversial adverse events that may arise from therapy. The most common adverse event associated with TTh is erythrocytosis, defined as a serum hematocrit >52%. In advanced cases, erythrocytosis can lead to hyperviscosity and theoretically lead to venothrombotic events such as myocardial infarction (MI), stroke, and deep venous thrombosis (CDVT) [102], which led the FDA in 2015 to require a black box warning of these risks. However, erythrocytosis has not been definitively shown to lead to cardiovascular complications in the setting of HG, with studies not conclusively demonstrating an association between TTh-induced erythrocytosis and venothrombotic events [7, 30, 103–106]. Another more controversial adverse event is a potential increased risk of independent cardiovascular events in men on TTh. Most studies have not shown an association between TTh and cardiac events; however, a few studies have suggested potential harm, throwing the safety profile of TTh into question. Many of the studies that have identified an increased risk of cardiovascular events in men on TTh have been heavily criticized for design flaws including poor patient selection and inclusion and exclusion criteria [41, 63, 102, 107–115], leading the FDA to state that the "evidence linking testosterone therapy to an increased risk of heart attack, stroke, and death is inconclusive" and to recommend more comprehensive studies to better assess the potential cardiovascular risks with TTh [116].

TTh was initially thought to increase the risk for development or progression of prostate cancer (CaP), given the cancer's androgen-responsive nature. Although the FDA requires a warning of the risk of developing CaP while on TTh, most available evidence suggests that TTh is not linked to development or progression of CaP [7, 28, 110]. Similarly, due to a lack of data demonstrating a risk for TTh in men with past or active CaP, it is currently recommended that these patients be evaluated on a case-by-case basis based on the potential benefits of treatment [75, 117–122].

TTh can also reduce fertility, and men should be counseled that TTh should be stopped in advance of any effort to conceive. While two-thirds of men recover sperm in their ejaculate within 6 months of stopping TTh, recovery is highly variable, as 10% may take up to 2 years to recover, and some may not recover spermatogenesis at all [123–125]. Lastly, the effects of TTh on men with preexisting cardiovascular disease such as hypertension, MI, and stroke are unclear, and thus, it is recommended that these patients are observed for 3–6 months before initiating TTh [126, 127].

CONCLUSION

Hypogonadism has tangible effects on men's health, and its treatment with TTh improves symptoms and metabolic parameters of conditions that are influenced by the disease. Although controversy remains on the diagnosis of HG and the long-term safety of TTh, most available studies support its safety and efficacy. Clearly, more prospective studies are needed to better define the morbidity and mortality of hypogonadism and the potential for TTh to safely improve quality of life in these patients.

References

[1] Khoo EM, Tan HM, Low WY. Erectile dysfunction and comorbidities in aging men: an urban cross-sectional study in Malaysia. J Sex Med 2008;5(12):2925–34. Epub 2008/09/03, https://doi.org/10.1111/j.1743-6109.2008.00988.x18761590.

[2] Ponholzer A, Madersbacher S, Rauchenwald M, Jungwirth S, Fischer P, Tragl KH. Vascular risk factors and their association to serum androgen levels in a population-based cohort of 75-year-old men over 5 years: results of the VITA study. World J Urol 2010;28(2):209–14. Epub 2009/06/30. https://doi.org/10.1007/s00345-009-0440-y19562348.

[3] Wong SY, Chan DC, Hong A, Woo J. Prevalence of and risk factors for androgen deficiency in middle-aged men in Hong Kong. Metabolism 2006;55(11):1488–94. Epub 2006/10/19. https://doi.org/10.1016/j.metabol.2006.06.01917046551.

[4] Mulligan T, Frick MF, Zuraw QC, Stemhagen A, McWhirter C. Prevalence of hypogonadism in males aged at least 45 years: the HIM study. Int J Clin Pract 2006;60(7):762–9. Epub 2006/07/19. https://doi.org/10.1111/j.1742-1241.2006.00992.x16846397.

[5] Millar AC, Lau AN, Tomlinson G, Kraguljac A, Simel DL, Detsky AS, et al. Predicting low testosterone in aging men: a systematic review. CMAJ 2016;188(13):E321–30. Epub 2016/06/22. https://doi.org/10.1503/cmaj.15026227325129.

[6] Bhasin S, Cunningham GR, Hayes FJ, Matsumoto AM, Snyder PJ, Swerdloff RS, et al. Testosterone therapy in men with androgen deficiency syndromes: an Endocrine Society clinical practice guideline. J Clin Endocrinol Metab 2010;95(6):2536–59. Epub 2010/06/09. https://doi.org/10.1210/jc.2009-235420525905.

[7] Mulhall JP, Trost LW, Brannigan RE, Kurtz EG, Redmon JB, Chiles KA, et al. Evaluation and management of testosterone deficiency: AUA guideline. J Urol 2018;200(2):423–32.

[8] Tajar A, Huhtaniemi IT, O'Neill TW, Finn JD, Pye SR, Lee DM, et al. Characteristics of androgen deficiency in late-onset hypogonadism: results from the European Male Aging Study (EMAS). J Clin Endocrinol Metab 2012;97(5):1508–16. Epub 2012/03/16. https://doi.org/10.1210/jc.2011-251322419720.

[9] Haddad RM, Kennedy CC, Caples SM, Tracz MJ, Bolona ER, Sideras K, et al. Testosterone and cardiovascular risk in men: a systematic review and meta-analysis of randomized placebo-controlled trials. Mayo Clin Proc 2007;82(1):29–39. Epub 2007/02/09. https://doi.org/10.4065/82.1.2917285783.

[10] Wu FC, von Eckardstein A. Androgens and coronary artery disease. Endocr Rev 2003;24(2):183–217. Epub 2003/04/18. https://doi.org/10.1210/er.2001-002512700179.

[11] Corona G, Mannucci E, Forti G, Maggi M. Hypogonadism, ED, metabolic syndrome and obesity: a pathological link supporting cardiovascular diseases. Int J Androl 2009;32(6):587–98. Epub 2009/02/20. https://doi.org/10.1111/j.1365-2605.2008.00951.x19226407.

[12] Corona G, Rastrelli G, Vignozzi L, Mannucci E, Maggi M. Testosterone, cardiovascular disease and the metabolic syndrome. Best Pract Res Clin Endocrinol Metab 2011;25(2):337–53. Epub 2011/03/15. https://doi.org/10.1016/j.beem.2010.07.00221397202.

[13] Behre HM, Kliesch S, Leifke E, Link TM, Nieschlag E. Long-term effect of testosterone therapy on bone mineral density in hypogonadal men. J Clin Endocrinol Metab 1997;82(8):2386–90. Epub 1997/08/01. https://doi.org/10.1210/jcem.82.8.41639253305.

[14] Pellitero S, Olaizola I, Alastrue A, Martinez E, Granada ML, Balibrea JM, et al. Hypogonadotropic hypogonadism in morbidly obese males is reversed after bariatric surgery. Obes Surg 2012;22(12):1835–42. Epub 2012/08/28. https://doi.org/10.1007/s11695-012-0734-922923309.

[15] Wang C, Jackson G, Jones TH, Matsumoto AM, Nehra A, Perelman MA, et al. Low testosterone associated with obesity and the metabolic syndrome contributes to sexual dysfunction and cardiovascular disease risk in men with type 2 diabetes. Diabetes Care 2011;34(7):1669–75. Epub 2011/06/29. https://doi.org/10.2337/dc10-233921709300.

[16] Atlantis E, Martin SA, Haren MT, O'Loughlin PD, Taylor AW, Anand-Ivell R, et al. Demographic, physical and lifestyle factors associated with androgen status: the Florey Adelaide Male Ageing Study (FAMAS). Clin Endocrinol 2009;71(2):261–72. Epub 2009/01/31. https://doi.org/10.1111/j.1365-2265.2008.03463.x19178527.

[17] Haring R, Ittermann T, Volzke H, Krebs A, Zygmunt M, Felix SB, et al. Prevalence, incidence and risk factors of testosterone deficiency in a population-based cohort of men: results from the study of health in Pomerania. Aging Male 2010;13(4):247–57. Epub 2010/05/28. https://doi.org/10.3109/13685538.2010.48755320504090.

[18] Wu FC, Tajar A, Pye SR, Silman AJ, Finn JD, O'Neill TW, et al. Hypothalamic-pituitary-testicular axis disruptions in older men are differentially linked to age and modifiable risk

factors: the European Male Aging Study. J Clin Endocrinol Metab 2008;93(7):2737–45. Epub 2008/02/14. https://doi.org/10.1210/jc.2007-197218270261.

[19] Malik RD, Wang CE, Lapin B, Lakeman JC, Helfand BT. Characteristics of men undergoing testosterone replacement therapy and adherence to follow-up recommendations in metropolitan multicenter health care system. Urology 2015;85(6):1382–8. Epub 2015/04/12. https://doi.org/10.1016/j.urology.2015.01.02725862121.

[20] Baillargeon J, Urban RJ, Kuo YF, Holmes HM, Raji MA, Morgentaler A, et al. Screening and monitoring in men prescribed testosterone therapy in the U.S., 2001-2010. Public Health Rep 2015;130(2):143–52. Epub 2015/03/03. https://doi.org/10.1177/003335491513 00020725729103.

[21] Malik RD, Lapin B, Wang CE, Lakeman JC, Helfand BT. Are we testing appropriately for low testosterone?: characterization of tested men and compliance with current guidelines. J Sex Med 2015;12(1):66–75. Epub 2014/11/11. https://doi.org/10.1111/jsm.1273025382540.

[22] Wang C, Nieschlag E, Swerdloff R, Behre HM, Hellstrom WJ, Gooren LJ, et al. Investigation, treatment and monitoring of late-onset hypogonadism in males: ISA, ISSAM, EAU, EAA and ASA recommendations. Eur J Endocrinol 2008;159(5):507–14. Epub 2008/10/29. https://doi.org/10.1530/eje-08-060118955511.

[23] Hellstrom WJ, Paduch D, Donatucci CF. Importance of hypogonadism and testosterone replacement therapy in current urologic practice: a review. Int Urol Nephrol 2012;44 (1):61–70. Epub 2010/12/15. https://doi.org/10.1007/s11255-010-9879-421152980.

[24] Rosner W, Auchus RJ, Azziz R, Sluss PM, Raff H. Position statement: utility, limitations, and pitfalls in measuring testosterone: an Endocrine Society position statement. J Clin Endocrinol Metab 2007;92(2):405–13. Epub 2006/11/09. https://doi.org/10.1210/jc.2006-186417090633.

[25] Anaissie J, DeLay KJ, Wang W, Hatzichristodoulou G, Hellstrom WJ. Testosterone deficiency in adults and corresponding treatment patterns across the globe. Transl Androl Urol 2017;6 (2):183–91. Epub 2017/05/26. 10.21037/tau.2016.11.1628540225.

[26] Lazarou S, Reyes-Vallejo L, Morgentaler A. Wide variability in laboratory reference values for serum testosterone. J Sex Med 2006;3(6):1085–9. Epub 2006/11/15. https://doi.org/10.1111/j.1743-6109.2006.00334.x17100942.

[27] Khera M, Broderick GA, Carson 3rd CC, Dobs AS, Faraday MM, Goldstein I, et al. Adult-onset hypogonadism. Mayo Clin Proc 2016;91(7):908–26. Epub 2016/06/28. https://doi.org/10.1016/j.mayocp.2016.04.02227343020.

[28] Snyder PJ, Bhasin S, Cunningham GR, Matsumoto AM, Stephens-Shields AJ, Cauley JA, et al. Effects of testosterone treatment in older men. N Engl J Med 2016;374(7):611–24. Epub 2016/02/18. https://doi.org/10.1056/NEJMoa150611926886521.

[29] Brock G, Heiselman D, Maggi M, Kim SW, Rodriguez Vallejo JM, Behre HM, et al. Effect of testosterone solution 2% on testosterone concentration, sex drive and energy in hypogonadal men: results of a placebo controlled study. J Urol 2016;195(3):699–705. Epub 2015/10/27. https://doi.org/10.1016/j.juro.2015.10.08326498057.

[30] Maggi M, Heiselman D, Knorr J, Iyengar S, Paduch DA, Donatucci CF. Impact of testosterone solution 2% on ejaculatory dysfunction in hypogonadal men. J Sex Med 2016;13(8):1220–6. Epub 2016/07/21. https://doi.org/10.1016/j.jsxm.2016.05.01227436077.

[31] Paduch DA, Polzer PK, Ni X, Basaria S. Testosterone replacement in androgen-deficient men with ejaculatory dysfunction: a randomized controlled trial. J Clin Endocrinol Metab 2015;100(8):2956–62. Epub 2015/07/15. https://doi.org/10.1210/jc.2014-443426158605.

[32] Svartberg J, Agledahl I, Figenschau Y, Sildnes T, Waterloo K, Jorde R. Testosterone treatment in elderly men with subnormal testosterone levels improves body composition and BMD in the hip. Int J Impot Res 2008;20(4):378–87. Epub 2008/05/16. https://doi.org/10.1038/ijir.2008.1918480825.

[33] Travison TG, Basaria S, Storer TW, Jette AM, Miciek R, Farwell WR, et al. Clinical meaningfulness of the changes in muscle performance and physical function associated with testosterone administration in older men with mobility limitation. J Gerontol A Biol Sci Med Sci 2011;66(10):1090–9. Epub 2011/06/24. https://doi.org/10.1093/gerona/glr10021697501.

[34] Orengo CA, Fullerton L, Kunik ME. Safety and efficacy of testosterone gel 1% augmentation in depressed men with partial response to antidepressant therapy. J Geriatr Psychiatry Neurol 2005;18(1):20–4. Epub 2005/02/01. https://doi.org/10.1177/089198870427176715681624.

[35] Vaughan C, Goldstein FC, Tenover JL. Exogenous testosterone alone or with finasteride does not improve measurements of cognition in healthy older men with low serum testosterone. J Androl 2007;28(6):875–82. Epub 2007/07/05. https://doi.org/10.2164/jandrol.107.00293117609296.

[36] Cherrier MM, Anderson K, Shofer J, Millard S, Matsumoto AM. Testosterone treatment of men with mild cognitive impairment and low testosterone levels. Am J Alzheimers Dis Other Demen 2015;30(4):421–30. Epub 2014/11/14. https://doi.org/10.1177/153331751455687425392187.

[37] Grossmann M, Hoermann R, Wittert G, Yeap BB. Effects of testosterone treatment on glucose metabolism and symptoms in men with type 2 diabetes and the metabolic syndrome: a systematic review and meta-analysis of randomized controlled clinical trials. Clin Endocrinol 2015;83(3):344–51. Epub 2015/01/06. https://doi.org/10.1111/cen.1266425557752.

[38] Corona G, Giagulli VA, Maseroli E, Vignozzi L, Aversa A, Zitzmann M, et al. Testosterone supplementation and body composition: results from a meta-analysis of observational studies. J Endocrinol Investig 2016;39(9):967–81. Epub 2016/06/01. https://doi.org/10.1007/s40618-016-0480-227241317.

[39] Gianatti EJ, Dupuis P, Hoermann R, Zajac JD, Grossmann M. Effect of testosterone treatment on constitutional and sexual symptoms in men with type 2 diabetes in a randomized, placebo-controlled clinical trial. J Clin Endocrinol Metab 2014;99(10):3821–8. . Epub 2014/07/01. https://doi.org/10.1210/jc.2014-187224978674.

[40] Page ST, Amory JK, Bowman FD, Anawalt BD, Matsumoto AM, Bremner WJ, et al. Exogenous testosterone (T) alone or with finasteride increases physical performance, grip strength, and lean body mass in older men with low serum T. J Clin Endocrinol Metab 2005;90(3):1502–10. Epub 2004/12/02. https://doi.org/10.1210/jc.2004-193315572415.

[41] Corona G, Maseroli E, Rastrelli G, Isidori AM, Sforza A, Mannucci E, et al. Cardiovascular risk associated with testosterone-boosting medications: a systematic review and meta-analysis. Expert Opin Drug Saf 2014;13(10):1327–51. Epub 2014/08/21. https://doi.org/10.1517/14740338.2014.95065325139126.

[42] Wang C, Ilani N, Arver S, McLachlan RI, Soulis T, Watkinson A. Efficacy and safety of the 2% formulation of testosterone topical solution applied to the axillae in androgen-deficient men. Clin Endocrinol 2011;75(6):836–43. Epub 2011/06/22. https://doi.org/10.1111/j.1365-2265.2011.04152.x21689131.

[43] Dobs AS, McGettigan J, Norwood P, Howell J, Waldie E, Chen Y. A novel testosterone 2% gel for the treatment of hypogonadal males. J Androl 2012;33(4):601–7. Epub 2011/10/08. https://doi.org/10.2164/jandrol.111.01430821979302.

[44] Kaufman JM, Miller MG, Fitzpatrick S, McWhirter C, Brennan JJ. One-year efficacy and safety study of a 1.62% testosterone gel in hypogonadal men: results of a 182-day open-label extension of a 6-month double-blind study. J Sex Med 2012;9(4):1149–61. Epub 2012/02/11. https://doi.org/10.1111/j.1743-6109.2011.02630.x22321357.

[45] Raynaud JP, Legros JJ, Rollet J, Auges M, Bunouf P, Sournac M, et al. Efficacy and safety of a new testosterone-in-adhesive matrix patch applied every 2 days for 1 year to hypogonadal men. J Steroid Biochem Mol Biol 2008;109(1–2):168–76. Epub 2008/03/08. https://doi.org/10.1016/j.jsbmb.2007.10.01018325757.

[46] Dobs AS, Meikle AW, Arver S, Sanders SW, Caramelli KE, Mazer NA. Pharmacokinetics, efficacy, and safety of a permeation-enhanced testosterone transdermal system in comparison with bi-weekly injections of testosterone enanthate for the treatment of hypogonadal men. J Clin Endocrinol Metab 1999;84(10):3469–78. Epub 1999/10/16. https://doi.org/10.1210/jcem.84.10.607810522982.

[47] Wang C, Swerdloff R, Kipnes M, Matsumoto AM, Dobs AS, Cunningham G, et al. New testosterone buccal system (Striant) delivers physiological testosterone levels: pharmacokinetics study in hypogonadal men. J Clin Endocrinol Metab 2004;89(8):3821–9. Epub 2004/08/05. https://doi.org/10.1210/jc.2003-03186615292312.

[48] Rogol AD, Tkachenko N, Bryson N. Natesto, a novel testosterone nasal gel, normalizes androgen levels in hypogonadal men. Andrology 2016;4(1):46–54. Epub 2015/12/24. https://doi.org/10.1111/andr.1213726695758.

[49] Morgentaler A, Dobs AS, Kaufman JM, Miner MM, Shabsigh R, Swerdloff RS, et al. Long acting testosterone undecanoate therapy in men with hypogonadism: results of a pharmacokinetic clinical study. J Urol 2008;180(6):2307–13. Epub 2008/10/22. https://doi.org/10.1016/j.juro.2008.08.12618930255.

[50] Fennell C, Sartorius G, Ly LP, Turner L, Liu PY, Conway AJ, et al. Randomized cross-over clinical trial of injectable vs. implantable depot testosterone for maintenance of testosterone replacement therapy in androgen deficient men. Clin Endocrinol (Oxf) 2010;73(1):102–9. Epub 2009/11/07. https://doi.org/10.1111/j.1365-2265.2009.03744.x19891698.

[51] Kaminetsky JC, Moclair B, Hemani M, Sand M. A phase IV prospective evaluation of the safety and efficacy of extended release testosterone pellets for the treatment of male hypogonadism. J Sex Med 2011;8(4):1186–96. Epub 2011/01/29. https://doi.org/10.1111/j.1743-6109.2010.02196.x21269402.

[52] Murray-Lyon IM, Westaby D, Paradinas F. Hepatic complications of androgen therapy. Gastroenterology 1977;73(6):1461. Epub 1977/12/01 199526.

[53] Moon DG, Park MG, Lee SW, Park K, Park JK, Kim SW, et al. The efficacy and safety of testosterone undecanoate (Nebido((R))) in testosterone deficiency syndrome in Korean: a multicenter prospective study. J Sex Med 2010;7(6):2253–60. Epub 2010/03/30. https://doi.org/10.1111/j.1743-6109.2010.01765.x20345732.

[54] Ong GS, Somerville CP, Jones TW, Walsh JP. Anaphylaxis triggered by benzyl benzoate in a preparation of depot testosterone undecanoate. Case Rep Med 2012;2012:384054. Epub 2012/01/25. https://doi.org/10.1155/2012/38405422272209.

[55] Cavender RK, Fairall M. Precocious puberty secondary to topical testosterone transfer: a case report. J Sex Med 2011;8(2):622–6. Epub 2010/12/01. https://doi.org/10.1111/j.1743-6109.2010.02082.x21114766.

[56] Merhi ZO, Santoro N. Postmenopausal virilization after spousal use of topical androgens. Fertil Steril 2007;87(4):976.e13–5. Epub 2007/01/27. https://doi.org/10.1016/j.fertnstert.2006.07.154717254577.

[57] Moran LJ, Brinkworth GD, Martin S, Wycherley TP, Stuckey B, Lutze J, et al. Long-term effects of a randomised controlled trial comparing high protein or high carbohydrate weight loss diets on testosterone, SHBG, erectile and urinary function in overweight and obese men. PLoS ONE 2016;11(9). Epub 2016/09/02. https://doi.org/10.1371/journal.pone.016129727584019.

[58] Boonchaya-Anant P, Laichuthai N, Suwannasrisuk P, Houngngam N, Udomsawaengsup S, Snabboon T. Changes in testosterone levels and sex hormone-binding globulin levels in extremely obese men after bariatric surgery. Int J Endocrinol 2016;2016:1416503. Epub 2016/10/12. https://doi.org/10.1155/2016/141650327725831.

[59] Armamento-Villareal R, Aguirre LE, Qualls C, Villareal DT. Effect of lifestyle intervention on the hormonal profile of frail, obese older men. J Nutr Health Aging 2016;20(3):334–40. Epub 2016/02/20. https://doi.org/10.1007/s12603-016-0698-x26982583.

[60] Braun M, Wassmer G, Klotz T, Reifenrath B, Mathers M, Engelmann U. Epidemiology of erectile dysfunction: results of the 'Cologne Male Survey'. Int J Impot Res 2000;12(6):305–11. Epub 2001/06/21. https://doi.org/10.1038/sj.ijir.390062211416833.

[61] Lewis RW, Fugl-Meyer KS, Corona G, Hayes RD, Laumann EO, Moreira Jr. ED, et al. Definitions/epidemiology/risk factors for sexual dysfunction. J Sex Med 2010;7:1598–607.

[62] Araujo AB, Travison TG, Ganz P, Chiu GR, Kupelian V, Rosen RC, et al. Erectile dysfunction and mortality. J Sex Med 2009;6(9):2445–54. Epub 2009/06/23. https://doi.org/10.1111/j.1743-6109.2009.01354.x19538544.

[63] Basaria S, Harman SM, Travison TG, Hodis H, Tsitouras P, Budoff M, et al. Effects of testosterone administration for 3 years on subclinical atherosclerosis progression in older men with low or low-normal testosterone levels: a randomized clinical trial. JAMA 2015;314(6):570–81. . Epub 2015/08/12. https://doi.org/10.1001/jama.2015.888126262795.

[64] Morales A, Black A, Emerson L, Barkin J, Kuzmarov I, Day A. Androgens and sexual function: a placebo-controlled, randomized, double-blind study of testosterone vs. dehydroepiandrosterone in men with sexual dysfunction and androgen deficiency. Aging Male 2009;12(4):104–12. Epub 2009/11/04. https://doi.org/10.3109/13685530903294388198832 95.

[65] Ng Tang Fui M, Hoermann R, Prendergast LA, Zajac JD, Grossmann M. Symptomatic response to testosterone treatment in dieting obese men with low testosterone levels in a randomized, placebo-controlled clinical trial. Int J Obes 2017;41(3):420–6. Epub 2016/12/29. https://doi.org/10.1038/ijo.2016.24228028318.

[66] Tan WS, Low WY, Ng CJ, Tan WK, Tong SF, Ho C, et al. Efficacy and safety of long-acting intramuscular testosterone undecanoate in aging men: a randomised controlled study. BJU Int 2013;111(7):1130–40. Epub 2013/05/09. https://doi.org/10.1111/bju.1203723651425.

[67] Chiang HS, Hwang TI, Hsui YS, Lin YC, Chen HE, Chen GC, et al. Transdermal testosterone gel increases serum testosterone levels in hypogonadal men in Taiwan with improvements in sexual function. Int J Impot Res 2007;19(4):411–7. Epub 2007/06/01. https://doi.org/10.1038/sj.ijir.390156217538639.

[68] Corona G, Isidori AM, Buvat J, Aversa A, Rastrelli G, Hackett G, et al. Testosterone supplementation and sexual function: a meta-analysis study. J Sex Med 2014;11(6):1577–92. Epub 2014/04/05. https://doi.org/10.1111/jsm.1253624697970.

[69] Haider A, Yassin A, Doros G, Saad F. Effects of long-term testosterone therapy on patients with "diabesity": results of observational studies of pooled analyses in obese hypogonadal men with type 2 diabetes. Int J Endocrinol 2014;2014:683515. Epub 2014/04/17. https://doi.org/10.1155/2014/68351524738000.

[70] Bolona ER, Uraga MV, Haddad RM, Tracz MJ, Sideras K, Kennedy CC, et al. Testosterone use in men with sexual dysfunction: a systematic review and meta-analysis of randomized placebo-controlled trials. Mayo Clin Proc 2007;82(1):20–8. Epub 2007/02/09. https://doi.org/10.4065/82.1.2017285782.

[71] Jarvis TR, Chughtai B, Kaplan SA. Testosterone and benign prostatic hyperplasia. Asian J Androl 2015;17(2):212–6. Epub 2014/10/23. https://doi.org/10.4103/1008-682x.14096625337845.

[72] Sasagawa I, Nakada T, Kazama T, Satomi S, Terada T, Katayama T. Volume change of the prostate and seminal vesicles in male hypogonadism after androgen replacement therapy. Int Urol Nephrol 1990;22(3):279–84.

[73] Behre HM, Bohmeyer J, Nieschlag E. Prostate volume in testosterone-treated and untreated hypogonadal men in comparison to age-matched normal controls. Clin Endocrinol 1994;40(3):341–9.

[74] Raynaud JP, Gardette J, Rollet J, Legros JJ. Prostate-specific antigen (PSA) concentrations in hypogonadal men during 6 years of transdermal testosterone treatment. BJU Int 2013;111(6):880–90. Epub 2013/01/09. https://doi.org/10.1111/j.1464-410X.2012.11514.x23294726.

[75] Marks LS, Mazer NA, Mostaghel E, Hess DL, Dorey FJ, Epstein JI, et al. Effect of testosterone replacement therapy on prostate tissue in men with late-onset hypogonadism: a randomized controlled trial. JAMA 2006;296(19):2351–61. Epub 2006/11/16. https://doi.org/10.1001/jama.296.19.235117105798.

[76] Morales A. Androgen replacement therapy and prostate safety. Eur Urol 2002;41(2):113–20. Epub 2002/06/2112074396.

[77] Jin B, Conway AJ, Handelsman DJ. Effects of androgen deficiency and replacement on prostate zonal volumes. Clin Endocrinol 2001;54(4):437–45. Epub 2001/04/2511318778.

[78] Lee JH, Kim Y, Park YW, Lee DG. Relationship between benign prostatic hyperplasia/lower urinary tract symptoms and total serum testosterone level in healthy middle-aged eugonadal men. J Sex Med 2014;11(5):1309–15. Epub 2014/03/13. https://doi.org/10.1111/jsm.1248924612680.

[79] Yassin DJ, El Douaihy Y, Yassin AA, Kashanian J, Shabsigh R, Hammerer PG. Lower urinary tract symptoms improve with testosterone replacement therapy in men with late-onset hypogonadism: 5-year prospective, observational and longitudinal registry study. World J Urol 2014;
32(4):1049–54. Epub 2013/10/19. https://doi.org/10.1007/s00345-013-1187-z24135918.

[80] Ko YH, Moon du G, Moon KH. Testosterone replacement alone for testosterone deficiency syndrome improves moderate lower urinary tract symptoms: one year follow-up. World J Mens Health 2013;31(1):47–52. Epub 2013/05/10. https://doi.org/10.5534/wjmh.2013.31.1.4723658865.

[81] Shigehara K, Sugimoto K, Konaka H, Iijima M, Fukushima M, Maeda Y, et al. Androgen replacement therapy contributes to improving lower urinary tract symptoms in patients with hypogonadism and benign prostate hypertrophy: a randomised controlled study. Aging Male 2011;14(1):53–8. Epub 2010/12/22. https://doi.org/10.3109/13685538.2010.51817821171937.

[82] Ford ES, Giles WH, Dietz WH. Prevalence of the metabolic syndrome among US adults: findings from the third National Health and Nutrition Examination Survey. JAMA 2002;287(3):356–9. Epub 2002/01/1611790215.

[83] Wilson PW, D'Agostino RB, Parise H, Sullivan L, Meigs JB. Metabolic syndrome as a precursor of cardiovascular disease and type 2 diabetes mellitus. Circulation 2005;112(20):3066–72. Epub 2005/11/09. https://doi.org/10.1161/circulationaha.105.53952816275870.

[84] Alberti KG, Eckel RH, Grundy SM, Zimmet PZ, Cleeman JI, Donato KA, et al. Harmonizing the metabolic syndrome: a joint interim statement of the International Diabetes Federation Task Force on Epidemiology and Prevention; National Heart, Lung, and Blood Institute; American Heart Association; World Heart Federation; International Atherosclerosis Society; and International Association for the Study of Obesity. Circulation 2009;120(16):1640–5. Epub 2009/10/07. https://doi.org/10.1161/circulationaha.109.19264419805654.

[85] Luconi M, Samavat J, Seghieri G, Iannuzzi G, Lucchese M, Rotella C, et al. Determinants of testosterone recovery after bariatric surgery: is it only a matter of reduction of body mass index?. Fertil Steril 2013;99(7):1872–1879.e1. Epub 2013/03/20. https://doi.org/10.1016/j.fertnstert.2013.02.03923507475.

[86] Traish AM. Testosterone and weight loss: The evidence. Curr Opin Endocrinol Diabetes Obes 2014;21(5):313–22. Epub 2014/08/12. https://doi.org/10.1097/med.0000000000000008625105998.

[87] Ding EL, Song Y, Malik VS, Liu S. Sex differences of endogenous sex hormones and risk of type 2 diabetes: a systematic review and meta-analysis. JAMA 2006;295(11):1288–99. Epub 2006/03/16. https://doi.org/10.1001/jama.295.11.128816537739.

[88] Yang YM, Lv XY, Huang WD, Xu ZR, Wu LJ. Study of androgen and atherosclerosis in old-age male. J Zhejiang Univ Sci B 2005;6(9):931–5. Epub 2005/09/01. https://doi.org/10.1631/jzus.2005.B093116130198.

[89] Kapoor D, Aldred H, Clark S, Channer KS, Jones TH. Clinical and biochemical assessment of hypogonadism in men with type 2 diabetes: correlations with bioavailable testosterone and visceral adiposity. Diabetes Care 2007;30(4):911–7. Epub 2007/03/30. https://doi.org/10.2337/dc06-142617392552.

[90] Osuna JA, Gomez-Perez R, Arata-Bellabarba G, Villaroel V. Relationship between BMI, total testosterone, sex hormone-binding-globulin, leptin, insulin and insulin resistance in obese men. Arch Androl 2006;52(5):355–61. Epub 2006/07/29. https://doi.org/10.1080/01485010600692017168 73135.

[91] Tanabe M, Akehi Y, Nomiyama T, Murakami J, Yanase T. Total testosterone is the most valuable indicator of metabolic syndrome among various testosterone values in middle-aged Japanese men. Endocr J 2015;62(2):123–32. Epub 2014/10/25. https://doi.org/10.1507/endocrj.EJ14-031325342164.

[92] Svartberg J, von Muhlen D, Schirmer H, Barrett-Connor E, Sundfjord J, Jorde R. Association of endogenous testosterone with blood pressure and left ventricular mass in men. The Tromso Study. Eur J Endocrinol 2004;150(1):65–71. Epub 2004/01/1014713281.

[93] Haffner SM, Mykkanen L, Valdez RA, Katz MS. Relationship of sex hormones to lipids and lipoproteins in nondiabetic men. J Clin Endocrinol Metab 1993;77(6):1610–5. Epub 1993/12/01. https://doi.org/10.1210/jcem.77.6.82631498263149.

[94] Maggio M, Basaria S. Welcoming low testosterone as a cardiovascular risk factor. Int J Impot Res 2009;21(4):261–4. Epub 2009/06/19. https://doi.org/10.1038/ijir.2009.2519536127.

[95] Anaissie J, Roberts NH, Wang P, Yafi FA. Testosterone replacement therapy and components of the metabolic syndrome. Sex Med Rev 2017;5(2):200–10. Epub 2017/02/17. https://doi.org/10.1016/j.sxmr.2017.01.00328202344.

[96] Corona G, Monami M, Rastrelli G, Aversa A, Tishova Y, Saad F, et al. Testosterone and metabolic syndrome: a meta-analysis study. J Sex Med 2011;8(1):272–83. Epub 2010/09/03. https://doi.org/10.1111/j.1743-6109.2010.01991.x20807333.

[97] Yassin DJ, Doros G, Hammerer PG, Yassin AA. Long-term testosterone treatment in elderly men with hypogonadism and erectile dysfunction reduces obesity parameters and improves metabolic syndrome and health-related quality of life. J Sex Med 2014;11(6):1567–76. Epub 2014/04/10. https://doi.org/10.1111/jsm.1252324712761.

[98] Traish AM, Haider A, Doros G, Saad F. Long-term testosterone therapy in hypogonadal men ameliorates elements of the metabolic syndrome: an observational, long-term registry study. Int J Clin Pract 2014;68(3):314–29. Epub 2013/10/17. https://doi.org/10.1111/ijcp.1231924127736.

[99] Jones TH, Arver S, Behre HM, Buvat J, Meuleman E, Moncada I, et al. Testosterone replacement in hypogonadal men with type 2 diabetes and/or metabolic syndrome (the TIMES2 study). Diabetes Care 2011;34(4):828–37. Epub 2011/03/10. https://doi.org/10.2337/dc10-123321386088.

[100] Aversa A, Bruzziches R, Francomano D, Rosano G, Isidori AM, Lenzi A, et al. Effects of testosterone undecanoate on cardiovascular risk factors and atherosclerosis in middle-aged men with late-onset hypogonadism and metabolic syndrome: results from a 24-month, randomized, double-blind, placebo-controlled study. J Sex Med 2010;7(10):3495–503. Epub 2010/07/22. https://doi.org/10.1111/j.1743-6109.2010.01931.x20646185.

[101] Aversa A, Bruzziches R, Francomano D, Spera G, Lenzi A. Efficacy and safety of two different testosterone undecanoate formulations in hypogonadal men with metabolic syndrome. J Endocrinol Invest 2010;33(11):776–83. Epub 2010/03/12. doi: 10.1007/bf03350341https://doi.org/10.3275/690320220293.

[102] Fernandez-Balsells MM, Murad MH, Lane M, Lampropulos JF, Albuquerque F, Mullan RJ, et al. Clinical review 1: adverse effects of testosterone therapy in adult men: a systematic

[103] Glueck CJ, Friedman J, Hafeez A, Hassan A, Wang P. Testosterone, thrombophilia, thrombosis. Blood Coagul Fibrinolysis 2014;25(7):683–7. Epub 2014/04/16. https://doi.org/10.1097/mbc.0000000000000126 24732175.

[104] Baillargeon J, Urban RJ, Morgentaler A, Glueck CJ, Baillargeon G, Sharma G, et al. Risk of venous thromboembolism in men receiving testosterone therapy. Mayo Clin Proc 2015; 90(8):1038–45. Epub 2015/07/25. https://doi.org/10.1016/j.mayocp.2015.05.012 26205547.

[105] Tsai AW, Cushman M, Rosamond WD, Heckbert SR, Polak JF, Folsom AR. Cardiovascular risk factors and venous thromboembolism incidence: the longitudinal investigation of thromboembolism etiology. Arch Intern Med 2002;162(10):1182–9. Epub 2002/05/22 12020191.

[106] Shibata J, Hasegawa J, Siemens HJ, Wolber E, Dibbelt L, Li D, et al. Hemostasis and coagulation at a hematocrit level of 0.85: functional consequences of erythrocytosis. Blood 2003;101(11):4416–22. Epub 2003/02/11. https://doi.org/10.1182/blood-2002-09-2814 12576335.

[107] Malkin CJ, Pugh PJ, Morris PD, Kerry KE, Jones RD, Jones TH, et al. Testosterone replacement in hypogonadal men with angina improves ischaemic threshold and quality of life. Heart 2004;90(8):871–6. Epub 2004/07/16. https://doi.org/10.1136/hrt.2003.021122 15253956.

[108] English KM, Steeds RP, Jones TH, Diver MJ, Channer KS. Low-dose transdermal testosterone therapy improves angina threshold in men with chronic stable angina: a randomized, double-blind, placebo-controlled study. Circulation 2000;102(16):1906–11. Epub 2000/10/18 11034937.

[109] Shores MM, Smith NL, Forsberg CW, Anawalt BD, Matsumoto AM. Testosterone treatment and mortality in men with low testosterone levels. J Clin Endocrinol Metab 2012;97 (6):2050–8. Epub 2012/04/13. https://doi.org/10.1210/jc.2011-2591 22496507.

[110] Calof OM, Singh AB, Lee ML, Kenny AM, Urban RJ, Tenover JL, et al. Adverse events associated with testosterone replacement in middle-aged and older men: a meta-analysis of randomized, placebo-controlled trials. J Gerontol A Biol Sci Med Sci 2005;60(11):1451–7. Epub 2005/12/13 16339333.

[111] Baillargeon J, Urban RJ, Kuo YF, Ottenbacher KJ, Raji MA, Du F, et al. Risk of myocardial infarction in older men receiving testosterone therapy. Ann Pharmacother 2014;48 (9):1138–44. Epub 2014/07/06. https://doi.org/10.1177/1060028014539918 24989174.

[112] Sharma R, Oni OA, Gupta K, Chen G, Sharma M, Dawn B, et al. Normalization of testosterone level is associated with reduced incidence of myocardial infarction and mortality in men. Eur Heart J 2015;36(40):2706–15. Epub 2015/08/08. https://doi.org/10.1093/eurheartj/ehv346 26248567.

[113] O'Connell MD, Roberts SA, Srinivas-Shankar U, Tajar A, Connolly MJ, Adams JE, et al. Do the effects of testosterone on muscle strength, physical function, body composition, and quality of life persist six months after treatment in intermediate-frail and frail elderly men?. J Clin Endocrinol Metab 2011;96(2):454–8. Epub 2010/11/19. https://doi.org/10.1210/jc.2010-1167 21084399.

[114] Srinivas-Shankar U, Roberts SA, Connolly MJ, O'Connell MD, Adams JE, Oldham JA, et al. Effects of testosterone on muscle strength, physical function, body composition, and quality of life in intermediate-frail and frail elderly men: a randomized, double-blind, placebo-controlled study. J Clin Endocrinol Metab 2010;95(2):639–50. Epub 2010/01/12. https://doi.org/10.1210/jc.2009-1251 20061435.

[115] Corona G, Maggi M. Perspective: regulatory agencies' changes to testosterone product labeling. J Sex Med 2015;12(8):1690–3. Epub 2015/08/21. https://doi.org/10.1111/jsm.12951 26289540.

[116] U.S. Food and Drug Administration. FDA Drug Safety Communication: FDA evaluating risk of stroke, heart attack and death with FDA-approved testosterone products. Maryland: US Department of Health and Human Services; 2014.

[117] Kaufman JM, Graydon RJ. Androgen replacement after curative radical prostatectomy for prostate cancer in hypogonadal men. J Urol 2004;172(3):920–2. Epub 2004/08/18. https://doi.org/10.1097/01.ju.0000136269.10161.3215310998.

[118] Khera M, Grober ED, Najari B, Colen JS, Mohamed O, Lamb DJ, et al. Testosterone replacement therapy following radical prostatectomy. J Sex Med 2009;6(4):1165–70. Epub 2009/02/12. https://doi.org/10.1111/j.1743-6109.2009.01161.x19207277.

[119] Pastuszak AW, Pearlman AM, Lai WS, Godoy G, Sathyamoorthy K, Liu JS, et al. Testosterone replacement therapy in patients with prostate cancer after radical prostatectomy. J Urol 2013;190(2):639–44. Epub 2013/02/12. https://doi.org/10.1016/j.juro.2013.02.00223395803.

[120] Sarosdy MF. Testosterone replacement for hypogonadism after treatment of early prostate cancer with brachytherapy. Cancer 2007;109(3):536–41. . Epub 2006/12/22. https://doi.org/10.1002/cncr.2243817183557.

[121] Pastuszak AW, Pearlman AM, Godoy G, Miles BJ, Lipshultz LI, Khera M. Testosterone replacement therapy in the setting of prostate cancer treated with radiation. Int J Impot Res 2013;25(1):24–8. Epub 2012/09/14. https://doi.org/10.1038/ijir.2012.2922971614.

[122] Rhoden EL, Morgentaler A. Testosterone replacement therapy in hypogonadal men at high risk for prostate cancer: results of 1 year of treatment in men with prostatic intraepithelial neoplasia. J Urol 2003;170(6 Pt 1):2348–51. Epub 2003/11/25. https://doi.org/10.1097/01.ju.0000091104.71869.8e14634413.

[123] Samplaski MK, Loai Y, Wong K, Lo KC, Grober ED, Jarvi KA. Testosterone use in the male infertility population: prescribing patterns and effects on semen and hormonal parameters. Fertil Steril 2014;101(1):64–9. Epub 2013/10/08. https://doi.org/10.1016/j.fertnstert.2013.09.00324094422.

[124] Liu PY, Swerdloff RS, Anawalt BD, Anderson RA, Bremner WJ, Elliesen J, et al. Determinants of the rate and extent of spermatogenic suppression during hormonal male contraception: an integrated analysis. J Clin Endocrinol Metab 2008;93(5):1774–83. Epub 2008/02/28. https://doi.org/10.1210/jc.2007-276818303073.

[125] Kohn TP, Louis MR, Pickett SM, Lindgren MC, Kohn JR, Pastuszak AW, et al. Age and duration of testosterone therapy predict time to return of sperm count after human chorionic gonadotropin therapy. Fertil Steril 2017;107(2):351–357.e1. Epub 2016/11/20. https://doi.org/10.1016/j.fertnstert.2016.10.00427855957.

[126] Basaria S, Coviello AD, Travison TG, Storer TW, Farwell WR, Jette AM, et al. Adverse events associated with testosterone administration. N Engl J Med 2010;363(2):109–22. Epub 2010/07/02. https://doi.org/10.1056/NEJMoa100048520592293.

[127] Kaufman JM, Miller MG, Garwin JL, Fitzpatrick S, McWhirter C, Brennan JJ. Efficacy and safety study of 1.62% testosterone gel for the treatment of hypogonadal men. J Sex Med 2011;8(7):2079–89. Epub 2011/04/16. https://doi.org/10.1111/j.1743-6109.2011.02265.x21492400.

CHAPTER 4.2

Other Hormonal Therapies and Men's Health

Dorota J. Hawksworth, Arthur L. Burnett, II
The James Buchanan Brady Urological Institute and Department of Urology, Johns Hopkins University School of Medicine, Baltimore, MD, United States

INTRODUCTION

Testosterone (T) is the most abundant androgen in blood and plays crucial roles in every phase of a man's life [1, 2]. During embryonal stages, it is involved in sexual organ differentiation; at the time of puberty, it further facilitates proper development into an adult male, while in adulthood, it is involved in several maintenance roles such as control of sexual function, modulation of central arousal, maintenance of the corpora cavernosa structural integrity, and regulation of psychological symptoms [3–5]. On a molecular level, T contributes to endothelial and smooth muscle homeostasis via reduction of pro-inflammatory markers in both penile and cardiac vascular beds [6]. Low T levels result in reduction of nocturnal and morning time sex-induced erections, decrease in ejaculate volume, delay in onset of ejaculation, and ultimately infertility [7, 8].

T production is governed by tight control of the hypothalamic-pituitary-gonadal axis (HPGA), ultimately regulated by multiple negative feedback mechanisms [9]. The majority of T (>95%) is synthesized in the Leydig cells of the testes, with cholesterol as a starting substrate for its synthesis. Only a very small amount of T remains stored in the testes as its majority is secreted into blood. The mechanism of T transport from the Leydig cells to plasma is not completely elucidated at this time. During transport, the majority of T is bound to albumin or sex hormone-binding globulin (SHBG). In normal men, only 2% of total T circulates as a free substance, 44% is bound to SHBG, and 54% is bound to albumin. The remaining (5%) T in men is synthesized by the adrenal glands. Excess T is estrogenized under the influence of aromatase into estradiol (E_2), the most predominant form of estrogen (E), ultimately pushing T balance into lower total levels. Both excess endogenous and exogenous T and E_2 levels negatively influence release of luteinizing hormone (LH) and follicle-stimulating hormone (FSH) from the pituitary and ultimately

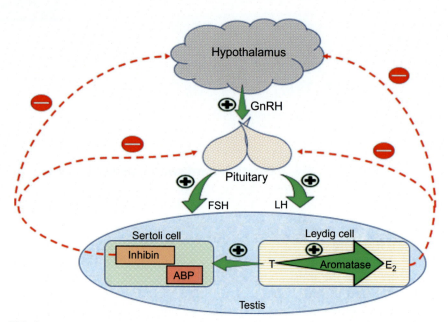

FIG. 1
Hypothalamic-pituitary-gonadal axis.

result in decreased testicular production of gonadotropins. Similarly, elevated levels of T and E_2 provide another negative feedback onto the hypothalamus itself, depressing the entire function of the HPGA (Fig. 1).

T deficiency, defined as "low T levels combined with symptoms or signs that are associated with low serum total testosterone," is prevalent, and its incidence continues to rise with age [10, 11]. Adult-onset hypogonadism (AOH) is a distinct entity from classical primary (testicular failure) or secondary (hypothalamic or pituitary failure) hypogonadism, and it occurs mainly in middle-aged or older men [12]. It is a clinical and biochemical syndrome encompassing biochemically assayed low T, associated signs and symptoms (reduced energy and stamina, depressed mood, increased irritability, difficulty concentrating, decreased libido, erectile dysfunction (ED), diminished penile sensation, difficulty attaining orgasm, and ejaculatory problems), and low or normal gonadotropin levels. Studies demonstrate AOH prevalence to be about 12%, which possibly further increases with increasing age [13, 14].

T deficiency has pro-inflammatory and proapoptotic effects on endothelial cells, and as a result, it has been demonstrated to contribute to increased risks of metabolic syndrome (MetS), type 2 diabetes (T2DM), and cardiovascular disease (CVD), independently of age and obesity [15, 16]. Although the association between T deficiency and MetS and T2DM is well established, the exact pathophysiological mechanisms are still somewhat unknown. The obesity-related decrease in SHBG and increase in aromatase activity, potentiated by insulin resistance (IR), provide some explanations. Increased levels of insulin, resulting from insulin resistance, suppress hepatic production of SHBG. This, in turn, results in decreased delivery of T to peripheral tissues and increased free T. Free T provides a negative feedback onto the HPGA, ultimately resulting in decreased release of gonadotropins. Increasing IR is also associated with a decrease in overall Leydig cell T secretion [17]. Moreover, excess T levels are converted locally by aromatase into E_2, which provides an additional negative feedback onto the already affected HPGA [18].

All of these molecular processes demonstrate that low T levels can further worsen a patient's metabolic profile, increase abdominal fat, and even exacerbate obesity-associated ED [19].

Treatment with exogenous T formulations is very effective in the management of T deficiency and significantly improves a patient's overall quality of life [8].

It is well known that T therapy (TT) can improve sexual desire and orgasm, nocturnal erections, frequency of intercourse, and overall sexual satisfaction and function [8]. Its effects on MetS have been studied particularly in combination with lifestyle modifications over the last decade. Studies report reductions in all-cause mortality and improvements in body composition (especially the loss of body fat with preservation of lean mass), lipid profiles, insulin resistance, and diastolic BP, further suggesting that TT should be considered as an adjunct to lifestyle interventions in patients with MetS and symptomatic AOH [20–23].

As with any therapy, TT has concerning side effects, to include a questionable relationship to CVD, prostate cancer, and thromboembolic events [24–26]. Furthermore, there are well-established, negative effects of TT on male fertility, testis size, and increase in breast tissue. These latter side effects are of significant concern to many younger, T-deficient patients. Together, with their providers, they seek alternative forms of TT in order to mitigate their risks of infertility.

The medical, nontestosterone-based alternatives to TT include Food and Drug Administration (FDA)-approved human chorionic gonadotropin (hCG), off-label selective estrogen receptor modulators (SERMs) such as clomiphene citrate (CC), and aromatase inhibitors (AIs) such as anastrozole and letrozole [11, 27]. All three of these groups of substances influence the HPGA at different levels of gonadotropin production, indirectly resulting in increased T production.

HUMAN CHORIONIC GONADOTROPIN

hCG

Introduction

Human chorionic gonadotropin (hCG) is a placental homologue of LH and is either derived from the urine of pregnant women or synthetized via recombinant DNA technology [28]. Its main physiological role is ascribed to maintenance of early pregnancy. Discovered in the early 1930s, hCG was initially used in women to stimulate follicular maturation [29]. In 1985, Finkel and colleagues examined men with pre- and postpubertal onset of hypogonadotropic hypogonadism (HH) and found that those with prior normal spermatogenesis (postpubertal HH) responded well to the hCG monotherapy and achieved normal sperm counts during this treatment [30]. A few years later, in 1992, Vicari and colleagues further confirmed that monotherapy with hCG induced spermatogenesis in men with isolated HH, and this occurred irrespective of their initial testicular volumes [31]. Furthermore, recent studies demonstrate that low-dose hCG (500 IU every other day) maintains intratesticular T levels and thus preserves spermatogenesis in men undergoing TT, while high-dose hCG (3000 IU every other day) can aid spermatogenic recovery in those with infertility resulting from prior TT [32, 33].

Mechanism of Action (Fig. 2)

Because of its structural similarity and virtually identical activity to LH, hCG acts as an LH agonist and stimulates T production from Leydig cells. Furthermore, through direct positive effects on the testes, hCG can also stimulate spermatogenesis.

Indications

FDA-approved for the treatment of HH and children with cryptorchidism [11].

Off-label treatment of T deficiency (other than HH) and idiopathic infertility.

Dosing

500–4000 IU units SQ or IM 2–3 times per week.

Side Effects and Safety Concerns

Currently, data on adverse effects of long-term hCG use are not available. The most commonly encountered side effects and reasons for discontinuation of therapy are listed below:

- Headache
- Irritability
- Depression
- Fatigue

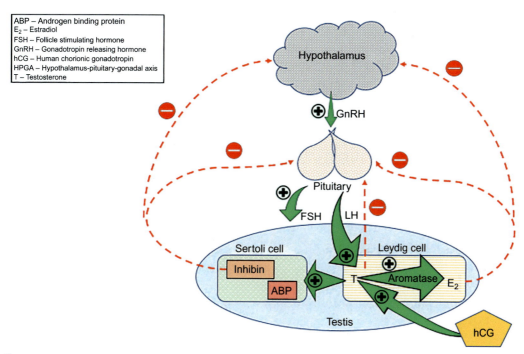

FIG. 2
Human chorionic gonadotropin and HPGA.

- Edema
- Gynecomastia
- Injection site pain

Benefits

In addition to beneficial hCG effects on increased intratesticular T levels and thus ultimate preservation and improvements in spermatogenesis, hCG can also reduce effects of standard TT, mainly preventing testicular atrophy. In patients, who are periodically "cycled off" their standard TT, hCG can maintain their TT effects [34].

Similar to TT, hCG therapy was recently demonstrated to have positive effects on metabolic indexes and body composition in T-deficient men. The Homeostasis Model Assessment of Insulin Resistance (HOMA-IR) index, triglyceride levels, body fat ratio, and waist-to-hip ratio decreased significantly in the study participants following 6 months of treatment [35].

Injections of hCG have been claimed to aid with weight loss via hunger reduction and changes in feelings of well-being [36]. Over the years, these claims have been critically evaluated in double-blind randomized placebo-controlled trials, and investigators found that placebo injections were as effective as hCG in the treatment of obesity [37]. A more recent meta-analysis of hCG effects in the treatment of obesity concluded that there was no scientific evidence for such claims [38]. However, in spite of this proved lack of efficacy, over-the-counter sales of various hCG formulations and direct-to-consumer advertisements continue to claim their weight loss effects.

SELECTIVE ESTROGEN RECEPTOR MODULATORS

Clomiphene citrate (Clomid)
Tamoxifen (Nolvadex, Soltamox)

Introduction

Selective estrogen receptor modulators (SERMs) are compounds that exhibit tissue-specific estrogen receptor agonist or antagonist activity. Although many SERMs have been developed, clomiphene citrate (CC) and tamoxifen are the ones most commonly used in men. CC has been used since the 1960s, initially as a treatment for female infertility, namely, oligomenorrhea, and then adapted as a treatment for anovulation and ovulation induction [39, 40].

The first use of CC in the treatment of male infertility was reported by Paulson et al. in 1976, who showed improved quantity and motility of spermatozoa were noted in 35 infertile men [41, 42]. Over time, the initial, exclusive use of CC for male infertility treatment evolved into another common current use, the treatment of T deficiency. Several studies evaluated CC therapy as an off-label alternative to standard TT, and they all found it to be safe and effective [43–45]. Low-dose CC not only raises T, LH, FSH, and E_2 levels and T/E_2 ratio but also leads to positive changes in sperm parameters [46, 47].

Tamoxifen was approved in the 1970s as an oral therapy for the treatment of hormone-responsive breast cancer, and credited to its mechanism of action, it has been also expanded into off-label treatment of male idiopathic infertility. The few randomized, placebo-controlled studies evaluating its use in men are less convincing than those evaluating the use of CC, however [48–50].

Mechanism of Action (Fig. 3)

CC is a mixture of two isomers, zuclomiphene and enclomiphene. As an E antagonist, it inhibits the negative feedback of E_2 at the level of the hypothalamus and the pituitary gland. This subsequently results in stimulation of LH and FSH release and downstream increase in spermatogenic and androgenic production.

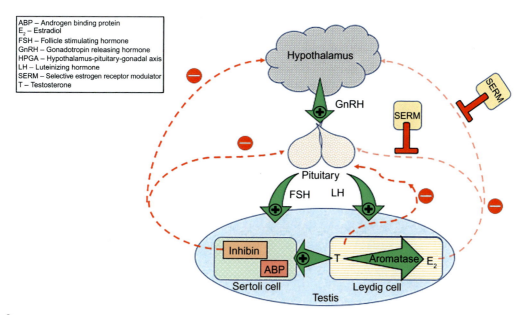

FIG. 3
Selective estrogen receptor modulators and HPGA.

Tamoxifen functions as an E receptor antagonist in breast tissue, and because of this property, it has been effectively used for the treatment of breast cancer. It also inhibits E_2 negative feedback mechanisms at the hypothalamus and the pituitary levels and, just like CC, results in increased LH and FSH stimulation and subsequent upregulation in both spermatogenesis and T synthesis.

Indications
Off-label treatment of T deficiency and idiopathic infertility.

Dosing
Clomiphene citrate: 25–50 mg orally, every 1–2 days.

Tamoxifen: 20 mg orally daily.

Side Effects and Safety Concerns
Overall, CC is established to be a safe and effective treatment of T deficiency.

The majority of symptoms resolve once therapy is terminated. However, a few cases of permanent visual derangements have been reported in literature [45].

Clomiphene citrate

- Vasomotor flushing
- Abdominal discomfort
- Nausea and vomiting
- Visual symptoms
- Headache

Tamoxifen

- Liver abnormalities/liver enzyme changes
- Ocular disturbances (including cataracts)
- Thromboembolic events (including deep vein thrombosis (DVT) and stroke)

Benefits

The use of SERMs does not suppress endogenous gonadotropin secretion, and thus, they do not impact spermatogenesis or testis size. This is a very attractive benefit, as T-deficient men can achieve appropriate T level normalization, in addition to preserving and/or improving their fertility potential. These benefits of CC use are less pronounced in men who already have elevated levels of LH prior to initiation of therapy, however. Mazzola and colleagues further identified men with low T and pretreatment LH levels ≤6 IU/mL to be the ideal candidates for this form of T replacement therapy [51].

Additionally, younger men may derive more benefit from this therapy. In the study by Tenover and colleagues, younger men (22–35 years of age) had a much better response to CC treatment than older (≥65 years of age) men, with their mean serum total T levels increasing by 100% and 32%, respectively [52].

Guay and colleagues evaluated the effects of raising endogenous T levels with CC therapy in men with ED and T deficiency and further demonstrated that only younger and healthier men had some improvements in sexual function parameters [53, 54]. Thus, CC therapy does not seem to be beneficial in older infertile men with AOH and ED.

As TT improves metabolic parameters in obese men, CC therapy has been evaluated for these same effects. Pelusi et al. examined possible benefits of CC in obese men with low T levels, impaired glucose tolerance, and T2DM, who were treated with metformin. This small, randomized, double-blind, placebo-controlled, crossover study demonstrated significant increases in serum total T levels in addition to mild improvements in serum fasting glucose and insulin levels [55].

The effects of CC on body mass index (BMI) and cholesterol levels remain controversial. While a retrospective review of data on 90 men who received CC for T deficiency demonstrated no significant alterations in BMI and cholesterol levels, a randomized, double-blind, placebo-controlled study in obese men demonstrated statistically significant improvements in lean body mass and muscle mass and reduction in high-density lipoprotein (HDL) cholesterol levels [56, 57].

AROMATASE INHIBITORS

Anastrozole (Arimidex)
Letrozole (Femara)

Introduction

Estradiol (E_2) is the most potent estrogen produced in the body. It is synthesized either from T via aromatase or from estrone (E_1) via 17β-hydroxysteroid dehydrogenase. While about 60% of circulating E_2 is derived from direct testicular secretion or from conversion of testicular androgens by aromatase, the remainder is derived from peripheral conversion of adrenal androgens [58]. Estrogen excess in males results in premature closure of epiphyseal plates, gynecomastia, and hypogonadism. Aromatase is a key enzyme in estrogen biosynthesis, and it has been identified in the gonads, brain, muscle, hair, bone, as well as in placental, adipose, and vascular tissues [59–63].

Aromatase inhibitors (AIs) have been extensively used and thoroughly studied in women with hormone-sensitive breast cancer [64]. More recently, AIs have become an attractive oral alternative to the treatment of T deficiency in men. AIs are classified into steroidal and nonsteroidal or into first, second, or third generation. Steroidal AIs differ from nonsteroidal ones by the way they inhibit aromatase function. Third-generation AIs (such as anastrozole and letrozole) are the most commonly used, as they are the most potent and the most specific inhibitors.

Mechanism of Action (Fig. 4)

AIs block aromatase enzyme activity that normally converts T to E_2. As result, they lower E_2 levels and thus limit the negative feedback of E_2 on the HPGA. These alterations consequently result in increased release of GnRH, LH, and FSH. The increased gonadotropins further stimulate testicular production of T, subsequently leading to elevation in T levels as well. With increase in FSH levels, improved sperm parameters have been demonstrated in patients on AI therapy as well [65, 66].

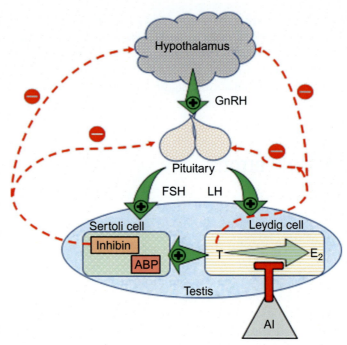

FIG. 4
Aromatase inhibitors and HPGA.

Indications
Off-label treatment of T deficiency and idiopathic infertility.

Dosing
0.05–1 mg orally every 1–3 days.

Side Effects and Safety Concerns
AIs are proved to be safe, convenient, and effective for the treatment of hormone-sensitive breast cancer in women [64]. Although their long-term use has not been ascribed to any major side effects, they have been associated with a moderate increase in bone resorption and a modest decrease in bone mineral density (BMD) [67]. Negative effects of AIs on bone metabolism are of the most concern in regard to their use in men.

Both androgens and estrogens influence bone health in men, and as men age, both of their levels decrease, resulting in an elevated E_2/T ratio, which, may in fact, have bone protective effects.

The common side effects encountered with AI therapy are listed below:

- Hot flashes
- Weight gain
- Insomnia
- Joint aches
- Hypertension
- Nausea
- Back pain
- Bone pain
- Decreased bone mineral density
- Dyspnea
- Peripheral edema

Benefits

The effects of AIs in men with T deficiency have been studied extensively. It is well established that AIs markedly increase LH and T levels in both the young and elderly, as well as in severely obese men [68–71]. The hormonal changes in older men, however, are not associated with improvements in body composition or strength, but rather lead to a decrease in BMD [68, 72]. Obese men on AI therapy demonstrate normalization of their T levels. However, the benefits of this treatment on overall body composition and insulin sensitivity are unclear at this time. A randomized controlled trial evaluating TT with T and Dutasteride (inhibiting conversion of T to dihydrotestosterone (DHT)) versus T and AIs in young obese men demonstrated no benefit to inhibiting aromatization of administered T to E_2, while the T and Dutasteride regimen improved insulin sensitivity [73]. Studies evaluating effects of AIs on cardiometabolic parameters, such as lipid and various inflammatory markers, do not demonstrate any advantageous or deleterious effects [74, 75].

SERM AND AI COMBINATION THERAPY (CLOMIPHENE CITRATE AND ANASTROZOLE)

Clomiphene citrate (CC) and anastrozole are the most common off-label medications used in the treatment of idiopathic infertility [76, 77]. As discussed earlier, these two medications are also commonly used in the treatment of T deficiency in men who wish to preserve their fertility.

Clomiphene citrate stimulates GnRH production at the level of the hypothalamus and leads to increases in T, LH, and FSH levels. Monotherapy with CC, however, may result in increased levels of E_2 and eventual disruption of the T/E_2 ratio, negatively affecting spermatogenesis [78]. Treatment with an aromatase inhibitor (AI) such as anastrozole may improve the decreased T/E_2 ratio and also improve sperm parameters [79].

The increasing incidence of obesity worldwide has been linked to many health conditions in men, to include decreased T levels and fertility problems [80]. Since adipose tissue is the essential site of aromatase activity, many obese men have lower baseline T/E_2 ratios. In practice, most infertile men are initially placed on a monotherapy with CC, and when they develop hyperestrogenemia, they require addition of an AI [79]. A recent study examining safety and efficacy of combined use of CC and AI (anastrozole) confirmed the association of BMI with E_2 levels. The authors further suggested that obese men may be more likely to develop hyperestrogenemia on CC monotherapy and may benefit from combination therapy. Their study, albeit small and retrospective in nature, found the combination therapy to be a safe and effective TT alternative [79].

CONCLUSIONS

As discussed in this manuscript, all nontestosterone hormonal therapeutic formulations are safe and efficacious in increasing T, LH, and FSH levels in T-deficient men. The fertility preservation and restoration effects make these options very attractive for those men who wish to father children. The overall health benefits of these medications, however, are still not well established, as there are no large, prospective, randomized, placebo-controlled trials comparing them with one another and with standard exogenous T formulations. With well-established T benefits in healthy men and in those men with obesity, metabolic syndrome, and cardiovascular disease, it is very likely that future research will prove the same for hCG, SERMs, and AIs.

For now, since hCG is the only FDA-approved medication and SERM and AI use continues to be off-label, careful clinical and laboratory evaluation, thorough patient counseling, and regular follow-up are recommended. Perhaps, rigorous professional societal guidelines will be established in the near future in order to further direct us in the use of these alternatives.

References

[1] Kelly DM, Jones TH. Testosterone: a vascular hormone in health and disease. J Endocrinol 2013;217(3):R47–71.

[2] Kelly DM, Jones TH. Testosterone: a metabolic hormone in health and disease. J Endocrinol 2013;217(3):R25–45.

[3] Aversa A, et al. Androgens and penile erection: evidence for a direct relationship between free testosterone and cavernous vasodilation in men with erectile dysfunction. Clin Endocrinol 2000;53(4):517–22.

[4] Rastrelli G, Corona G, Maggi M. Testosterone and sexual function in men. Maturitas 2018;112:46–52.

[5] Traish AM. Role of androgens in modulating male and female sexual function. Horm Mol Biol Clin Invest 2010;4(1):521–8.

[6] Mohamad NV, et al. The relationship between circulating testosterone and inflammatory cytokines in men. Aging Male 2018;1–12.

[7] Carnahan RM, Perry PJ. Depression in aging men: the role of testosterone. Drugs Aging 2004;21(6):361–76.

[8] Traish AM. Benefits and health implications of testosterone therapy in men with testosterone deficiency. Sex Med Rev 2018;6(1):86–105.

[9] Caroppo E. Male hypothalamic-pituitary-gonadal axis. In: Lipshultz LI, editor. Infertility in the male, 4th ed. New York, NY: Cambridge University Press; 2009.

[10] Gray A, et al. Age, disease, and changing sex hormone levels in middle-aged men: results of the Massachusetts Male Aging Study. J Clin Endocrinol Metab 1991;73(5):1016–25.

[11] Mulhall JP, et al. Evaluation and management of testosterone deficiency: AUA guideline. J Urol 2018;200(2):423–32.

[12] Sexual Medicine Society of North America (SMSNA). Consensus Statement and White Paper Executive Summary: Adult Onset Hypogonadism (AOH); 2015.

[13] Guay A, Seftel AD, Traish A. Hypogonadism in men with erectile dysfunction may be related to a host of chronic illnesses. Int J Impot Res 2010;22(1):9–19.

[14] Tajar A, et al. Characteristics of secondary, primary, and compensated hypogonadism in aging men: evidence from the European Male Ageing Study. J Clin Endocrinol Metab 2010;95(4):1810–8.

[15] Wang C, et al. Low testosterone associated with obesity and the metabolic syndrome contributes to sexual dysfunction and cardiovascular disease risk in men with type 2 diabetes. Diabetes Care 2011;34(7):1669–75.

[16] Allan CA, McLachlan RI. Androgens and obesity. Curr Opin Endocrinol Diabetes Obes 2010;17(3):224–32.

[17] Pitteloud N, et al. Increasing insulin resistance is associated with a decrease in Leydig cell testosterone secretion in men. J Clin Endocrinol Metab 2005;90(5):2636–41.

[18] Saboor Aftab SA, Kumar S, Barber TM. The role of obesity and type 2 diabetes mellitus in the development of male obesity-associated secondary hypogonadism. Clin Endocrinol 2013;78(3):330–7.

[19] Corona G, et al. Hypogonadism as a possible link between metabolic diseases and erectile dysfunction in aging men. Hormones (Athens) 2015;14(4):569–78.

[20] Corona G, et al. Obesity and late-onset hypogonadism. Mol Cell Endocrinol 2015;418(Pt 2):120–33.

[21] Ng Tang Fui M, et al. Symptomatic response to testosterone treatment in dieting obese men with low testosterone levels in a randomized, placebo-controlled clinical trial. Int J Obes 2017;41(3):420–6.

[22] Ng Tang Fui M, et al. Effects of testosterone treatment on body fat and lean mass in obese men on a hypocaloric diet: a randomised controlled trial. BMC Med 2016;14(1):153.

[23] Hackett G, et al. Serum testosterone, testosterone replacement therapy and all-cause mortality in men with type 2 diabetes: retrospective consideration of the impact of PDE5 inhibitors and statins. Int J Clin Pract 2016;70(3):244–53.

[24] Baillargeon J, et al. Risk of venous thromboembolism in men receiving testosterone therapy. Mayo Clin Proc 2015;90(8):1038–45.

[25] Morgentaler A. Controversies and advances with testosterone therapy: a 40-year perspective. Urology 2016;89:27–32.

[26] Morgentaler III A, Conners WP. Testosterone therapy in men with prostate cancer: literature review, clinical experience, and recommendations. Asian J Androl 2015;17(2):206–11.

[27] Lo EM, et al. Alternatives to testosterone therapy: a review. Sex Med Rev 2018;6(1):106–13.

[28] Practice Committee of American Society for Reproductive Medicine, Birmingham, Alabama. Gonadotropin preparations: past, present, and future perspectives. Fertil Steril 2008;90(5 Suppl):S13–20.

[29] Lunenfeld B. Historical perspectives in gonadotrophin therapy. Hum Reprod Update 2004;10(6):453–67.

[30] Finkel DM, Phillips JL, Snyder PJ. Stimulation of spermatogenesis by gonadotropins in men with hypogonadotropic hypogonadism. N Engl J Med 1985;313(11):651–5.

[31] Vicari E, et al. Therapy with human chorionic gonadotrophin alone induces spermatogenesis in men with isolated hypogonadotrophic hypogonadism–long-term follow-up. Int J Androl 1992;15(4):320–9.

[32] Hsieh TC, et al. Concomitant intramuscular human chorionic gonadotropin preserves spermatogenesis in men undergoing testosterone replacement therapy. J Urol 2013;189(2):647–50.

[33] Wenker EP, et al. The use of HCG-based combination therapy for recovery of spermatogenesis after testosterone use. J Sex Med 2015;12(6):1334–7.

[34] Lee YJ, et al. Human chorionic gonadotropin-administered natural cycle versus spontaneous ovulatory cycle in patients undergoing two pronuclear zygote frozen-thawed embryo transfer. Obstet Gynecol Sci 2018;61(2):247–52.

[35] Bayram F, et al. The effects of gonadotropin replacement therapy on metabolic parameters and body composition in men with idiopathic hypogonadotropic hypogonadism. Horm Metab Res 2016;48(2):112–7.

[36] Simeons AT. The action of chorionic gonadotrophin in the obese. Lancet 1954;267(6845):946–7.

[37] Greenway FL, Bray GA. Human chorionic gonadotropin (HCG) in the treatment of obesity: a critical assessment of the Simeons method. West J Med 1977;127(6):461–3.

[38] Lijesen GK, et al. The effect of human chorionic gonadotropin (HCG) in the treatment of obesity by means of the Simeons therapy: a criteria-based meta-analysis. Br J Clin Pharmacol 1995;40(3):237–43.

[39] Johnson Jr. JE, et al. The efficacy of clomiphene citrate for induction of ovulation. A controlled study. Int J Fertil 1966;11(3):265–70.

[40] Kistner RW. Use of clomiphene citrate, human chorionic gonadotropin, and human menopausal gonadotropin for induction of ovulation in the human female. Fertil Steril 1966;17(5):569–83.

[41] Paulson DF, et al. Clomiphene citrate: pharmacologic treatment of hypofertile male. Urology 1977;9(4):419–21.

[42] Paulson DF, Wacksman J. Clomiphene citrate in the management of male infertility. J Urol 1976;115(1):73–6.

[43] Katz DJ, et al. Outcomes of clomiphene citrate treatment in young hypogonadal men. BJU Int 2012;110(4):573–8.

[44] Patel DP, et al. The safety and efficacy of clomiphene citrate in hypoandrogenic and subfertile men. Int J Impot Res 2015;27(6):221–4.

[45] Moskovic DJ, et al. Clomiphene citrate is safe and effective for long-term management of hypogonadism. BJU Int 2012;110(10):1524–8.

[46] Kaminetsky J, et al. Oral enclomiphene citrate stimulates the endogenous production of testosterone and sperm counts in men with low testosterone: comparison with testosterone gel. J Sex Med 2013;10(6):1628–35.

[47] Shabsigh A, et al. Clomiphene citrate effects on testosterone/estrogen ratio in male hypogonadism. J Sex Med 2005;2(5):716–21.

[48] Brake A, Krause W. Treatment of idiopathic oligozoospermia with tamoxifen–a follow-up report. Int J Androl 1992;15(6):507–8.

[49] Kotoulas IG, et al. Tamoxifen treatment in male infertility. I. Effect on spermatozoa. Fertil Steril 1994;61(5):911–4.

[50] Krause W, Holland-Moritz H, Schramm P. Treatment of idiopathic oligozoospermia with tamoxifen–a randomized controlled study. Int J Androl 1992;15(1):14–8.

[51] Mazzola CR, et al. Predicting biochemical response to clomiphene citrate in men with hypogonadism. J Sex Med 2014;11(9):2302–7.

[52] Tenover JS, et al. The effects of aging in normal men on bioavailable testosterone and luteinizing hormone secretion: response to clomiphene citrate. J Clin Endocrinol Metab 1987;65(6):1118–26.

[53] Guay AT, Bansal S, Heatley GJ. Effect of raising endogenous testosterone levels in impotent men with secondary hypogonadism: double blind placebo-controlled trial with clomiphene citrate. J Clin Endocrinol Metab 1995;80(12):3546–52.

[54] Guay AT, et al. Clomiphene increases free testosterone levels in men with both secondary hypogonadism and erectile dysfunction: who does and does not benefit? Int J Impot Res 2003;15(3):156–65.

[55] Pelusi C, et al. Clomiphene citrate effect in obese men with low serum testosterone treated with metformin due to dysmetabolic disorders: a randomized, double-blind, placebo-controlled study. PLoS ONE 2017;12(9).

[56] Soares AH, et al. Effects of clomiphene citrate on male obesity-associated hypogonadism: a randomized, double-blind, placebo-controlled study. Int J Obes 2018;42(5):953–63.

[57] Lee D, Wosnitzer M, Winter A, Paduch D. Influence of clomiphene citrate on BMI and cholesterol levels in the treatment of hypogonadism. Fertil Steril 2014;102:E269–70.

[58] de Ronde W, et al. A direct approach to the estimation of the origin of oestrogens and androgens in elderly men by comparison with hormone levels in postmenopausal women. Eur J Endocrinol 2005;152(2):261–8.

[59] Harada N, et al. Localized expression of aromatase in human vascular tissues. Circ Res 1999;84(11):1285–91.

[60] Longcope C, et al. Aromatization of androgens by muscle and adipose tissue in vivo. J Clin Endocrinol Metab 1978;46(1):146–52.

[61] Schweikert HU, Milewich L, Wilson JD. Aromatization of androstenedione by isolated human hairs. J Clin Endocrinol Metab 1975;40(3):413–7.

[62] Schweikert HU, Wolf L, Romalo G. Oestrogen formation from androstenedione in human bone. Clin Endocrinol 1995;43(1):37–42.

[63] Stoffel-Wagner B, et al. Expression of CYP19 (aromatase) mRNA in different areas of the human brain. J Steroid Biochem Mol Biol 1999;70(4–6):237–41.

[64] Breast International Group 1-98 Collaborative Group, et al. A comparison of letrozole and tamoxifen in postmenopausal women with early breast cancer. N Engl J Med 2005;353(26):2747–57.

[65] Pavlovich CP, et al. Evidence of a treatable endocrinopathy in infertile men. J Urol 2001;165(3):837–41.

[66] Raman JD, Schlegel PN. Aromatase inhibitors for male infertility. J Urol 2002;167 (2 Pt 1):624–9.

[67] Perez EA, et al. Effect of letrozole versus placebo on bone mineral density in women with primary breast cancer completing 5 or more years of adjuvant tamoxifen: a companion study to NCIC CTG MA.17. J Clin Oncol 2006;24(22):3629–35.

[68] Burnett-Bowie SA, et al. Effects of aromatase inhibition in hypogonadal older men: a randomized, double-blind, placebo-controlled trial. Clin Endocrinol 2009;70(1):116–23.

[69] Leder BZ, et al. Effects of aromatase inhibition in elderly men with low or borderline-low serum testosterone levels. J Clin Endocrinol Metab 2004;89(3):1174–80.

[70] T'Sjoen GG, et al. Comparative assessment in young and elderly men of the gonadotropin response to aromatase inhibition. J Clin Endocrinol Metab 2005;90(10):5717–22.

[71] de Boer H, et al. Letrozole normalizes serum testosterone in severely obese men with hypogonadotropic hypogonadism. Diabetes Obes Metab 2005;7(3):211–5.

[72] Burnett-Bowie SA, et al. Effects of aromatase inhibition on bone mineral density and bone turnover in older men with low testosterone levels. J Clin Endocrinol Metab 2009;94 (12):4785–92.

[73] Juang PS, et al. Testosterone with dutasteride, but not anastrazole, improves insulin sensitivity in young obese men: a randomized controlled trial. J Sex Med 2014;11(2):563–73.

[74] Dias JP, et al. Testosterone vs. aromatase inhibitor in older men with low testosterone: effects on cardiometabolic parameters. Andrology 2017;5(1):31–40.

[75] Dougherty RH, et al. Effect of aromatase inhibition on lipids and inflammatory markers of cardiovascular disease in elderly men with low testosterone levels. Clin Endocrinol 2005;62(2):228–35.

[76] Rambhatla A, Mills JN, Rajfer J. The role of estrogen modulators in male hypogonadism and infertility. Rev Urol 2016;18(2):66–72.

[77] Wheeler K, et al. Clomiphene citrate for the treatment of hypogonadism. Sex Med Rev 2018. in press.

[78] Itoh N, et al. Changes in the endocrinological milieu after clomiphene citrate treatment for oligozoospermia: the clinical significance of the estradiol/testosterone ratio as a prognostic value. J Androl 1994;15(5):449–55.

[79] Alder NJ, et al. Combination therapy with clomiphene citrate and anastrozole is a safe and effective alternative for hypoandrogenic subfertile men. BJU Int 2018;122(4):688–94.

[80] Bieniek JM, et al. Influence of increasing body mass index on semen and reproductive hormonal parameters in a multi-institutional cohort of subfertile men. Fertil Steril 2016;106(5):1070–5.

Supplements

CHAPTER 5.1

Chinese Medicine and Men's Health

Barbara M. Chubak, Jillian Capodice
Department of Urology, Icahn School of Medicine at Mount Sinai, New York, NY, United States

In the Western world, "traditional Chinese medicine" (TCM) is commonly considered synonymous with percutaneous procedural interventions, such as acupuncture, moxibustion, and cupping. This limited appreciation of Chinese medicine is in part a result of its introduction to the American public media, through a reporter who experienced acupuncture as part of his surgical care during Nixon's visit to China in 1971 [1]. It has since been reinforced by the prominent use of cupping by elite athletes [2] and by provision of acupuncture as a part of cancer care in leading academic medical centers throughout the United States [3]. Widespread acknowledgment of the appeal and therapeutic value of certain Chinese medicine procedures is reflected in the willingness of many international health insurance providers to defray the cost of acupuncture treatments.

However, these procedures are only a minority of what Chinese medicine as a whole system offers patients. In the Eastern world, TCM therapy is dominated by the provision of herbal remedies from an extensive pharmacopeia. These herbal remedies are compounded and used in various forms, including oral (teas, tablets, capsules, tinctures, etc.), topical (poultices, salves, patches, etc.), and intravenous formulations. Despite its Chinese name and origin, the theory and practice of TCM reach beyond national borders, drawing its materials and evidence for their use from across a wide swath of space and time. Thus, practical differences can be found between practitioners trained in the Chinese, Japanese, and Korean TCM traditions, despite their shared reliance on the same underlying medical concepts [4]. These concepts date back to the ancient world, and TCM practitioners continue to rely heavily on texts such as the *Inner Canon of the Yellow Emperor*, dated to the 1st–3rd centuries BCE [5].

Although this and other *yang shen* (nourishing life) texts dated to the Han Dynasty (206 BCE–CE 220) do refer to health problems specific to men,

---

☆Note that all Chinese words transliterated using English script (or pin yin) are rendered in italics to demonstrate their foreign-language status.

offering guidelines for optimizing sexual development and productivity throughout the lifespan, they do not consider men's health as a specific medical field for concern. Nor does most of the extensive medical literature that followed, in which TCM disease etiologies, mechanisms, and treatments have been investigated and elaborated through the centuries into the present [6]. The Chinese medical subspecialty of *nanke* (andrology) developed during the late 20th century in the crucible of the Communist revolution, and was shaped by compromises wrought between TCM practitioners and Nationalist government factions that favored Western biomedicine [7].

Nanke, the men's health of Chinese medicine, is thus exemplary of *zhong xi yi jie he* (integrated Chinese-Western medicine), and in China, clinical and academic departments of *Nanke* are found equally within institutions dedicated to TCM and to Western biomedicine. The balance of biomedicine and TCM principles and practice applied in *Nanke* clinics varies on a case-by-case basis, with diagnostic and therapeutic decisions motivated by a combination of provider and patient preference and perceived need [8]. This idiosyncrasy of Chinese medicine has great appeal to patients in Western medical systems, where financial and temporal pressures on the health-care encounter can leave them feeling "like widgets in a production line" [9]. Recent years have witnessed growing sales of TCM-based nutraceuticals and interventions worldwide [10].

This chapter describes the TCM aspects of *Nanke*, including their underlying principles, characteristic diagnoses, and some common treatments. Although *Nanke* is inclusive of Western biomedicine, we assume that the reader is familiar with this pathophysiology and therapy—if not in general, then from the preceding chapters of this book—and therefore limit our discussion of biomedicine to the ways in which it compares and contrasts with TCM.

It is essential to appreciate how profoundly different TCM is from Western biomedicine. Whereas Western biomedicine principally relies on the sciences of biology and chemistry (and their derivative disciplines of anatomy, physiology, and pathophysiology) for its logic, TCM conceives of the body as governed by a kind of physics, with optimal health characterized by smooth flows of energy and entrained vibrations between body parts, in resonance with the natural and social structures that surround them [11]. The basic material of all that exists, which is moved by these flows and vibrations, is *qi*. *Qi* manifests in various forms, with movements and transformations that are described in terms of *yin* and *yang* phases. An exception to this rule is *jing* (vital essence), a form of *qi* that is neither *yin* nor *yang* in character, but liminal between them. *Jing* is the material representation of the body's capacity for creativity, growth, and reproduction: in men, it materializes as semen, and in women as menstrual blood [3].

Although the TCM body, like the biomedical body, is constituted of parts, rather than focusing on their individual anatomy and physiology, TCM is concerned with the ways that they relate to each other and the surrounding world. The body's viscera and fluids (not only the aforementioned *jing* but also *xue* (which is most often translated as blood but may also include lymph and cerebrospinal fluid)) are understood to function in a network, connected by channels through which *qi* circulates. These channels are named for the anatomy with which they are associated, but their influence is not limited to the organs for which they are named. Thus, the kidney channel runs through the genitalia and exerts a powerful effect on sexual function; so too do the liver, heart, and midline governing *du mai* and conception *ren mai* channels frequently contribute to sexual health. These channels do not function in isolation from each other, but rather run along *yin* or *yang* meridians to interact with the others in a network.

In addition to this channel or network vessel-based body concept, TCM also comprehends the body in health and illness through *zang xiang xue shuo* (visceral manifestation theory.) This concept considers the body in terms of *zang* and *fu* organs: The former are the heart, pericardium, lung, spleen, liver, and kidney; the latter are the bowels (including stomach, small intestine, and large intestine), gallbladder, urinary bladder, and triple burner. The absence of any clear correlate for the triple burner *San jiao* organ in Western anatomy emphasizes how despite their similar terminologies, the TCM body is very different from the biomedical version [12]. Organs with familiar names serve some unfamiliar functions, which manifest in signs and symptoms that, while entirely logical and coherent to the TCM practitioner, often seem bizarre to biomedical healers, who presume similarities based on translated nomenclature where none may exist.

Biomedicine and TCM do have in common an aversion to extremes: normal good health in each is characterized by moderation, an avoidance of hypo- or hyperfunction and extremes of behavior and lifestyle. The healthy TCM body possesses an appropriate amount of *qi* that moves in a balanced fashion: words such as "vacuity," "repletion," "stasis," and "insufficiency" are used to explain the etiology of disease. This is equally true of psychological and somatic problems, as all of these are both cause and effect of improper flow of *qi*. Unfortunately, this mechanistic vocabulary loses some of its subtle meaning in translation, and the conventional English language terminology may seem troublingly vague to those more used to the biomedical preference for precise quantitative description. For example, "overtaxation" may refer to general overwork, sleep deprivation, and dietary irregularities, but it may also be specific to excessive sexual activity. Seminal retention practices, in which ejaculation is purposefully delayed or avoided as a form of personal discipline and to prevent the loss of *jing* and a deficit of *qi*, are an age-old TCM tradition. But the exchange of energy between sexual partners and even the release of *qi* from stasis achieved by solitary masturbation are also salubrious. Thus, in sexual activity, as in other parts of life, TCM encourages moderation [6].

Ideal moderation can be difficult to achieve, particularly in the present age of fast food, distant travel, and constant communication, which encourage extremes of behavior, consumption, and stress. In the paragraphs that follow, we describe andrological diseases according to their Chinese diagnoses, only some of which have biomedical diagnostic correlates. This lack of clear correlation, while unsurprising given the fundamental differences of anatomy and mechanism that we have already established, challenges integrative practice and comparative research.

A prototypical penile disease is *yang wei* (*yang* wilt) or erectile dysfunction (ED). Like the biomedical diagnosis of ED, yang wilt is most commonly seen in older men; unlike ED, which is considered a psychogenic, hormonal, and/or neurovascular disease, *yang* wilt not only is commonly attributed to kidney channel pathology, sexual overtaxation, or diminished life-gate fire with aging, but also generates the following differential diagnosis in TCM:

> Thought and fright damaging the heart and spleen, liver depression, liver channel damp-heat, yin phlegm-dampness obstructing yang qi, and blood stasis. In vacuity patterns, the pathomechanism is that the penis lacks warmth, power, or nourishment for erection. In repletion patterns, there may be sufficient warmth and power, but qi stagnation and/or substantial evils obstruct the free flow of qi and blood to the penis and hence interfere with normal erectile response.
>
> **Ref. [6]**

The patterns referred to above and the broad diagnostic differential given are narrowed by a consideration of patient context, history, and physical exam, the last with a particular emphasis on the qualities of tongue and pulse.

It is possible to perceive connections between TCM pathomechanisms and Western biomedical pathophysiology, to link the aforementioned heart damage, "blood stasis," and obstruction of "the free flow of qi and blood to the penis" to a diagnosis of vasculogenic ED. However, such connections must be made with caution, as they are necessarily inaccurate, reducing TCM's complexly interactive body networks to a single vascular system. Patients who are familiar with and committed to TCM health-care practices may find such reductions off-putting and even dangerous: this is a reason why Viagra, a financial blockbuster drug in the West, did not sell as successfully in China, despite growing concern about an "impotence epidemic" in that country [6].

Other TCM diagnoses are less readily compatible with biomedical equivalents. For example, *yang quiang* (rigid *yang*) is a condition of persistent penile rigidity, with an inability to ejaculate, which is followed later by a spontaneous seminal discharge: is this akin to priapism? Or perhaps delayed ejaculation? Is it possible to include both, and if so, what if any diagnostic or therapeutic insights might be gleaned from making a connection between these otherwise disparate biomedical diagnoses? Table 1 lists common TCM andrological diseases with

Table 1 Andrological Diagnoses in TCM and Proposed Biomedical Equivalents

TCM Diagnosis	Western Biomedical Diagnosis
Yang wei (*Yang* wilt)	Erectile dysfunction
Yang quiang (rigid *yang*)	Priapism
	Delayed ejaculation
Jing zhong tong (pain in the penis)	Urethritis
	Prostatitis (acute or chronic)
	Cystitis (acute or chronic)
	Chronic pelvic pain syndrome
	Bladder pain syndrome
Yin jing duan xiao (small penis)	Micropenis
	Hypogonadism
Zao xie (premature ejaculation)	Premature ejaculation
Yi jing (seminal emission, outside of conscious sexual activity)	Urethritis
	Chronic prostatitis
	Persistent genital arousal disorder
Xue jing (hematospermia)	Hematospermia
Shen nang feng (kidney sac wind)	Scrotal dermatosis
Shan qi (Mounting qi)	Scrotal enlargement NOS
• *Hu shan* (foxy mounting)	• Inguinal hernia
• *Shui shan* (water mounting)	• Hydrocele
• *Xue shan* (blood mounting)	• Scrotal hematoma
• *Tui shan* (prominent mounting)	• Lymphedema
Nang yong (scrotal welling abscess)	Scrotal abscess
Zi yong (testicular welling abscess)	Testicular abscess
Zi tan (testicular phlegm)	Epididymitis
	Orchitis
	(including tubercular infection)
Yin yang (genital itching)	Neuropathic scrotal pruritus
	Tinea cruris
Yin suo (retracted genitals)	?
• *Nang suo* (scrotal retraction)	
• *Jing suo* (penile retraction)	
Infertility	Infertility
Long bi (dribbling urinary block)	Benign prostatic hyperplasia (BPH)
	Prostatitis
	Prostate cancer
	Urethral stricture
Lin zheng (strangury)	BPH
	Prostatitis
	Cystitis

their proposed biomedical equivalents; the presence of multiple possible biomedical diagnoses for some TCM conditions and the absence of any for others again highlight the challenge of integrating these very different health-care models.

Because a single TCM diagnosis (e.g., *yang* wilt) may be a consequence of one or more underlying dysfunctions (e.g., kidney vs liver pathology), treatment is tailored less to the diagnosis than to its etiologic pattern or imbalance. For example, when sexual dysfunction is attributable to a deficit of kidney *qi*, treatment is aimed at warming and supplementing that energy, particularly kidney *yang*, which when vacuous engenders cold. Correlations between this and other diseases and treatment patterns, based predominantly on *Zang Fu* organ theory, are outlined in Table 2 [6]. Various medicinal compounds are used to address each of these treatment patterns in the TCM historical and scientific literature, often in combinations that include between 6 and 20 separate herbs. For patients with comorbid conditions (e.g., liver depression in addition to kidney vacuity), standard of care dictates that these compounds be tweaked if possible: this is feasible for older, published recipes but impossible for more recently developed TCM medications, which are often proprietary, and their components kept secret.

The remarkable complexity of TCM herbal compounds poses multiple challenges to integrative and evidence-based biomedical clinical practice. For historical medications whose components are known, some may be derived from endangered species and their use no longer permitted by TCM's professional regulating bodies [13]; it is uncertain to what extent these recipe changes may affect efficacy. Other ingredients, such as *Zi He Che* (Placenta hominis, or dried human placenta), pose obvious risks of disease transmission if not safely sourced and processed. Ensuring the safety and investigating the claims of newer proprietary compounds imported from abroad is a persistent struggle: the US Food and Drug Administration (FDA) website perennially updates its list of nutraceuticals. Of particular concern for andrologists and their patients is the frequency with which dietary supplements and herbal formulas sold as TCM medications are tainted with sildenafil [14].

This is not to suggest that the herbs in TCM compounds lack evidence of efficacy. Beyond the iterative evidence accrued over millennia of accumulated case reports, which while meaningful is vulnerable to accusations of bias, recent scientific literature explores some component herbs taken individually, to better determine the underlying mechanisms for their effects. For example, *Tu Si Zi* (Cuscutae semen, the seed of the Dodder plant) contains flavonoids that when administered to male rats resulted in increased LH secretion and testosterone production [15]. Korean red ginseng (*Panax ginseng*) has shown efficacy

Table 2 Some Correlations Between Treatment Patterns and Diagnoses

Treatment Principal	Pathology Addressed
Inhibition of damp heat, which may obstruct the kidney, bladder, and triple burner	Lower urinary tract symptoms Dysuria Infertility
Course the liver and rectify depressed flow of *qi*	ED Premature ejaculation (PE) Urinary retention Mounting *qi* (i.e., scrotal enlargement) Age-related hypogonadism
Warm the liver and dissipate cold	Pelvic pain Testicular pain Mounting *qi* (i.e., scrotal enlargement) Retracted genitals
Quicken the blood and transform stasis	Mounting *qi* (i.e., scrotal enlargement) Penile pain BPH ED Infertility
Transform phlegm and soften hardness	Any genital mass or swelling Peyronie's disease BPH
Supplement the spleen and boost *qi*	ED Infertility PE Age-related hypogonadism Chronic prostatitis
Warm and supplement kidney *qi* and *yang*	ED Retracted genitals Small penis Testicular atrophy Infertility Nocturia Urinary frequency
Enrich and supplement kidney *yin*	PE Delayed ejaculation Ejaculatory obstruction Nocturnal emission Age-related hypogonadism
Nourish the heart and calm the spirit	ED Infertility Anxiety re: masculinity

compared with placebo for treatment of ED in both human [16] and animal studies [17]; in vitro studies suggest a dual mechanism, with ginseng acting as both an antioxidant and a stimulus to smooth muscle relaxation, not unlike sildenafil [18].

The active ingredients of many biomedicines are also derived from plants (e.g., paclitaxel from the Yew tree, digoxin from *Digitalis* or foxglove, and quinine from Cinchona bark,), so the notion that individual herbs can contain therapeutic alkaloids, saponins, antioxidants, and other agents is as familiar in the West as it is in the East. Whereas most Western medications have only one or two active ingredients, TCM medications embrace a multiplicity of active herbals. While studying each in isolation can be informative in theory, in practice, this reductionist lens both poorly reflects consumers' real-life experience and defies the foundational tenets of TCM.

For example, even for *Panax ginseng*, one of the more rigorously investigated TCM herbs, the literature prompts more questions than it provides answers: what preparation should be used for maximal efficacy, an extract from the traditional root or from the berry? Is the 1 g 3 × daily dose used in most human-subject studies truly the most appropriate, or would another be better? What are the side effects of ingestion, and how ought patients to be counseled about them, given the relatively unregulated market for nutraceuticals and the ubiquity (and exotic appeal) of foreign-language labeling? If *Panax ginseng* acts as a smooth muscle relaxant similar to sildenafil, is it similarly unsafe for patients taking nitrate medications? What of other pharmaceutical and nutraceutical interactions? And what of the potential promise of other, less studied ingredients in the TCM pharmacopeia? [19].

Given this abundance of questions and few, if any, answers, systematic reviews of TCM medications tend to emphasize the risk of bias in extant randomized and controlled clinical trials and to encourage further study [20]. While appropriate, for the millions of men who are using TCM-based medicinals and other therapeutic interventions today, this caution might be compared with buying a ticket for a train that has long departed the station. For them, the most important resources are thoughtful health-care providers, both TCM and biomedically trained, who can help them to navigate their options safely, along the Chinese *Nanke* integrative model.

References

[1] Li Y. Acupuncture journey to America: a turning point in 1971. J Tradit Chin Med Sci 2014;1:81–3.

[2] Peter J. How cupping works and why Olympic athletes use it. USA Today August 8, 2016. https://www.usatoday.com/story/sports/olympics/rio-2016/2016/08/08/how-cupping-works-and-why-olympic-athletes-use/88410804/. Accessed 1 September 2018.

[3] Academic Consortium for Integrative Medicine and Health, https://imconsortium.org. [Accessed 30 August 2018].

[4] Kohn L. Health and long life: the Chinese way. Magdalena, NM: Three Pines Press; 2005.

[5] Sivin N. Huang ti nei ching. In: Loewe M, editor. Early Chinese texts: a bibliographical guide. Berkeley: University of California Press; 1993. p. 196–215.

[6] Damone B. Principles of Chinese medical andrology: an integrated approach to male reproductive and urological health. Boulder, CO: Blue Poppy Press; 2008.

[7] Lei SH. Neither donkey nor horse: medicine in the struggle over China's modernity. Chicago, IL: University of Chicago Press; 2014.

[8] Zhang EY. The impotence epidemic: men's medicine and sexual desire in contemporary China. Durham, NC: Duke University Press; 2015.

[9] Gillick MR. Old and sick in America: the journey through the health care system. Chapel Hill, NC: University of North Carolina Press; 2017.

[10] Zhang F, et al. What is important during the selection of traditional Chinese medicine (TCM) in a health care reimbursement or insurance system? Critical issues of assessment from the perspective of TCM practitioners. Value Health Reg Issues 2013;2(1):141–6.

[11] Kaptchuk TJ. The web that has no weaver: understanding Chinese medicine. Chicago, IL: Contemporary Books; 2000.

[12] Liu Z, Liu L, editors. Essentials of Chinese medicine. Springer; 2010.

[13] Campbell C. Traditional Chinese medical authorities are unable to stop the booming trade in rare animal parts. Time November 22, 2016. http://time.com/4578166/traditional-chinese-medicine-tcm-conservation-animals-tiger-pangolin/. Accessed 27 August 2018.

[14] FDA. Tainted sexual enhancement products, Last updated August 10, 2018. https://www.fda.gov/drugs/resourcesforyou/consumers/buyingusingmedicinesafely/medicationhealthfraud/ucm234539.htm; 2018. Accessed 27 August 2018.

[15] Qin DN, et al. Effects of flavonoids from semen cuscutae on the reproductive system in male rats. Asian J Androl 2000;2(2):99–102.

[16] Hong B, et al. A double-blind, crossover study evaluating the efficacy of Korean red ginseng in patients with erectile dysfunction: a preliminary report. J Urol 2002;168(5):2070–3.

[17] Kim SD, et al. Improvement of erectile dysfunction by Korean red ginseng in a rat model of metabolic syndrome. Asian J Androl 2013;15(3):395–9.

[18] Cho KS, et al. Effects of Korean ginseng berry extract (GB0710) on penile erection: evidence from in vitro and in vivo studies. Asian J Androl 2013;15(4):503–7.

[19] Park NC, Okuyama A. Modern oriental phytotherapy in sexual medicine. Seoul: Koonja Publishing; 2010.

[20] Xiong G, et al. Chinese herb formulae for treatment of erectile dysfunction: a systematic review of randomised controlled clinical trials. Andrologia 2014;46(3):201–23.

CHAPTER 5.2

Over-the-Counter Supplements and Men's Health

Farouk M. El-Khatib*, Natalie R. Yafi[†], Faysal A. Yafi*

*Department of Urology, University of California Irvine, Newport Beach, CA, United States,
[†]Independent Registered Dietitian, Costa Mesa, CA, United States

INTRODUCTION

Scope of OTC

Over-the-counter medicine, also known as OTC or nonprescription medicine, refers to medicine that you can buy without a prescription. It is the US Food and Drug Administration (FDA) that determines whether medicines are prescription or nonprescription. The term prescription refers to medicines that are safe and effective when used under a doctor's care, whereas nonprescription or OTC drugs are medicines the FDA determines are safe and effective for use without a doctor's prescription [1]. OTC medicine consists mainly of vitamins, minerals, herbs, supplements, and certain medications that are safe and effective when patients follow the directions on the label or those given to them by their health-care professional.

OTCs are used to treat an expanding range of conditions, whether these are considered illnesses or deficiencies. OTCs often do more than just relieve aches, pains, and itches. Some can prevent diseases like tooth decay [2], cure diseases like athlete's foot [3], and help manage recurring conditions like vaginal yeast infections [4] or migraines [5].

Market Size of OTC

The OTC market is growing in an exponential manner. Between 30% and 50% of the adult population in industrialized nations use some form of complementary and alternative medicine to prevent or treat a variety of health-related problems [6]. A recent article in "Global Market Insights" reported that growth driven by massive demand for health supplements and analgesics is expected to exceed $178 billion in revenues by 2024 [6]. Comparing the previous number with the gross domestic product (GDP) of all the countries worldwide, the OTC drugs market will be higher than 50% of some countries' GDP [7].

NATUROPATHIC MEDICINE

As per the American Association of Naturopathic Physicians (AANP), "naturopathic medicine is a distinct primary health-care profession, emphasizing prevention, treatment, and optimal health through the use of therapeutic methods and substances that encourage individuals' inherent self-healing process" [8]. It embraces many therapies, including herbs, massage, acupuncture, exercise, and nutritional counseling.

Naturopathic medicine mainly targets the body as a whole and aims at preventing illnesses. It involves a wellness-oriented diet, lifestyle changes, and supplements that support the body. Naturopathic medicine is often considered for chronic illnesses, such as arthritis or osteoporosis [9], because these conditions improve when diet, lifestyle, and nutritional deficiencies are addressed. It can also be considered for nonemergency acute illnesses, such as colds and flus, because it has been shown to have an effect on cytokines in the body that could help alter the immune system [10].

Comparing conventional medicine with naturopathic medicine can help explain the push by some physicians toward combining both approaches together for better overall benefit. Conventional medicine treats symptoms directly, usually with drugs or surgery. On the other hand, naturopathic medicine attempts to treat the root cause of the health issue so that the body can ultimately heal itself. The results can be measured as a reduction or elimination of drugs, improved vitality, and/or a complete reversal of disease. This does not eliminate the importance and inevitability of many features of conventional medicine, as many of its aspects are not only beneficial but also essential for patient's care. Some diagnostic tools such as laboratory values and imaging are very crucial for establishing the appropriate diagnosis of diseases and monitoring prognosis. As such, it is believed that combining both fields can have a synergistic effect toward achieving optimum care.

OTC AND SEXUAL FUNCTION

Erectile Dysfunction

Erectile dysfunction (ED) is defined as "the inability to attain and/or maintain penile erection sufficient for satisfactory sexual performance" [11]. The prevalence of ED is positively related to age [12]. A comprehensive review of the literature pertaining to ED showed that the prevalence of ED in young men to be as high as 30% [13] that equates to almost 13.5 million men. ED has several etiologies that can be grouped into vasculogenic (arteriogenic and veno-occlusive dysfunction), neurogenic, psychogenic (stress, anxiety, etc.), endocrinologic, or medication induced. Treatments for ED differ based on the etiology and range from minimally invasive approaches such as oral medications (phosphodiesterase type 5 inhibitors (PDE5i)) to medical devices (vacuum erection device), intracavernous injections, intraurethral suppositories, and even surgery (penile implants).

Oral phosphodiesterase type 5 inhibitors (PDE5i) such as sildenafil, tadalafil, and vardenafil work through the L-arginine-nitric oxide-guanylyl cyclase-cyclic guanosine monophosphate (cGMP) pathway and are considered highly effective and safe for the management of ED [14]. In some countries, these pills can now be purchased directly by patients without a prescription [15]. Below is a list of the most commonly used OTC supplements for the management of ED.

Korean Ginseng

Ginseng has been used to restore and enhance the normal well-being of the body [16]. Its active compounds are ginsenosides (triterpene saponins) that have an effect on the central nervous system, metabolism, immune function, and cardiovascular system. Ginsenosides induce the activation of large-conductance KCa channels in human corporal smooth muscle cells leading to hyperpolarization of the cell membrane and, as a consequence, to muscle relaxation [17]. Since erections require two main processes, one of them being cavernous artery smooth muscle relaxation, researchers have suggested that ginsenosides may potentially help in inducing erections. No human studies are available, but data from animal studies have suggested that this action is nitric oxide (NO)-dependent [18].

Korean Red Ginseng (KRG) (Panax Ginseng)

KRG belongs to the Araliaceae family, and its major active ingredients are also ginsenosides (triterpene saponins) [19]. The active ingredient works by relaxing the penile smooth muscle through the NO-cyclic guanosine monophosphate (cGMP) pathway, by increasing the production of NO leading to improved erectile and vascular endothelial function [20]. In a clinical trial, 60% of patients who used ginseng reported a therapeutic effect versus 30% of the patients who were on placebo [21]. In a randomized control trial, ginseng demonstrated improvements in erectile function and reduction in fatigue, insomnia, and depression [22]. In a double-blind crossover study, administration of 1 g of oral ginseng over a period of 8 weeks significantly improved International Index of Erectile Function (IIEF) scores in a control group compared with placebo (16.4 ± 2.9 to 21.0 ± 6.3 vs 17.0 ± 3.1 to 17.7 ± 5.6) [23].

Korean Mountain Ginseng

In a similar manner, 1 g of an extract of tissue-cultured Korean mountain ginseng given orally twice daily over a period of 8 weeks has been shown to significantly improve IIEF scores from 29.78 ± 13.14 to 39.86 ± 15.29.

Furthermore, significant improvements were also noted in orgasmic function and libido in the group that used the Korean mountain ginseng [24]. The only adverse effect that was reported among the experimental group was minor headaches with a prevalence of 4.6% [24].

Gingko biloba

Ginkgo biloba is a popular supplement and one of the top-selling herbal medicines because of its use for a variety of reasons [25]. It is speculated that *G. biloba* can improve vascular perfusion and, in a German study, was found to enhance chronic cerebrovascular insufficiency [26]. The active components of *G. biloba*, which are flavonoid glycosides, have a complicated mechanism of action through which they improve the damage in vessel walls and dysfunction in the tension of blood vessels [26]. In a clinical study, 60 mg of *G. biloba* was given over a period of 12–18 months to a group of patients suffering from antidepressant-induced sexual dysfunction predominately caused by selective serotonin reuptake inhibitors (SSRI). After treatment, 50% of patients who suffered from ED regained their erectile function following 6 months of use [27]. *G. biloba* also improved other aspects of sexual function including enhanced desire, excitement, and orgasm [28]. Reported adverse effects included nausea, diarrhea, dizziness, headaches, stomach aches, restlessness, and vomiting [29].

Rubus coreanus

Unlike other herbal supplements, the season during which the *R. coreanus* is harvested and its ripeness degree are crucial to its efficacy [20]. Korean natives have been using the dry unripe fruit for centuries as a herbal medicine without any major reported safety issues [20]. Crude *R. coreanus* has been described for the management of impotence, spermatorrhea, enuresis, asthma, allergic reaction, and stomachache [30]. No human studies to date have assessed its efficacy in vivo on erectile function, but an animal study showed that extracts of *R. coreanus* can induce penile corpus cavernosum relaxation in a concentration-dependent manner and may even enhance sildenafil's action [31]. In another animal study, the extract has been shown to act by activating the NO-cGMP system, and it may improve cases of ED that do not completely respond to sildenafil citrate through endothelium-independent and endothelium-dependent pathways that relax the corpus cavernosum [32].

Schisandra chinesis

The berries of this plant have been used by Korean natives for ED, enuresis, frequency, spontaneous sweating, night sweats, cough, asthma, sputum, wheezing, jaundice, and diabetes. Recent pharmacological studies have shown a dose-dependent relaxation effect on vascular smooth muscle that was mediated by an endothelium-dependent NO pathway and exerted via dephosphorylation of the myosin light chain [33]. Animal studies have depicted this mechanism of action in more detail. Extracellular application of *S. chinensis* significantly increases whole-cell calcium-sensitive K^+ channel currents (117.4%) and inward rectifier K^+ channel currents (110.0%). This activation of K^+ channels and the inhibition of canonical transient receptor potential Ca^{2+} channel help promote this relaxatory effect [34]. In another study, chemical compounds called schisandrol A & B (part of lignans) were isolated from the fruit. These compounds were found to enhance sildenafil citrate-induced relaxation [35].

Maca (Lepidium meyenii)

L. meyenii, known as maca or Peruvian ginseng, is an edible herbaceous biennial plant that is native to the high Andes mountains of Peru. For centuries, this plant has been used in the Andes as a fertility enhancer in both humans and animals [36]. A systematic review was performed by Byung-Cheul et al. to assess the clinical evidence for or against the effectiveness of the maca plant as a treatment for sexual dysfunction. In this systematic review, several databases (Medical Literature Analysis and Retrieval System Online (Medline), Allied and Complementary Medicine Database (AMED), Cumulative Index to Nursing and Allied Health Literature (CINAHL), Excerpta Medica Database (EMBASE), PsycINFO, etc.) were checked and the inclusion criteria for studies were only trials involving people with normal sexual function and those with sexual dysfunction [37]. Out of the four RCTs that met the inclusion criteria, two suggested a significant positive effect of maca on sexual dysfunction or sexual desire in men ($P < .001$ and $P < .006$), while one RCT failed to show any effects [37]. In patients with sexual dysfunction, one RCT compared the effects of maca with placebo treatments in patients with ED and showed positive effects of maca on the IIEF-5 in patients with mild ED compared with the placebo control (MD, 1.10, 95% CIs, 0.61–1.59, $P < .001$) [38].

Ferula harmonis "Zallouh"

Zallouh is the dried root of the herb *F. hermonis* that grows at high altitude in the Middle East region [39]. Although it is widely used among natives in that region as an aphrodisiac and a male sexual behavior enhancer, there is currently no supportive scientific evidence. In one animal study, the oil extracted from the seeds of *F. hermonis* was tested for its efficacy in enhancing erectile function in male rats and demonstrated that it could potentially enhance erectile function in a dose-dependent fashion [39]. Further studies and additional human studies are needed for validation.

Yohimbine

The bark of the yohimbine tree has long been recognized as an aphrodisiac, and recent studies have shown that it can be effective in the symptomatic treatment of ED [40]. In a double-blind, partial crossover study of 82 men with ED, the therapeutic effects of yohimbine were examined [41]. Patients were given a

Table 1 Effect of Various Herbal Medicines on Erectile Function

Substance	Effect
Korean ginseng	Corporal smooth muscle relaxation
Korean red ginseng (KRG) (Panax ginseng)	Increases the production of NO leading to improved erectile and vascular endothelial function
Korean mountain ginseng	Improves orgasmic function and libido
Ginkgo biloba	Improves antidepressant-induced sexual dysfunction
Rubus coreanus	Corporal smooth muscle relaxation
Schisandra chinesis	Corporal smooth muscle relaxation
Maca (Lepidium meyenii)	Improves erectile dysfunction
Ferula harmonis "zallouh"	Aphrodisiac and enhances male sexual behavior
Yohimbine	Restores full and sustains erections

daily oral maximum dose of 42 mg over a period of 1 month. Results showed that 14% of patients experienced restoration of full and sustained erections, 20% reported a partial response to the therapy, and 65% reported no improvement. It took 2–3 weeks to reach maximal effect (Table 1). Although few side effects were recorded, it is crucial to check for any health contraindications before administration as yohimbine can cause severe cardiac issues in some scenarios [42].

Ejaculatory and Orgasmic Function

Ejaculation and orgasm are two separate physiological processes that are sometimes difficult to distinguish. Ejaculation is a complex physiological process that is composed of three events (emission, bladder neck contracture, and expulsion) and is influenced by intricate neurological and hormonal pathways. Orgasm is an intense transient peak sensation of intense pleasure creating an altered state of consciousness associated with ejaculation [43]. As ejaculation and orgasm are associated with each other, several supplements have been tested to check if they can affect both phenomena. The most commonly used ones are reviewed below.

Epimedium koreanum

E. koreanum is a traditional Korean herbal medicine that has icariin as the active ingredient. It is believed to have many biological effects targeting cardiovascular function, hormone regulation, and immunologic function [44]. A study targeting sexual behavior in rats used two different doses of *E. koreanum*. Group A was on 300 mg/kg, whereas group B was on 750 mg/kg. The supplement was administered daily for 10 days. Results showed an increase in the acts of intromission to 23.3 ± 2.6 in group A and to 20.1 ± 2.3 in group B. Concerning ejaculation, the daily number increased to 2.6 ± 0.4 in group A versus 1.1 ± 0.3

in the placebo group. The latency period of ejaculation was also affected; it decreased to 9.8 ± 1.5 minutes in group A versus 14.2 ± 1.8 minutes in the placebo group. To date, no human studies have been reported [44].

L. meyenii

L. meyenii is a root vegetable that belongs to the *Brassicaceae* family. In its dry form, the root is rich in amino acids, iodine, iron, and magnesium. Traditionally, it has been used as an aphrodisiac and for its fertility-enhancing properties [20]. In an animal study, the latency period of ejaculation in male rats was tested. The control group had a latency period of 112 ± 13 seconds when on regular diet without the supplement. The oral supplementation of a purified form of *L. meyenii* with dosages of 45, 180, or 1800 mg/kg per day for 22 days reduced the latency period of ejaculation to 71 ± 12 seconds, 73 ± 12 seconds, and 41 ± 13 seconds, respectively [45]. In a double-blind clinical trial, the experimental group received a dose of 1.5 or 3 g daily for 12 weeks. Data demonstrated that the group *L. meyenii* had a significant effect on improving sexual desire at 8 and 12 weeks of treatment. This effect was not thought to be due to changes in neither Hamilton scores for depression or anxiety nor serum testosterone and estradiol levels. [46].

Tribulus terrestris and *Cornus officinalis*

Severance secret (SS) cream is a topical agent derived from the extracts of nine natural products (ginseng radix alba, Angelicae Gigantis Radix, Cistanches Herba, *Zanthoxylum* fruits, Torilis semen, Asiasari radix, Caryophylli flos, Cinnamomi cortex, and Bufonis venenum) and has been used for the treatment of premature ejaculation [20]. In an animal study, SS cream was used to check what effects it may have on the mean latency of somatosensory evoked potential (SEP). SEPs are electric signals generated by the nervous system in response to sensory stimuli [47]. When SS cream was applied to the experimental group, the mean amplitude of SEP decreased to 1.83 ± 0.07 µV and 1.72 ± 0.05 µV at 30 and 60 minutes, respectively. This decrease directly led to a prolongation in the mean latency of SEP from 21.57 ± 1.86 milliseconds to 23.09 ± 0.85 milliseconds and 27.49 ± 2.4 milliseconds at 30 and 60 minutes, respectively [48]. These results suggest that this inherent ability to delay SEP may translate to potential ability to delay latency period between ejaculations. In a phase III clinical study, 106 volunteers who suffered from premature ejaculation were given SS cream, and both efficacy and safety were checked [49]. The mean ejaculatory latency time was prolonged to 10.92 ± 0.92 minutes in the SS cream group compared with 2.45 ± 0.29 minutes in the placebo group. The percentage of patients who prolonged their ejaculatory latency time by over two minutes in the placebo and SS cream groups was 15.09% and 79.81%, respectively [49]. Mild local burning and pain were the only adverse effects reported in the treatment group.

Table 2 Effect of Various Herbal Medicines on Ejaculatory and Orgasmic Function

Substance	Effect
Epimedium koreanum	Increases the act of intromission and decrease latency period
Lepidium meyenii	Improves sexual desire
Tribulus terrestris and *Cornus officinalis*	Delays latency period between ejaculations
Cabergoline	Improves orgasm

Cabergoline

Cabergoline is a dopamine agonist that acts on the central nervous system to normalize serum prolactin. Prolactin is a hormone secreted from the pituitary gland and plays a regulatory role in both ejaculation and orgasm pathways. Hyperprolactinemia is associated with an inhibitory effect on male sexual desire [50], and data have shown that an increase in serum prolactin levels (15–20 ng/mL) has been detected in men after orgasm and could be contributing to the after-orgasm refractory period [51]. One study showed that cabergoline administration induced a marked fall in serum prolactin level that began within 3 hours and continued for 7 days [52]. In a retrospective pilot analysis, 131 men suffering from delayed orgasm or anorgasmia were treated with cabergoline 0.5 mg twice weekly. Results showed that 87 men (66.4%) reported subjective improvement in orgasm, and 44 (33.6%) reported no change in orgasm ($P=.0004$). Neither age ($P=.90$) nor prior prostatectomy ($P=.41$) influenced the outcome of cabergoline treatment. Serum testosterone levels before ($P=.26$) and after ($P=.81$) treatment were not significantly different in responders versus nonresponders (Table 2). As such, it is hypothesized that cabergoline may be a potentially effective treatment for male orgasmic disorder, independent of patient age or orgasmic disorder etiology [53].

OTC AND HORMONES

Testosterone

Testosterone (T) is an androgen that is mostly produced in the Leydig cells of the testicles and stimulates the development of male characteristics. It also initiates the development of the male internal and external reproductive organs during fetal development and is essential for puberty and development of secondary male traits. This hormone also has an anabolic function, which means it signals the body to produce new cells and ensures growth of muscle mass and increase in bone density and strength [54].

Anabolic steroids have been popular now for over a decade as a mean to increase muscle mass, decrease body fat, and enhance athletic performance and body appearance [55]. Recent evidence suggests that anabolic steroid use may be one of the top causes of hypogonadism in men, as multiple surveys have shown that these supplements are highly used by bodybuilders to achieve their goals [56]. Despite recent regulatory efforts that have banned specific compounds, many anabolic-androgenic steroids (AAS) remain available in over-the-counter dietary supplements that are legally sold in the United States [57]. The main concern from the use of these supplements without prescription or without the supervision of a specialist is the dangerous side effects that include hepatotoxicity [58], cholestasis [58], renal failure [59], hypogonadism [60], gynecomastia [61], and infertility. While some of these side effects may be reversible, more aggressive use may result in more permanent end-organ damage [56].

Androstenedione
Due to its anabolic effects, many individuals use androstenedione as a dietary supplement. Though widely available over the counter or by mail order in America and Europe, androstenedione is easily accessible [62]. Androstenedione is an androgen steroid hormone that is an intermediate in the biosynthesis of testosterone from dehydroepiandrosterone. In one study, to check the volume of androstenedione usage among athletes, anonymous questionnaires were distributed to five gymnasiums. Among men, 18% reported the use of androstenedione within the last 3 years [62]. Despite their known adverse effects, unknown long-term risks, and possible potential for causing abuse or dependence, a large volume of men are using these supplements without proper guidance.

Dehydroepiandrosterone (DHEA)
Dehydroepiandrosterone (DHEA) is another androgenic steroid hormone that can be found as an OTC supplement. *T. terrestris* Linn (TT) is a plant that has been used for a long time in both the Indian and Chinese systems of medicine. Air-dried aerial parts of this plant contain a major saponin named protodioscin (PTN) [63]. In one study, after bolus intravenous administration of the TT extract at doses of 7.5, 15, and 30 mg/kg body weight in primates, PTN produced a moderate increase in testosterone, dihydrotestosterone, and dehydroepiandrosterone sulfate levels. This increase in DHEA level improved libido and sexual activity [63]. Human studies that were performed using DHEA have unfortunately not looked at sexual benefits. A recent double-blind, placebo-controlled human study showed that DHEA exerts an immunomodulatory action in the elderly [64].

Male Silkworm Extract
Male silkworm extract is a natural herbal combination produced by the *Bombyx mori* moth and is believed to enhance masculinity without side effects. It has been used in both young and adult men who suffer from reduced physical

strength, weak urinary function, or fatigue [20]. In a clinical study with 168 volunteers, two capsules of the extract were administered three times daily over a period of 4 weeks, after which three capsules were administered three times daily for an additional 2 weeks to the volunteers who did not respond to the initial treatment. Results showed that 48% of volunteers who followed this protocol demonstrated increased testosterone levels from 4.4 to 4.9 nmol/L [65]. Furthermore, a statistically significant improvement was noted in symptoms such as fatigue, ED, weak urinary stream, muscular atrophy, and sleep disturbances beginning 2 weeks after the initiation of therapy [65]. Mild adverse effects were reported along with these improvements such as digestive difficulty, sputum production, temporary chest palpitations, drowsiness, and increased stool frequency [65].

Estrogen

Estrogen has been long considered to be the "female sex hormone"; however, estradiol, the predominant form of estrogen, plays a critical role in male sexual function. Estradiol in men is essential for modulating libido, erectile function, and spermatogenesis and is essential to bone density [66]. Areas related to sexual arousal that are found in the brain show increased synthesis of estradiol [67]. The corpus cavernosum has a high concentration of estradiol receptors around the neurovascular bundle that are important for erectile function [68]. Most estrogen supplementation is naturally performed by women, and accordingly, there aren't enough data reporting the use of estrogen as an OTC supplement to enhance male sexual health. There was, however, one study that targeted isoflavone-rich soy protein isolate to determine its effects on circulating hormone profiles and hormone receptor expression patterns in men at high risk for developing advanced prostate cancer [69]. Overall, 58 volunteers were randomly assigned to consume 1 of 3 protein isolates containing 40g/d protein: (1) soy protein isolate (SPI+), (2) alcohol-washed soy protein isolate (SPI-), or (3) milk protein isolate. Serum samples collected at 0, 3, and 6 months were analyzed for circulating estradiol, estrone, sex hormone-binding globulin, androstenedione, androstanediol glucuronide, dehydroepiandrosterone sulfate, dihydrotestosterone, testosterone, and free testosterone concentrations. Results showed that consumption of SPI− significantly increased estradiol and androstenedione concentrations and tended to suppress androgen receptor expression ($P=.09$). Although the effects of SPI− consumption on estradiol and androstenedione are difficult to interpret and the clinical relevance is uncertain, these data show that androgen receptor expression in the prostate is suppressed by soy protein isolate consumption, which may be potentially beneficial in preventing prostate cancer [69].

Calcium-D-Glucarate

Calcium-D-glucarate is a substance made by combining calcium and glucaric acid. It is a compound found naturally in the body and in several types of fruits and vegetables [70] or can be chemically synthesized and purchased without prescription. Walaszek et al. [71] showed that oral supplementation of calcium-D-glucarate inhibited beta-glucuronidase, which is an enzyme produced by colonic microflora. Since elevated activity of beta-glucuronidase is associated with an increased risk of estrogen-dependent cancers (breast, prostate, and colon), inhibiting this enzyme by calcium-D-glucarate intake can help decrease the incidence of these types of cancers [71]. In an animal study, serum levels of estrogen decreased by 23% when calcium-D-glucarate was administered in large doses [72]. Clinical trials on humans are still pending for final outcomes, and preliminary results have not shown adverse effects to its use [73].

OTC AND URINARY FUNCTION

Benign Prostatic Hyperplasia

Benign prostatic hyperplasia (BPH) is an enlargement of the prostate gland that occurs in the transition zone caused by a benign overgrowth of chiefly glandular tissue [12]. As men age, the prostate undergoes two major modifications that lead to clinically lower urinary tract symptoms (LUTS) due to BPH. First, there is an increase in the amount of prostate stroma in the lumen of the prostatic urethra. Second, there is also an increase in the number of alpha 1 receptors in the stroma, which in turn causes an increase in the smooth muscle tone within the prostate and the bladder neck. Both of these modifications lead to obstruction of urine flow [74]. Five-alpha-reductase inhibitors treat BPH by shrinking the prostate stroma, whereas alpha 1 blockers treat BPH by relaxing the prostatic and bladder neck smooth muscle. Many men unfortunately do not discuss their LUTS with their health-care practitioners or seek treatment due to embarrassment and thus remain undiagnosed, untreated, or both. Accordingly, many men will seek self-treatment with OTC drugs, as, to date, alpha blockers have not been approved for OTC use due to concerns about the possibility of inappropriate self-diagnosis and/or self-treatment by patients and the potential for missing more serious underlying disease [75]. The following are the most commonly reported OTC supplements for the management of LUTS/BPH.

Saw Palmetto

In 1995, a large randomized multicenter trial ($n = 176$) investigated "saw palmetto (*Serenoa repens*)" extract (SPE) for any beneficial effects for the management of LUTS. Following the usage of SPE over a 2-month period, the experimental group reported significant increase in urinary flow and decrease in nocturia [76]. To understand the mechanism of action of SPE, an in vitro experiment was performed and showed similarities between the action of SPE and finasteride, a 5-alpha-reductase inhibitor. SPE blocked the conversion of testosterone to dihydroxytestosterone (a major growth stimulator of the prostate) by inhibiting the 5-alpha-reductase enzyme [77]. SPE also blocked the uptake of testosterone and dihydroxytestosterone by the prostate without affecting serum testosterone levels [77].

Pygeum africanum

Pygeum is a supplement made from the bark of the African plumb tree that has been used for many years to treat enlarged prostate [78]. To investigate whether extracts of *P. africanum* are more effective than placebo in the treatment of BPH, Wilt et al. performed a systematic review of trials in computerized general and specialized databases [78]. A total of 18 RCTs involving 1562 men who met the inclusion criteria were analyzed. Results showed that men who used *P. africanum* were more than twice as likely to report an improvement in overall symptoms (RR = 2.1, 95% CI = 1.4, 3.1). Nocturia was reduced by 19%, residual urine volume was reduced by 24%, and peak urine flow was increased by 23%. Adverse effects due to *P. africanum* were mild and comparable with placebo [78].

Rye-Grass Pollen Extract (Cernilton)

Cernilton is a flower (Graminaceae) pollen extract (ryegrass pollen extract) that has been used in Europe to treat nonbacterial prostatitis and benign prostatic hyperplasia (BPF) for more than 35 years [79]. Cernilton contains hydroxamic acid, which has been shown to inhibit the growth of a common human prostate cancer cell line, in vitro [80]. Results of clinical studies using this supplement have demonstrated a marked reduction in residual urine volume, prostate volume, and substantial improvement in urinary flow rate in patients with BPH [81]. MacDonald et al. conducted a systematic review to check the clinical effects and safety of the Cernilton in men with symptomatic BPH [82]. Results in three RCTs showed that Cernilton improved self-rated urinary symptoms when compared with placebo and another plant product (Tadenan). The weighted mean (95% confidence interval) risk ratio (RR) for self-rated improvement versus placebo was 2.40 (1.21–4.75), and the weighted RR versus Tadenan was 1.42 (1.21–4.75). Cernilton reduced nocturia compared with placebo or Paraprost (a mixture of amino acids); against placebo, the weighted RR was 2.05 (1.41–3.00), and against Paraprost, the weighted mean difference for nocturia was −0.40 times per evening (−0.73 to 0.07) (Table 3). Cernilton did not improve urinary flow rates, residual volume, or prostate size compared with placebo or the comparative study agents [82]. Adverse events were rare and mild; the withdrawal rate for Cernilton was 4.8%, compared with 2.7% for placebo and 5.2% for Paraprost [82].

Table 3 Effect of Various Herbal Medicines on Benign Prostatic Hyperplasia

Substance	Effect
Saw palmetto	Blocks the conversion of testosterone to dihydroxytestosterone
Pygeum africanum	Improves lower urinary tract symptoms
Ryegrass pollen extract (Cernilton)	Reduces residual urine volume and prostate volume and improves urinary flow rate

Prostatitis

Chronic prostatitis (CP), or inflammation of the prostate, sometimes also known as chronic pelvic pain syndrome (CPPS), comprises a group of syndromes that affect almost 50% of men at least once in their lifetime and make up the majority of visits to urology clinics [83]. Bacterial infection is one of the top causes of prostatitis and usually starts when bacteria in the urine leak into the prostate [84]. Antibiotics are used to treat the infection, but treatment might not be totally achieved, which results in recurrent episodes of prostatitis (chronic bacterial prostatitis) [85]. Frequent use of antibiotics may also lead to more problems down the road such as antibiotic resistance and gut microbiome dysbiosis. The treatment of prostatitis/CPPS should be tailored individually for patients based on their symptoms and the suspected cause for their condition. Since several etiologies can cause urological pelvic pain syndromes, the urinary, psychosocial, organ-specific, infection, neurological/systemic and tenderness (UPOINT) system was designed to classify patients with an established diagnosis of CP/CPPS into a clinically relevant phenotype that can rationally guide therapy [86]. As mentioned in previous sections, naturopathic medicine is one of the alternative pharmacotherapies proposed to abate inflammatory processes such as prostatitis/CPPS.

Pollen Extract

A randomized double-blind trial assessed the efficacy and safety of a pollen extract preparation for the treatment of patients with chronic nonbacterial prostatitis [87]. Sixty adult patients who had been symptomatic for at least 6 months without response to any given therapy were enrolled in the study. Patients were evaluated at the start and after 6 months of treatment with the help of a symptom questionnaire, similar to the one devised by Krieger et al. [88]. Following 6 months of treatment, more patients in the treatment group were "cured" or improved compared with placebo [87]. In a larger double-blind randomized controlled trial that involved multiple centers, the safety and efficacy of a standardized pollen extract in men with inflammatory chronic prostatitis CPPS were assessed. Evaluations were made based on symptomatic improvement in the pain domain of the NIH Chronic Prostatitis Symptom Index (NIH-CPSI), the number of leukocytes in postprostatic massage urine (VB3), the International Prostate Symptom Score (IPSS), and the sexuality domain of a life satisfaction questionnaire. Overall, 139 participants were randomized to receive two capsules of the pollen extract orally every 8 hours ($n=70$) or placebo ($n=69$) over a period of 12 weeks. Results showed an improvement in pain ($P=.0086$), quality of life ($P=.0250$), and NIH-CPSI score ($P=.0126$) in the pollen extract group [89]. Both these studies highlighted the fact that pollen extract may be potentially effective in the treatment of chronic prostatitis with minimal reported side effects.

Quercetin

This supplement is found in several plants such as green tea, onions, and red wine. It belongs to the family of bioflavonoids that are known for their antioxidant and antiinflammatory effects [90]. In a prospective double-blind placebo-controlled trial, treatment with quercetin for 1 month resulted in mean improvement of the NIH-CPSI score from 21.0 to 13.1 ($P=.003$) compared with the placebo group [91]. One patient taking quercetin developed a headache after the first few doses, which resolved, and 1 patient taking quercetin noted mild tingling of the extremities after each dose. All these side effects resolved after cessation of therapy [87]. In another study, a quercetin formulation mixed with the digestive enzymes bromelain and papain similarly demonstrated an 82% improvement in mean NIH symptom scores from 25.1 to 14.6 [92]. Numerous quercetin supplements are currently available in the market and are often mixed with the aforementioned pollen extract.

Saw Palmetto Extract (SPE)

As mentioned previously, SPE has been studied for the treatment of LUTS by mimicking the function of the 5-alpha-reductase inhibitors [92]. It has similarly been investigated for the management of CPPS. A prospective randomized study that included 64 patients with CPPS compared SPE with finasteride (5-alpha-reductase inhibitor). After 1 year of treatment, the mean NIH-CPSI score in the finasteride group significantly decreased from 23.9 to 18.1 ($P=.003$) and not significantly from 24.7 to 24.6 ($P=.41$) in SPE group. Well-designed multicentered, randomized, prospective trials are, however, needed before any recommendations can be put forth (Table 4).

Table 4 Effect of Various Herbal Medicines on Chronic Prostatitis

Substance	Effect
Pollen extract	Improves pain, quality of life, and NIH-CPSI score
Quercetin	Improves the NIH-CPSI score
Saw palmetto extract (SPE)	Need more studies

RISKS OF OTC

Whether OTCs are beneficial in certain conditions or not, taking supplements without prescription will remain a contentious topic due to the lack of formal agency regulation and potential side effects. Overdosing by taking a larger dose or by taking the supplement for a longer period of time also remains of concern. Mineral supplements can be dangerous; for example, selenium, boron, and iron supplements can be toxic in large amounts [93]. Many people who are willing to take OTCs might already be on certain medications to treat their chronic conditions. Bypassing their primary care physician can put them in danger of various drug-drug interactions. Since a vast majority of the OTC supplements are herbs, some people might have serious allergic reactions to them. While there is some scientific evidence supporting the use of a large number of OTC supplements, more studies continue to be needed to clearly ensure safety and efficacy [94].

TAKE HOME MESSAGES

1. Herbal supplements that have been used across generations tend to be safer to use and tend to be beneficial: Korean ginseng has been used for erectile dysfunction, *L. meyenii* improves sexual desire, and saw palmetto reduces symptoms of BPH and chronic prostatitis.
2. Seeking guidance from the primary care physician is very important to detect subclinical conditions and prevent major complications prior to starting OTC supplements.
3. Future studies are needed to better assess for the true safety and efficacy of most OTC supplements.

References

[1] Center for Drug Evaluation and Research. Questions & answers - prescription drugs and over-the-counter (OTC) drugs: questions and answers. US Food and Drug Administration Home Page, Center for Drug Evaluation and Research; 2017.

[2] Benson, et al. Fluorides for the prevention of early tooth decay (demineralised white lesions) during fixed brace treatment. White Rose Research Online, Sage; 1 January, 1970.

[3] Zatcoff RC, et al. Treatment of tinea pedis with socks containing copper-oxide impregnated fibers. The Foot 2008;18(3):136–41. https://doi.org/10.1016/j.foot.2008.03.005.

[4] Pirotta MV, Garland SM. Genital Candida species detected in samples from women in Melbourne, Australia, before and after treatment with antibiotics. J Clin Microbiol 2006;44(9):3213–7. https://doi.org/10.1128/jcm.00218-06.

[5] Wenzel RG, et al. Over-the-counter drugs for acute migraine attacks: literature review and recommendations. Pharmacotherapy 2003;23(4):494–505. https://doi.org/10.1592/phco.23.4.494.32124.

[6] Astin JA. Why patients use alternative medicine: results of a national study. Complement Ther Med 1999;7(1):51. https://doi.org/10.1016/s0965-2299(99)80067-3.

[7] OTC drugs market revenue to exceed $178B by 2024. Contract Pharma.

[8] "Projected GDP ranking (2018-2023)." Sector-wise contribution of GDP of India - StatisticsTimes.com, statisticstimes.com/economy/projected-world-gdp-ranking.php.

[9] Oleson CV. Osteoporosis in rheumatologic conditions and inflammatory disorders. Osteopor Rehabil 2017. p. 225–49. https://doi.org/10.1007/978-3-319-45084-1_12.

[10] Armstrong AR, et al. Australian adults use complementary and alternative medicine in the treatment of chronic illness: a national study. Aust NZ J Public Health 2011;35(4):384–90. https://doi.org/10.1111/j.1753-6405.2011.00745.x.

[11] Spelman K, et al. Modulation of cytokine expression by traditional medicines: a review of herbal immunomodulators. Altern Med Rev 2006;11(2):128–50.

[12] Selvin E, et al. Prevalence and risk factors for erectile dysfunction in the US. Am J Med 2007;120(2):151–7.

[13] Bennett N. Faculty of 1000 evaluation for prevalence and risk factors for erectile dysfunction in the US. F1000—Post-Publication Peer Review of the Biomedical Literature; 2016. https://doi.org/10.3410/f.725054348.793515475.

[14] Corbin JD. Mechanisms of action of PDE5 inhibition in erectile dysfunction. Int J Impot Res 2004;16(S1). https://doi.org/10.1038/sj.ijir.3901205.

[15] Sharlip ID. Difference of opinion–is there a space to improve the treatment of erectile dysfunction in the next years? Opinion: no. Ten reasons that there will be no new pharmacologic therapies for erectile dysfunctionin the foreseeable future. Int Braz J Urol 2015;41(5):832–4.

[16] Leung KW, Wong AST. Ginseng and male reproductive function. Spermatogenesis 2013;3(3):e26391.

[17] Sung HH, et al. Effects of ginsenoside on large-conductance K(Ca) channels in human corporal smooth muscle cells. Int J Impot Res 2011;23(5):193–9. https://doi.org/10.1038/ijir.2011.25.

[18] Chen X, Lee TJ. Ginsenosides-induced nitric oxide-mediated relaxation of the rabbit corpus cavernosum. Br J Pharmacol 1995;115(1):15–8. https://doi.org/10.1111/j.1476-5381.1995.tb16313.x.

[19] Kim YM, Namkoong S, Yun YG, Hong HD, Lee YC, Ha KS, et al. Water extract of Korean red ginseng stimulates angiogenesis by activating the PI3K/Akt-dependent ERK1/2 and eNOS pathways in human umbilical vein endothelial cells. Biol Pharm Bull 2007;30(9):674–1679. https://doi.org/10.1248/bpb.30.1674.

[20] Shin YS, et al. Current status and clinical studies of oriental herbs in sexual medicine in Korea. World J Men's Health 2015;33(2):62–72.

[21] Choi HK, Seong DH, Rha KH. Penile blood change after oral medication of Korean Red Ginseng in erectile dysfunction patients. J Ginseng Res 2003;27(4):165–70. https://doi.org/10.5142/jgr.2003.27.4.165.

[22] Choi YD, Rha KH, Choi HK. In vitro and in vivo experimental effect of Korean red ginseng on erection. J Urol 1999. p. 1508–11. https://doi.org/10.1097/00005392-199910000-00093.

[23] Hong B, Ji YH, Hong JH, Nam KY, Ahn TY. A double-blind crossover study evaluating the efficacy of Korean red ginseng in patients with erectile dysfunction: a preliminary report. J Urol 2002. p. 2070–73. https://doi.org/10.1097/00005392-200211000-00041.

[24] Kim TH, Jeon SH, Hahn EJ, Paek KY, Park JK, Youn NY, et al. Effects of tissue-cultured mountain ginseng (Panax ginseng CA Meyer) extract on male patients with erectile dysfunction. Asian J Androl 2009;11(3):356–61. https://doi.org/10.1038/aja.2008.32.

[25] Ginkgo Biloba Extract Market: Global Industry Analysis, Size, Share, Growth, Trends, and Forecasts 2016–2024, *Industry News Today* 11 Feb. 2019, industrynewstoday.com/19026/

ginkgo-biloba-extract-market-global-industry-analysis-size-share-growth-trends-and-forecasts-2016-2024/.

[26] Kleijnen J, Knipschild P. Ginkgo biloba. Lancet 1992;340(8828):1136–9.

[27] Ashton AK. Antidepressant-induced sexual dysfunction and ginkgo biloba. Am J Psychiatry 2000;157(5):836–7. https://doi.org/10.1176/appi.ajp.157.5.836.

[28] Paick JS, Lee JH. An experimental study of the effect of ginkgo biloba extract on the human and rabbit corpus cavernosum tissue. J Urol 1996. p. 1876–80. https://doi.org/10.1097/00005392-199611000-00113.

[29] Nordqvist J. Ginkgo biloba: health benefits, side effects, risks, and history. Medical News Today MediLexicon International, 18 December 2017.

[30] Shin TY, et al. Action of Rubus coreanus extract on systemic and local anaphylaxis. Phytother Res 2002;16(6):508–13. https://doi.org/10.1002/ptr.925.

[31] Zhao C, Kim HK, Kim SZ, Chae HJ, Cui WS, Lee SW, et al. What is the role of unripe Rubus coreanus extract on penile erection? Phytother Res 2011;25(7):1046–53. https://doi.org/10.1002/ptr.3393.

[32] Lee JH, Chae MR, Sung HH, Ko M, Kang SJ, Lee SW. Endothelium-independent relaxant effect of Rubus coreanus extracts in corpus cavernosum smooth muscle. J Sex Med 2013;10(7):1720–9. https://doi.org/10.1111/jsm.12183.

[33] Park JY, Shin HK, Lee YJ, Choi YW, Bae SS, Kim CD. The mechanism of vasorelaxation induced by Schisandra chinensis extract in rat thoracic aorta. J Ethnopharmacol 2009;121(1):69–73. https://doi.org/10.1016/j.jep.2008.09.031.

[34] Han DH, Lee JH, Kim H, Ko MK, Chae MR, Kim HK, et al. Effects of Schisandra chinensis extract on the contractility of corpus cavernosal smooth muscle (CSM) and Ca2 + homeostasis in CSM cells. BJU Int 2011;109(9):1404–13. https://doi.org/10.1111/j.1464-410x.2011.10567.x.

[35] Kim HK, Bak YO, Choi BR, Zhao C, Lee HJ, Kim CY, et al. The role of the lignan constituents in the effect of Schisandra chinensis fruit extract on penile erection. Phytother Res 2011.

[36] Hudson T. Maca: new insights on an ancient plant. Integr Med 2008;7(6):54–7.

[37] Shin B-C, et al. Maca (L. Meyenii) for improving sexual function: a systematic review. BMC Complement Altern Med 2010;10:44.

[38] Zenico T, Cicero AF, Valmorri L, Mercuriali M, Bercovich E. Subjective effects of Lepidium meyenii (Maca) extract on well-being and sexual performances in patients with mild erectile dysfunction: a randomised, double-blind clinical trial. Andrologia 2009;41(2):95–9. https://doi.org/10.1111/j.1439-0272.2008.00892.x.

[39] El-Thaher TS, et al. Ferula harmonis 'zallouh' and enhancing erectile function in rats: efficacy and toxicity study. Int J Impot Res 2001;13(4):247–51. https://doi.org/10.1038/sj.ijir.3900706.

[40] Pittler MH. Yohimbine in therapy of erectile dysfunction. Curr Neurol Neurosci Rep BMJ 1998;317(7156):478. https://doi.org/10.1136/bmj.317.7156.478.

[41] Susset JG, et al. Effect of yohimbine hydrochloride on erectile impotence: a double-blind study. J Urol 1989;141(6):1360–3. https://doi.org/10.1016/s0022-5347(17)41308-5.

[42] Yohimbe: uses, side effects, interactions, dosage, and warning. WebMD, 2017.

[43] Alwaal A, Breyer BN, Lue TF. Normal male sexual function: emphasis on orgasm and ejaculation. Fertil Steril 2015;104(5):1051–60.

[44] Makarova MN, Pozharitskaya ON, Shikov AN, Tesakova SV, Makarov VG, Tikhonov VP. Effect of lipid-based suspension of Epimedium koreanum Nakai extract on sexual behavior in rats. J Ethnopharmacol 2007;114(3):412–6. https://doi.org/10.1016/j.jep.2007.08.021.

[45] Zheng BL, He K, Kim CH, Rogers L, Shao Y, Huang ZY, et al. Effect of a lipidic extract from lepidium meyenii on sexual behavior in mice and rats. Urology 2000;55(4):598–602. https://doi.org/10.1016/s0090-4295(99)00549-x.

[46] Gonzales GF, Córdova A, Vega K, Chung A, Villena A, Góñez C, et al. Effect of Lepidium meyenii (MACA) on sexual desire and its absent relationship with serum testosterone levels in adult healthy men. Andrologia 2002;34(6):367–72. https://doi.org/10.1046/j.1439-0272.2002.00519.x.

[47] Alan D, et al. General principles of somatosensory evoked potentials: overview, electrical stimulation parameters, recording parameters. In: Background, pathophysiology, etiology, 2 july, 2018.

[48] Xin ZC, et al. Sensory evoked potential and effect of SS-cream in premature ejaculation. Yonsei Med J 1995;36(5):397. https://doi.org/10.3349/ymj.1995.36.5.397.

[49] Choi HK, Jung GW, Moon KH, Xin ZC, Choi YD, Lee WH, et al. Clinical study of SS-cream in patients with lifelong premature ejaculation. Urology 2000;55(2):257–61. https://doi.org/10.1016/s0090-4295(99)00415-x.

[50] Buvat J. Hyperprolactinemia and sexual function in men: a short review. Int J Impot Res 2003;15(5):373–7. https://doi.org/10.1038/sj.ijir.3901043.

[51] Exton MS, Krüger TH, Koch M, Paulson E, Knapp W, Hartmann U, et al. Coitus-induced orgasm stimulates prolactin secretion in healthy subjects. Psychoneuroendocrinology 2001;26(3):287–94. https://doi.org/10.1016/s0306-4530(00)00053-6.

[52] Ferrari C, et al. Long-lasting prolactin-lowering effect of cabergoline, a new dopamine agonist, in hyperprolactinemic patients. J Clin Endocrinol Metab 1986;63(4):941–5. https://doi.org/10.1210/jcem-63-4-941.

[53] Hollander AB, et al. Cabergoline in the treatment of male orgasmic disorder—a retrospective pilot analysis. Sex Med 2016;4(1):e28–33.

[54] Bain J. The many faces of testosterone. Clin Interv Aging 2007;2(4):567–76.

[55] Wedro B. Steroid abuse symptoms, side effects & treatment. MedicineNet; 2017.

[56] Rahnema CD, et al. Anabolic steroid-induced hypogonadism: diagnosis and treatment. Fertil Steril 2014;101(5):1271–9. https://doi.org/10.1016/j.fertnstert.2014.02.002.

[57] Rahnema CD, et al. Designer steroids – over-the-counter supplements and their androgenic component: review of an increasing problem. Andrology 2015;3(2):150–5. https://doi.org/10.1111/andr.307.

[58] Brown AC. Heart toxicity related to herbs and dietary supplements: online table of case reports. Part 4 of 5. J Dietary Suppl 2017;15(4):516–55. https://doi.org/10.1080/19390211.2017.1356418.

[59] Krishnan PV, et al. Prolonged intrahepatic cholestasis and renal failure secondary to anabolic androgenic steroid-enriched dietary supplements. J Clin Gastroenterol 2009;43(7):672–5. https://doi.org/10.1097/mcg.0b013e318188be6d.

[60] Gill GV. Anabolic steroid induced hypogonadism treated with human chorionic gonadotropin. Postgrad Med J 1998;74(867):45–6.

[61] Maravelias C, et al. Adverse effects of anabolic steroids in athletes. Toxicol Lett 2005;158(3):167–75. https://doi.org/10.1016/j.toxlet.2005.06.005.

[62] Kanayama G, et al. Over-the-counter drug use in gymnasiums: an underrecognized substance abuse problem? Psychother Psychosomat 2001;70(3):137–40. https://doi.org/10.1159/000056238.

[63] Bowers LD. Oral dehydroepiandrosterone supplementation can increase the testosterone/epitestosterone ratio. Clin Chem 1999;45(2):295–7.

[64] Rutkowski K, et al. Dehydroepiandrosterone (DHEA): hypes and hopes. Drugs 2014;74(11):1195–207. https://doi.org/10.1007/s40265-014-0259-8.

[65] Kim DC, Kim YW, Park MS, Suh JK, Lee DS, Lee SH, et al. Effects of the Nuegra from male silkworm extract on enhancement of the masculine function and activation of overall physical function. Int J Ind Entomol 2002;5:109–22.

[66] Schulster M, Bernie AM, Ramasamy R. The role of estradiol in male reproductive function. Asian J Androl 2016;18(3):435–40.

[67] Roselli CE, Abdelgadir SE, Resko JA. Regulation of aromatase gene expression in the adult rat brain. Brain Res Bull 1997;44(4):351–7. https://doi.org/10.1016/s0361-9230(97)00214-1.

[68] Dean RC, Lue TF. Physiology of penile erection and pathophysiology of erectile dysfunction. Urol Clin North Am 2005;32(4):379–95. https://doi.org/10.1016/j.ucl.2005.08.007.

[69] Hamilton-Reeves JM, et al. Isoflavone-rich soy protein isolate suppresses androgen receptor expression without altering estrogen receptor-beta expression or serum hormonal profiles in men at high risk of prostate cancer. J Nutr 2007;137(7):1769–75. https://doi.org/10.1093/jn/137.7.1769.

[70] Dwivedi C, Heck WJ, Downie AA, et al. Effect of calcium glucarate on beta-glucuronidase activity and glucarate content of certain vegetables and fruits. Biochem Med Metab Biol 1990;43:83–92.

[71] Walaszek Z, Szemraj J, Narog M, et al. Metabolism, uptake, and excretion of a D-glucaric acid salt and its potential use in cancer prevention. Cancer Detect Prev 1997;21:178–90.

[72] Walaszek Z, Hanausek-Walaszek M, Minto JP, Webb TE. Dietary glucarate as anti-promoter of 7,12-dimethylbenz[a]anthracene-induced mammary tumorigenesis. Carcinogenesis 1986; 7:1463–6.

[73] Calcium-D-glucarate. Altern Med Rev 2002, www.ncbi.nlm.nih.gov/pubmed/12197785.

[74] Wieder JA. Pocket guide to urology. J. Wieder Medical; 2010.

[75] Roehrborn CG, et al. Are over-the-counter alpha blockers in the best interest of men with lower urinary tract symptoms? Urol Pract 2017;4(5):395–404. https://doi.org/10.1016/j.urpr.2016.09.006.

[76] Descotes J, Rambeaud J, Deschaseaux P, Faure G. Placebo-controlled evaluation of the efficacy and tolerability of permixon in benign prostatic hyperplasia after exclusion of placebo responders. Clin Drug Invest 1995;9(5):291–7. https://doi.org/10.2165/00044011-199509050-00007.

[77] Sultan C, Terraza A, Devillier C. Inhibition of androgen metabolism and binding by a lipidosterolic extract of "Serenoa repens B" in human foreskin fibroblasts. J Steroid Biochem 1984; 20(1):515–9. https://doi.org/10.1016/0022-4731(84)90264-4.

[78] Ishani A, et al. Pygeum africanum for the treatment of patients with benign prostatic hyperplasia: a systematic review and quantitative meta-analysis. Am J Med 2000;109(8):654–64. https://doi.org/10.1016/s0002-9343(00)00604-5.

[79] Yasumoto R, et al. Clinical evaluation of long-term treatment using cernitin pollen extract in patients with benign prostatic hyperplasia. Clin Ther 1995;17:82–6.

[80] Habib ZX, et al. Isolation and characteristics of a cyclic hydroxamic acid from a pollen extract, which inhibits cancerous cell growth in vitro. J Med Chem 1994;38(4):735–8.

[81] Qian X, et al. Therapeutic efficacy of Cernilton in benign prostatic hyperplasia patients with histological prostatitis after transurethral resection of the prostate. Int J Clin Exp Med 2015; 8(7):11268–75.

[82] MacDonald R, et al. A systematic review of cernilton for the treatment of benign prostatic hyperplasia. BJU Int 2001;85(7):836–41. https://doi.org/10.1046/j.1464-410x.2000.00365.x.

[83] Khan A, Murphy AB. Updates on therapies for chronic prostatitis/chronic pelvic pain syndrome. World J Pharmacol 2015;4(1):1. https://doi.org/10.5497/wjp.v4.i1.1.

[84] Balentine JR. What is prostatitis? symptoms, treatment, & causes. MedicineNet; 2018.

[85] Prostatitis. Mayo Clinic, Mayo Foundation for Medical Education and Research, 16 May, 2018. 16 May.

[86] Zhao Z, et al. Clinical utility of the UPOINT phenotype system in Chinese males with chronic prostatitis/chronic pelvic pain syndrome (CP/CPPS): a prospective study. PloS ONE 2013; 8(1):e52044.

[87] Elist J. Effects of pollen extract preparation prostat/poltit on lower urinary tract symptoms in patients with chronic nonbacterial prostatitis/chronic pelvic pain syndrome: a randomized, double-blind, placebo-controlled study. Urology 2006;67(1):60–3. https://doi.org/10.1016/j.urology.2005.07.035.

[88] Krieger JN, Egan KJ, Ross SO, et al. Chronic pelvic pains represent the most prominent urogenital symptoms of "chronic prostatitis". Urology 1996;48:715–21.

[89] Wagenlehner FM, Schneider H, Ludwig M, Schnitker J, Brähler E, Weidner W. A pollen extract (Cernilton) in patients with inflammatory chronic prostatitis-chronic pelvic pain syndrome: a multicentre, randomised, prospective, double-blind, placebo-controlled phase 3 study. Eur Urol 2009;56(3):544–51. https://doi.org/10.1016/j.eururo.2009.05.046.

[90] Shoskes DA, Nickel JC. Quercetin for chronic prostatitis/chronic pelvic pain syndrome. Urol Clin North Am 2011;38(3):279–84. https://doi.org/10.1016/j.ucl.2011.05.003.

[91] Shoskes DA, Zeitlin SI, Shahed A, Rajfer J. Quercetin in men with category III chronic prostatitis: a preliminary prospective, double-blind, placebo-controlled trial. Urology 1999;54(6):960–3. https://doi.org/10.1016/s0090-4295(99)00358-1.

[92] Suh LK, Lowe FC. Alternative therapies for the treatment of chronic prostatitis. Curr Urol Rep 2011;12(4):284–7. https://doi.org/10.1007/s11934-011-0188-y.

[93] Aldosary BM, et al. Case series of selenium toxicity from a nutritional supplement. Clin Toxicol 2011;50(1):57–64. https://doi.org/10.3109/15563650.2011.641560.

[94] Davidson MH, Geohas CT. Efficacy of over-the-counter nutritional supplements. Curr Atheroscler Rep 2003;5(1):15–21. https://doi.org/10.1007/s11883-003-0063-5.

Further Reading

[95] Adimoelja A. Phytochemicals and the breakthrough of traditional herbs in the management of sexual dysfunctions. Int J Androl 2000;23(S2):82–4. https://doi.org/10.1046/j.1365-2605.2000.00020.x.

Recreational Habits

CHAPTER 6.1

Smoking and Men's Health

U. Milenkovic, M. Albersen
Department of Urology, University Hospitals Leuven, Leuven, Belgium

INTRODUCTION

In the United States, the principal source of preventable morbidity and mortality is cigarette smoking (CS). It accounts for about half million deaths each year, which is about 20% of deaths [1]. Fortunately, smoking habits have seen a decline over recent years. In 2016, 15.5% or c. 37.8 million (down from 20.9% in 2005) of all US adults are currently smoking cigarettes. However, more than 16 million Americans must still live with smoking-related comorbidities [1]. Over 60% of noncommunicable diseases list smoking as a possible risk factor [2]. Approximately one-third of all male adults worldwide and 30% of women of reproductive age smoke cigarettes, with Europe still being the leading continent [3].

Cigarettes contain tobacco, which is derived from a dicotyledonous plant from the species *Nicotiana tabacum*, belonging to the Solanaceae family (together with coffee beans, potato, and tomato plants). Its sprouting is stimulated by being cut and is planted throughout the year. Its growth may take several years and may grow up to 2 m (c. 6 ft) high. Nicotine, the substance responsible for the addictive component of CS, is an alkaloid residing in the plant under the form of an organic salt. It has the highest concentration in the highest leaves [4].

CS is notorious for causing a multitude of health hazards such as cardiovascular disease, respiratory disease, cancer, and birth defects in infants [1]. Traditionally, cigarette smoke can be separated in two distinct phases; the tar (or particulate) and gas phase [5]. The free radicals inside the tar phase have high longevity (hours to several months), while gaseous free radicals have a life span of a few seconds [6–8]. Mainstream smoke, that is, smoke that is inhaled through the tobacco into a smoker's mouth, contains 8% tar and 92% gas phase components. Nicotine is a component of the tar phase [9].

In this chapter, an overview of the effect of CS on male (sexual) health will be discussed with a focus on male fertility, testosterone, erectile dysfunction, nephrolithiasis, and urological cancers.

SMOKING AND MALE FERTILITY

Infertility is defined by the American Society for Reproductive Medicine as the absence of pregnancy after 12 months of regular, unprotected sexual intercourse. Literature shows that approximately 10%–15% of couples wanting to have children suffer from infertility in industrialized nations [10]. Nearly 30% is caused by a sole male factor, and an additional 20% has a combined male-female factor. Thus, male factors play a role in around 50% of all cases of infertility [11]. Moreover, men of reproductive age (20–39 years old) make up for nearly half of all smokers (46%) [12], so there has been great interest in the effects of CS on male reproductive function. By tackling the preventable environmental and lifestyle factors influencing fertility, there is potential to lower the social burden and public health cost of male infertility [3].

It has proved difficult to provide a definition of male infertility, and diagnosis has largely been based on comparison of semen analysis results with the World Health Organization (WHO) standards, and since 2010, there are new criteria for the laboratory examination of human semen [13]. Worryingly so, there has been a steady decline of sperm concentration over the past 35 years [14].

Studying the link between smoking and male infertility has been fraught by contrasting results. Most studies show a negative impact of smoking on semen analysis and male infertility; however, opposing results were found regarding its effect on sperm motility and the extent of DNA damage. This has been reviewed extensively by Harlev et al. and Sharma et al. [12, 15].

Due to obvious ethical concerns, the effects of recreational drug use, alcohol consumption, secondhand smoking, and CS cannot be investigated in randomized, interventional clinical trials. Consequently, most of the studies concerning these subjects have been retrospective with a large amount of confounding factors (alcohol use, medical illness, and socioeconomic status), for which it is difficult to control [16]. A partial solution to this can be offered by way of animal studies; however, exposure in these studies is much higher and over a more limited amount of time in comparison with the human situation; thus, these studies should be interpreted with caution.

Presently, more than 4700 different chemicals have been found in tobacco smoke [17], ranging from heavy metals, to polycyclic aromatic hydrocarbons, to mutagenic chemicals. Metal micronutrients such as arsenic, cadmium, and lead are routinely inhaled during the combustion process of tobacco or cigarette paper [17–19]. These are all involved in the generation of oxidative stress and mutagenesis in the spermatogenetic process. They are associated with increased risk of male infertility, despite not affecting semen volume, concentration, and motility [20, 21]. Oxidative injury can also be caused by the increased concentration of benzo(a)pyrene in sperm from smoking males. Benzo(a)pyrene is a highly mutagenic carcinogen that binds to the DNA. Smoking also lowers the serum antioxidant activity by attenuation of superoxide dismutase [22, 23].

Clinical studies have shown adverse effects on several parameters of semen analysis. Two thousand five hundred forty-two healthy men were analyzed in a cross-sectional fashion from 1987 to 2004 [24]. Cigarette smoking caused lower semen volumes and oligoasthenozoospermia (lower sperm count and decreased motility), and this happened in a dose-dependent manner. Smoking more than 20 cigarettes a day caused a nearly 20% worse semen quality than in other smokers. Other authors have had similar findings [25–29]. Moreover, in 1997, the Ontario Farm Family Health Study reported decreased fecundability rates for smoking males in a retrospective cohort of 2607 cases of planned pregnancy [30].

In addition to observational studies, a meta-analysis that included 57 observational studies and nearly 30,000 men found that, both in infertile and fertile men, CS had a significant impact on all sperm-related parameters (volume, density, total counts, and motility). Similarly, Kunzle and colleagues have found a significant association between smoking and reduced sperm concentration in 2100 men presenting for fertility evaluation [25].

Definite evidence involving the exact mechanisms of impaired sperm quality has yet to be provided. Ultrastructural abnormalities involving microtubules, tail alterations, acrosome reaction, and capacitation have been most commonly reported [31–33]. The latter two processes are essential for the fertilization processes. Increased oxidative stress has been put forward as one of the possible mechanisms of impaired sperm function. Hypoxia might also be associated with impaired spermatogenesis [16, 34].

In 2010, the relationship between zinc levels and semen quality was investigated [35]. It seems that zinc levels corresponded with seminal parameters. In short, smokers had an overall lower zinc concentration in the semen, with an associated decrease in sperm concentration, motility, and morphology. However, smokers that had normal zinc levels did not experience the same degree of semen abnormality compared with the smokers with low zinc.

Epigenetic and DNA methylation pattern changes and certain microRNAs (miRNAs) can impact semen quality as well. Histones are proteins that package DNA in somatic cells and oocytes. This role is taken over by protamines during advanced stages of spermatogenesis and is regulated epigenetically [12]. Protamines offer protection of the genetic material from oxidation and other deteriorating substances in the female genital tract. Twenty-eight differentially expressed miRNAs between semen from nonsmokers and smokers have been found. Four of these play a role in proliferation, differentiation, and apoptosis in spermatozoa [36].

Semen parameters and sperm function tests are 22% poorer in smokers, and its effects are dose-dependent according to the 2012 American Society of Reproductive Medicine [37], but unambiguous results have not been provided yet. Cigarette smoking has a significant effect on spermatogenesis, even though its precise mechanisms are yet to be elucidated. Considering the available literature and despite the lack of randomized controlled trials, smoking cessation and reduction of secondhand smoking in both sexes should be recommended to couples attempting to conceive [38].

SMOKING AND TESTOSTERONE

Levels of total testosterone (TT) have been reported in many studies to be related to a variety of disorders and comorbidities including cardiovascular disease, type 2 diabetes, metabolic syndrome, erectile dysfunction, depression, obesity, stroke, and osteoporosis [39–45]. Total testosterone includes sex hormone-binding globulin (SHBG)-bound testosterone, albumin-bound testosterone, and free testosterone (FT).

Hormone levels fluctuate with several factors including age, cigarette smoking, body mass index (BMI), alcohol consumption, and physical activity [46]. However, the reported effects of smoking on FT or TT in men have been controversial; some report no association [47, 48]; others report higher [49, 50] or even lower [51, 52] levels of TT among smokers. There was a similarly conflicting report regarding the connection between smoking and FT and SHBG. As mentioned before, due to the retrospective study design, there are several confounding factors that should be accounted for, but this is rarely possible.

Hypothetically, chronic cigarette smoking increases liver metabolism of testosterone while at the same time inducing secretory dysfunction of Leydig and Sertoli cells [53]. However, there seems to be no consensus on the effects of smoking on follicle-stimulating hormones (FSH) and luteinizing hormone (LH) production: some studies have shown lower or no different levels of both gonadotropins among smokers (54), whereas different researchers have observed increased concentration of LH and/or FSH following tobacco consumption [54, 55]. Considering all the possible confounding factors, testosterone concentration is remarkably difficult to ascertain among smokers: some studies have reported increased serum testosterone and dehydroepiandrosterone levels in smokers [54, 56], whereas others have suggested that mean levels of testosterone are not significantly different [57] between smokers and nonsmokers.

The literature does not provide us with clear answers surrounding the relationship between smoking and testosterone levels, and thus, no formal recommendations can be made.

SMOKING AND ERECTILE DYSFUNCTION

Normal erections are a hormonally regulated, combined neurovascular tissue phenomenon. Arterial dilatation, trabecular smooth muscle relaxation, and activation of the corporal veno-occlusive mechanisms are essential components of the process. The NIH Consensus Development Panel on Impotence defined erectile dysfunction (ED) as the persistent inability to attain and maintain an erection sufficient for satisfactory sexual intercourse. ED is the preferred term since sexual desire and the ability to achieve orgasm may as well be intact, in the absence of an erection.

The Massachusetts Male Aging Study (MMAS) [58] was the first large prospective observational longitudinal study performed in healthy, randomly selected men from all over the globe. It was revealed that ED has a high prevalence and incidence worldwide. The highest number reported includes a prevalence as high as 52% in the United States among noninstitutionalized men between 40 and 70 years old (of which 17% was minimal, 25% moderate, and 10% complete ED). Even though ED cannot be viewed as an immediate danger to the patients' health status per se, it is still considered to have important effects on the quality of life and personal well-being of patients and partners alike. Patients with ED have increased rates of decreased self-esteem, depression, anxiety, anger, and relationship dissatisfaction.

Additionally, ED is an important independent predictor for cardiovascular morbidity and mortality. An ED complaint should always warrant further investigation, as the urological dogma states "the penis is the antenna of the heart." Underlying hypertension, hyperlipidemia, metabolic syndrome, diabetes mellitus, cardiac disease, and smoking are major predictors of ED and vice versa and should be identified in a patient presenting with ED. In short, it is an important health concern and could be an augur for other serious medical conditions.

Briefly, there are four important criteria that need to be fulfilled to allow for penile erection:

- (i) Intact neuronal transmission of proerectile responses, that is, efficacious delivery of the neurotransmitter nitric oxide (NO) to the smooth muscle cells (SMC) in the corpus cavernosum (CC). During sexual stimulation, NO is set free from nerve endings and vascular endothelial cells in the penis, diffusing into vascular and corporal SMCs. Subsequent

stimulation of guanylyl cyclase (GC) raises the amount of intracellular cGMP. The second messenger cascade causes a lowering of intracytoplasmic calcium, resulting in SMC relaxation, vasodilatation, and penile erection.
(ii) Furthermore, an intact arterial blood supply is essential to fill the cavernous erectile tissue with oxygenated blood.
(iii) The CC needs to be in good health and SMC elastic to successfully expand and compress the subtunical venous plexus during the rigid phase of the erection.
(iv) Lastly, hormonal balance should be maintained as testosterone not only is the main driver of libido but also keeps peripheral effectors healthy through genomic effects. ED develops if one of these systems fails, but often, there is a multilevel failure [59].

Many cross-sectional studies have established the correlation between CS and ED [60–64]. These studies have been performed in populations all over the world, and odds ratios have ranged between 1.4 and 3.1. Many of these studies accounted for most known comorbid vascular factors (age, hypertension, obesity, and diabetes); however, it is still quite difficult to determine the significance of any one isolated factors, since many exist together and cannot be separated.

Even in a specific cohort including young men under 40 years old, there was a significant correlation between CS and ED. Multivariate analysis in this study did not show any other vascular risk factor, which strongly suggests an independent, direct pathophysiological role of CS in ED [65].

A recent systematic review compiling eight case-control and cohort studies totaling 28,586 men showed that there was an increased risk with OR of 1.81 for smokers to develop ED [66].

CS has been suggested to act as an independent, dose-dependent risk factor for heart disease and ED. Smoking a greater number than 10 cigarettes a day increased the OR [60]. Additionally, in a younger, less comorbid patient population, those who were heavy smokers (>20 cigarettes per day) had twice the likelihood of developing severe ED compared with those who smoked less [67]. Moreover, cumulative pack-years correlated with higher probability for ED. For example, it was found that 29 pack-years cumulative history was associated with a much higher risk for ED compared with <12 pack-years. Surprisingly, the latter carried the same risk as nonsmokers [68]. Diabetes is also an important contributor to ED through both micro- and macrovascular damages [69]. More than 50% of patients with diabetes have ED, as shown in some studies [70, 71]. Additionally, a Chinese study of over 51,000 men has shown that smokers with type 2 diabetes mellitus have 1.25 increased risk of developing ED than nonsmokers [72].

The molecular mechanism of smoking-related ED has been well studied. Both constituent forms of NO synthase (NOS) (endothelial (eNOS) and neuronal NOS (nNOS)) are attenuated by CS [73]. nNOS activity is diminished in both in vitro and in vivo models of smoking through enzymatic blockade. Moreover, it has been well established in the vascular literature that CS impairs vasodilatation through endothelial injury and eNOS inhibition [74–77]. Additionally, smoke metabolites such as superoxide anions actively lower the availability of free NO in the CC through the activation of the NADH oxidase enzyme [78]. This causes the vasoactive NO to be shunted into the peroxynitrite pathway [79]. Aside from the direct vasodilation effects, intracellular NO functions to inhibit the Rho-associated coiled-coil kinase (ROCK). ROCK is known to maintain the flaccid state [80]. Thus, if intracellular NO is lowered, ROCK is activated, leading to a worsening of ED [81].

Smoking also instigates intrinsic vessel injury, thus preventing elastic dilatation despite present paracrine signaling. CS alters the elastin contents of the extracellular matrix and induces calcification leading to arterial stiffness [82].

Current literature and available studies have yet to provide a definitive answer if smoking cessation influences ED specifically. Former smokers (defined as >1 year smoking cessation) still have a higher risk of developing ED compared with people who had never smoked [60, 68, 83]. In a different study that had excluded patients with cardiovascular disease, former smokers had the same amount of ED compared with current smokers. Thus, it appears that even without apparent vasculopathy, there is still a silent vascular damage that does not subside spontaneously. However, not all the damage appears to be permanent. There is currently mounting evidence that some damage is reversible if smoking is stopped prior to middle age [84, 85]. In summary, large epidemiological studies could not show a benefit regarding ED after quitting smoking. However, smoking cessation at an earlier age might increase the chances of ED improvement. Not only does CS affect ED directly, but also it is associated with numerous other comorbidities that cause ED. Atherosclerosis and cardiovascular disease are also known to disturb erectile function by impairing penile perfusion pressures leading to increased time to maximum erection and decreasing the peak rigidity [86, 87].

Peyronie's Disease

Peyronie's disease (PD) is a connective tissue disease situated on the tunica albuginea of the penis. Prevalence ranges from 1% to 3% of males; however, screening studies have shown that this might only be the tip of the iceberg. There is an increase with age, diabetes, and preexisting ED, up to 9%–13%

[88, 89]. It occurs most frequently in men 50–60 years old and has a devastating impact on mental health, as it was shown that around 50% of men with PD develop a clinically meaningful depression [90]. Currently available treatment options recommended by international guidelines include either surgery or local injection of collagenase [91]. Depending on the severity or complexity of the curvature and preexisting ED, different types of surgery can be recommended: either a Nesbit plication procedure, or plaque incision with replacement of the defect by various graft materials (bovine epicardium (Tutoplast) and porcine small intestinal submucosa (SIS) being the most popular), or implantation of a penile prosthesis device [92]. However, even with surgery, patients with severe deformities or curvatures remain hard to treat. Additionally, intralesional injections can be paired with side effects (pain, corporal rupture, and allergic reaction) and only cause a limited improvement in curvature (up to 30%) [93].

Smoking has not been identified as an independent risk factor PD, mostly due to the small study populations and a limited proportion of patients consulting a physician for this problem. CS however is heavily associated with the development of ED, hypertension, atherosclerosis, cardiovascular disease, etc. [94]. Up to 54% of patients with PD report ED, and the two disorders may influence one another [95, 96]. Diabetes and atherosclerosis (due to penile fibrosis) are associated with both PD and ED [97–100]. Thus, even though CS is not an independent risk factor, it appears to contribute toward most other risk factors of PD. It can be hypothesized that smoking cessation would lead to lower incidence and prevalence of PD as well.

NEPHROLITHIASIS

Nephrolithiasis (NL) or renal stone formation is a commonplace phenomenon with a wide array of risk factors, including dietary contents, genetic composition, obesity, medication use, and geography [101]. However, the role of CS in NL remains ill-investigated. Tamadon et al. [102] reviewed 102 stone-forming patients and compared them with age-matched healthy controls. They showed that CS increased the risk of stone formation, as 27% of stone disease patients were smokers, while this was only 15% in the control group. Liu and colleagues [103] also showed that CS increased the formation of calcium stones. Several hypotheses for increased NL have been proposed: decrease of urinary output; oxidative stress; and an increased nucleation, aggregation, and crystal formation in the kidneys [104, 105].

However, the investigated links between smoking and NL are weak, and no real recommendations can be made based on the currently available data.

UROLOGICAL CANCERS

Bladder Cancer

Bladder cancer (BC) is the ninth most common cancer worldwide, with more than 430,000 new cases diagnosed in 2012 [106]. Most BC (about 75%) patients present with nonmuscle-invasive BC (NMIBC). NMIBC is the cancer with the highest recurrence rate of all cancers (50%–70%), and 15%–20% evolves to muscle-invasive disease [107]. Additionally, due to frequent need for resection or follow-up, it is also the cancer with one of the highest costs for society. The remaining one-fourth of BC patients suffer from muscle-invasive BC (MIBC). MIBC has a high mortality rate of 50% over 5 years. Moreover, standard of care treatment includes neoadjuvant chemotherapy followed by radical cystectomy, pelvic lymph node dissection, and urinary diversion; surgery with high morbidity; and impact on quality of life. The most well-known risk factor remains CS and is directly related to its frequency and duration. Consequently, smoking cessation reduces BC incidence [108–112]. The most important carcinogenic substances are aromatic nitrates, formed during the pyrolysis (or tobacco combustion) [111–113]. The precise mechanism of carcinogenicity is unknown, but genotoxic effects have been put forward as the most commonly occurring [114]. Carcinogenic substances in cigarette smoke are aromatic amines, polycyclic aromatic hydrocarbons (PAHs, of which benzo(a)pyrene is an example), and tobacco-specific nitrosamines. The formation of repair-mechanism resistant DNA damage in essential cancer-related genes is a well-known phenomenon in tobacco-related cancers [114].

Exposure to these chemicals also occurs in several industries, such as rubber and textile manufacturing; aluminum transformation; and gas, coal, pesticide, and cosmetic production. Other sources include hair dye use, engine exhaust, and paint fumes. Additionally, dietary (e.g., pesticides in food), lifestyle (e.g., hair dye use), and environmental (e.g., engine exhaust) sources also play a significant role [115–117].

In an Italian study from 1991 [118] and similar findings in 2011 [119], there was a distinct difference in the sex distribution of BC. Only 13.9% of BC cases were present in nonsmoking males, while a staggering 63.9% occurred in females with no prior smoking habits. This suggests a significant sex difference in BC risk. Smoking cessation could reduce the amount of new BC by 30,000 in the United States.

Prostate Cancer

The etiology of prostate cancer remains an elusive concept. Several factors are thought to contribute to its development: genetic predisposition, diet, infections, hormonal imbalance, and exposure to toxins (like those found in CS) [120].

In 2015, Carter et al. published data from a longitudinal cohort study encompassing almost 1 million patients (both male and female) across the United States. This revealed the significance of smoking in prostate cancer. Not only does smoking increase the incidence of prostate cancer, but also it heightens the risk of advanced-stage disease, and mortality was 43% higher in smokers [121]. There is not much known about the possible etiopathology. One North Indian study in 2011 evaluated 578 men with prostate cancer. They found increased levels of the inflammatory cytokine IL-18 in men who smoke. Moreover, it was also associated with higher-stage disease. Chronic inflammation has been proposed to contribute significantly to the development of prostate cancer [122].

Kidney Cancer (Renal Cell Carcinoma [RCC])

Renal cell carcinoma (RCC) is responsible for 90% of all kidney cancers. About 20%–30% of patients are diagnosed with metastatic disease (mRCC) [123]. Additionally, 20% of patients undergoing nephrectomy for localized disease will relapse and develop metastases [124]. Patients diagnosed with mRCC have a very poor prognosis, which makes RCC a significant problem for health care around the globe. Kidney cancer is the 13th most common malignancy worldwide and has the highest rates in Europe, North America, and Australia [125]. A meta-analysis of 24 studies in 2005 by Hunt et al. [126] showed only a weak association between smoking and RCC. However, the International Agency for Research on Cancer (IARC) Monograph series in the United Kingdom reported the association between smoking and a myriad of cancers [127, 128]. It was revealed that the relative risk (RR) for the development of RCC in smoking males was 2.5 (smoking females had an RR of 1.5). By comparison, these numbers are not much less than the reported RR for smoking and BC (3.0). It is estimated that around 29% of RCC cases in men and 15% in women are caused by CS, and this effect is dose-dependent [126, 127]. Patel et al. [129] showed that smoking is mostly a risk factor for clear-cell and papillary RCC, but not for the chromophobe phenotype.

Other risk factors for RCCs include lifestyle (obesity) and iatrogenic (hypertension and the use of antihypertensive medication and acquired renal cystic disease), occupational (trichloroethylene, lead, glass fibers, mineral wool fibers, and brick dust), and genetic load (von Hippel-Lindau disease, hereditary papillary renal cell carcinoma, familial leiomyomatosis RCC, and Birt-Hogg-Dubé syndrome) [130].

References

[1] U.S. Department of Health and Human Services. The health consequences of smoking—50 years of progress: a report of the surgeon general. Atlanta: U.S. Department of Health and Human Services, Centers for Disease Control and Prevention, National Center for Chronic Disease Prevention and Health Promotion, Office on Smoking and Health; 2014. [Accessed 22 February 2018].

[2] Yao FD, Mills JN. Male infertility: lifestyle factors and holistic, complementary, and alternative therapies. Asian J Androl 2016;18(3):410–8.

[3] Sansone A, Di Dato C, de Angelis C, et al. Smoke, alcohol and drug addiction and male fertility. Reprod Biol Endocrinol 2018;16:3.

[4] Méndez-Rubio S, Salinas-Casado J, Esteban-Fuertes M, et al. Urological disease and tobacco. A review for raising the awareness of urologists. Actas Urol Esp 2016;40(7):424–33.

[5] Ambrose JA, Barua RS. The pathophysiology of cigarette smoking and cardiovascular disease: an update. J Am Coll Cardiol 2004;43(10).

[6] Pryor WA, Stone K. Oxidants in cigarette smoke: radicals, hydrogen peroxide, peroxynitrate, and peroxynitrite. Ann N Y Acad Sci 1993;686:12–28.

[7] Smith CJ, Fischer TH. Particulate and vapor phase constituents of cigarette mainstream smoke and risk of myocardial infarction. Atherosclerosis 2001;158:257–67.

[8] Pryor WA, Stone K, Zang LY, Bermudez E. Fractionation of aqueous cigarette tar extracts: fractions that contain the tar radical cause DNA damage. Chem Res Toxicol 1998;11:441–8.

[9] Powell JT. Vascular damage from smoking: disease mechanisms at the arterial wall. Vasc Med 1998;3:21–8.

[10] World Health Organization. Infections, pregnancies, and infertility: perspectives on prevention. Fertil Steril 1987;47(6):964–8.

[11] Kovac JR, Pastuszak AW, Lamb DJ. The use of genomics, proteomics, and metabolomics in identifying biomarkers of male infertility. Fertil Steril 2013;99(4):998–1007.

[12] Harlev A, Agarwal A, Gunes SO, et al. Smoking and male infertility: an evidence-based review. World J Mens Health 2015;33(3):143–60.

[13] WHO laboratory manual for the examination and processing of human semen. 5th ed. Geneva, Switzerland: WHO Press; 2010.

[14] Sengupta P, Dutta S, Krajewska-Kulak E. The disappearing sperms: Analysis of reports published between 1980 and 2015. Am J Mens Health 2017;11(4):1279–304.

[15] Sharma R, Harlev A, Agarwal A, Esteves SC. Cigarette smoking and semen quality: a new meta-analysis examining the effect of the 2010 WHO Laboratory methods for the examination of human semen. Eur Urol 2016;70(4):635–45.

[16] Gabrielsen JS, Tanrikut C. Chronic exposures and male fertility: the impacts of environment, diet, and drug use on spermatogenesis. Andrology 2016;4(4):648–61.

[17] Hammond D, Fong GT, Cummings KM, O'Connor RJ, Giovino GA, McNeill A. Cigarette yields and human exposure: a comparison of alternative testing regimens. Cancer Epidemiol Biomarkers Prev 2006;15:1495–501.

[18] Benoff S, Centola GM, Millan C, Napolitano B, Marmar JL, Hurley IR. Increased seminal plasma lead levels adversely affect the fertility potential of sperm in IVF. Hum Reprod 2003;18(2):374–83.

[19] Jurasovic J, Cvitkovic P, Pizent A, Colak B, Telisman S. Semen quality and reproductive endocrine function with regard to blood cadmium in Croatian male subjects. Biometals 2004;17(6):735–43.

[20] Wang X, Zhang J, Xu W, Huang Q, Liu L, Tian M, Xia Y, Zhang W, Shen H. Low-level environmental arsenic exposure correlates with unexplained male infertility risk. Sci Total Environ 2016;571:307–13.

[21] de Angelis C, Galdiero M, Pivonello C, Salzano C, Gianfrilli D, Piscitelli P, Lenzi A, Colao A, Pivonello R. The environment and male reproduction: the effect of cadmium exposure on reproductive functions and its implication in fertility. Reprod Toxicol 2017;73:105–27.

[22] Pasqualotto FF, Umezu FM, Salvador M, et al. Effect of cigarette smoking on antioxidant levels and presence of leukocytospermia in infertile men: a prospective study. Fertil Steril 2008;90(2):278–83.

[23] Zenzes MT, Bielecki R, Reed TE. Detection of benzo(a)pyrene diol epoxide-DNA adducts in sperm of men exposed to cigarette smoke. Fertil Steril 1999;72(2):330–5.

[24] Ramlau-Hansen CH, Thulstrup AM, Aggerholm AS, Jensen MS, Toft G, Bonde JP. Is smoking a risk factor for decreased semen quality? A cross-sectional analysis. Hum Reprod 2007;22(1):188.

[25] Kunzle R, Mueller MD, Hanggi W, Birkhauser MH, Drescher H, Bersinger NA. Semen quality of male smokers and nonsmokers in infertile couples. Fertil Steril 2003;79(2):287.

[26] Saaranen M, Suonio S, Kauhanen O, Saarikoski S. Cigarette smoking and semen quality in men of reproductive age. Andrologia 1987;19(6):670.

[27] Zhang JP, Meng QY, Wang Q, Zhang LJ, Mao YL, Sun ZX. Effect of smoking on semen quality of infertile men in Shandong, China. Asian J Androl 2000;2(2):143.

[28] Murawski M, Saczko J, Marcinkowska A, Chwilkowska A, Grybos M, Banas T. Evaluation of superoxide dismutase activity and its impact on semen quality parameters of infertile men. Folia Histochem Cytobiol 2007;45(Suppl. 1):S123–6.

[29] Sanocka D, Miesel R, Jedrzejczak P, Kurpisz MK. Oxidative stress and male infertility. J Androl 1996;17(4):449.

[30] Curtis KM, Savitz DA, Arbuckle TE. Effects of cigarette smoking, caffeine consumption, and alcohol intake on fecundability. Am J Epidemiol 1997;146(1):32–41.

[31] Zavos PM, Correa JR, Karagounis CS, Ahparaki A, Phoroglou C, Hicks CL, Zarmakoupis-Zavos PN. An electron microscope study of the axonemal ultrastructure in human spermatozoa from male smokers and nonsmokers. Fertil Steril 1998;69(3):430–4.

[32] Yeung CH, Tuttelmann F, Bergmann M, Nordhoff V, Vorona E, Cooper TG. Coiled sperm from infertile patients: characteristics, associated factors and biological implication. Hum Reprod 2009;24(6):1288–95.

[33] Zalata AA, Ahmed AH, Allamaneni SS, Comhaire FH, Agarwal A. Relationship between acrosin activity of human spermatozoa and oxidative stress. Asian J Androl 2004;6(4):313–8.

[34] Taha EA, Ezz-Aldin AM, Sayed SK, Ghandour NM, Mostafa T. Smoking influence on sperm vitality, DNA fragmentation, reactive oxygen species and zinc in oligoasthenoteratozoospermic men with varicocele. Andrologia 2014;46(6):687–91.

[35] Liu RZ, Gao JC, Zhang HG, Wang RX, Zhang ZH, Liu XY. Seminal plasma zinc level may be associated with the effect of cigarette smoking on sperm parameters. J Int Med Res 2010;38(3):923.

[36] Marczylo EL, Amoako AA, Konje JC, Gant TW, Marczylo TH. Smoking induces differential miRNA expression in human spermatozoa: a potential transgenerational epigenetic concern? Epigenetics 2012;7:432–9.

[37] Practice Committee of the American Society for Reproductive Medicine. Smoking and infertility: a committee opinion. Fertil Steril 2012;98(6):1400–6.

[38] Yao DF, Mills JN. Male infertility: lifestyle factors and holistic, complementary, and alternative therapies. Asian J Androl 2016;18(3):410–8.

[39] Rhoden EL, Ribeiro EP, Teloken C, Souto CA. Diabetes mellitus is associated with subnormal serum levels of free testosterone in men. BJU Int 2005;96:867–70.

[40] Laaksonen DE, Niskanen L, Punnonen K, et al. Sex hormones, inflammation and the metabolic syndrome: a population-based study. Eur J Endocrinol 2003;149:601–8.

[41] Shores MM, Sloan KL, Matsumoto AM, et al. Increased incidence of diagnosed depressive illness in hypogonadal older men. Arch Gen Psychiatry 2004;61:162–7.

[42] Isidori AM, Giannetta E, Gianfrilli D, et al. Effects of testosterone on sexual function in men: results of a meta-analysis. Clin Endocrinol (Oxf) 2005;63:381–94.

[43] Orwoll ES, Belknap JK, Klein RF. Gender specificity in the genetic determinants of peak bone mass. J Bone Miner Res 2001;16:1962–71.

[44] Corona G, Monami M, Boddi V, et al. Low testosterone is associated with an increased risk of MACE lethality in subjects with erectile dysfunction. J Sex Med 2010;7:1557–64.

[45] Basaria S, Coviello AD, Travison TG, et al. Adverse events associated with testosterone administration. N Engl J Med 2010;363:109–22.

[46] Tajar A, Forti G, O'Neill TW, et al. Characteristics of secondary, primary, and compensated hypogonadism in aging men: evidence from the European Male Ageing Study. J Clin Endocrinol Metab 2010;95:1810–8.

[47] Harman SM, Metter EJ, Tobin JD, Pearson J, Blackman MR. Longitudinal effects of aging on serum total and free testosterone levels in healthy men. Baltimore Longitudinal Study of Aging. J Clin Endocrinol Metab 2001;86:724–31.

[48] Barrett-Connor E, Khaw KT. Cigarette smoking and increased endogenous estrogen levels in men. Am J Epidemiol 1987;126:187–92.

[49] Shiels MS, Rohrmann S, Menke A, et al. Association of cigarette smoking, alcohol consumption, and physical activity with sex steroid hormone levels in US men. Cancer Causes Control 2009;20:877–86.

[50] Svartberg J, Midtby M, Bonaa KH, et al. The associations of age, lifestyle factors and chronic disease with testosterone in men: the Tromso Study. Eur J Endocrinol 2003;149:145–52.

[51] Briggs MH. Cigarette smoking and infertility in men. Med J Aust 1973;1:616–7.

[52] Shaarawy M, Mahmoud KZ. Endocrine profile and semen characteristics in male smokers. Fertil Steril 1982;38:255–7.

[53] Dai JB, Wang ZX, Qiao ZD. The hazardous effects of tobacco smoking on male fertility. Asian J Androl 2015;17(6):954–60.

[54] Blanco-Munoz J, Lacasana M, Aguilar-Garduno C. Effect of current tobacco consumption on the male reproductive hormone profile. Sci Total Environ 2012;426:100–5. https://doi.org/10.1016/j.scitotenv.2012.03.071.

[55] Ochedalski T, Lachowicz-Ochedalska A, Dec W, Czechowski B. Examining the effects of tobacco smoking on levels of certain hormones in serum of young men. Ginekol Pol 1994;65(2):87–93.

[56] Lotti F, Corona G, Vitale P, Maseroli E, Rossi M, Fino MG, Maggi M. Current smoking is associated with lower seminal vesicles and ejaculate volume, despite higher testosterone levels, in male subjects of infertile couples. Hum Reprod 2015;30(3):590–602.

[57] Pasqualotto FF, Sobreiro BP, Hallak J, Pasqualotto EB, Lucon AM. Cigarette smoking is related to a decrease in semen volume in a population of fertile men. BJU Int 2006;97:324–6.

[58] Feldman HA, Goldstein I, Hatzichristou DG, et al. Impotence and its medical and psychosocial correlates: results of the Massachusetts Male Aging Study. J Urol 1994;151(1):54–61.

[59] Shamloul R, Ghanem H. Erectile dysfunction. Lancet 2013;381:153–65.

[60] Austoni EMV, Parazzini F, Fasolo CB, Turchi P, Pescatori ES, Ricci E, Gentile V. Andrology prevention week centres; Italian Society of Andrology. Eur Urol 2005;48:810–8.

[61] Kupelian V, Link CL, McKinlay JB. Association between smoking, passive smoking, and erectile dysfunction: results from the Boston Area Community Health (BACH) Survey. Eur Urol 2007;52:416–22.

[62] Chew KK, Bremner A, Stuckey B, Earle C, Jamrozik K. Is the relationship between cigarette smoking and male erectile dysfunction independent of cardiovascular disease? Findings from a population-based cross-sectional study. J Sex Med 2009;6:222–31.

[63] Ghalayini IFA-GM, Al-Azab R, Bani-Hani I, Matani YS, Barham AE, Harfeil MN, Haddad Y. Erectile dysfunction in a Mediterranean country: results of an epidemiological survey of a representative sample of men. Int J Impot Res 2010;22:196–203.

[64] Wu C, Zhang H, Gao Y, Tan A, Yang X, Lu Z, Zhang Y, Liao M, Wang M, Mo Z. The association of smoking and erectile dysfunction: results from the Fangchenggang Area Male Health and Examination Survey (FAMHES). J Androl 2012;33:59–65.

[65] Elbendary MA, El-Gamal OM, Salem KA. Analysis of risk factors for organic erectile dysfunction in Egyptian patients under the age of 40 years. J Androl 2009;30:520–4.

[66] Cao S, Yin X, Wang Y, Zhou H, Song F, Lu Z. Smoking and risk of erectile dysfunction: systematic review of observational studies with meta-analysis. PLoS ONE 2013;8.

[67] Natali A, Mondaini N, Lombardi G, Del Popolo G, Rizzo M. Heavy smoking is an important risk factor for erectile dysfunction in young men. Int J Impot Res 2005;17:227–30.

[68] Gades NM, Nehra A, Jacobson DJ, McGree ME, Girman CJ, Rhodes T, Roberts RO, Lieber MM, Jacobsen SJ. Association between smoking and erectile dysfunction: a population-based study. Am J Epidemiol 2005;161:346–51.

[69] Maiorino MI, Bellastella G, Esposito K. Diabetes and sexual dysfunction: current perspectives. Diabetes Metab Syndr Obes 2014;7:95–105.

[70] Giuliano FA, Leriche A, Jaudinot EO, de Gendre AS. Prevalence of erectile dysfunction among 7689 patients with diabetes or hypertension, or both. Urology 2004;64:1196–201.

[71] Thorve VS, Kshirsagar AD, Vyawahare NS, Joshi VS, Ingale KG, Mohite RJ. Diabetes-induced erectile dysfunction: epidemiology, pathophysiology and management. J Diabetes Complications 2011;25:129–36.

[72] Shi L, Shu XO, Li H, Cai H, Liu Q, Zheng W, Xiang YB, Villegas R. Physical activity, smoking, and alcohol consumption in association with incidence of type 2 diabetes among middle-aged and elderly Chinese men. PLoS ONE 2013;8.

[73] Kovac JR, Labbate C, Ramasamy R, et al. Effects of cigarette smoking on erectile dysfunction. Andrologia 2015;47(10):1087–92.

[74] Xie Y, Garban H, Ng C, Rajfer J, Gonzalez-Cadavid NF. Effect of long-term passive smoking on erectile function and penile nitric oxide synthase in the rat. J Urol 1997;157:1121–6.

[75] Demady DR, Lowe ER, Everett AC, Billecke SS, Kamada Y, Dunbar AY, Osawa Y. Metabolism-based inactivation of neuronal nitric-oxide synthase by components of cigarette and cigarette smoke. Drug Metab Dispos 2003;31:932–7.

[76] Celermajer DS, Sorensen KE, Georgakopoulos D, Bull C, Thomas O, Robinson J, Deanfield JE. Cigarette smoking is associated with dose-related and potentially reversible impairment of endothelium-dependent dilation in healthy young adults. Circulation 1993;88:2149–55.

[77] Butler R, Morris AD, Struthers AD. Cigarette smoking in men and vascular responsiveness. Br J Clin Pharmacol 2001;52:145–9.

[78] Orosz Z, Csiszar A, Labinskyy N, Smith K, Kaminski PM, Ferdinandy P, Wolin MS, Rivera A, Ungvari Z. Cigarette smoke-induced proinflammatory alterations in the endothelial phenotype: Role of NAD(P)H oxidase activation. Am J Physiol Heart Circ Physiol 2007;292:H130–9.

[79] Peluffo G, Calcerrada P, Piacenza L, Pizzano N, Radi R. Superoxide-mediated inactivation of nitric oxide and peroxynitrite formation by tobacco smoke in vascular endothelium: studies in cultured cells and smokers. Am J Physiol Heart Circ Physiol 2009;296:H1781–92.

[80] Mills TM, Chitaley K, Wingard CJ, Lewis RW, Webb RC. Effect of rho-kinase inhibition on vasoconstriction in the penile circulation. J Appl Physiol 2001;91:1269–73.

[81] Chitaley K, Wingard CJ, Clinton Webb R, Branam H, Stopper VS, Lewis RW, Mills TM. Antagonism of rho-kinase stimulates rat penile erection via a nitric oxide-independent pathway. Nat Med 2001;7:119–22.

[82] Guo X, Oldham MJ, Kleinman MT, Phalen RF, Kassab GS. Effect of cigarette smoking on nitric oxide, structural, and mechanical properties of mouse arteries. Am J Physiol Heart Circ Physiol 2006;291:H2354–61.

[83] Bacon CG, Mittleman MA, Kawachi I, Giovannucci E, Glasser DB, Rimm EB. A prospective study of risk factors for erectile dysfunction. J Urol 2006;176:217–21.

[84] Pourmand GAM, Rasuli S, Maleki A, Mehrsai A. Do cigarette smokers with erectile dysfunction benefit from stopping? A prospective study. BJU Int 2004;94:1310–3.

[85] Sighinolfi MC, Mofferdin A, De Stefani S, Micali S, Cicero AF, Bianchi G. Immediate improvement in penile hemodynamics after cessation of smoking: previous results. Urology 2007;69:163–5.

[86] Shabsigh R, Fishman IJ, Chum C, Dunn JK. Cigarette smoking and other vascular risk factors in vasculogenic impotence. Urology 1991;38:227–31.

[87] Sullivan ME, Thompson CS, Dashwood MR, Khan MA, Jeremy JY, Morgan RJ, Mikhailidis DP. Nitric oxide and penile erection: is erectile dysfunction another manifestation of vascular disease? Cardiovasc Res 1999;43:658–65.

[88] Mulhall JP, Creech SD, Boorjian SA, et al. Subjective and objective analysis of the prevalence of Peyronie's disease in a population of men presenting for prostate cancer screening. J Urol 2004;171(6):2350–3.

[89] Al-Thakafi S, Al-Hathal N. Peyronie's disease: a literature review on epidemiology, genetics, pathophysiology, diagnosis and work-up. Transl Androl Urol 2016;5(3):280–9.

[90] Nelson CJ, Diblasio C, Kendirci M, Hellstrom W, Guhring P, Mulhall JP. The chronology of depression and distress in men with peyronie's disease. J Sex Med 2008;5(8):1985–90.

[91] Hatzimouratidis K, Eardley I, Giuliano F, et al. EAU guidelines on penile curvature. Eur Urol 2012;62(3):543–52.

[92] Hatzichristodoulou G. Grafting techniques for Peyronie's disease. Transl Androl Urol 2016;5(3):334–41.

[93] Hellstrom WJG, Feldman RA, Coyne KS, et al. Self-report and clinical response to Peyronie's disease treatment: Peyronie's disease questionnaire results from 2 large double-blind, randomized, placebo-controlled phase 3 studies. Urology 2015;86(2):291–8.

[94] Serefoglu EC, Smith TM, Kaufman GJ, Liu G, Yafi FA, Hellstrom WJG. Factors associated with erectile dysfunction and the Peyronie's disease questionnaire in patients with Peyronie disease. Urology 2017;107:155–60.

[95] Rhoden EL, Teloken C, Ting HY, Lucas ML, Teodósio da Ros C, Ary Vargas Souto C. Prevalence of Peyronie's disease in men over 50-y-old from southern Brazil. Int J Impot Res 2001;13:291–3.

[96] El-Sakka AI. Prevalence of Peyronie's disease among patients with erectile dysfunction. Eur Urol 2006;49:564–9.

[97] Schwarzer U, Sommer F, Klotz T, Braun M, Reifenrath B, Engelmann U. The prevalence of Peyronie's disease: results of a large survey. BJU Int 2001;88:727–30.

[98] Kadioglu A, Tefekli A, Erol B, Oktar T, Tunc M, Tellaloglu S. A retrospective review of 307 men with Peyronie's disease. J Urol 2002;168:1075–9.

[99] Sommer F, Schwarzer U, Wassmer G, et al. Epidemiology of Peyronie's disease. Int J Impot Res 2002;14:379–83.

[100] Gonzalez-Cadavid NF. Mechanisms of penile fibrosis. J Sex Med 2009;6:353–62.

[101] Hall PM. Nephrolithiasis: treatment, causes and prevention. Cleve Clin J Med 2009;76:583–91.

[102] Tamadon MR, Nassaji M, Ghorbani R. Cigarette smoking and nephrolitiasis in adult individuals. Nephrourol Mon 2013;5(1):702–5.

[103] Liu CC, Huang SP, Wu WJ, et al. The impact of cigarette smoking, alcohol drinking and betel quid chewing on the risk of calcium urolithiasis. Ann Epidemiol 2009;19:539–45.

[104] Robert M, Roux JO, Bourelly F, et al. Circadian variations in the risk of urinary calcium oxalate stone formation. Br J Urol 1994;74:294–7.

[105] Obata T, Tomaru K, Nagakura T, et al. Smoking and oxidative stress: assay of isoprostane in human urine by gas chromatography-mass spectrometry. J Chromatogr B Biomed Sci Appl 2000;746:11–5.

[106] GLOBOCAN. v1.0, cancer incidence and mortality worldwide: IARC CancerBase no. 11. International Agency for Research on Cancer; 2012.http://globocan.iarc.fr.

[107] Clark PE, Agarwal N, Biagioli MC, et al. National Comprehensive Cancer Network (NCCN). Bladder cancer. J Natl Compr Canc Netw 2013;11:446–75.

[108] U.S. Department of Health and Human Services. The health consequences of smoking: a report of the surgeon general. Atlanta, GA, USA: U.S. Department of Health and Human Services, Centers for Disease Control and Prevention, National Center for Chronic Disease Prevention and Health Promotion, Office on Smoking and Health; 2004.

[109] Boffetta P. Tobacco smoking and risk of bladder cancer. Scand J Urol Nephrol Suppl 2008;42:45–54.

[110] Clavel J, Cordier S, Boccon-Gibod L, Hemon D. Tobacco and bladder cancer in males: Increased risk for inhalers and smokers of black tobacco. Int J Cancer 1989;44:605–10.

[111] D'Avanzo B, Negri E, La Vecchia C, Gramenzi A, Bianchi C, Franceschi S, Boyle P. Cigarette smoking and bladder cancer. Eur J Cancer 1990;26:714–8.

[112] Vineis P, Esteve J, Hartge P, Hoover R, Silverman DT, Terracini B. Effects of timing and type of tobacco in cigarette-induced bladder cancer. Cancer Res 1988;48:3849–52.

[113] Bartsch H, Malaveille C, Friesen M, Kadlubar FF, Vineis P. Black (air-cured) and blond (flue-cured) tobacco cancer risk. IV: molecular dosimetry studies implicate aromatic amines as bladder carcinogens. Eur J Cancer 1993;29A:1199–207.

[114] Besaratinia A, Tommasi S. Genotoxicity of tobacco smoke-derived aromatic amines and bladder cancer: current state of knowledge and future research directions. FASEB J 2013;27:2090–100.

[115] International Agency for Research on Cancer. 2-Naphthylamine. In: Some aromatic amines, hydrazine and related substances, N-nitroso compounds and miscellaneous alkylating agents. Lyon, France: International Agency for Research on Cancer; 1974. p. 97–111.

[116] Parkes HG, Evans AEJ. Epidemiology of aromatic amine cancers. In: Searle CE, editor. Chemical carcinogens. vol. 1. Washington DC: American Chemical Society; 1984. p. 277–301.

[117] Steineck G, Plato N, Norell SE, Hogstedt C. Urothelial cancer and some industry-related chemicals: an evaluation of the epidemiologic literature. Am J Ind Med 1990;17:371–91.

[118] La Vecchia C, Negri E, D'Avanzo B, et al. Genital and urinary tract diseases and bladder cancer. Cancer Res 1991;51:629–31.

[119] Freedman ND, Silverman DT, Hollenbeck AR, et al. Association between smoking and risk of bladder cancer among men and women. JAMA 2011;306:737–45.

[120] Dwivedi S, Goel A, Mandhani A, et al. Tobacco exposure may enhance inflammation in prostate cancer patients: an exploratory study in north Indian population. Toxicol Int 2012;19:310–8.

[121] Carter BD, Abnet CC, Feskanich D, et al. Smoking and mortality-beyond established causes. N Engl J Med 2015;372:631–40.

[122] Dwivedi S, Goel A, Mandhani A, et al. Tobacco exposure may enhance inflammation in prostate cancer patients: an exploratory study in north Indian population. Toxicol Int 2012;19:310–8.

[123] Gupta K, Miller JD, Li JZ, Russell MW, Charbonneau C. Epidemiologic and socioeconomic burden of metastatic renal cell carcinoma (mRCC): a literature review. Cancer Treat Rev 2008;34:193–205.

[124] Athar U, Gentile TC. Treatment options for metastatic renal cell carcinoma: a review. Can J Urol 2008;15:3954–66.

[125] Ferlay J, Shin HR, Bray F, Forman D, Mathers C, Parkin DM. Estimates of worldwide burden of cancer in 2008: GLOBOCAN 2008. Int J Cancer 2010;15:2893–917.

[126] Hunt JD, van der Hel OL, McMillan GP, Boffetta P, Brennan P. Renal cell carcinoma in relation to cigarette smoking: meta-analysis of 24 studies. Int J Cancer 2005;114:101–8.

[127] Cogliano VJ, Baan R, Straif K, et al. Preventable exposures associated with human cancers. J Natl Cancer Inst 2011;103:1827–39.

[128] Parkin DM. 2. Tobacco-attributable cancer burden in the UK in 2010. Br J Cancer 2011;195 (Suppl. 2):S6–S13.

[129] Patel NH, Attwood KM, Hanzly M, et al. Comparative analysis of smoking as risk factor among renal cell carcinoma histological subtypes. J Urol 2015;194:640–6.

[130] Ljungberg B, Campbell SC, Cho HY, Jacqmin D, Lee JE, Weikert S, Kiemeney LA. The epidemiology of renal cell carcinoma. Eur Urol 2011;60(4):615–21.

CHAPTER 6.2

The Opioid Epidemic and Men's Sexual Health

Hossein Sadeghi-Nejad*, Radhika Ragam[†]

*Professor of Surgery, Rutgers New Jersey Medical School and Hackensack University Medical Center, Hackensack, NJ, United States,
[†]Rutgers New Jersey Medical School, Newark, NJ, United States

INTRODUCTION

The opioid epidemic, a problem that spans across various races, age groups, and socioeconomic layers, is currently the worst drug crisis in US history. This national emergency has accounted for decreased average life expectancy for 2 consecutive years with quadruple the number of deaths secondary to heroin abuse since 2000 and billions spent on opioid-related health and social costs per year. Despite an overall decline in illicit drug use in the United States, the incidence of prescription opioid abuse has drastically increased since the 1990s, with the global number of users reaching 12–21 million people [1]. The number of heroin users has increased from 1.6 to 2.5 per 1000 persons between 2002 and 2013, with heroin overdose death rates increasing fivefold from 2013 to 2016 [2, 3].

The prescription opioid epidemic has developed into a significant health-care issue affecting the general population, displayed by the increase in deaths and imprisonments related to heroin abuse and overall fatal outcomes associated with opioid addiction. Along with the lethal sequelae of opioid abuse, long-term users have also been subjected to an increase in the incidence of opioid-induced endocrinopathy, mostly manifesting as androgen deficiency [4]. Opioid-induced androgen deficiency (OPIAD) affects sexual functioning in males while simultaneously eliciting various central and peripheral effects. This chapter highlights the importance of opioid-induced androgen deficiency, details treatment options, and presents a comprehensive review of the current opioid epidemic and its consequential effects on sexual health in males.

BACKGROUND AND EPIDEMIOLOGY

Heroin use while historically associated with lower socioeconomic groups has shifted to become more prevalent among different demographics. Non-

Hispanic white men, those living in nonurban communities, and younger users have exhibited large increases in heroin use; users aged 18–44 years old have the highest rates of heroin-related deaths at 7.0 per 100,000 persons [5]. It is proposed that heroin-related overdoses in new, younger heroin users might have been catalyzed by the transition from prescription opioid use to heroin [2]; approximately three out of four new heroin users report prescription opioid abuse prior to heroin use [6].

The prevalence of OPIAD in symptomatic men and asymptomatic men on chronic opioid therapy has been estimated to be 90% and 53%, respectively [7, 8]. The nonspecific clinical manifestations of OPIAD are often unrecognized as opioid-related symptoms, making the incidence of the condition difficult to determine. Providers should hold a high index of suspicion for OPIAD among chronic opioid users and be prepared to confirm the diagnosis with the appropriate laboratory tests.

PATHOPHYSIOLOGY OF OPIOID-INDUCED HYPOGONADISM

Opioids are mostly known for their analgesic effects, which occur through binding and activation of μ-receptors in the presynaptic regions of the periaqueductal gray region, ventromedial hypothalamus, and superficial dorsal horn of the spinal cord. Common side effects of opioid use such as constipation, urinary retention, respiratory depression, dizziness, and nausea also result from activation of μ-receptors in other areas of the body [9]. Hypogonadism, lost within a constellation of aforementioned side effects, has not received much attention or investigation within the medical community despite increasing reports of high prevalence among chronic opioid users. Current literature reports the incidence of hypogonadism in chronic opioid users to be between 50% and 90% [10].

Opioid use leads to hypogonadism by affecting the hypothalamic-pituitary-gonadal axis (HPGA), the main pathway involved in the development and the hormonal regulation of the reproductive system. Gonadotropin-releasing hormone (GnRH), secreted by the hypothalamus in a pulsatile manner, stimulates the pituitary gland to release luteinizing hormone (LH) and follicle-stimulating hormone (FSH). In men, LH regulates the production of testosterone through Leydig cells, and FSH regulates spermatogenesis through Sertoli cells. Alterations in this pathway lead to an imbalance in hormone production and ensuing side effects. Opioids bind to μ-receptors in the hypothalamus, disrupting the pulsatile release of GnRH, causing diminished production of testosterone and subsequent effects such as erectile dysfunction, decreased muscle mass, reduced body hair, and the loss of libido [11].

Dehydroepiandrosterone (DHEA) and DHEA sulfate (DHEAS) are both endogenous precursors to the androgens testosterone and dihydrotestosterone and are produced from the adrenal glands. Opioids have been shown to decrease DHEAS production by inhibiting corticotropin-releasing hormone (CRH) and disrupting the hypothalamic-adrenal axis, leading to side effects that include fatigue, depression, weakness, and sexual dysfunction from DHEA/DHEAS deficiency. Therefore, it has been hypothesized that the nature of OPIAD may additionally be derived from decreased CRH secondary to opioid use [12].

The prevalence of μ-receptors extends to the human spermatozoa as well, specifically in the plasma membrane of the sperm head and middle tail regions; activation of these receptors by opioids leads to decreased sperm motility [13]. Studies have used DNA fragmentation index to determine that chronic opioid users have worse semen parameters when compared with their healthy age-matched counterparts, shown by significantly decreased sperm concentration (22 vs 66 million/mL), increased DNA fragmentation (37% vs 27%), and decreased catalase-like and superoxide-dismutase activity [14].

Opioids appear to affect sexual behavior in different species, even at the genomic level. Aloisi et al. showed that morphine-treated mice have increased expression of testosterone-catabolizing enzymes such as 5-alpha reductase and P-450 aromatase in the liver, testes, and brain that may persist long term [15]. Songbirds generally have increased testosterone production and androgen receptor expression within the medial preoptic nucleus during springtime, with accompanying increase in sexually motivated songs [16]. Cordes et al. showed that these sexually motivated songs are stunted as a result of opioid release and μ-receptor expression in these areas, possibly inhibiting seasonal sexual behavior in songbirds [17].

Although opioid use has rapid effects on testosterone levels, cessation of use allows levels to return back to baseline [18, 19]. Researchers have proposed the nature of opioid-induced hypogonadism to be related to sustained serum opioid levels reached with long-acting opioids causing further cumulative GnRH suppression as opposed to intermittent GnRH suppression with shorter-acting opioids. Rubinstein et al. supported this explanation by showing a significant association of hypogonadism with long-acting opioids (74% vs 34%), but not total daily opioid dose [8]. Therefore, the duration of opioid action seems to be a fundamental factor in the nature of OPIAD.

DIAGNOSIS

The nonspecific symptoms of OPIAD make the diagnosis of this condition difficult. Though there are no definitive guidelines for OPIAD diagnosis, a thorough history should initially be obtained, as with other medical conditions. The symptoms accompanying hypogonadism may include fatigue, reduced sense of vitality, depressed mood, weight gain, and decreased muscle mass, which can be attributed to other causes like natural aging or chronic pain states, even in patients with OPIAD. Further, patients may be hesitant to discuss with their physician more personal and specific symptoms of low testosterone including anorgasmia, erectile dysfunction, and diminished ejaculatory volume. Thus, the true prevalence of men with hypogonadism secondary to chronic opioid use is likely understated due to the aforementioned difficulties in identifying men with OPIAD. Clinical suspicion for this condition should be heightened in the setting of men presenting with the aforementioned symptoms and should prompt taking a thorough history that includes questions on opioid use. Patients who are maintained on opioid treatments should also be asked about hypogonadism symptoms.

While a thorough history is an appropriate initial step in identifying patients with hypogonadism, serum testosterone levels are more specific and should be subsequently obtained. Low testosterone has been linked with aging, diabetes, obesity, and poor physical activity but must be identified in order to diagnose hypogonadism. Bhasin et al. suggested that free testosterone levels should be measured in obese or diabetic patients in order to determine if they are truly hypogonadal [20]. Samples should be collected in the early morning and repeated if abnormal. The current accepted lower limit of normal testosterone is 300–350 ng/dL [21]. Different organizations have varying testosterone thresholds for the diagnosis of hypogonadism, with the lower levels reaching 200 ng/dL in some publications. However, it is accepted that patients with testosterone levels below 230 ng/dL will benefit from testosterone replacement therapy [9]. Bawor et al. performed a meta-analysis that showed that about 50% of opioid-using men have significantly suppressed testosterone levels, with the opioid-user group having mean total testosterone levels 165 ng/dL lower than that of nonopioid users ($P<.0001$) [22]. If abnormal testosterone levels persist even after being repeated, further serum analyses of SHBG, LH, FSH, and prolactin may be performed to further elucidate the cause of the hypogonadism and identify which part of the HPGA is disrupted.

These additional serum hormone levels can help differentiate between the three categories of hypogonadism—primary, secondary, and tertiary hypogonadism. Primary hypogonadism is the most common type of hypogonadism due to testicular failure. Since the testes are unable to produce testosterone, the lack of negative feedback on the hypothalamus results in increased levels of LH and FSH produced by the pituitary gland [11]. Secondary and tertiary hypogonadism are caused by defects in the hypothalamic and pituitary portions of the HPGA and as a result often have low levels of both testosterone and gonadotropins. Secondary hypogonadism differs from tertiary hypogonadism in that it stems from defects at the hypophysis, instead of the hypothalamus. Opioids cause hypogonadism mainly at the level of the hypothalamus, making OPIAD a form of tertiary hypogonadism.

TREATMENT

Patients with OPIAD must first address the problem by way of diet and lifestyle modification. After conservative measures are attempted, other options for treatment include stopping opioid use, decreasing opioid dose or type, substituting opioids with nonnarcotic painkillers such as nonsteroidal antiinflammatory drugs (NSAIDs), or hormone replacement therapy [23]. Understandably, treatments such as halting opioid use or reducing opioid dose may not always be realistic in patients that require (or have become accustomed to) high-dose opioids to manage severe pain. In these circumstances, patients may consider cycling different opioids to avoid using long-acting opioids, which have been associated with an increased incidence of OPIAD compared with short-acting opioids [8]. If patients have hypogonadism symptoms, androgen replacement therapy is also an appropriate option. A subset of patients, specifically those with risk factors such as younger age, history of depression, length of hospital stay, and major renal surgery, will occasionally develop opioid dependence and overdose after urological surgery. Shah et al. recently demonstrated that 0.09% of urological patients experience opioid-related complications within 1 year of a procedure [24]. The patients who are more likely to develop opioid dependence should be identified, counseled on risks with narcotic use, and managed postoperatively with nonopioid painkillers if feasible.

Testosterone replacement therapy (TRT) is the core treatment for men with symptomatic hypogonadism who want to restore normal testosterone levels. Optimal replacement therapy has not yet been defined for men with OPIAD, although studies have trialed the effect of replacement therapy with transdermal gels, patches, and injections in this population. Basaria et al. conducted a randomized, double-blind, parallel-group, placebo-controlled trial, which found that hypogonadal men on opioids for chronic noncancer pain had improvement with sexual desire, body composition, and sensitivity, with 5 g transdermal testosterone gel when compared with placebo. There was no significant difference in self-reported pain between treatment and control groups [25]. Other studies have also reported significant increases in testosterone levels and improvement in sexual function using questionnaires such as the Brief Male Sexual Function Inventory and the Watts Sexual Function Questionnaire for Men [26, 27]. Additional proved benefits include increased muscle mass and improvement in mood. Currently, a multitude of testosterone replacement therapies are available and endorsed by the endocrine society. The optimal treatment should be tailored to each patient based on factors like cost, preference, side effects, and pharmacokinetics [20].

TREATMENT OPTIONS

Intramuscular Testosterone Enanthate or Cypionate, 75–100 mg Every Week or 150–200 mg Every 2 Weeks

While this treatment is the least expensive option, it can cause testosterone levels to rise to supraphysiological levels and drop to hypogonadal levels at the end of the dosing schedule. Swings in mood and libido are associated with these "roller coaster" fluctuations in levels and may be bothersome to some patients. Others, however, get used to (or may actually prefer) the highs and lows of injections. Among the TRT options, this is the most likely to cause polycythemia in the first few months of treatment [28]. Clinicians administering this treatment should also be cognizant of needle use in patients who are recovering from heroin addiction, as injection may pose as a trigger for relapse. Different treatments with other routes of administration should be considered for this subgroup.

Transdermal Testosterone Patches, One or Two 5 mg Applied Nightly

Testosterone patches more closely mimic normal circadian serum testosterone levels. However, patch sites may cause irritation and/or allergic reactions and should be regularly examined, at times requiring topical steroid therapy [29].

1.625% Testosterone Gel 20.25–81 mg Daily

Testosterone gels are supplied as metered-dose pumps and therefore allow for flexible titrations of dosing. Adjustments are usually made based on serum testosterone levels measured 2 weeks after treatment has started. Patients should be informed that that testosterone gel is susceptible to secondary transfer by skin-to-skin contact and is flammable when wet [30].

Subcutaneous Testosterone Pellets, Varied Dosing, Interval Injection of 3–6 Months [31]

Subcutaneous implantation of testosterone pellet eliminates the risk of secondary transfer seen with testosterone gel and also increases treatment compliance. Uncommon but possible side effects of this treatment include pellet extrusion and local site infection, which can be readily treated.

Prior to initiation of TRT, baseline testosterone, free testosterone, complete blood count (CBC), prostate-specific antigen (PSA), LH/FSH, and prolactin levels should be obtained. The risks of exogenous testosterone, specifically its detrimental effects on semen quality and future fertility, should be explained to patients starting treatment. After starting, patients should follow up frequently to monitor for metabolic or prostate-specific events. In order to ensure adequate response to therapy with increasing testosterone levels, repeat labs should be drawn 3–6 months after TRT initiation and then proceed to every 3–12 months [20]. The potential adverse events may impact quality of life for patients and therefore should be regularly assessed through quality-of-life questionnaires.

TRT is not without shortcomings, with reported adverse effects such as polycythemia, sleep apnea, reductions in HDL, and acne in both males and females. Oral testosterone has been associated with hepatotoxicity due to its susceptibility to hepatic first-pass effect. Some male-specific adverse events include gynecomastia, azoospermia, and priapism. Specific patients may be at risk for additional side effects. Patients with benign prostatic hyperplasia should be warned that their obstructive urinary symptoms might worsen and lead to acute urinary retention while being on treatment. Although TRT has not been proved to increase the risk of prostate cancer in hypogonadal men and serum concentration of sex hormones have not been associated with risk of prostate cancer, TRT is not generally recommended for patients with known history of prostate cancer. In a carefully monitored setting, more trends toward testosterone therapy for those with successfully treated prostate cancer are being reported worldwide [32, 33]. Those with high suspicion of having prostate cancer should be managed cautiously on TRT [4, 34].

The growing use of opioids for cancer and noncancer pain has raised questions about the relationship between TRT and the risk of cardiovascular events. A study by Basaria et al. showed that hypogonadal men aged 60–63 who had previously undergone coronary angiography and later started TRT were more likely to experience cardiovascular adverse events in 3 years than those without TRT. However, study limitations included small size and small number and variability of reported adverse events [35]. According to a meta-review of 51 studies analyzing adverse events with TRT, there was an associated increase in hematocrit and hemoglobin, decrease in HDL, and no reported difference in mortality or major cardiovascular events between the testosterone and control groups [36]. Huang et al. investigated the effects of TRT on metabolic and inflammatory markers in men with OPIAD. This study found that lipid profile, fasting glucose, fasting insulin, and C-reactive protein outcomes were similar between treatment and placebo groups. Body composition was increased in the testosterone group as well. This study was reassuring in showing that TRT may increase libido, body composition, and overall quality of life while avoiding deleterious metabolic consequences. Though, the studies that exist concerning the risk of cardiovascular events and TRT in men with OPIAD are limited by short follow-up and therefore are unable to make long-term conclusive statements on this exact risk. Further, longer-term studies, specifically double-blind, randomized, multicentered trials, may be beneficial in fully defining this association [37].

Opioids work within multiple signaling pathways and potentially have beneficial effects in other disease processes. Studies have shown that opioids and estrogen activate common signaling pathways, which can additionally lead to lower testosterone levels by inducing downstream effects of estrogen. Naloxone, an opioid receptor antagonist, has been shown to modulate estrogen receptors and inhibit breast cancer growth in mice [38, 39].

Clomiphene citrate (CC) is a selective estrogen receptor modulator that blocks estrogen binding at the hypothalamic level, increasing central production of GnRH, which stimulates LH/FSH release from the pituitary, leading to increased endogenously produced testosterone. The use of CC in hypogonadal men has shown effectiveness in increasing testosterone levels with minimal side effects [40, 41]. Hussein et al. have even demonstrated increased sperm count and success rate of testis sperm retrieval in patients with nonobstructive azoospermia using CC [42]. Although urologists have long used CC to treat hypogonadal men, in particular those who wish to preserve their fertility, it has not been FDA approved for this population and has predominantly been used to treat fertility in women. Anastrozole, a reversible inhibitor of the aromatase enzyme, has also been used with occasional success to treat men with hypogonadism and increased estradiol levels. The inhibition of aromatase reduces the conversion of peripheral androgens to estrogens and induces the aforementioned HPGA. While CC and anastrozole may become more commonly used treatments for men with OPIAD, more clinical and molecular studies are needed to determine the true benefit of these treatments in this population and the definite relationship between opioid use and estrogen levels.

CONCLUSION

The rising incidence of opioid use has developed into a global crisis and has simultaneously generated a largely neglected complication—hypogonadotropic hypogonadism. Hypogonadism may often be overlooked or thought to be unrelated to opioid use, and thus, any patient being evaluated for sexual dysfunction or infertility should be thoroughly inquired about medications and drug use, including history of prescription narcotic use. Conversely, men on opioid treatment should be monitored for development of OPIAD. Clinical suspicion of OPIAD can be high with reported symptoms, but the diagnosis is made through testosterone levels while factoring in other possible causes of secondary hypogonadism.

Guidelines for management of OPIAD are generally limited by the lack of prospective randomized control trials on treatment in this population and have room for further improvement with development of these studies. Fortunately, current recommendations for this condition may be modeled after treatment guidelines for secondary hypogonadism. While the best treatment regimen for each patient differs, the options for OPIAD treatment include opioid discontinuation with or without pain control substitution, use of short-acting opioids, and testosterone replacement therapy when indicated. Testosterone replacement therapy has multiple forms of administration such as injections, gels, and patches, which all have similar efficacy in men with OPIAD. Patients receiving testosterone replacement therapy should be counseled on the risks and benefits of each form of treatment and participate in shared decision-making and comprehensive counseling prior to choosing therapy. Though testosterone replacement therapy appears to not affect metabolic parameters, evidence on its association with cardiovascular events is inconsistent and therefore must be discussed with patients prior to initiation. Treatments that regulate estrogen levels such as clomiphene citrate and anastrozole have been used off-label to treat low testosterone levels in men with hypogonadism and may have some promise in treating men with OPIAD. However, more data are needed to establish the proved benefit of these therapies. Increased incidence of chronic opioid use and thus OPIAD necessitates large, prospective randomized studies that will elucidate further details of this disease process and generate the most comprehensive practice guidelines for the diagnosis and management of patients with OPIAD.

References

[1] Kolodny A, Courtwright DT, Hwang CS, Kreiner P, Eadie JL, Clark TW, et al. The prescription opioid and heroin crisis: a public health approach to an epidemic of addiction. Annu Rev Public Health 2015;36:559–74. Epub 2015/01/13, https://doi.org/10.1146/annurev-publhealth-031914-12295725581144.

[2] Unick GJ, Ciccarone D. US regional and demographic differences in prescription opioid and heroin-related overdose hospitalizations. Int J Drug Policy 2017;46:112–9. Epub 2017/07/10, https://doi.org/10.1016/j.drugpo.2017.06.00328688539.

[3] Cicero TJ, Ellis MS, Kasper ZA. Increased use of heroin as an initiating opioid of abuse. Addict Behav 2017;74:63–6. Epub 2017/06/06, https://doi.org/10.1016/j.addbeh.2017.05.03028582659.

[4] Smith HS, Elliott JA. Opioid-induced androgen deficiency (OPIAD). Pain Physician 2012;15 (3 Suppl):ES145–56. Epub 2012/07/20 22786453.

[5] Lucyk SN, Nelson LS. Toxicosurveillance in the US opioid epidemic. Int J Drug Policy 2017;46:168–71. Epub 2017/07/25, https://doi.org/10.1016/j.drugpo.2017.05.05728735771.

[6] Cicero TJ, Ellis MS, Surratt HL, Kurtz SP. The changing face of heroin use in the United States: a retrospective analysis of the past 50 years. JAMA Psychiatry 2014;71(7):821–6. Epub 2014/05/30, https://doi.org/10.1001/jamapsychiatry.2014.36624871348.

[7] Rajagopal A, Vassilopoulou-Sellin R, Palmer JL, Kaur G, Bruera E. Symptomatic hypogonadism in male survivors of cancer with chronic exposure to opioids. Cancer 2004;100(4):851–8. Epub 2004/02/11, https://doi.org/10.1002/cncr.2002814770444.

[8] Rubinstein AL, Carpenter DM, Minkoff JR. Hypogonadism in men with chronic pain linked to the use of long-acting rather than short-acting opioids. Clin J Pain 2013;29(10):840–5. Epub 2014/01/05, https://doi.org/10.1097/AJP.0b013e31827c7b5d24384986.

[9] Ali K, Raphael J, Khan S, Labib M, Duarte R. The effects of opioids on the endocrine system: an overview. Postgrad Med J 2016;92(1093):677–81. Epub 2016/09/21, https://doi.org/10.1136/postgradmedj-2016-13429927647927.

[10] Schneider J. Hypogonadism in men treated with chronic opioids. Arch Phys Med Rehabil 2008;89(7):1414. Epub 2008/07/05, https://doi.org/10.1016/j.apmr.2008.05.001.18586148.

[11] Khera M, Broderick GA, Carson III CC, Dobs AS, Faraday MM, Goldstein I, et al. Adult-onset hypogonadism. Mayo Clin Proc 2016;91(7):908–26. Epub 2016/06/28, https://doi.org/10.1016/j.mayocp.2016.04.02227343020.

[12] Zhang GF, Ren YP, Sheng LX, Chi Y, Du WJ, Guo S, et al. Dysfunction of the hypothalamic-pituitary-adrenal axis in opioid dependent subjects: effects of acute and protracted abstinence. Am J Drug Alcohol Abuse 2008;34(6):760–8. Epub 2008/11/19, https://doi.org/10.1080/00952990802385781190161818181.

[13] Ragni G, De Lauretis L, Bestetti O, Sghedoni D, Gambaro V. Gonadal function in male heroin and methadone addicts. Int J Androl 1988;11(2):93–100. Epub 1988/04/01, 3372047.

[14] Safarinejad MR, Asgari SA, Farshi A, Ghaedi G, Kolahi AA, Iravani S, et al. The effects of opiate consumption on serum reproductive hormone levels, sperm parameters, seminal plasma antioxidant capacity and sperm DNA integrity. Reprod Toxicol (Elmsford, NY) 2013;36:18–23. Epub 2012/12/05, https://doi.org/10.1016/j.reprotox.2012.11.01023207164.

[15] Aloisi AM, Ceccarelli I, Fiorenzani P, Maddalena M, Rossi A, Tomei V, et al. Aromatase and 5-alpha reductase gene expression: modulation by pain and morphine treatment in male rats. Mol Pain 2010;6:69. Epub 2010/10/28, https://doi.org/10.1186/1744-8069-6-6920977699.

[16] Cordes MA, Stevenson SA, Driessen TM, Eisinger BE, Riters LV. Sexually-motivated song is predicted by androgen-and opioid-related gene expression in the medial preoptic nucleus of male European starlings (Sturnus vulgaris). Behav Brain Res 2015;278:12–20. Epub 2014/09/30, https://doi.org/10.1016/j.bbr.2014.09.02925264575.

[17] Ventura-Aquino E, Paredes RG. Animal models in sexual medicine: the need and importance of studying sexual motivation. Sex Med Rev 2017;5(1):5–19. Epub 2016/08/28, https://doi.org/10.1016/j.sxmr.2016.07.00327566910.

[18] Woody G, McLellan AT, O'Brien C, Persky H, Stevens G, Arndt I, et al. Hormone secretion in methadone-dependent and abstinent patients. NIDA Res Monogr 1988;81:216–23. Epub 1988/01/01, 3136363.

[19] Mendelson JH, Meyer RE, Ellingboe J, Mirin SM, McDougle M. Effects of heroin and methadone on plasma cortisol and testosterone. J Pharmacol Exp Ther 1975;195(2):296–302. Epub 1975/11/01, 1185598.

[20] Bhasin S, Cunningham GR, Hayes FJ, Matsumoto AM, Snyder PJ, Swerdloff RS, et al. Testosterone therapy in men with androgen deficiency syndromes: an Endocrine Society clinical practice guideline. J Clin Endocrinol Metab 2010;95(6):2536–59. Epub 2010/06/09, https://doi.org/10.1210/jc.2009-2354, 20525905.

[21] Montorsi F, Adaikan G, Becher E, Giuliano F, Khoury S, Lue TF, et al. Summary of the recommendations on sexual dysfunctions in men. J Sex Med 2010;7(11):3572–88. Epub 2010/11/03, https://doi.org/10.1111/j.1743-6109.2010.02062.x21040491.

[22] Bawor M, Bami H, Dennis BB, Plater C, Worster A, Varenbut M, et al. Testosterone suppression in opioid users: a systematic review and meta-analysis. Drug Alcohol Depend 2015;149:1–9. Epub 2015/02/24, https://doi.org/10.1016/j.drugalcdep.2015.01.03825702934.

[23] Daniell HW. Hypogonadism in men consuming sustained-action oral opioids. J Pain 2002;3(5):377–84. Epub 2003/11/19, 14622741.

[24] Shah AS, Blackwell RH, Kuo PC, Gupta GN. Rates and risk factors for opioid dependence and overdose after urological surgery. J Urol 2017;198(5):1130–6. Epub 2017/05/17, https://doi.org/10.1016/j.juro.2017.05.037 28506855.

[25] Basaria S, Travison TG, Alford D, Knapp PE, Teeter K, Cahalan C, et al. Effects of testosterone replacement in men with opioid-induced androgen deficiency: a randomized controlled trial. Pain 2015;156(2):280–8. Epub 2015/01/20, https://doi.org/10.1097/01.j.pain.0000460308.86819.aa 25599449.

[26] Blick G, Khera M, Bhattacharya RK, Nguyen D, Kushner H, Miner MM. Testosterone replacement therapy outcomes among opioid users: the Testim Registry in the United States (TRiUS). Pain Med (Malden, MA) 2012;13(5):688–98. Epub 2012/04/28, https://doi.org/10.1111/j.1526-4637.2012.01368.x22536837.

[27] Daniell HW, Lentz R, Mazer NA. Open-label pilot study of testosterone patch therapy in men with opioid-induced androgen deficiency. J Pain 2006;7(3):200–10. Epub 2006/03/07, https://doi.org/10.1016/j.jpain.2005.10.00916516826.

[28] Dobs AS, Meikle AW, Arver S, Sanders SW, Caramelli KE, Mazer NA. Pharmacokinetics, efficacy, and safety of a permeation-enhanced testosterone transdermal system in comparison with bi-weekly injections of testosterone enanthate for the treatment of hypogonadal men. J Clin Endocrinol Metab 1999;84(10):3469–78. Epub 1999/10/16, https://doi.org/10.1210/jcem.84.10.607810522982.

[29] Ullah MI, Riche DM, Koch CA. Transdermal testosterone replacement therapy in men. Drug Des Devel Ther 2014;8:101–12. https://doi.org/10.2147/DDDT.S43475.

[30] Shoskes JJ, Wilson MK, Spinner ML. Pharmacology of testosterone replacement therapy preparations. Transl Androl Urol 2016;5(6):834–43. https://doi.org/10.21037/tau.2016.07.10.

[31] Kaminetsky JC, Moclair B, Hemani M, Sand M. A phase IV prospective evaluation of the safety and efficacy of extended release testosterone pellets for the treatment of male hypogonadism. J Sex Med 2011;8(4):1186–96. Epub 2011/01/29, https://doi.org/10.1111/j.1743-6109.2010.02196.x21269402.

[32] Morgentaler III A, Conners WP. Testosterone therapy in men with prostate cancer: literature review, clinical experience, and recommendations. Asian J Androl 2015;17(2):206–11. Epub 2015/02/06, https://doi.org/10.4103/1008-682X.14806725652633.

[33] Kaplan AL, Hu JC, Morgentaler A, Mulhall JP, Schulman CC, Montorsi F. Testosterone therapy in men with prostate cancer. Eur Urol 2016;69(5):894–903. Epub 2016/01/01, https://doi.org/10.1016/j.eururo.2015.12.005, 26719015.

[34] Ory J, Flannigan R, Lundeen C, Huang JG, Pommerville P, Goldenberg SL. Testosterone therapy in patients with treated and untreated prostate cancer: impact on oncologic outcomes. J Urol 2016;196(4):1082–9. Epub 2016/05/02, https://doi.org/10.1016/j.juro.2016.04.06927131465.

[35] Basaria S, Coviello AD, Travison TG, Storer TW, Farwell WR, Jette AM, et al. Adverse events associated with testosterone administration. N Engl J Med 2010;363(2):109–22. Epub 2010/07/02, https://doi.org/10.1056/NEJMoa1000485 20592293.

[36] Fernandez-Balsells MM, Murad MH, Lane M, Lampropulos JF, Albuquerque F, Mullan RJ, et al. Clinical review 1: adverse effects of testosterone therapy in adult men: a systematic review and meta-analysis. J Clin Endocrinol Metab 2010;95(6):2560–75. Epub 2010/06/09, https://doi.org/10.1210/jc.2009-257520525906.

[37] Huang G, Travison T, Maggio M, Edwards RR, Basaria S. Effects of testosterone replacement on metabolic and inflammatory markers in men with opioid-induced androgen deficiency. Clin Endocrinol 2016;85(2):232–8. Epub 2016/03/02, https://doi.org/10.1111/cen.1304926928845.

[38] Lee CW-S, Ho I-K. Sex differences in opioid analgesia and addiction: interactions among opioid receptors and estrogen receptors. Mol Pain 2013;9:45. https://doi.org/10.1186/1744-8069-9-45.

[39] Farooqui M, Geng ZH, Stephenson EJ, Zaveri N, Yee D, Gupta K. Naloxone acts as an antagonist of estrogen receptor activity in MCF-7 cells. Mol Cancer Ther 2006;5(3):611–20. Epub 2006/03/21, https://doi.org/10.1158/1535-7163.Mct-05-001616546975.

[40] Katz DJ, Nabulsi O, Tal R, Mulhall JP. Outcomes of clomiphene citrate treatment in young hypogonadal men. BJU Int 2012;110(4):573–8. Epub 2011/11/03, https://doi.org/10.1111/j.1464-410X.2011.10702.x22044663.

[41] Shabsigh A, Kang Y, Shabsign R, Gonzalez M, Liberson G, Fisch H, et al. Clomiphene citrate effects on testosterone/estrogen ratio in male hypogonadism. J Sex Med 2005;2(5):716–21. Epub 2006/01/21, https://doi.org/10.1111/j.1743-6109.2005.00075.x16422830.

[42] Hussein A, Ozgok Y, Ross L, Niederberger C. Clomiphene administration for cases of nonobstructive azoospermia: a multicenter study. J Androl 2005;26(6):787–91. discussion 92–3. Epub 2005/11/18, https://doi.org/10.2164/jandrol.0418016291975.

CHAPTER 6.3

Alcohol and Men's Health

Brian Dick, Scott Brimley, Peter Tsambarlis, Wayne Hellstrom
Tulane University School of Medicine, Department of Urology, New Orleans, LA, United States

INTRODUCTION

Ethanol (Fig. 1), the active chemical in alcohol, is mainly absorbed in the small intestines and metabolized in the liver [1]. Excess alcohol consumption is a problem that disproportionately affects men. In 2010, the World Health Organization reported 3.14 million men, nearly double the 1.72 million women, died of causes related to excessive use of alcohol. Alcohol is linked to physical and social consequences in three ways: intoxication, dependence, and direct biological effects [2]. These direct biological effects lead to many disease states including oral and esophageal malignancies, liver cancer, unipolar major depression, epilepsy, hypertension, hemorrhagic stroke, cirrhosis, and coronary heart disease.

The effects of alcohol, however, are not uniformly negative. Several investigations have reported mild consumption of alcohol to be beneficial to cardiovascular health [3]. When reviewing studies examining the relationship between alcohol and sexual dysfunction, it becomes clear that the effects depend on both the frequency and volume of alcohol consumed. In this review, we examine how acute and chronic alcohol consumption affects erections, ejaculation, and men's health.

> …it provokes the desire, but it takes away the performance
> — William Shakespeare

ACUTE EFFECTS OF ALCOHOL

Erectile Dysfunction (ED)

It is a commonly held belief that acute alcohol intake promotes sexual desire and arousal in men [3, 4]. For adolescent men, sex and regular alcohol use both mark transitions into adulthood, and drinking and sexual intercourse are often

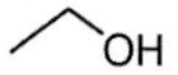

FIG. 1
Ethanol.

combined [5]. Researchers disagree on alcohol's effects on erections. Historically, it was postulated that high blood alcohol content (BAC) attenuated sustained erections [6]. The popular adage from Shakespeare's Macbeth captures the cultural idea of what alcohol does before copulation: "it provokes the desire, but it takes away the performance."

Many consumers believe that acute alcohol intake will promote a man's erectile function and sexual performance. One proposed mechanism is through the relaxation of the smooth muscle of the corpus cavernosum (CC). Choi et al. observed this effect in rats. In their experiments, they separated 40 male rats into eight groups depending on the level of alcohol in each rat's blood. They observed that increasing concentrations of alcohol led to decreased contraction of the CC when stimulated with phenylephrine [6, 7]. Acute alcohol intake resulted in significantly increased cyclic AMP (cAMP) levels in the rats' CC as compared with the control group. CC relaxation did not take place, however, if the rats were pretreated with propranolol (an adrenergic B-receptor blocker). The authors concluded that activation of the cAMP pathway is beneficial to healthy erections.

More recent data contradict this precept. In a study from 2011, enough alcohol was administered to a group of 65 male volunteers to achieve a BAC of .10%, and participants were then shown erotic videos to induce sexual arousal. Little physiological difference was observed between this group and the controls, indicating that, at least at .10% BAC, alcohol had a limited effect on erectile function. Additionally, when researchers asked subjects to rate their levels of sexual arousal on a scale from 1 to 7 (1= "no sexual arousal at all" and 7= "extremely sexually aroused"), there was no significant difference between the two groups [8, 9]. In another study, Morlet et al. followed 11 males for three nights to determine alcohol's effect on erections. During the first two nights, the subjects were observed while asleep and monitored for erections for control purposes. On the third night, the subjects were administered enough alcohol to reach $0.12\,g/100\,mL$ BAC. The researchers then measured penile circumference at two points, proximal to the glans and at the base of the shaft. The researchers reported no difference in size, duration, or the number of erections in the 11 subjects they followed. Interestingly, the authors suggest that acute alcohol use does little to physiologically change erectile function but can cause a psychological and social increase in desire and likely helps relax men who use it. They suggest that altered perception due to alcohol may be the underlying etiology of decreased erectile function [10].

George et al. hypothesized that whether the BAC was ascending or descending may have confounded the previous research, explaining the discrepancies seen in some results. Their study aimed to test this theory; 78 heterosexual men were exposed to erotica as their BAC descended from .08%. They observed that alcohol slightly inhibited erectile capacity and control compared with sober controls [11]. Though more information on the effects alcohol has on the mind and on sexual desire is needed, practical and ethical limitations make finding conclusive data on this topic difficult.

Ejaculatory Dysfunction

Delayed Ejaculation

Delayed ejaculation (DE) is defined as the delay or absence of male ejaculation and subsequent orgasm that results in distress [10]. The cause of DE is often not immediately understood, as the complex nature of reaching orgasm includes genetics, neurophysiology, and social variables. Alcohol, like many other pharmacological agents (including selective serotonin reuptake inhibitors (SSRIs), diuretics, and narcotics), interferes with peripheral and central nerve receptors controlling ejaculation [12]. However, there is little evidence that acute alcohol intake causes DE. As of now, there is no FDA-approved medication for the treatment of DE, as the drugs that have been tested for its treatment have little to no supporting evidence [12].

Premature Ejaculation

Premature ejaculation (PE) generally consists of three constructs: a short ejaculatory latency, an inability to control or delay ejaculation, and a distress to the individual and/or partner [13]. Reportedly, between 20% and 30% of men complain of some form of PE [14]. PE is recognized to be the most common sexual dysfunction among males [13]. Although chronic alcohol abuse is linked to PE, acute intake of alcohol seems to have limited effect. As previously mentioned, intoxication may have a slight beneficial effect with regard to PE [14]. Laumann et al. reported a small but significant ($P=.045$) negative association of alcohol with PE symptoms. However, because of low statistical power, when the patient sample is compared with an age-matched control group, the significant association between alcohol consumption and PE is no longer observed. The authors recommend changing specific lifestyle factors, such as physical activity, that may contribute to the PE rather than relying on alcohol as a treatment [15].

CHRONIC EFFECTS OF ALCOHOL

Chronic alcohol use is commonly measured in terms of drinks per day (d/d), drinks per week (d/w), or grams of ethanol per day (g/d). Approximately 10 g of ethanol is equal to one drink [16]. Moderate alcohol use will be defined as <4 d/d, which is <40 g/d, or <28 d/w. Heavy alcohol use will be defined as anything >4 d/d.

Erectile Dysfunction

The relationship between chronic alcohol use and ED varies from study to study; some report causative effects with consumption of as little as one drink per week [17], while others report protective effects when consuming as many as four drinks per day [18]. However, for the most part, moderate chronic alcohol use appears to promote erectile function, while heavy chronic alcohol use appears to be associated with ED.

Moderate Alcohol Consumption

Most studies suggest that drinking a moderate amount of alcohol preserves erectile function. A study of 8347 Australian men found drinking 1–4 d/d to be significantly associated with healthy erectile function [18]. Statistically significant associations between healthy erectile function and moderate alcohol consumption were also noted in a study of 799 Belgian men consuming 20–40 g/d [16], in a study of 342 Brazilian men consuming ≤3 d/d [19], and in a study of 2301 Bostonian men consuming 1–3 d/d [20]. In one of the largest studies performed, Weber et al. evaluated 101,674 Australian men and reported that men who drank 6–10 d/w were less likely to have ED than those who drank 1–5 d/w. Furthermore, men who drank 0 d/w were more likely to have ED than those consuming 1–5 d/w [21]. In line with these findings, Cheng et al. performed a meta-analysis in 2007 evaluating the relationship between alcohol and erectile function and reported a significant association between healthy erections and consuming >8 d/w. They also found an association between good erectile function and consuming 1–7 d/w, but this was not statistically significant [22]. Several other studies report an association, albeit not a statistically significant one, between moderate alcohol consumption and good erections; in three US studies from 2000, 2003, and 2006, men who drank 1–3 d/d [23], 5–29.9 g/d [24], and 5–14.9 g/d [25], respectively, were found to be at lower risk of developing ED. Austoni et al. also found a nonsignificant correlation between healthy erectile function and having ≤3 d/d in their study of 16,724 Italian men [26].

Conversely, a nonsignificant association between moderate alcohol consumption and erectile *dysfunction* was reported in a study of 2120 Danish men drinking >8 d/w [27], a Korean study of 3501 men consuming >4 d/w [28], and a study of 2412 men from different nations who consumed >8 d/w [29].

There is one study that reports a statistically significant association between moderate alcohol consumption and ED. Polsky et al. studied 335 Canadian men and found consuming more than 1 d/w to be linked to ED [17].

Heavy Alcohol Consumption

While moderate alcohol consumption is predominantly associated with healthy erections, heavy alcohol consumption is more complex. Many studies do not define a level that corresponds to excessive consumption, often grouping together all subjects who consume more than 3–4 d/d or more than 20 d/w. This is problematic when trying to determine the effect that drinking copious amounts of alcohol, that is, >8 d/d, has on erectile function. Because few participants drink at this level, their results will be masked by the majority of participants drinking in the 4–5 d/d range. However, these few participants who do drink copious amounts of alcohol will exhibit a trend toward a positive association between heavy alcohol use and ED causing many results to become nonsignificant. Thus, there are no studies with statistically significant associations between heavy alcohol use and healthy erectile function.

A nonsignificant association between healthy erectile function and heavy alcohol consumption is reported in the study of 8347 Australian men consuming >4 d/d [18], in the study of 342 Brazilian men consuming >3 d/d [19], in the study of 2301 Bostonian men consuming >3 d/d [20], in the study of 513 Bostonian men consuming >3 d/d [23], and in a study of 2101 Italian men consuming >21 d/w [30].

Conversely, a nonsignificant association between erectile *dysfunction* and heavy alcohol use is reported in the study of 799 Belgian men consuming ≥40 g/d [16], in the study of 2120 Danish men consuming ≥21 d/w [27], and in the study of 2412 men from different nations consuming >21 d/w [29].

A statistically significant association between ED and heavy alcohol use can be observed in studies evaluating patients who consume excessive amounts of alcohol. A 2010 study of 1956 Italian men found a statistically significant association between consuming >6 d/d and ED [31]. This was seen again in the large study of 101,674 Australian men, where consumption of >30 d/w was associated with ED [21].

A summary of these results can be viewed in Table 1, which includes studies that report alcohol consumption in d/d or g/d, and Table 2, which includes studies that report alcohol consumption in d/w.

Mechanism

To better understand how alcohol causes ED, one must understand the physiology of a normal erection. Healthy males achieve erection through vasodilation of arterioles in the corpus spongiosum and corpora cavernosa. Blood flows into these spaces, causing expansion of the cavernosal spaces and an increase in pressure that occludes subtunical venules, which reduces the outflow of blood from the penis [32]. The leading cause of age-related ED is occluded vessels due to

Table 1 Alcohol Consumption and Erectile Dysfunction (d/d or g/d)

Year	Authors	Age	Country	0 d/d	1 d/d	2 d/d	3 d/d	4 d/d	5 d/d	6 d/d	>6 d/d	ED Assessment Method
2005	Austoni E et al.	16+	Italy	0 d/d		≤3 d/d				>3 d/d		Single-question assessment
2006	Millet C et al.	16–59	Australia	0 d/d		1–4 d/d					>4 d/d	Single-question assessment
2008	Kupelian V et al.	30–79	Boston, United States	0 d/d	<1 d/d	1–3 d/d				>3 d/d		IIEF-5, ED considered ≤16
2002	Mak R et al.	40–69	Belgium		<20 g/d					≥40 g/d		Single-question assessment
2002	Moreira ED et al.	40–70	Brazil	1		≤3 d/d				>3 d/d		Single-question assessment
2000	Feldman HA et al.	40–70	Massachusetts	<1 d/d		1–3 d/d			≥4 d/d			23-item questionnaire
2006	Bacon CG et al.	40–75	United States	0 d/d	0.1–4.9 g/d	5–14.9 g/d	15–29.9 g/d					Single-question assessment
2003	Bacon CG et al.	50+	United States	0 d/d	0.1–4.9 g/d	5–14.9 g/d	15–29.9 g/d			≥30 g/d		Single-question assessment
2010	Boddi V et al.	N/A	Italy			<3 d/d			4–6 d/d		>6 d/d	Structured interview on ED (SIEDY)

Table 2 Alcohol Consumption and Erectile Dysfunction (d/w)

Year	Authors	Age	Country	0 d/w	1–4 d/w	5–7 d/w	8–14 d/w	15–21 d/w	21–25 d/w	26–30 d/w	>30 d/w	ED Assessment Method
2011	Christensen BS et al.	16–97	Denmark	0 d/w	1–7 d/w		8–21 d/w			≥21 d/w		Single-question assessment
2000	Parazzini F et al.	18+	Italy	0 d/w	1–7 d/w		8–14 d/w	15–21 d/w		≥22 d/w		Single-question assessment
2003	Okulate G et al.	18–63	Nigeria	0 d/w			Any amount no association P = .48					IIEF
2003	Cho BL et al.	20+	Korea	<3 d/w	3–4 d/w			>4 d/w				IIEF-5
2009	Chew KK et al.	20+	Australia	0 d/w			Any amount					IIEF-5 <22
2003	Nicolosi A et al.	40–70	Multiple	0 d/w	1–7 d/w		8–21 d/w			≥21 d/w		Single-question assessment
2013	Weber MF et al.	45+	Australia	0 d/w	1–5 d/w	6–10 d/w		11–25 d/w		26–30 d/w	>30 d/w	Single-question assessment
2005	Polsky JY et al.	50–80	Canada	<1 d/w	1–7 d/w				>8 d/w			ED diagnosed in hospital
2007	Cheng JYW et al.	N/A	Multiple	0 d/w	1–7 d/w			>8 d/w				IIEF or questions

OR <1, sig; OR <1, not sig; OR >1, not sig; OR >1, sig; d/d, drinks per day; d/w, drinks/week; g/d, grams/day; g/w, grams/week.

atherosclerosis from conditions such as diabetes and hypertension [33]. Additionally, inflammation in these vessels can impair blood flow [34]. By moderating vessel obstruction or inflammation, alcohol may influence erectile function.

One mechanism through which alcohol can affect both pathways is through the PON1 gene. PON1, also known as serum paraoxonase/arylesterase 1, is an enzyme associated with HDL that destroys oxidized LDL to nonharmful products and inhibits the oxidation of LDL [35]. Oxidized LDL is typically phagocytosed by macrophages forming foam cells and initiating atherosclerotic plaque formation (Fig. 2). Additionally, oxidized LDL has been demonstrated to increase reactive oxygen species, with associated inflammation, in endothelial cells [36]. If alcohol can upregulate the expression of the PON1 gene, it would be able to inhibit some of the leading causes of age-related ED. This is exactly what Rao et al. observed when comparing serum PON1 levels in nondrinkers, light drinkers who had consumed 1–3 d/d for the past 6 months of longer, and heavy drinkers who had consumed ≥ 6 d/d for the last 6 months or longer. Compared with nondrinkers, light drinkers had a 395% increase in serum PON1, while heavy drinkers had a 45% decrease [35]. A separate study analyzing PON1 activity in drinkers versus nondrinkers also reported an association between alcohol and PON1 activity [37]. However, this association was nonsignificant and nonlinear suggesting that while moderate alcohol use may upregulate PON1, the amount of alcohol to achieve this effect varies from person to person. That said, alcohol's ability to alter PON1 activity offers a possible

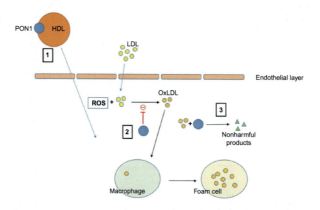

FIG. 2

(1) PON1 circulates through the body attached to HDL. It leaves the bloodstream and passes through endothelial cells to enter the tunica intima. (2) Once in the tunica intima, PON1 inhibits the formation of oxidized LDL (OxLDL) by blocking the reaction between reactive oxygen species (ROS) and LDL. (3) PON1 is also able to break down preexisting OxLDL into nonharmful products.

explanation as to why moderate chronic alcohol consumption is correlated with healthy erectile function but heavy chronic alcohol consumption is correlated with ED.

A physiological representation of decreased PON1 activity was witnessed in a 2010 study by Boddi et al. When compared with nondrinkers, men who consumed >6 d/d had significantly reduced dynamic peak systolic velocity (measured by penile duplex Doppler ultrasound) suggesting vascular impairment and increased likelihood of developing ED [31].

Ejaculatory Dysfunction

After ED, ejaculatory dysfunctions are reported to be some of the most common sexual dysfunctions in alcohol-dependent men [38, 39]. However, there are fewer studies examining the effect alcohol has on ejaculatory function than those examining its effect on erectile function. Additionally, these studies report increased rates of both PE and DE, which is counterintuitive as these are driven by opposing physiological mechanisms.

A study of 2120 Danish men noted that any alcohol consumption is positively associated with PE. However, this association only reaches statistical significance when men drank >21 d/w [27]. A study of 96 men drinking an average of 20 drinks per day over the last 8 years revealed 37% to have PE and 10% to have DE [40]. Lastly, a 2010 paper found a positive association between consuming >6 alcoholic d/d and DE [31].

Mechanism

PE and DE are opposite clinical complaints. It follows that they are caused by different physiological mechanisms. PE is associated with hyperandrogenism, hypoprolactinemia, and hyperthyroidism, while DE is associated with hypoandrogenism, hyperprolactinemia, and hypothyroidism [41]. In 2010, Boddi et al. reported that, compared with nondrinkers, men who consumed alcohol had decreased levels of both prolactin and thyroid-stimulating hormone (TSH) [31]. Decreased TSH levels are associated with hyperthyroidism, suggesting chronic alcohol consumers are more likely to exhibit hypoprolactinemia and hyperthyroidism. These two conditions are associated with PE, so it follows that many chronic consumers of alcohol experience PE.

Despite chronic alcohol consumption leading to hormonal changes associated with PE, some studies found increased rates of DE. Arackal et al. suggest that this may be the product of psychogenic sexual dysfunction stemming from marital strife often seen in alcoholic families [40].

Hypogonadism

Excess alcohol consumption is linked to hypogonadism and decreased sexual function due to its role in decreasing testosterone (T) levels [31]. T is intimately involved in each step of male sexual function, and therefore, low T leads to other sexual dysfunctions, such as ED. Alcohol can lead to testicular damage by decreasing T production and release [42]. This is due to the direct and indirect actions of alcohol—alcohol intake is correlated with dysregulated luteinizing hormone-releasing hormone (LHRH) and luteinizing hormone (LH), the hormones responsible for T production signaling [43]. Alcohol also decreases the secretion of T from the testicles [44].

Both acute and chronic exposures to ethanol decrease serum T [45], but the mechanisms differ; acute alcohol intake increases LHRH and LH but lowers T; chronic alcohol use results in a decrease in LHRH, LH, and T [46]. Along with lower T, alcohol consumption can lead to increased estrogen levels, possibly due to increased aromatization in the liver [47]. Subsequently, feminization, that is, swelling of breast tissue, decrease in prostate size, and diminished pubic and facial hair, is a side effect of chronic alcohol abuse [48].

Although a major lack of T has a significant deleterious effect on libido, chronic alcoholism may not produce this effect because T is produced elsewhere in the male body [49]. Of note, because of the important role T has on the development of the male, the use of alcohol during puberty may have significant adverse consequences on reproductive function [49]. T supplementation can be helpful to improve sexual function but only in men with overt ED and hypogonadism [47].

POTENTIAL CONFOUNDING FACTORS

While most studies suggest a benefit of moderate consumption on erectile function and an association between heavy alcohol use and ED, there are several exceptions. Explanations for these discrepancies may include differences in how the studies assessed erectile function and the different attitudes toward alcohol in different cultures.

Assessing Erectile Dysfunction

The International Society for Sexual Medicine (ISSM) defines ED as the consistent inability to achieve or maintain an erection sufficient for sexual satisfaction [50]. Several ways to assess erectile function exist including the International Index of Erectile Function (IIEF), the Brief Male Sexual Function Inventory (BMSFI), and a clinical examination. The IIEF is a 15-question survey about the patients' sexual experiences over the preceding 4 weeks. Questions are ranked on a 5-point scale, and patient erectile function is found by totaling the number of points in the survey [51]. The IIEF-5 is a shorter, 5-question, version of the IIEF-15 that gives similar results [52]. The BMSFI is an 11-question survey to assess a patient's sexual drive, ability, and satisfaction over the preceding 30 days [53].

In 2005, O'Donnell et al. were dissatisfied with the high nonresponse rate of these surveys and tried to develop a single question to assess male erectile function. Subjects were instructed to answer whether their erection was *always*, *usually*, *sometimes*, or *never* hard enough for satisfactory sexual activity. Based on this answer, they were categorized as not impotent, minimally impotent, moderately impotent, or completely impotent [54]. The results of this question correlated not only with the IIEF-15 and BMSFI but also with the results of an independently performed urological examination [54]. Tables 1 and 2 include the erectile function assessment method used by the studies cited in this chapter.

In 2007, Cheng et al. performed a meta-analysis of ED risk and alcohol consumption and studied the odds ratio of developing ED when studies were grouped according to how they assessed ED. The group of studies using a single-question ED assessment method noted a significant correlation between healthy erectile function and alcohol use [22]. The group of studies that assessed ED using the IIEF also found a correlation between the two, but it was not significant [22]. This suggests that different assessment methods may be the reason 1–3 d/d is associated with healthy erections in one study but has no effect in another.

Societal Difference in Alcohol Consumption

Another factor that changes from study to study is the location in which participants grew up or currently reside. Consuming 1–3 d/d may be considered "light" drinking in some areas, while it is considered "heavy" in others. This is important to consider because "light drinkers" may, overall, be more health conscious than "heavy drinkers." The healthier lifestyle rather than the alcohol consumption could be what is influencing erectile function [55]. This is a possible explanation for why intermediate alcohol use can be linked to healthy erectile function or ED depending on the country in which the study took place. In short, other lifestyle differences may be masking the subtle effects of alcohol at this "in-between" level of consumption. An example of this is an Italian study reporting a nonsignificant association between healthy erectile function and consuming 8–22 d/w and ED [30], while a Danish study reports a nonsignificant association between ED and consuming 8–22 d/w [27].

CONCLUSION

In summary, the effect of acute and chronic alcohol use on men's health is directly related to the amount and frequency with which it is consumed. Acute alcohol consumption has historically been associated with ED, but more recent studies suggest that this may not be the case. Reported effects range from causing no difference in erectile function and sexual satisfaction to producing better erections through protective vascular effects. Additionally, chronic consumption of moderate amounts of alcohol may have a protective effect on erectile function through upregulation of the PON1 gene with its corresponding cardiovascular benefits. However, chronic consumption of heavy amounts of alcohol is associated with the development of ED, potentially due to downregulation of the PON1 gene. Studies vary in their reported "cutoff" points for what constitutes heavy versus moderate alcohol intake. There is less research examining the relationship between alcohol and ejaculatory dysfunction than alcohol and ED, but chronic alcohol use appears to induce hormonal changes leading to PE. Lastly, alcohol abuse is linked to the development of hypogonadism. The negative effects of heavy alcohol use outweigh any possible benefits that may be present, and patients should be instructed to avoid excess consumption.

References

[1] Jung YC, Namkoong K. Alcohol: intoxication and poisoning - diagnosis and treatment. Handb Clin Neurol 2014;125:115–21.

[2] Baker P, Dworkin SL, Tong S, Banks I, Shand T, Yamey G. The men's health gap: men must be included in the global health equity agenda. Bull World Health Organ 2014;92(8):618–20.

[3] Rehm J, Room R, Graham K, Monteiro M, Gmel G, Sempos CT. The relationship of average volume of alcohol consumption and patterns of drinking to burden of disease: an overview. Addiction 2003;98(9):1209–28.

[4] Zakhari S. Alcohol and the cardiovascular system: molecular mechanisms for beneficial and harmful action. Alcohol Health Res World 1997;21(1):21–9.

[5] McLean AL, Flanigan BJ. Transition-marking behaviors of adolescent males at first intercourse. Adolescence 1993;28(111):579–95.

[6] Choi SM, Seo DH, Lee SW, Lee C, Jeh SU, Kam SC, et al. The effect of alcohol administration on the corpus cavernosum. World J Mens Health 2017;35(1):34–42.

[7] Maisto SA, Simons JS. Research on the effects of alcohol and sexual arousal on sexual risk in men who have sex with men: implications for HIV prevention interventions. AIDS Behav 2016;20(Suppl. 1):S158–72.

[8] Wilson GT. Alcohol and human sexual behavior. Behav Res Ther 1977;15(3):239–52.

[9] George WH, Davis KC, Norris J, Heiman JR, Schacht RL, Stoner SA, et al. Alcohol and erectile response: the effects of high dosage in the context of demands to maximize sexual arousal. Exp Clin Psychopharmacol 2006;14(4):461–70.

[10] Morlet A, Watters GR, Dunn J, Keogh EJ, Creed KE, Tulloch AG, et al. Effects of acute alcohol on penile tumescence in normal young men and dogs. Urology 1990;35(5):399–404.

[11] George WH, Cue Davis K, Schraufnagel TJ, Norris J, Heiman JR, Schacht RL, et al. Later that night: descending alcohol intoxication and men's sexual arousal. Am J Mens Health 2008;2(1):76–86.

[12] Sadowski DJ, Butcher MJ, Kohler TS. A review of pathophysiology and management options for delayed ejaculation. Sex Med Rev 2016;4(2):167–76.

[13] Althof SE, McMahon CG, Waldinger MD, Serefoglu EC, Shindel AW, Adaikan PG, et al. An update of the International Society of Sexual Medicine's guidelines for the diagnosis and treatment of premature ejaculation (PE). Sex Med 2014;2(2):60–90.

[14] Santtila P, Sandnabba NK, Jern P. Prevalence and determinants of male sexual dysfunctions during first intercourse. J Sex Marital Ther 2009;35(2):86–105.

[15] Laumann EO, Paik A, Rosen RC. Sexual dysfunction in the United States: prevalence and predictors. JAMA 1999;281(6):537–44.

[16] Mak R, De Backer G, Kornitzer M, De Meyer JM. Prevalence and correlates of erectile dysfunction in a population-based study in Belgium. Eur Urol 2002;41(2):132–8.

[17] Polsky Jane Y, Aronson Kristan J, Heaton Jeremy PW, Adams MA. Smoking and other lifestyle factors in relation to erectile dysfunction. BJU Int 2005;96(9):1355–9.

[18] Millett C, Wen LM, Rissel C, Smith A, Richters J, Grulich A, et al. Smoking and erectile dysfunction: findings from a representative sample of Australian men. Tob Control 2006;15(2):136.

[19] Moreira Jr. ED, Bestane WJ, Bartolo EB, Fittipaldi JA. Prevalence and determinants of erectile dysfunction in Santos, southeastern Brazil. Sao Paulo Med J 2002;120(2):49–54.

[20] Kupelian V, Link CL, Rosen RC, McKinlay JB. Socioeconomic status, not race/ethnicity, contributes to variation in the prevalence of erectile dysfunction: results from the Boston Area Community Health (BACH) Survey. J Sex Med 2008;5(6):1325–33.

[21] Weber MF, Smith DP, O'Connell DL, Patel MI, de Souza PL, Sitas F, et al. Risk factors for erectile dysfunction in a cohort of 108 477 Australian men. Med J Aust 2013;199(2):107–11.

[22] Cheng JYW, Ng EML, Chen RYL, Ko JSN. Alcohol consumption and erectile dysfunction: meta-analysis of population-based studies. Int J Impot Res 2007;19:343.

[23] Feldman HA, Johannes CB, Derby CA, Kleinman KP, Mohr BA, Araujo AB, et al. Erectile dysfunction and coronary risk factors: prospective results from the Massachusetts Male Aging Study. Prev Med 2000;30(4):328–38.

[24] Bacon CG, Mittleman MA, Kawachi I, Giovannucci E, Glasser DB, Rimm EB. Sexual function in men older than 50 years of age: results from the health professionals follow-up study. Ann Intern Med 2003;139(3):161–8.

[25] Bacon CG, Mittleman MA, Kawachi I, Giovannucci E, Glasser DB, Rimm EB. A prospective study of risk factors for erectile dysfunction. J Urol 2006;176(1):217–21.

[26] Austoni E, Mirone V, Parazzini F, Fasolo CB, Turchi P, Pescatori ES, et al. Smoking as a risk factor for erectile dysfunction: data from the andrology prevention weeks 2001-2002 a study of the Italian Society of Andrology (s.I.a.). Eur Urol 2005;48(5):810–7. discussion 7–8.

[27] Christensen BS, Grønbæk M, Pedersen BV, Graugaard C, Frisch M. Associations of unhealthy lifestyle factors with sexual inactivity and sexual dysfunctions in Denmark. J Sex Med 2011;8(7):1903–16.

[28] Cho BL, Kim YS, Choi YS, Hong MH, Seo HG, Lee SY, et al. Prevalence and risk factors for erectile dysfunction in primary care: results of a Korean study. Int J Impot Res 2003;15:323.

[29] Nicolosi A, Glasser DB, Moreira ED, Villa M. Prevalence of erectile dysfunction and associated factors among men without concomitant diseases: a population study. Int J Impot Res 2003;15:253.

[30] Parazzini F, Menchini Fabris F, Bortolotti A, Calabrò A, Chatenoud L, Colli E, et al. Frequency and determinants of erectile dysfunction in Italy. Eur Urol 2000;37(1):43–9.

[31] Boddi V, Corona G, Monami M, Fisher AD, Bandini E, Melani C, et al. Priapus is happier with Venus than with Bacchus. J Sex Med 2010;7(8):2831–41.

[32] Giuliano F. Neurophysiology of erection and ejaculation. J Sex Med 2011;8(Suppl. 4):310–5.

[33] Solomon H, Man JW, Jackson G. Erectile dysfunction and the cardiovascular patient: endothelial dysfunction is the common denominator. Heart 2003;89(3):251–3.

[34] Gareri P, Castagna A, Francomano D, Cerminara G, De Fazio P. Erectile dysfunction in the elderly: an old widespread issue with novel treatment perspectives. Int J Endocrinol 2014;2014:878670.

[35] Rao MN, Marmillot P, Gong M, Palmer DA, Seeff LB, Strader DB, et al. Light, but not heavy alcohol drinking, stimulates paraoxonase by upregulating liver mRNA in rats and humans. Metabolism 2003;52(10):1287–94.

[36] Cominacini L, Garbin U, Pasini AF, Davoli A, Campagnola M, Pastorino AM, et al. Oxidized low-density lipoprotein increases the production of intracellular reactive oxygen species in endothelial cells: inhibitory effect of lacidipine. J Hypertens 1998;16(12 Pt 2):1913–9.

[37] Schwedhelm C, Nimptsch K, Bub A, Pischon T, Linseisen J. Association between alcohol consumption and serum paraoxonase and arylesterase activities: a cross-sectional study within the Bavarian population. Br J Nutr 2016;115(4):730–6.

[38] Grover S, Mattoo SK, Pendharkar S, Kandappan V. Sexual dysfunction in patients with alcohol and opioid dependence. Indian J Psychol Med 2014;36(4):355–65.

[39] Pendharkar S, Mattoo SK, Grover S. Sexual dysfunctions in alcohol-dependent men: a study from north India. Indian J Med Res 2016;144:393–9.

[40] Arackal BS, Benegal V. Prevalence of sexual dysfunction in male subjects with alcohol dependence. Indian J Psychiatry 2007;49(2):109–12.

[41] Dick B, Reddy A, Gabrielson AT, Hellstrom WJ. Organic and psychogenic causes of sexual dysfunction in young men. Int J Med Rev 2017;4(4):102–11.

[42] Ylikahri RH, Huttunen MO, Harkonen M. Hormonal changes during alcohol intoxication and withdrawal. Pharmacol Biochem Behav 1980;13(Suppl. 1):131–7.

[43] Cicero TJ. Alcohol-induced deficits in the hypothalamic-pituitary-luteinizing hormone axis in the male. Alcohol Clin Exp Res 1982;6(2):207–15.

[44] Emanuele MA, Emanuele N. Alcohol and the male reproductive system. Alcohol Res Health 2001;25(4):282–7.

[45] Little PJ, Adams ML, Cicero TJ. Effects of alcohol on the hypothalamic-pituitary-gonadal axis in the developing male rat. J Pharmacol Exp Ther 1992;263(3):1056–61.

[46] Noth RH, Walter Jr. RM. The effects of alcohol on the endocrine system. Med Clin North Am 1984;68(1):133–46.

[47] Rachdaoui N, Sarkar DK. Pathophysiology of the effects of alcohol abuse on the endocrine system. Alcohol Res 2017;38(2):255–76.

[48] Purohit V. Can alcohol promote aromatization of androgens to estrogens? A review. Alcohol 2000;22(3):123–7.

[49] Abel EL. A review of alcohol's effects on sex and reproduction. Drug Alcohol Depend 1980;5(5):321–32.

[50] McCabe MP, Sharlip ID, Atalla E, Balon R, Fisher AD, Laumann E, et al. Definitions of sexual dysfunctions in women and men: a consensus statement from the Fourth International Consultation on Sexual Medicine 2015. J Sex Med 2016;13(2):135–43.

[51] Rosen RC, Riley A, Wagner G, Osterloh IH, Kirkpatrick J, Mishra A. The international index of erectile function (IIEF): a multidimensional scale for assessment of erectile dysfunction. Urology 1997;49(6):822–30.

[52] Rosen RC, Cappelleri JC, Smith MD, Lipsky J, Pena BM. Development and evaluation of an abridged, 5-item version of the International Index of Erectile Function (IIEF-5) as a diagnostic tool for erectile dysfunction. Int J Impot Res 1999;11(6):319–26.

[53] O'Leary MP, Fowler FJ, Lenderking WR, Barber B, Sagnier PP, Guess HA, et al. A brief male sexual function inventory for urology. Urology 1995;46(5):697–706.

[54] O'Donnell AB, Araujo AB, Goldstein I, McKinlay JB. The validity of a single-question self-report of erectile dysfunction. Results from the Massachusetts Male Aging Study. J Gen Intern Med 2005;20(6):515–9.

[55] Lee ACK, Ho LM, Yip AWC, Fan S, Lam TH. The effect of alcohol drinking on erectile dysfunction in Chinese men. Int J Impot Res 2010;22:272.

CHAPTER 6.4

Exercise, Sports, and Men's Health

Joshua T. Randolph*, Lindsey K. Burleson*, Alyssa Sheffield[†], Johanna L. Hannan*

*Department of Physiology, Brody School of Medicine, East Carolina University, Greenville, NC, United States, [†]Department of Public Health, Brody School of Medicine, East Carolina University, Greenville, NC, United States

INTRODUCTION

Erectile dysfunction (ED) is a common ailment among men, affecting as many as 10% of men by age 40, increasing to approximately 80% of men after age 70 [1]. ED is defined as "the inability to attain and/or maintain penile erection sufficient for satisfactory sexual performance" [2]. Studies have repeatedly demonstrated an intimate relationship between cardiovascular health and erectile function, suggesting that the diagnosis of ED may indicate the presence of underlying cardiovascular disease (CVD) or increased CVD risk [3]. Exercise has shown to be beneficial to both cardiovascular health and erectile function. This chapter will address evidence that increased physical activity benefits erectile function, examine if different types of exercise can recover ED, and explain how exercise impacts fertility and libido. Additionally, we will discuss the impact of specific sports on erectile function, and whether sexual activity affects athletic performance.

EXERCISE AND MEN'S HEALTH

Exercise Benefits Erectile Function

Many association studies have demonstrated that keeping a moderate diet and regular exercise habits might be associated with better erectile function. In a study of 180 Brazilian men aged 40–75 years, physical fitness and ED were estimated by max O_2 intake capacity during exercise and the International Index of Erectile Function (IIEF-5), respectively [4]. The researchers found that the odds of having ED were 10-fold higher in sedentary men than in physically active men. Additionally, men who were in "good" physical condition were almost five times less likely to have ED compared with men who were in only "tolerable" physical condition. Another association study looked at ED and physical activity in 510 diabetic and nondiabetic men of European and South Asian origin [5]. After controlling for age, diabetic status, and the use of antihypertensive drugs, the authors found that physical inactivity was associated with increased incidence of ED for European men, but not South Asian men. It was suggested that this observation may be the result of differences in metabolism between Caucasians and South Asians that engage in similar exercise intensities. An additional possible influence may have been that Asian individuals tend to be more disinclined to discuss matters of sexual dysfunction compared with other cultures. Another self-reported study of 460 Japanese men with type II diabetes found an independent inverse relationship between average intensity of regular exercise habits and ED [6]. Furthermore, in this population of Japanese type II diabetic men, there was an association between self-reported sitting time and ED indicating that sedentary men are likely to have ED [7].

A systematic review of 19 similar studies including men from 8 different countries came to several conclusions [8]. The first was that low activity or inactivity was associated with greater prevalence of ED. The second conclusion was that physical activity is protective of erectile function, since it occurs less in active individuals. Additionally, the study reiterated the previously mentioned finding that there is a negative correlation between intensity of physical activity and prevalence of ED. Only two out of the 19 studies included in the review did not find an association between self-reported physical activity habits and ED [8]. Taken together, these studies provide compelling evidence that a connection exists between physical inactivity and erectile dysfunction.

Can Erectile Function be Recovered With Exercise?

The next question is as follows: can implementation of a regular exercise regimen improve currently existing ED? A study from 2011 took 50 men with ED and put them on a Mediterranean diet and a moderate-intensity exercise routine for 3 months [9]. These men, whose average age was 57, were made to work out for 30 minutes daily for 5 days each week at 40%–60% of their max heart rate. Erectile function was assessed by IIEF-5 and by Doppler ultrasound of penile arterial blood flow. At the end of the program, both measures of erectile function were significantly improved in the exercise group compared with controls. Additionally, the exercise group had significantly reduced levels of circulating endothelial progenitor cells (EPCs) and endothelial microparticles (EMPs), both of which are markers of endothelial dysfunction. The authors suggested that the improvement of endothelial function, signified by reduced endothelial apoptosis, may serve as a mechanism by which erectile function is improved in the exercise group [9]. In fact, this interpretation is supported by a previously published review that summarized several mechanisms by which physical activity and diet have been implicated in improving ED, including reduction of inflammation, improved insulin sensitivity, decreased visceral adiposity, and improved endothelial function [10]. Another review summarized the importance of endothelial nitric oxide (NO) release in both ED and cardiovascular health, which is enhanced in the setting of exercise [11]. This adds yet another possible connection by which exercise may impact erectile function.

Although the previous study represents a single sample, there have been meta-analyses of several similar interventional experiments. One such study analyzed seven randomized controlled trials from 2004 to 2013, involving 478 total men with ED [12]. The interventions administered in these experiments included aerobic exercise, aerobic and resistance exercises, pelvic floor muscle-specific exercises, or exercise with pharmacological treatment. The analysis concluded a statistically significant 3.85 point average improvement in IIEF-5 scores in the intervention groups compared with controls across all trials [12]. A recent systematic review of 10 intervention studies between 2009 and 2016, involving men from six countries, reported 14%–86% improvement in erectile function in intervention groups that underwent continuous or interval-based aerobic training at moderate to vigorous intensity [13]. The review also concluded that resistance training can complement the erectile function benefits conferred by aerobic exercise. Finally, the authors recommended 40 minutes of supervised aerobic exercise of moderate intensity four times weekly for 6 months in men with arterial ED, especially for those individuals who are physically inactive, obese, and hypertensive or suffer from metabolic syndrome or cardiovascular disease [13].

Exercise and Erectile Function in Prostate Cancer Survivors

Although exercise clearly represents a viable treatment choice for men suffering from impotence due to underlying endothelial dysfunction or cardiovascular disease, exercise may not be effective for all ED etiologies. Prostate cancer is a very common disease that will affect approximately one in nine men, and among those who choose to undergo treatment, rates of posttherapeutic ED have been reported to be at least 60% at 2 years post intervention [14]. Clinicians and researchers have experimented with various methods of prophylaxis or treatment for ED in prostate cancer survivors, including nerve-sparing procedures, phosphodiesterase-5 inhibitor regimens, and exercise programs. In a trial from 2014 involving 50 men status-postradical retropubic prostatectomy, subjects were made to walk five times weekly for 30–45 minutes per day at 55%–100% peak O_2 consumption [15]. After 6 months, the exercise group demonstrated significantly better cardiovascular function than the control group as measured by brachial artery flow-mediated dilation (FMD) and peak

O_2 consumption, but there was no difference in self-reported erectile dysfunction [15]. Another study in 2015 enrolled 85 men that had undergone prostatectomy just 6–12 weeks prior and started them on a program mandating moderate-intensity exercise once weekly for 60 minutes [16]. Again, researchers reported an improvement in cardiovascular health over control subjects, measured by peak O_2 consumption, but no difference in erectile function. Although it is fair to point out the small sample size in both of these studies, it is important to point out the distinction in the arteriopathic ED that tends to occur as men age and cardiovascular health declines and the neurogenic ED that occurs as a result of nerve trauma brought about during prostate cancer treatment. This is one reason why PDE-5 inhibitors are largely ineffective at restoring erectile function following treatment, since they target the vascular mechanisms of the erectile response and not the neural mechanisms. Similarly, if exercise is meant to improve erectile function by reversing endothelial/cardiovascular dysfunction, then physical activity may not have a significant impact on the neurogenic ED that is seen following prostate cancer treatment.

Pelvic Floor Muscle Rehabilitation and Men's Sexual Health

Pelvic floor exercises, such as Kegel exercises, are commonly performed in women to strengthen their pelvic floor to improve continence and sexual function. Recent studies suggest that pelvic floor exercises can also benefit erectile function. Ischiocavernosus muscles surrounding the crus and bulb of the penis are contracted during the rigidity phase of an erection and contribute to increased intracavernous pressure [17]. One study examined 122 men with symptoms of ED that included difficulty to maintain an erection or inability to penetrate who underwent 20 sessions of 30 minutes of voluntary contractions coupled with ischiocavernosus muscle electric stimulation [18]. Indirect measurements of intracavernous pressure and ischiocavernosus muscle force were increased; however, this study did not assess if sexual function improved for these men. Another study examined 16 patients with persistent ED 1 year after radical prostatectomy who received 3 months of individual pelvic floor muscle training with a therapist [19]. Patients receiving pelvic floor muscle training had a better recovery of erectile function compared with patients without pelvic floor therapy as measured by the IIEF-EF questionnaire. Authors also speculated that combination of pelvic floor muscle training and PDE5-I may further improve erectile function. These studies are carried out in a very small number of patients and, although promising, need to be repeated in a much larger cohort.

Exercise and Fertility in Men

There is strong evidence of the benefit of exercise on sexual function in men; however, its effects on fertility and reproduction have been less researched.

There is evidence that endurance and resistance training can have both positive and negative effects on hormonal profiles and sperm morphology in men.

Exercise can be an aggravating factor for existing fertility problems, while conversely, it may benefit the hormonal and semen profiles of sedentary men. Endurance aerobic training can acutely increase androgenic hormones, and men that practice mild to moderate aerobic exercise will have higher baseline levels of free testosterone, dihydrotestosterone, and sex hormone-binding globulin [20]. Long-term, intense endurance training can lead to chronic suppression of both gonadotropin-releasing hormone (GnRH) and serum testosterone. Another study has found that ultraendurance athletes who run or cycle more than 10–20 hours a week can experience hypogonadal state [21]. A hypogonadal state in men can alter certain masculine characteristics and lower sex drive.

Resistance training has been difficult to research, given the frequent parallel use of anabolic steroids. Some studies have suggested that resistance training provokes an increase in free and total testosterone levels after one training session, although it varies with intensity, volume, recovery, age, experience, performance level, and amount of muscle mass involved [22]. Frequent or excessive resistance training can also have the opposite effect and lead to decreased testosterone levels due to persistent fatigue from overtraining [21, 23]. When comparing aerobic exercise with resistance exercise in patients with ED, researchers concluded that increasing cardiorespiratory fitness and lowering body fat percentage by aerobic exercise will superiorly raise serum testosterone levels compared with resistance training [24].

Long-term exhaustive exercise can negatively affect sperm quality and reproductive potential [25]. The difficulty in measuring endurance exercise is that high-performance athletes have been training for many years and it is difficult to establish a threshold that induces sperm changes—especially in men that have been training since prepubescence. Several studies list a threshold of running 100 km per week to induce changes in sperm density, motility, and morphology [26]. Motility impairment may inhibit the sperm's ability to penetrate cervical mucus and fertilize an egg. There is a high inverse correlation between sperm morphology and weekly cycling volumes greater than 300 km per week [27]. Sperm morphology is a high predictor of fertility rates; therefore, men who practice high-volume cycling may be at higher risk for developing infertility. It's not only sperm that can be injured with overtraining but also semen collected from men who enlist in a 16-week intensive cycling programs contained inflammatory interleukins (IL) such as IL-β, IL-6, IL-8, and tumor necrosis factor alpha (TNF-α) [28]. It should be noted that these studies that investigate the effects of cycling on fertility factors are unable to distinguish the difference between the cardiovascular effects and mechanical effects of the outcomes. Similar to excessive exercise being linked to low libido and ED, long-term high-intensity exercise will also impair male fertility [21].

Exercise and Libido in Men

Exercise has been shown in animal studies to increase libido and improve sexual performance in rats. Rats with established baseline activity that were subjected to 1 hour of swimming had an increased intromission rate and reduced mount latency when compared with sedentary controls, indicating both improved sexual desire and performance [29]. The swimming rats also had significantly higher levels of serum testosterone. In a prospective study, men were followed over 9 months and kept a daily diary over the first and last month of the study recording exercise, diet, and smoking habits. Men who exercised for 1 hour a day, 3 days a week, reported significantly better sexual function, measured in frequency; reliability of erections; and satisfying orgasms [30].

Similar to the trends seen in fertility studies, strenuous exercise can also be linked to lower libido in men. A survey study conducted in approximately 300 running clubs found that high levels of chronic endurance running were correlated with a low libido state [31]. The mechanism of how chronic high-intensity exercise affects libido is unclear, but a combination of lower testosterone levels and physical exhaustion plays a role.

SPORTS AND MEN'S HEALTH

Sports and Genitourinary Trauma

Although male genital trauma from sports is infrequent, the impact of genitourinary tract traumatic sport injuries on ED has been well studied. These sport injuries typically occur from collisions, kicks, falls, straddle injuries, or equipment malfunctions and involve scrotal injuries more frequently than penile structures [32]. Persec et al. note that up to 80% of penile trauma is caused by blunt mechanisms, often resulting in penile fracture with rupture of the cavernosal tunica albuginea and associated injuries to the corpus spongiosum and urethra [33]. Complications of such trauma include ED, urethral stricture, and coital difficulty, often the result of distortion of the vascular anatomy. The incidence of ED after penile injury is 0.6%–9% in the literature, and in this particular study, only one patient (4.5%) did not report complete erectile function recovery by 12 months post injury [33].

For patients sustaining trauma resulting in pelvic fracture, the incidence of erectile dysfunction ranges from 11% to 24%, with a higher rate suspected for those with associated urethral injuries [33]. The mechanism of ED in these patients is thought to be due to neurogenic etiology, although other mechanisms include strictures due to arterial or venous insufficiency. Chung et al. performed a retrospective review of 50 males treated for pelvic fractures at a level 1 trauma center over an 11-year period, utilizing the International Index of Erectile Function (IIEF-5) questionnaire [34]. The authors concluded that ED was more severe in patients who sustained pelvic fracture with urethral injury. While the urethral injury was not the cause of ED, it was likely a surrogate for more extensive neurovascular injury.

Continued Controversy—Does Cycling Lead to ED?

There has been extensive controversy in the literature regarding the association between cycling and ED since the 1990s. Michiels et al. performed a large systematic review to further examine this relationship and found that to date, there is no evidence to indicate that the prevalence of ED is higher in cyclists than the general population [35]. A cross-sectional population study by Hollingworth et al. included 5282 male cyclists in the United Kingdom and found that there was no relationship between time cycling and ED or infertility [36]. They concluded that there was no correlation between cycling hours per week or years of cycling and ED or infertility. Another study surveyed 3932 low or high cyclists, runners, and swimmers for sexual function using the Sexual Health Inventory for Men (SHIM) questionnaire [37]. Multivariate analysis indicated that runners/swimmers had significantly lower SHIM scores than low- and high-intensity cyclists. Overall, these data indicate that cyclists have no higher risk for ED.

While it is hypothesized that cycling generates excessive perineal pressure and compression of the pudendal neurovascular bundle leading to ED, there have been no studies to date that demonstrate such causality [35]. The biggest contributors to increased perineal pressure are the bicycle seat or saddle, the position of the saddle, what it is made of, the cycling position, and the workload. Studies have shown that there is less pressure created by using modified seat without the narrow nose at the end of the seat and a wider seat [38]. The seats with the perineal hole in the center do not appear to lessen pressure but to increase pressure [39]. Additionally, the forward lean position of the rider is believed to be a main contributing factor to increased perineal pressure [40]. Overall, the literature seems to support the warning of Sommer et al. in which patients with genital numbness are a risk for the development of ED, even though genital numbness may occur unrelated to ED [40]. To date, there is not enough evidence to support a causal relationship between cycling and ED.

Sex Before Sports—Does It Impact Athletic Performance?

Since ancient Greece and Rome, athletes have been advised to abstain from sexual intercourse prior to important athletic events. Claims have been made that abstaining from sex before sports will increase aggression, boost energy, and testosterone levels. However, to date, there have been very few studies examining the effect of sexual activity on athletic performance. A systematic review performed by Stefani et al. revealed that sexual activity the day before competition generally does not negatively influence sports performance but that any association between sexual activity and negative lifestyle choices such as smoking or alcohol abuse may promote worse performance [41]. One study examined the effect of sexual activity in 15 young male high-level athletes on their performance on a graded cycle ergometer stress test and arithmetic mental concentration test while exercising and examined testosterone levels and cardiac activity via ECG [42]. Sexual activity did not impact the maximum workload, mental concentration, or testosterone levels of the athletes. However, 2 hours post sexual activity, stress test posteffort heart rate values were higher suggesting decreased heart rate recovery capacity if intercourse took place within 2 hours of competition. Another study examined whether sexual intercourse altered lower-extremity muscle force in 12 physically active men [43]. Knee flexion or extension muscle force was measured before or 12 hours after sexual intercourse, and no difference was seen in lower-extremity muscle force generation. As a result, the authors concluded that restriction of sexual activity prior to athletic activities requiring high lower-extremity muscle force is not necessary [43]. These studies are all very small sample sizes and needed to be repeated on a larger scale to confirm that sexual intercourse will not impact athletic performance.

OVERALL CONCLUSIONS

Overall, exercise is beneficial to men's sexual health, erectile function, libido, and fertility. Many studies recommend that men undergo 40 minutes of moderate aerobic exercise four times weekly in order to improve or maintain erectile function. It is beneficial to have this exercise supervised by a personal trainer or with a friend to ensure that exercise is being performed at an appropriate intensity. Chronic high-intensity endurance exercise can have detrimental effects on sex hormone levels, fertility, and libido. Sport injury-related ED seems to be more closely associated with extensive disruption of the physical mechanisms and/or associated perineum anatomy but is often without permanent negative effects. While perineal numbness can be an issue with cycling, the studies evaluating cycling-induced ED are hindered by small sample sizes and results that have not been reproducible. Thus, evidence indicates that erectile function can be preserved or improved with moderate-intensity exercise on a regular basis.

References

[1] O'Leary MP, Rhodes T, Girman CJ, Jacobson DJ, Roberts RO, Lieber MM, Jacobsen SJ. Distribution of the Brief Male Sexual Inventory in community men. Int J Impot Res 2003;15(3):185–91.

[2] NIH Consensus Conference on Impotence. NIH Consensus Development Panel on impotence. JAMA 1993;270:83–90.

[3] Nehra A, Jackson G, Miner M, Billups KL, Burnett AL, Buvat J, Carson CC, Cunningham GR, Ganz P, Goldstein I, Guay AT, Hackett G, Kloner RA, Kostis J, Montorsi P, Ramsey M, Rosen R, Sadovsky R, Seftel AD, Shabsigh R, Vlachopoulos C, Wu FCW. The Princeton III Consensus recommendations for the management of erectile dysfunction and cardiovascular disease. Mayo Clin Proc 2012;87(8):766–78.

[4] Agostini LC, Netto JM, Miranda Jr MV, Figueiredo AA. Erectile dysfunction association with physical activity level and physical fitness in men aged 40-75 years. Int J Impot Res 2011;23(3):115–21.

[5] Malavige LS, Wijesekara P, Ranasinghe P, Levy JC. The association between physical activity and sexual dysfunction in patients with diabetes mellitus of European and South Asian origin: The Oxford Sexual Dysfunction Study. Eur J Med Res 2015;20:90.

[6] Minami H, Furukawa S, Sakai T, Niiya T, Miyaoka H, Miyake T, Yamamoto S, Kanzaki S, Maruyama K, Tanaka K, Ueda T, Senba H, Torisu M, Tanigawa T, Matsuura B, Hiasa Y, Miyake Y. Physical activity and prevalence of erectile dysfunction in Japanese patients with type 2 diabetes mellitus: The Dogo Study. J Diabetes Investig 2018;9(1):193–8.

[7] Furukawa S, Sakai T, Niiya T, Miyaoka H, Miyake T, Yamamoto S, Kanzaki S, Maruyama K, Tanaka K, Ueda T, Senba H, Torisu M, Minami H, Tanigawa T, Matsuura B, Hiasa Y, Miyake Y. Self-reported sitting time and prevalence of erectile dysfunction in Japanese patients with type 2 diabetes mellitus: The Dogo Study. J Diabetes Complicat 2017;31(1):53–7.

[8] Araujo C, Souza M, Fernandes A, Pelegrini A, Andrade A, Coutinho de Azevedo Guimarães A. Physical activity and erectile dysfunction: a systematic review. Rev Bras Ativ Fis Saúde 2015;20(1):3–16.

[9] La Vignera S, Condorelli R, Vicari E, D'Agata R, Calogero A. Aerobic physical activity improves endothelial function in the middle-aged patients with erectile dysfunction. Aging Male 2011;14(4):265–72.

[10] Hannan JL, Maio MT, Komolova M, Adams MA. Beneficial impact of exercise and obesity interventions on erectile function and its risk factors. J Sex Med 2009;6(Suppl. 3):254–61.

[11] Meldrum DR, Gambone JC, Morris MA, Meldrum DA, Esposito K, Ignarro LJ. The link between erectile and cardiovascular health: the canary in the coal mine. Am J Cardiol 2011;108(4):599–606.

[12] Silva AB, Sousa N, Azevedo LF, Martins C. Physical activity and exercise for erectile dysfunction: systematic review and meta-analysis. Br J Sports Med 2017;51(19):1419–24.

[13] Gerbild H, Larsen CM, Graugaard C, Areskoug Josefsson K. Physical activity to improve erectile function: a systematic review of intervention studies. Sex Med 2018;6(2):75–89.

[14] Alemozaffar M, Regan MM, Cooperberg MR, Wei JT, Michalski JM, Sandler HM, Hembroff L, Sadetsky N, Saigal CS, Litwin MS, Klein E, Kibel AS, Hamstra DA, Pisters LL, Kuban DA, Kaplan ID, Wood DP, Ciezki J, Dunn RL, Carroll PR, Sanda MG. Prediction of erectile function following treatment for prostate cancer. JAMA 2011;306(11):1205–14.

[15] Jones LW, Hornsby WE, Freedland SJ, Lane A, West MJ, Moul JW, Ferrandino MN, Allen JD, Kenjale AA, Thomas SM, Herndon II JE, Koontz BF, Chan JM, Khouri MG, Douglas PS,

[15] Eves ND. Effects of nonlinear aerobic training on erectile dysfunction and cardiovascular function following radical prostatectomy for clinically localized prostate cancer. Eur Urol 2014;65 (5):852–5.

[16] Zopf EM, Bloch W, Machtens S, Zumbé J, Rübben H, Marschner S, Kleinhorst C, Schulte-Frei B, Herich L, Felsch M, Predel HG, Braun M, Baumann FT. Effects of a 15-month supervised exercise program on physical and psychological outcomes in prostate cancer patients following prostatectomy: The ProRehab Study. Integr Cancer Ther 2015;14(5):409–18.

[17] Cohen D, Gonzalez J, Goldstein I. The role of pelvic floor muscles in male sexual dysfunction and pelvic pain. Sex Med Rev 2016;4(1):53–62.

[18] Lavoisier P, Roy P, Dantony E, Watrelot A, Ruggeri J, Dumoulin S. Pelvic-floor muscle rehabilitation in erectile dysfunction and premature ejaculation. Phys Ther 2014;94(12):1731–43.

[19] Geraerts I, Van Poppel H, Devoogdt N, De Groef A, Fieuws S, Van Kampen M. Pelvic floor muscle training for erectile dysfunction and climacturia 1 year after nerve sparing radical prostatectomy: a randomized controlled trial. Int J Impot Res 2016;28(1):9–13.

[20] Safarinejad MR, Azma K, Kolahi AA. The effects of intensive, long-term treadmill running on reproductive hormones, hypothalamus-pituitary-testis axis, and semen quality: a randomized controlled study. J Endocrinol 2009;200(3):259–71.

[21] Vaamonde D, Garcia-Manso JM, Hackney AC. Impact of physical activity and exercise on male reproductive potential: a new assessment questionnaire. Rev Andal Med Deport 2017;10 (2):79–93.

[22] Maïmoun L, Lumbroso S, Manetta J, Paris F, Leroux JL, Sultan C. Testosterone is significantly reduced in endurance athletes without impact on bone mineral density. Horm Res 2003;59 (6):285–92.

[23] Cadore EL, Lhullier FL, Brentano MA, da Silva EM, Ambrosini MB, Spinelli R, et al. Hormonal responses to resistance exercise in long-term trained and untrained middle-aged men. J Strength Cond Res 2008;22(5):1617–24.

[24] Yeo JK, Cho SI, Park SG, Jo S, Ha JK, Lee JW, Cho SY, Park MG. Which exercise is better for increasing serum testosterone levels in patients with erectile dysfunction? World J Mens Health 2018;36(2):147–52.

[25] Vaamonde D, Da Silva-Grigoletto ME, García-Manso JM, Vaamonde-Lemos R, Swanson RJ, Oehninger SC. Response of semen parameters to three training modalities. Fertil Steril 2009;92(6):1941–6.

[26] Arce JC, De Souza MJ, Pescatello LS, Luciano AA. Subclinical alterations in hormone and semen profile in athletes. Fertil Steril 1993;59(2):398–404.

[27] Jóźków P, Rossato M. The impact of intense exercise on semen quality. Am J Mens Health 2017;11(3):654–62.

[28] Hajizadeh Maleki B, Tartibian B. Long-term low-to-intensive cycling training: impact on semen parameters and seminal cytokines. Clin J Sport Med 2015;25(6):535–40.

[29] Allouh MZ. Effects of swimming activity on the copulatory behavior of sexually active male rats. Int J Impot Res 2015;27(3):113–7.

[30] Simon RM, Howard L, Zapata D, Frank J, Freedland SJ, Vidal AC. The association of exercise with both erectile and sexual function in black and white men. J Sex Med 2015;12 (5):1202–10.

[31] Hackney AC, Lane AR, Register-Mihalik J, O'leary CB. Endurance exercise training and male sexual libido. Med Sci Sports Exerc 2017;49(7):1383–8.

[32] Hunter SR, Lishnak TS, Powers AM, Lisle DK. Male genital trauma in sports. Clin Sports Med 2013;32(2):247–54.

[33] Persec Z, Persec J, Puskar D, Sovic T, Hrgovic Z, Fassbender WJ. Penile injury and effect on male sexual function. Andrologia 2011;43(3):213–6.

[34] Chung PH, Gehring C, Firoozabadi R, Voelzke BB. Risk stratification for erectile dysfunction after pelvic fracture urethral injuries. Urology 2018;115:174–8.

[35] Michiels M, Van der Aa F. Bicycle riding and the bedroom: can riding a bicycle cause erectile dysfunction? Urology 2015;85:725–30.

[36] Hollingworth M, Harper A, Hamer M. An observational study of erectile dysfunction, infertility, and prostate cancer in regular cyclists: cycling for health UK study. J Men's Health 2014;11(2).

[37] Awad MA, Gaither TW, Murphy GP, Chumnarnsongkhroh T, Metzler I, Sanford T, Sutcliffe S, Eisenberg ML, Carroll PR, Osterberg EC, Breyer BN. Cycling, and male sexual and urinary function: results from a large, multinational, cross-sectional study. J Urol 2018;199(3):798–804.

[38] Spears IR, Cummins NK, Brenchley Z, Donohue C, Turnbull C, Burton S, Macho GA. The effects of saddle design on stresses in the perineum during cycling. Med Sci Sports Exerc 2003;35:1620–5.

[39] Carpes FP, Dagnese F, Kleinpaul JF, Martins Ede A, Mota CB. Effects of workload on seat pressure while cycling with two different saddles. J Sex Med 2009;6(10):2728–35.

[40] Sommer F, Goldstein I, Korda JB. Bicycle riding and erectile dysfunction: a review. J Sex Med 2010;7(7):2346–58.

[41] Stefani L, Galanti G, Padulo J, Bragazzi NL, Maffulli N. Sexual activity before sports competition: a systematic review. Front Physiol 2016;7:246.

[42] Sztajzel J, Périat M, Marti V, Krall P, Rutishauser W. Effect of sexual activity on cycle ergometer stress test parameters, on plasmatic testosterone levels and on concentration capacity. A study in high-level male athletes performed in the laboratory. J Sports Med Phys Fitness 2000;40(3):233–9.

[43] Valenti LM, Suchil C, Beltran G, Rogers RC, Massey EA, Astorino TA. Effect of sexual intercourse on lower extremity muscle force in strength-trained men. J Sex Med 2018;15:888–93.

PART 7

External factors

CHAPTER 7.1

Environmental Toxins and Men's Health

J. Marinaro, C. Tanrikut
Department of Urology, Georgetown University School of Medicine, Washington, DC, United States

The general population is exposed to a variety of potential toxins through daily activities: the food we eat, pollution in our living and working environments, and consumer products we use for convenience. As society develops at an ever-increasing pace, humans encounter new and potentially harmful environmental exposures. These exposures may contribute to global decreases in semen quality reported over the past several decades [1]. More specifically, several studies have demonstrated significant decreases in mean sperm count, concentration, motility, and normal morphology over the last half-century [2–5]. In addition, environmental toxicants with endocrine-disrupting activity may further impact male development, endocrine profile, and risk for malignancy. This chapter reviews several environmental factors that have been implicated in negatively effecting male fertility and discusses the role of endocrine-disrupting chemicals (EDCs) on men's health.

ENVIRONMENTAL TOXINS AND MEN'S HEALTH

Air Pollution

It has been well established that air pollution is detrimental to human health [6]. While air pollution is best known for being harmful to the cardiovascular and respiratory systems, it has also been implicated in affecting male fertility. To fully understand the literature regarding the impact of air pollution on men's health, it is important to understand how air pollution exposure is reported. Air pollution is described with respect to particulate matter (PM) size, typically either $PM_{2.5}$ (particles <2.5 μm) or PM_{10} (particles <10 μm). While particles smaller than 10 μm are able to enter the respiratory tract, the proportion of coarse (2.5–10 μm) and fine particles (<2.5 μm) varies substantially around the world, likely leading to some variation in health effects [7]. While coarse particles are typically produced by mechanical processes (e.g., construction, road dust, and wind), fine particles are typically produced by combustion sources (e.g., automobiles) [7]. However, both coarse and fine particles have demonstrated detrimental effects on human health, even at concentrations below current air quality standards [7]. As noted in the 2005 World Health Organization guidelines, both $PM_{2.5}$ and PM_{10} particulates have been found to impact rates of cardiovascular disease, respiratory disease, and lung cancer at concentrations 50% lower than the acceptable legal limits in many developed countries [6, 7].

Looking specifically at increased risk of death from long-term exposure to these smaller $PM_{2.5}$ particles, the findings are impressive. In a systematic review of European and American studies from 1950 to 2007, long-term exposure to air pollution was found to increase the risk of all-cause mortality by 6% for each increase of 10 $\mu g/m^3$ in atmospheric $PM_{2.5}$ [8]. More specifically, long-term exposure to $PM_{2.5}$ particles was associated with a large increase in mortality from both lung cancer (15%–21% increase per 10 $\mu g/m^3$ increase in atmospheric $PM_{2.5}$) and cardiovascular events (12%–14% increase per 10 $\mu g/m^3$ increase in atmospheric $PM_{2.5}$). Pope et al. reported similar increases in mortality: in this study of 500,000 adults living in urban areas throughout the United States, each 10 $\mu g/m^3$ increase in atmospheric $PM_{2.5}$ was associated with an approximately 4% increase in all-cause mortality, 6% increase in cardiopulmonary mortality, and 8% increase in lung cancer mortality [9].

Even long-term exposure to $PM_{2.5}$ particles below the recommended limits has been demonstrated to impact mortality rates. A recent study by Di et al. estimated the risk of all-cause mortality associated with long-term exposures to $PM_{2.5}$ concentrations, including those below the current annual National Ambient Air Quality Standard of 12 $\mu g/m^3$. This group found that a 10 $\mu g/m^3$ increase in $PM_{2.5}$ concentrations was associated with a 7.3% increase in all-cause mortality [10]. These findings corroborate other data that have been previously reported. However, when looking specifically at people with exposure to $PM_{2.5}$ of less than the recommended 12 $\mu g/m^3$, the same increase of 10 $\mu g/m^3$ in $PM_{2.5}$ was found to carry with it an even greater risk of all-cause mortality (13.6%). Additionally, subgroup analysis demonstrated that *men* had a higher estimated risk of death from any cause in association with $PM_{2.5}$ exposure, compared with the general population [10].

Interestingly, short-term exposure to $PM_{2.5}$ also has been shown to carry with it an increase in mortality. In one study of nine counties in California, an increase in $PM_{2.5}$ concentration of 10 $\mu g/m^3$ over only 2 days demonstrated a 0.6% increase in all-cause mortality rate [11].

In addition to negatively impacting cardiopulmonary conditions and mortality, air pollution has also been found to impair male fertility. In several animal studies, air pollution was found to be detrimental to both sperm quality and testicular structure. With respect to sperm quality, studies have shown both a decrease in the production of spermatozoa [12–14] and an increase in abnormal sperm morphology [15] among mice and rats exposed to car exhaust. This decrease in sperm production was found regardless of age at time of exposure: rats exposed to exhaust starting at birth [12] and mice exposed to exhaust starting at 6 weeks [14] were both found to have decreased sperm production. More specifically, those animals with exposure starting at 6 weeks were found to have a dose-dependent decrease in sperm production with increasing diesel exhaust exposure [14].

Even prenatal exposure to exhaust has proved to be detrimental. In animal models, prenatal exposure to diesel exhaust has been found to decrease sperm production and raise follicle-stimulating hormone (FSH) levels in male offspring [13, 16]. Remarkably, prenatal exposure to only ambient levels of $PM_{2.5}$ in some geographic locations has been found to impair several stages of spermatogenesis and decrease both testicular weight and volume, emphasizing the potent reproductive toxicity of such air pollutants [17].

Of the sperm that are produced by rodents with exposure to air pollution, there is an increase in morphological abnormalities [15] and DNA alterations. In one study of mice exposed to ambient levels of air pollution in an industrial area in the Great Lakes region, paternal germ line mutations were 1.6 times more common in the exposed group versus unexposed controls; however, there was no difference in maternal mutation rates [18]. Additionally, Yauk et al. observed a significant increase in sperm DNA breakage and sperm DNA hypermethylation in mice exposed to ambient air pollution in an industrial Canadian city [19]. Germ line changes were again demonstrated in this study, confirming the findings of Somers et al. It is hypothesized that the epigenetic changes caused by air pollutants may lead to germ line mutations. If true, this theory suggests that air pollution may have both durable and intergenerational health effects.

At the testicular level, as previously mentioned, prenatal exposure to air pollution is associated with decreased testicular weight and volume [17]. From a histopathologic perspective, there is evidence of both structural changes in Leydig cells [14] and a decrease in the number of Sertoli cells [13] in rats exposed to air pollutants.

The accumulated data regarding the effect of air pollution on male rodent development and spermatogenesis are significant and alarming. Interpreting the body of literature available on humans is somewhat more difficult, given that most studies are retrospective and not always comparable. Often, these studies differ in terms of study populations, type of exposure, and duration of exposure. However, despite these investigative challenges, most studies have confirmed that there are alternations in sperm quality with exposure to air pollution. Abnormalities in human sperm motility [20–23] and morphology

[23, 24] have been consistently reported. Abnormalities in sperm counts, on the other hand, have been variable. A number of studies have found no significant change in sperm count or concentration with exposure to ambient air pollution [23, 25, 26], although there is some evidence that increased exposure to car exhaust through occupational exposure can reduce sperm counts. In a study performed by Guven et al., 38 men working as highway toll collectors were compared with 35 men working in an office setting. Semen analyses demonstrated significantly lower sperm counts ($P=.002$), decreased motility ($P=.003$), and increased abnormal morphology ($P=.001$), among the toll collectors compared with controls [21].

As seen in animal models, exposure to air pollution can increase the percentage of spermatozoa with DNA abnormalities in men. In two studies of men living in an industrialized district in the Czech Republic, exposure to ambient levels of air pollutants was associated with an increased percentage of sperm with abnormal chromatin [23, 26].

Overall, the individual role that specific air pollutants play in the alteration of semen parameters is difficult to isolate, and the pathophysiology responsible for these changes in semen quality remains under investigation. Hormonal disturbances, oxidative stress induction, cellular DNA alterations, and epigenetic alterations are four proposed mechanisms, likely working in combination, to explain these negative changes [27]. However, further investigation is needed to elucidate the specific mechanisms of impact. Given the fact that air pollution is a global problem with wide-reaching health impacts, public awareness and governmental attempts to limit pollution may be prudent.

Water Pollution

Humans can be exposed to water pollutants in a variety of ways, including ingestion, handwashing, bathing, and boiling. Often, these pollutants include disinfection by-products (DBPs) as well as pharmaceuticals, pesticides, and other chemicals [1]. These substances and their effects on men's health are further discussed below.

Disinfection Byproducts

While water disinfection has served as an important public health advance, many of the disinfectants used—such as chlorine, ozone, chlorine dioxide, and chloramines—are powerful oxidants. When these disinfecting agents interact with naturally occurring substances in drinking water (i.e., organic matter, bromides, and iodides), they form disinfection by-products (DBPs) [28].

Scientists first became aware of DBPs in the early 1970s. Since then, more than 600 different types of DBPs have been identified in chlorinated drinking water, with the most common types being trihalomethanes and haloacetic acids [28].

In animal models, several of these DBPs have been found to exhibit testicular toxicity and impair sperm quality. In one study by Linder et al., daily exposure of male rats to dichloroacetic acid produced alterations in spermiation, sperm morphology, and sperm motility [29]. Similarly, exposure of rats to dibromoacetic acid also reduced epididymal sperm motility and sperm counts [30] and increased abnormal sperm morphology [31]. Dibromoacetic acid has also been associated with delayed pubertal development, decreased testicular and epididymal weights, and atrophic seminiferous tubules [31, 32].

In humans, where exposures to DBPs are typically lower and less regimented than in animal models, the data are mixed. In one study of over 2000 Chinese men, urinary trichloroacetic acid (TCAA) levels were measured as a marker of exposure to DBPs in drinking water. Increased urinary TCAA concentrations were negatively associated with several semen quality parameters, including sperm concentration and count, as well as motility [33]. However, another study of 418 men did not confirm these findings [34]. In contrast, this smaller study suggested that elevated urinary TCAA concentrations were associated only with decreased sperm motility.

Several other studies have used serum trihalomethane (THM) levels as a marker of exposure to drinking water DBPs and again found mixed results in regard to male fertility. In one study of over 400 Chinese men, Zeng et al. observed a suggestive dose-response relationship between elevated serum THM concentration and decreased sperm concentration [35]. Similarly, in a prospective study of 324 Chinese men living in the same water supply district, THM exposure via ingestion of contaminated tap water was associated with decreased sperm concentration and count [36]. However, several other studies have not validated this relationship between THM exposure and semen parameters. For example, in one study of 157 men, total THM levels were not associated with any changes in semen quality [37]. This finding was reproduced in a larger study of over 1500 men from 13 fertility clinics across the United Kingdom, which again found no evidence that THM exposure had any impact on semen quality [38]. Finally, one study of 228 US men did not identify a significant negative relationship between DBP exposure and semen parameters; interestingly, however, subjects in the top quartile of exposure to THMs actually had higher sperm counts and better morphology than those in the bottom two quartiles [39]. Given the conflicting evidence, the relationship between DBP exposure and human male fertility remains unclear.

Other Contaminants
In addition to DBPs, there is an increasing presence of detectable levels of pharmaceuticals, pharmaceutical by-products, pesticides, and other chemicals in the water supply [1]. Even after water treatment, finished drinking water has

been found to contain detectable levels of prescription and nonprescription drugs, fragrance compounds, flame retardants and plasticizers, cosmetic compounds, and solvents [40]. Most of these compounds do not have established drinking water standards or health advisories, making the potential health risks associated with exposure unknown [40]. However, many of these chemicals have been associated with endocrine-disrupting activity, which will be discussed in detail later in the chapter. Further research is needed to elucidate the health effects of chronic, low-dose ingestion of these chemicals and the possible cumulative or synergistic effects that exposure to multiple compounds may play in human health.

Heat

While mild elevations in ambient temperature are largely inconsequential for other aspects of men's health, such temperature changes can have significant adverse effects on spermatogenesis. For spermatogenesis to occur properly, testicular temperatures must be maintained 2–4°C cooler than core body temperature [41]. Fortunately, there are several mechanisms that permit the testes to remain significantly cooler than body temperature. For example, the scrotal sac itself aids thermoregulation of the testis in several ways, including the thin skin with minimal subcutaneous fat, dense sweat glands, and scant hair distribution [42]. Additionally, the cremaster muscle that surrounds the testes and the dartos muscle that lies beneath the scrotal skin both have the capacity to relax. This relaxation allows the testes to hang away from the abdomen, for the scrotal skin to slacken and for the total surface area available for heat dissipation to increase in order to maintain thermal homeostasis [42].

Additionally, the testes are thermoregulated by the countercurrent heat exchange function of the blood vessels in the spermatic cord [42, 43]. In normal testicular anatomy, the internal spermatic artery is surrounded by several internal spermatic veins, which form the pampiniform plexus. This arterial and venous juxtaposition promotes efficient heat exchange and thus allows cooling of the arterial blood before it reaches the testes [42]. In the presence of a varicocele, this method of countercurrent heat exchange is compromised, as will be discussed later in the chapter.

When the testes are exposed to elevated temperatures, spermatogenesis is inhibited in a number of ways. One of the most significant consequences of heat stress on the testes is the loss of germ cells via apoptosis [44]. Even just short exposure of the testes to elevated temperatures has been found to induce apoptosis. In several rodent models, exposure of the scrotum to a hot water bath of 43°C for just 15–20 minutes resulted in marked activation of germ cell apoptosis [45, 46]. This cell death can be so significant that it subsequently causes a measurable decrease in testicular weight [47].

The precise mechanisms at play in this programmed cell death remain under investigation. However, evidence has shown that spermatocyte apoptosis may be facilitated by a group of proteins called the "heat-shock proteins" (HSPs) [48]. There are two types of HSP: constitutive and inducible [42]. Under normal circumstances, constitutively produced HSPs act as a quality-control mechanism. Essentially, they are molecular chaperones that ensure that other polypeptides are assembled and transported correctly [42]. However, when elevated temperatures cause increased cellular stress and protein denaturation, these altered proteins activate transcription factors that, in turn, initiate the transcription of the other inducible HSPs [49]. One such transcription factor is heat-shock factor 1 (HSF1). HSF1 appears to have two paradoxical but unique roles in male germ cells. First, it promotes apoptotic cell death of spermatocytes in response to thermal stress [48]. However, it also acts as a cell survival factor for more immature germ cells when the testes are exposed to high temperatures [48]. Ultimately, it has been postulated that HSF1 serves to both eliminate injured spermatocytes and preserve undifferentiated germ cells that maintain the potential to develop normally after the period of thermal stress has subsided [48].

In addition to causing apoptosis, thermal stress also induces sperm DNA damage [44]. In fact, just a single, mild, transient scrotal heat stress in male mice has been found to induce DNA strand breaks and defects in DNA synapses in spermatocytes [50]. It is suggested that heat stress induces this DNA damage in germ cells by increasing the number of reactive oxygen species (ROS) present and suppressing gene expression associated with defense mechanisms against DNA damage [44].

As demonstrated in these animal models, heat poses a threat to spermatogenesis in a number of ways. In humans, there are both endogenous and exogenous sources of heat that have been demonstrated to impair male fertility.

Endogenous Sources of Testicular Heating
In regard to endogenous sources, the two most common pathologies associated with testicular heat exposure include cryptorchidism and varicocele [41]. However, fever and obesity can also contribute to testicular warming and have a negative impact on spermatogenesis.

Cryptorchidism
Cryptorchidism is the most frequent congenital abnormality found in newborn boys and affects about 2%–4% of all full-term male births [51]. In cryptorchidism, one or both testes may be retained within the abdominal cavity or inguinal canal, rather than descending normally into the scrotum. This can be due to a variety of factors, and in most cases, the etiology for cryptorchidism remains unknown [51]. Regardless of the cause, the consequences of cryptorchidism are well established and include both reduced fertility and increased risk of testicular cancer [41].

The main reason for reduced fertility in males with cryptorchidism is the thought to be related to elevated testicular temperatures [51]. At the histological level, these cryptorchid testes appear to have a loss of germ cells, delayed maturation of germ cells, and progressive interstitial fibrosis, similar to testes treated with experimental hyperthermia [52]. In males with cryptorchidism, these effects are likely progressive. For example, it has been reported that newborns with intra-abdominal testes have a normal number of germ cells at birth [53]. These intra-abdominal testes remain histologically normal for up to 6 months, but by 6–8 months of age, they show a sharp decline in germ cell development and maturation [51]. By the time of puberty, 69% of males with cryptorchidism have a complete absence of germ cells [54]. Of the 31% with evidence of germ cells, 2% demonstrate mature germ cells in low numbers (hypospermatogenesis), and none exhibit normal spermatogenesis [54]. While the severity of infertility secondary to cryptorchidism is multifactorial—including whether one or both testes fail to descend, the position of the testes, the cause of the incomplete descent, and the length of time before surgical correction—it is generally accepted that early surgical correction may help to preserve fertility outcomes for these patients [51]. Therefore, it is recommended that males undergo orchiopexy before 12 months of age to minimize the progression of these histological changes [52, 55].

With respect to malignancy, cryptorchidism is an established risk factor for developing testicular cancer, with the reported relative risk ranging between 3.7 and 7.5 [52]. The risk has been shown to increase with increasing age at correction, with the highest risk being among those men whose cryptorchid testes were never corrected [56]. While the risk of developing testicular tumors is not eliminated following orchiopexy [57], several groups have shown decreased risk of malignancy after surgical correction. Halme et al. found that orchiopexy significantly decreased the risk of seminoma development, which is the most common type of malignant tumor associated with cryptorchidism [58]. Interestingly, Herrington et al. found that men with a history of cryptorchidism who either had spontaneous testicular descent or successful orchiopexy before age 11 did not have any increased risk of testicular cancer compared with controls [59]. Therefore, early orchiopexy remains the standard or care for cryptorchidism, along with appropriate counseling regarding the risk of malignancy and long-term follow-up [60].

Varicocele

Varicocele is the most common identifiable, surgically correctible condition associated with male infertility [61]. While varicoceles are found in approximately 15% of the general population [62], they are found in up to 40% of men with primary infertility and 80% of men with secondary infertility [63]. Varicoceles are dilated veins in the pampiniform plexus of the spermatic cord that alter the efficiency of heat transfer within the testes [61]. As previously mentioned, the testis is thermoregulated in part by a countercurrent heat exchange method. In normal testicular anatomy, several internal spermatic veins surround the internal spermatic artery. This arterial and venous juxtaposition promotes efficient heat exchange and ensures that the arterial blood that reaches the testes is relatively cooler than core body temperature [42]. In the case of a varicocele, this method of countercurrent heat exchange is compromised. In animal models, the presence of a varicocele results in an approximately 1°C increase in intratesticular temperature [64, 65]. In human studies, this temperature increase has been found to be even more pronounced. In a study by Goldstein and Eid, the intratesticular temperature of testes with varicoceles was 2.43–2.72°C higher than controls [66]. This is well within the range of the 2–4°C increase in testicular temperature at which spermatogenesis becomes impaired.

As one would expect given the thermal dysregulation associated with varicoceles, they are associated with impaired sperm counts, motility, and morphology and with testicular atrophy [61]. There is some evidence that this is a progressive process. As Chehval and Purcell demonstrated, patients with varicoceles have a progressive decline in semen parameters over time, including sperm concentration and motility [67]. Similarly, in a study by Lipshultz and Corriere, it was noted that sperm concentration and testicular volume are decreased in older patients with uncorrected varicoceles [68].

Given the association between varicoceles and male-factor infertility, the American Society for Reproductive Medicine (ASRM) practice guidelines indicate that varicoceles should be treated when most or all of the following criteria are met: (1) The varicocele is palpable on physical examination; (2) the couple has known infertility; (3) the female partner has normal fertility or a potentially treatable cause of infertility, and the time to conception is not a concern; and (4) the male partner has abnormal semen parameters [69]. The goal of varicocele repair is to eliminate the pooling and reflux of warm venous blood into the scrotum while preserving the other key structures in the spermatic cord, including the internal spermatic artery, lymphatic vessels, and vas deferens. The two possible approaches for varicocele treatment include surgical varicocelectomy and percutaneous embolization. The microsurgical varicocelectomy approach is generally preferred, given its increased cost-effectiveness, lower morbidity, and lower rates of recurrence compared with the other techniques [70, 71].

Improvement in semen parameters after varicocele repair is somewhat difficult to measure, given that there are no standard definitions for what exactly constitutes improvement. However, most studies have observed relative improvements in semen parameters and fertility after varicocele repair [69]. For example, in a meta-analysis of infertile men who underwent varicocele repair, sperm concentration increased by an average of 12 million sperm/mL, while motility increased by an average of 11% [72]. Similarly, in one randomized controlled trial of men 20–39 years of age with palpable varicoceles, all semen parameters studied improved significantly within 12 months of microsurgical varicocelectomy, including sperm concentration, progressive motility, and normal morphology [73]. The spontaneous pregnancy rate was also found to increase from 13.9% in the untreated group to 32.9% in the treated group [73]. This increase in pregnancy rate after varicocelectomy repair has been replicated in several other studies. Remarkably, in one randomized controlled trial, the natural pregnancy rate rose to nearly 60% among treated patients within 1 year of surgery, while it remained at only 10% for untreated patients [74].

In addition to negatively impacting semen parameters, varicoceles have also been consistently associated with decreased testosterone production. In one study of 325 men with varicoceles, this cohort was found to have a significantly lower mean testosterone level compared with controls (416 vs 469 ng/dL, respectively) [75]. Even controlling for age, this difference in testosterone level remained significant. After microsurgical varicocelectomy, however, testicular function improved. Of the 200 men in the varicocele group who had documented pre- and postoperative testosterone levels, average serum testosterone level increased from 358 to 454 ng/dL within 3–12 months postoperatively.

This improvement in serum testosterone level after varicocelectomy has been replicated in several other studies. In one study by Hsiao et al., 272 men with a clinically palpable varicocele underwent microsurgical varicocelectomy by a single surgeon. Postoperatively, these men were found to have significant increases in sperm concentration, total sperm count, and serum testosterone level regardless of age [76]. Similarly, in a smaller study by Su et al., 53 patients with varicoceles presenting to a single infertility center underwent microsurgical inguinal varicocelectomy. Postoperatively, these patients were found to have a significant increase in serum testosterone level (319 ± 12 ng/dL preoperatively vs 409 ± 23 ng/dL postoperatively, $P < .0004$) [77]. As anticipated, these patients also experienced a significant increase in sperm concentration and motility. Interestingly, however, there was no correlation between magnitude of change in serum testosterone level and improvement in semen parameters following surgery, suggesting that improvement in testicular hormonal function may be independent of improvements in spermatogenesis.

Ultimately, the evidence strongly suggests that varicoceles play an important role in male infertility and hormone production. While the surgical treatment options available have demonstrated encouraging results in improving semen parameters, pregnancy rates, and testosterone levels, further research is needed to fully understand the mechanisms behind these improvements in testicular function.

Other Endogenous Heat Sources

Aside from cryptorchidism and varicocele, other endogenous sources of testicular warming include fevers and obesity.

Fevers that last 1 day or longer can have a negative impact on spermatogenesis [42]. Depending on the stage of spermatogenesis affected, fevers can lead to decreases in sperm concentration of up to 35%, decreases in normal sperm morphology of up to 7.4%, and increases in the number of immotile sperm of up to 20% [78]. In addition to altering these typical semen parameters, high fevers can also damage and denature the DNA contained within the sperm that are produced [79].

Similarly, obesity has also been implicated in causing elevated testicular temperatures. In one study by Garolla et al., there was a significantly greater 24 hour mean scrotal temperature among 20 obese men (35.38°C) compared with controls (34.73°C) [80]. These obese men also had lower sperm counts, decreased motility, and decreased normal morphology compared with the control group [80]. The measured elevated temperatures and concordant compromised semen parameters may be related to the deposition of fat around the spermatic cord and compromised thermoregulation [81]. In one study, removing some of this excess scrotal and suprapubic fat demonstrated improvement in semen parameters, including sperm count and motility; unfortunately, testicular temperature was not measured in this study, so the results remain inconclusive [82].

Exogenous Sources of Testicular Heating

While there are many possible exogenous sources of testicular heating, several sources that have been studied include clothing, sitting, laptops, occupational exposures, and saunas.

Generally, scrotal temperature is lower on an unclothed, upright body [42]. Clothing contributes to an approximately 1.2–1.5°C increase in scrotal temperature [83]. As one might expect, in some studies, tighter-fitting underwear has been associated with a relatively greater increase in scrotal temperature compared with loose underwear or no underwear [84].

This association between type of underwear and testicular heating may have wider impacts on testicular function and fertility. In one recent study of 656 men presenting to a single fertility center, type of underwear was significantly associated with alterations in several semen parameters. Men who reported primarily wearing boxer shorts (loose-fitting underwear) had a 25% higher sperm concentration, 17% higher sperm count, and 33% higher total motile sperm count compared with those who wore tighter-fitting underwear [85]. Additionally, men who wore boxer shorts were found to have 14% lower serum FSH levels; there was no difference in other reproductive hormones or sperm

DNA integrity parameters. Interestingly, the differences in sperm concentration and total sperm count according to type of underwear became insignificant when models were adjusted for serum FSH level. This suggests that the testicular injury caused by tight underwear may lead to a compensatory increase in gonadotropin secretion, which may in turn lead to impaired fertility outcomes. However, these results have not been reliably reproduced: other studies have shown that there is no significant difference in either scrotal temperature or semen parameters related to type of underwear worn [86]. Ultimately, the effect of clothing on scrotal temperature and fertility remains unclear and the subject of further investigation.

Prolonged sitting has also been associated with increased testicular temperatures. For example, men who spent at least 2 hours per day driving were found to have a significant increase in scrotal temperature of 1.7–2.2°C [87]. Similar results were seen among men who reported a predominantly sedentary position at work [88]. These sedentary men were found to have significantly higher scrotal temperatures than controls, with a clear dose-dependent relationship between scrotal temperature and time spent at work [88]. When examining the relationship between semen quality and sedentary work position, sedentary workers were noted to have relatively lower sperm concentrations; however, these results were not statistically significant [89]. Like clothing, the effect of a sedentary position on fertility outcomes remains unclear.

While sitting may cause elevations in scrotal temperature, sitting with a laptop computer may elevate scrotal temperatures even more. In one study by Sheynkin et al., sitting for 60 minutes increased scrotal temperature by 2.1°C. However, sitting while using a laptop computer on the lap for 60 minutes increased scrotal temperature by 2.6–2.8°C: a statistically significant increase ($P < .0001$) [90]. Again, however, the effect of this transient increase in scrotal temperature on fertility outcomes remains uncertain.

While a sedentary position can perhaps be considered an occupational hazard for those with office jobs, there are a number of other occupations that are associated with intense heat and subsequent testicular hyperthermia. For example, welders are exposed to both high levels of heat and toxic metals and fumes. In looking at welders with only heat exposure, groin temperatures increased by an average of 1.4°C during periods of radiant heat exposure. After 6 weeks of exposure, sperm count, sperm concentration, and the number of motile sperm were reduced, but not significantly. The proportion of sperm with normal morphology, however, did decrease significantly ($P < .01$) [91]. Similar to welders, bakers and ceramic oven operators are also subject to occupational heat exposure. Both of these groups have been found to have longer times to pregnancy compared with controls [92], further suggesting that occupational heat exposure may be a risk factor for male subfertility.

Finally, sauna exposure has also been studied as a potential risk factor for male infertility. The temperature of the scrotum has been found to increase significantly both during sauna exposure (2.4°C) and after exposure (1.7°C) [93]. This change in temperature is likely the cause of the alterations in semen parameters seen among men with sauna exposure. In one study by Garolla et al., men who underwent two 15-minute sauna sessions per week for 3 months had a significant decrease in both sperm count and motility ($P < .001$) [94]. Even just one sauna exposure of 85°C for 20 minutes has been shown to decrease sperm count [95]. Both of these changes in semen parameters were ultimately reversible, but it took up to 6 months in the group with repeat exposure and up to 5 weeks in the group with one-time exposure to return to baseline [94, 95].

Radiation

The two primary men's health concerns primarily associated with radiation exposure include reduced fertility and increased risk of malignancy.

Alterations in Spermatogenesis

Ionizing radiation has been well established as a cause of cellular damage. Specifically, radiation is known to induce a broad spectrum of DNA lesions, which subsequently trigger cellular pathways leading to apoptosis or carcinogenesis [96]. In males, radiation is also well known to impair spermatogenesis.

There are several mechanisms for this alteration in sperm production. First, as previously mentioned, radiation directly induces DNA damage. Once this damage is detected by the body, it can trigger cell-signaling cascades that induce spermatogonial apoptosis [96] and oxidizing events [97]. These oxidizing events increase the total number of reactive oxygen species (ROS) present within the cell, which in turn can further precipitate DNA damage, trigger cellular apoptosis, and result in abnormal or inhibited spermatogenesis [96]. Finally, there is also evidence that cells that have not been directly exposed to radiation can experience adverse effects. Through radiation-induced bystander signaling, factors released by irradiated cells can interact with unexposed cells and activate the same deleterious pathways as those in the exposed cells [96]. Given the fact that radiation can have adverse effects on both exposed and unexposed cells alike, one would expect that even low-dose radiation exposures would have a negative impact on spermatogenesis.

As several early studies demonstrated, testicular irradiation causes a significant, dose-related decrease in sperm production [98]. In these studies, healthy subjects from the prison population were exposed to between 0 and 600 rad of testicular irradiation [98]; the greater the radiation exposure, the more significant the decrease in sperm production. Sperm production was notably suppressed at dosages as low as 15 rad and sometimes completely inhibited at dosages of

50 rad [98]. While sperm production eventually recovered in most patients, at least one patient exposed to 600 rad became azoospermic and did not recover within the 2 years of follow-up, suggesting that this damage can be both severe and permanent [98].

Outside of the experimental laboratory setting, men exposed to large-scale nuclear accidents have also demonstrated significant impairment in spermatogenesis. In a small study of men exposed to ionizing radiation both during and after the Chernobyl power plant accident, 50% were found to be either azoospermic or oligospermatic [99]. Even men only involved in the Chernobyl cleanup efforts were found to have abnormalities in semen parameters. In a study of 18 male liquidators who lived in a nonradioactive area, there was no significant difference in sperm concentration, viability, or morphology compared with controls. However, there was a significant decrease in sperm motility and a significant increase in amorphous head shapes when these sperm were examined under electron microscopy [100], suggesting that even low-dose exposure can have a potentially negative impact on sperm quality.

Other studies have confirmed that this low-dose radiation exposure can have deleterious effects on spermatogenesis. In one study of health-care workers in China, those exposed to low-dose ionizing radiation were found to have significantly lower sperm motility, lower normal morphology, and higher DNA fragmentation index compared with controls [101]. While there was no change in sperm concentration, these findings suggest a negative effect on sperm quality.

Similarly, elevated levels of natural background radiation have also been found to increase the number of random microdeletions in the azoospermia factor a, b, and c regions on the long arm of the Y chromosome [102]. While semen parameters were not reported in this study, these findings again suggest that even low-dose radiation may affect sperm production and quality.

Low-dose electromagnetic field (EMF) radiation exposure has also been controversially linked to alterations in spermatogenesis. Of particular interest is the EMF radiation exposure transmitted by cellular phones. Ultimately, the evidence has been inconclusive regarding potential impacts on male fertility potential. In one short-term study of rats exposed to mobile phone radiation for either 30 or 60 minutes per day for 3 months, there were no abnormal findings for the group with only 30 minutes of daily exposure; however, among the rats with 60 minutes of daily exposure, 18.75% were found to have hypospermatogenesis, and 18.75% were found to have maturation arrest [103]. When cell phone exposure was extended to 3 hours per day for 4 months in a mouse model, more significant changes in semen parameters were noted. These exposed mice were found to have decreased sperm count, decreased sperm viability, increased germ cell apoptosis, decreased seminal testosterone, and decreased seminiferous tubule diameter compared with controls [104]. However, these results were not replicated in a more long-term study. Among rats exposed to cell phone radiation for 3 hours per day for 1 year, there were no differences in sperm concentration, motility, or seminiferous tubule diameter compared with controls [105].

In humans, the data have been similarly unconvincing. In four recent meta-analyses of cell phone usage and semen quality, two studies found significant deleterious effects on semen parameters [106, 107]: One was equivocal [108], and one found significant decreases in sperm counts in rats but not humans [109]. Of the two studies that found deleterious effects on human semen parameters, sperm concentration, motility, and viability were all significantly decreased [106, 107]. A greater percentage of abnormal morphology was noted in one analysis [106], but not the other [107]. In the analysis with more equivocal outcomes, exposure to cell phones was found to have an inconsistent effect on sperm concentration; however, sperm motility and viability were consistently decreased [108]. Finally, in the last meta-analysis, there was consistent evidence that mobile phone exposure decreased sperm concentration and motility in animal models [109]. However, when looking at human studies, this decrease in semen parameters was not present [109]. When human sperm were studied in vitro, though, radio-frequency radiation was indeed found to have a detrimental effect on sperm motility and viability [109].

Collectively, these studies suggest that sperm motility and viability may be sensitive to mobile phone exposure; however, the results are variable and inconsistent. Of note, the majority of these meta-analyses contained a significant amount of heterogeneity. This high degree of heterogeneity and relatively low number of studies make it difficult to assess publication bias [108]. Further research will be needed to more definitively assess the impact of mobile phones on semen parameters and men's health.

Risk of Testicular Cancer
Radiation is well known to induce a broad spectrum of DNA lesions, leading to an activation of cellular pathways inducing to apoptosis or carcinogenesis [96]. The risk of solid organ malignancy increases in a dose-dependent fashion after radiation exposure of 100 rad or higher [110]. This radiation-related cancer risk varies between organs and tissues. For example, in a study of atomic bomb survivors, only the bladder, female breast, lung, brain, thyroid, colon, esophagus, ovary, stomach, liver, and skin (excluding melanoma) were found to have an increased risk of malignancy; other organs did not demonstrate a similar risk [110].

In looking specifically at the risk of ionizing radiation on testicular cancer, most studies do not suggest that ionizing radiation increases the risk of testicular cancer [111]. In one meta-analysis, there was no relationship between occupational ionizing radiation exposure and testicular cancer [112]. Similarly, in a study of over 90,000 US radiation technologists, there was no significant increase in testicular cancer with occupational radiation exposure; in these exposed males, only the risks of melanoma and thyroid cancer were increased [113].

In regard to the relationship between nonionizing radiation and the risk of testicular cancer, the data are less clear. In the same meta-analysis by Yousif et al., six of nine studies analyzed demonstrated an increased risk of testicular cancer with EMF exposure [112]. However, even in one of the largest studies with this conclusion, a dose-response relationship could not be established [114], suggesting that further research is needed to assess the significance of EMF exposure on testicular malignancy.

Overall, even though the incidence of testicular cancer has increased in the past several decades, this increase is unlikely to be attributable to ionizing radiation exposure; the effect of nonionizing radiation exposure remains unclear and worthy of further study [111].

Hypoxia

Spermatocytes are known to be quite sensitive to hypoxia. Perhaps, the most severe cause of testicular hypoxia is testicular torsion. Torsion is considered to be a surgical emergency, since ischemic injury of as few as 4 hours can threaten the viability of the affected testis [115]. In animal models, even just 1 hour of experimental torsion and reperfusion was found to cause apoptosis of germ cells [116]. In humans, unilateral testicular torsion has been found to seriously interfere with spermatogenesis in about 50% of patients and produce borderline impairment in another 20% [115]. In one study of patients with a history of unilateral torsion, 39% had sperm counts less than 20 million/mL, and about two-thirds had abnormal sperm motility or morphology [117]. Overall, total motile sperm count was found to be significantly correlated with duration of torsion; however, serum levels of luteinizing hormone (LH), follicle-stimulating hormone (FSH), and testosterone remained unchanged [117]. It is thought that this posttorsion, germ cell-specific apoptosis is likely induced by reactive oxidative species [116]; however, methods for reducing this oxidative injury remain under investigation.

Though much less severe than torsion, altitude-associated hypobaric hypoxemia has also been noted to impair spermatogenesis. More specifically, it has been shown to cause alterations in testicular morphology, the loss of germ cells, and increased metabolic stress in spermatogenic cells [118]. In rat models, animals raised in hypobaric chambers simulating conditions at approximately

5000 m were found to have significant germ cell apoptosis and sloughing of seminiferous tubule epithelium after only 30 days of exposure [119]. Similarly, Farias et al. demonstrated that male rats exposed to a simulated 4600 m altitude demonstrated decreased testicular mass, decreased seminiferous tubule epithelium height, and the loss of spermatogenic cells in all stages of the spermatogenic cycle [118]. In a follow-up study, this group also found that, during times of hypobaric hypoxemia, the testes underwent increased vascularization. This increase in warm blood flow subsequently increased testicular temperature by approximately 1.5°C [120], suggesting that both hyperthermia and increased oxidative stress may contribute to reduced spermatogenesis.

In humans, similar impairments in spermatogenesis have been noted. Men with only about 4 weeks of high-altitude exposure have been shown to have significantly decreased sperm counts [121–123], decreased sperm motility [121, 123], and increased abnormal morphology [121–123]. However, these changes in semen parameters were found to be reversible and returned to normal values within 6 months [123] to 2 years [122] of returning to sea level. These men were also found to have slightly decreased testosterone levels, but again, these normalized with time [122]. These changes in testicular function appear to be only evident with acute exposure. Among the human populations that live at high altitudes permanently, there is no significant change in fertility rates [124]. This may suggest that there is some compensatory mechanism at play, the details of which have not yet been elucidated.

Finally, hypoxia has been implicated in the impaired spermatogenesis associated with cigarette smoking, iron deficiency, and varicocele [1], which are further discussed in other sections of this book.

Physical and Mental Stress

Occupational stress and exertion have also been implicated in altering semen parameters. In one prospective study of men attempting to conceive, work-related heavy exertion was consistently associated with lower sperm concentration and count [125]. Similarly, in one study of 202 consecutive male patients presenting to an infertility clinic, having a physically demanding job in either industry or construction was found to be an independent risk factor for male infertility [126].

Similarly, having a mentally demanding job may also be to be a risk factor for infertility. In one study by Sheiner et al., 106 men presenting to a single fertility center because of a male-factor fertility problem were found to have significantly greater job burnout, tension, listlessness, and cognitive weariness compared with controls [126]. These results are similar to that of El-Helaly et al., who found that work-related stress was significantly greater among men presenting to an infertility clinic compared with controls [127]. However, it is unclear in this population whether infertility is a consequence of stress or perhaps a contributor to it. Further studies are needed to assess the contributions of both mental and occupational stresses to fertility outcomes.

In addition to strenuous occupational activity, vigorous exercise has also been shown to negatively affect semen parameters. In one study of 286 subjects assigned to moderate-intensity versus high-intensity exercise regimens, those in the high-intensity group were found to have significantly decreased sperm concentration, motility, and morphology at 24 weeks [128]. These semen parameters continued to decline even more significantly at 36 weeks [128]. Similarly, Tartibian and Maleki found that elite athletes had significantly lower semen volume, sperm counts, motility, and percentage of normal sperm morphology compared with recreationally active men [129].

However, these results have not been universally reproduced. In fact, some studies have found that a higher level of physical activity is associated with increased sperm count and concentration [130]. Generally, the collective body of literature suggests that moderate recreational exercise is associated with either positive or neutral effects on semen parameters; however, professional athletes should be aware that intense training has been implicated in reducing sperm concentration, motility, and morphology [131].

ENDOCRINE DISRUPTORS AND MEN'S HEALTH

What Are Endocrine Disrupting Chemicals?

An EDC is defined by the US Environmental Protection Agency as "an exogenous agent that interferes with the production, release, transport, metabolism, binding action, or elimination of natural hormones in the body responsible for the maintenance of homeostasis and the regulation of developmental processes" [132]. Put more simply, EDCs are chemicals, or chemical mixtures, that interfere with normal hormone function. These compounds are pervasive, found in the air, water, and soil around us [133]. This ubiquity is concerning, given that exposure at nearly any point in life may have a negative impact on a man's health.

Types of Endocrine Disrupting Compounds

The list of potential endocrine-disrupting chemicals is extensive. According to The Endocrine Disruption Exchange—a nonprofit research institute that aims to make evidence-based data on endocrine-disrupting chemicals readily available—there are over 1450 different chemicals that have been identified as potential endocrine disruptors [134]. While it is beyond the scope of this chapter to review all 1450 different chemicals, several high-priority compounds will be discussed in greater detail.

Heavy Metal Exposure
Several heavy metals have been reported to have endocrine-disrupting activity, particularly cadmium and lead.

Cadmium
Cadmium is a metal that is typically found in low concentrations in nature. In humans, the two most significant sources of cadmium exposure are diet and smoking [135]. Human activities—such as combustion of fossil fuels, industrial emissions, and application of fertilizers and sewage sludge to farmlands—lead to cadmium soil contamination and subsequent uptake by crops and livestock [135, 136]. This cadmium is passed up the food chain and eventually accumulates in the tissues of humans who consume these crops and meats.

In terms to cadmium exposure via smoking, one cigarette contains about 1–2 μg of cadmium [135]. About 10% of this cadmium is inhaled during smoking, and about half of the cadmium inhaled is subsequently absorbed by the body [135]. Industrial workers who work as smelters, in pigment plants or in battery factories, have also historically had high occupational exposures to cadmium; however, in many countries, improvements in industrial hygiene have improved workplace air quality and reduced this risk [135].

Regardless of the mechanism by which cadmium exposure occurs, this heavy metal has the capacity to accumulate within the body and cause a number of adverse effects. It is a known carcinogen [137] and a reproductive toxicant. In animal models, cadmium induces significant edema and hemorrhage in the testes within 24 hours of exposure [138]. Over time, this leads to a dose-dependent decrease in sperm concentrations and serum testosterone levels [139]. At very high levels of exposure, the weights of the testes, seminal vesicles, and epididymides are reduced by nearly 50% [139]. These findings were reproduced by El-Ashmawy and Youssef, who found that administering a single dose of cadmium to rats decreased sperm concentration; sperm motility; and the weights of the testes, epididymides, and accessory sex organs within 2 months [140]. Histologically, these testes were found to have degenerated seminiferous tubules, abnormal Leydig cells, and significant fibrosis in addition to reduced size [138].

Although these animal models clearly demonstrate a negative relationship between cadmium exposure and reproductive parameters, it is important to remember that most of these studies follow the acute effects of a single, high bolus dose. This is likely not the main mechanism of exposure in most humans.

In human studies, the effect of low-dose cadmium exposure remains unclear. Several studies have demonstrated a significant negative correlation between cadmium exposure and semen parameters, such as sperm concentration [141–143] and motility [141, 142]. Among male infertility patients, both blood and seminal levels of cadmium have been found to be elevated compared with sperm donors or the general population [144]. However, other studies did not identify any impact of cadmium on semen parameters [145, 146].

Though the overall effect of cadmium on semen parameters remains debatable, it is suspected that cadmium acts as a reproductive toxicant both via direct injury to cells and by indirectly disturbing the hypothalamus-pituitary-testis axis [147]. In regard to direct injury, cadmium has been shown to induce oxidative DNA damage in human spermatozoa [143]. However, it also accumulates in the hypothalamus, where it alters neurotransmission and subsequent gonadotropin secretion [148]. Cadmium has also been found to have androgenic effects through its high-affinity binding to the androgen receptor and activation of downstream signaling. Cadmium both mimics the actions of androgens on cell growth and gene expression in vitro and prevents other androgens from binding to their receptor [149].

Lead

Environmental exposure to lead can derive from a variety of sources, including paint, brass water fixtures, ceramic coatings, and smoking [1]. In animal studies, lead exposure has been found to decrease sperm density, decrease sperm activity, and increase sperm malformation. Histologically, lead disrupts germ cell alignment, reduces the number of germ cell layers, and decreases the number of spermatozoa in the testes [150].

In humans, occupational lead exposure has also been associated with impaired spermatogenesis. In one study of 119 men employed by a lead smelting plant, blood lead level was negatively correlated with sperm count and concentration. Those men with blood lead level concentrations of 40 μg/dL or more had a particularly increased risk of below-normal sperm concentration and count [151]. Similarly, in a study of industrial workers in Croatia, lead exposure was associated with decreased sperm concentration, count, and motility and increased abnormal morphology [152].

In the general population, the relationship between seminal lead levels and standard semen parameters is unclear. In one study, only moderate levels of seminal lead were found to be significantly associated with decreased sperm concentration and motility [141]. However, this finding has not been universally reproduced. Another study of 56 nonsmokers found no significant difference between seminal lead concentration and semen quality [143]. In a study of couples undergoing in vitro fertilization (IVF), elevated seminal lead levels were insignificantly associated with decreased semen parameters (concentration, motility, and normal morphology) but significantly associated with reduced fertilization [153]. This suggests that seminal lead levels may have a role in fertility outcomes, even without significantly altering semen parameters.

Lead exposure has also been associated with erectile dysfunction (ED). In one study of 34 men with ED in Egypt, men with ED had a significantly higher blood lead level compared with a control group [154]. There was also a significant positive correlation between blood lead level and cavernous tissue lead level, with deposits of gray lead granules being seen within cavernosal tissue histologically [154]. Another study of 28 men with a history of chronic lead exposure was also found to have a statistically significant relationship between elevated blood lead levels and worsening erectile dysfunction [155].

Phytoestrogens

Genistein and daidzein are plant-derived phytoestrogens that have the potential to act as endocrine disruptors in several ways. First, in utero exposure to genistein has been found to contribute to the development of hypospadias. In one animal study, pregnant mice supplemented with genistein gave birth to offspring with significantly altered gene expression, particularly those genes and signaling pathways implicated in urethral development [156]. By altering these pathways, genistein appears to be intimately involved with organ morphogenesis and apoptosis, ultimately contributing to hypospadias.

Phytoestrogens have also been found to alter semen parameters and sex hormone levels. In one study of rabbits fed a soybean meal isoflavone-containing diet, sperm concentration and testosterone concentrations were significantly decreased compared with controls [157]. Similarly, in a study of juvenile rats treated with daidzein, those who received medium (20 mg/kg) or high (100 mg/kg) doses of daidzein had lower testosterone levels and fewer erections [158]. Even in rats with only prenatal exposure to genistein, these animals had a significant decrease in testicular weight and a dose-dependent decrease in sperm production and plasma testosterone levels [159].

In humans, the data on phytoestrogen exposure and semen parameters have been inconsistent. In one study of 99 male partners of subfertile couples presenting to a single fertility center, there was a significant negative association between soy food intake and sperm concentration. Strikingly, in multivariate analyses, men with the highest soy intake had 41 million sperm per milliliter less than men who did not consume any soy foods [160]. However, these results have not been consistently reproduced. In one study of healthy volunteers, daily supplementation with isoflavones for 2 months had no effect on semen parameters or sex hormones [161]. Even more surprisingly, one small study by Song et al. found that higher levels of isoflavone consumption were actually associated with better semen parameters, including higher sperm count, better motility, and lower sperm DNA fragmentation index [162]. Ultimately, further research is needed to fully understand the impact of phytoestrogens on semen parameters and overall fertility outcomes.

Synthetic Chemicals and Solvents
Diethylstilbestrol (DES)

Diethylstilbestrol (DES) is an orally active synthetic estrogen that historically was thought to prevent complications of pregnancy. Between the late 1940s and early 1970s, it was prescribed to 2–3 million pregnant women in the United States before ultimately being banned [163]. Like many endocrine-disrupting chemicals, the negative effects of DES vary by developmental stage of exposure and by dose. In utero exposure has been associated with several different types of structural abnormalities in boys and men, such as varicoceles, epididymal cysts, cryptorchidism, and testicular hypoplasia [164–167]. DES exposure has also been linked to several abnormalities in the penile urethra, including hypospadias, urethral stenosis, and microphallus [166, 168, 169].

While these structural abnormalities are a well-established consequence of in utero DES exposure, there has been some debate regarding whether this exposure increases the risk of testicular cancer. In one animal study, mice exposed to DES in utero were found to have an increased rate of testicular tumors compared with controls [170]. Though several tumors were benign, the majority of the lesions were malignant interstitial cell tumors [170]. In humans, however, the data have been less clear. In one study of 3613 men with prenatal DES exposure, the relative rates for developing testicular cancer were 3.05 times higher than those of unexposed men [171], yet this increase was compatible with a chance observation and remains in question. Similarly, in another study of over 250 men with a history of DES exposure, there was no increased risk of testicular cancer compared with controls [172].

Neonatal exposure to DES has also been associated with a number of negative effects. In one study of rats that were administered DES between the ages of 2 and 12 days, these animals were found to have decreased numbers of Sertoli cells and suppressed testicular growth. They were also found to have an increased proportion of apoptotic germ cells, decreased sperm production, and an increased number of Sertoli-cell-only tubules, suggesting that DES exposure has an effect on spermatogenesis. Additionally, these animals were found to have significantly lower FSH levels and lower inhibin B levels. With all of these findings combined, there is evidence that DES has a significant effect in directly modulating Sertoli cell development [173]. Similarly, another study of neonatal male mice exposed to DES found that several of these animals had Sertoli-cell-only tubules and significant inflammation in the epididymis and testis. It is thought that epididymal inflammation may be a consequence of

DES triggering inflammatory pathways via estrogen or testosterone receptors; however, further study is needed to fully elucidate these pathways [174].

Additionally, as a reproductive toxicant with endocrine activity, DES can alter gene expression in target tissues and has the potential for transgenerational effects. Animal studies have shown that either prenatal or neonatal exposure to DES increases the susceptibility for tumors in male and female descendants [175].

Clearly, DES is a potent endocrine-disrupting chemical associated with significant adverse effects on male development at multiple stages in life.

Phthalates
Phthalates and phthalate esters are a large group of compounds used as liquid plasticizers. They are found in a wide range of products, including plastics, coatings, cosmetics, personal care products, and toys [176]. Because they are not chemically bound to plastic, they can leach into and contaminate the surrounding environment [176]. Since they are commonly used in food and water containers, the majority of human exposure comes from dietary contamination, although exposure can also occur via dermal contact or inhalation [1]. Exposure is nearly ubiquitous. In one study of 2540 participants in the National Health and Nutrition Examination Survey, metabolites of phthalates were found in nearly every urine sample collected: Monoethyl phthalate (MEP), monobutyl phthalate (MBP), and monobenzyl phthalate (MBzP) were detected in >97% of the samples, and mono-(2-ethylhexyl) phthalate (MEHP) was detected in >75% of the samples [177].

Phthalates are known developmental and reproductive toxicants in animal models. In utero exposure of males to phthalates produces a syndrome of effects on the reproductive tract that parallels testicular dysgenesis syndrome. More specifically, phthalate-induced changes in male offspring include fetal Leydig cell aggregation, gonocyte multinucleation, focal testis dysgenesis, a decrease in INSL3 levels (a peptide hormone crucial for normal testicular descent), and a decrease in testosterone production, as well as gene expression for androgen biosynthesis [178]. It has been suggested that phthalates cause this syndrome by acting as antiandrogens, reducing tissue testosterone levels and testosterone production during a critical period of sexual differentiation in the male rat fetus [179].

This in utero exposure can have significant impacts on spermatogenesis and fertility into adulthood. One proposed mechanism suggests that fetal Leydig cell aggregation induces structural changes in the testis that inhibit spermatogenesis. In one study of rats exposed to dibutyl phthalate (DBP), testosterone levels and Leydig cell size were significantly reduced [180]. Additionally, abnormal aggregates of Leydig cells were found to trap Sertoli cells within them, forming misshapen, dysgenic seminiferous tubules. While these dysgenic tubules initially contained germ cells, by adulthood, germ cells were absent, and only Sertoli cells were identified, implying that the process of Leydig cell aggregation may interfere with normal spermatogenesis [180].

Phthalate exposure in adult rats has also been found to impair spermatogenesis. In one study of adult male rats treated with oral dibutyl phthalate (DBP), serum FSH levels, testosterone levels, testicular weights, and sperm counts all decreased in a dose-dependent manner. These rats also had evidence of increased oxidative stress, including decreased serum total antioxidant capacity and decreased testicular antioxidant enzyme activity. Therefore, it has been suggested that in addition to having antiandrogenic effects, phthalates also induce oxidative stress that further contributes to testicular toxicity [181].

In humans, phthalate exposure is generally negatively associated with serum testosterone and sperm counts, but this has not been found in all studies or for all types of phthalates. In in vitro studies, high-dose phthalate exposure has been found to cause a dose-dependent decrease in sperm motility and cytotoxicity after prolonged, high-dose exposure [182]. In in vivo studies, the results are more heterogeneous. For example, in one study of 463 male partners of subfertile couples who presented for semen analysis, several urinary phthalate levels were measured. These men were found to have a negative dose-response relationship between urinary monobutyl phthalate (MBP) levels and both sperm count and motility. However, there was no significant relationship between other phthalates—monobenzyl phthalate (MBzP) and di (2-ethylhexyl) phthalate (DEHP) metabolites—on semen parameters [183].

Similarly, in a study of 60 males in India, seminal levels of several phthalates were measured, including di(2-ethylhexyl) phthalate (DEHP), dibutyl phthalate (DBP), and diethyl phthalate (DEP). After adjusting for potential confounders, only DEHP levels were significantly negatively associated with sperm motility, concentration, and normal morphology [184].

Finally, another study of serum levels of six phthalate metabolites found a negative association between these phthalate levels and serum testosterone, semen volume, and total sperm count. However, there were no significant associations between serum phthalate metabolite concentrations and sperm concentration, morphology, or motility [185].

Undoubtedly, these studies are somewhat discordant. The heterogeneity between these study outcomes is likely related to the different mechanisms by which the phthalates were measured and the specific phthalate/metabolite investigated [1].

In addition to evidence that some phthalates alter semen parameters and decrease testosterone, there is also evidence that the phthalates are positively associated with increased estradiol. In one study of 82 men working in a polyvinyl chloride (PVC) plant who had been exposed to di(2-ethylhexyl) phthalate (DEHP) via inhalation of contaminated air, urinary concentrations of DEHP were significantly associated with increased serum estradiol concentrations and an increased estradiol/testosterone ratio [186]. This further emphasizes the significant endocrine-disrupting effect of phthalates on men's health.

Bisphenol A (BPA)

Bisphenol A (BPA) is a widely used plasticizer used in the production of polycarbonate plastics, epoxy resins, and many common items, including toys, water pipes, drinking containers, eyeglass lenses, sport safety equipment, medical equipment and tubing, and consumer electronics. It is one of the highest volume chemicals produced worldwide, with over 6 billion pounds produced each year [187]. Given its universal presence, there are many opportunities for human exposure; however, it most commonly enters human consumption by leaching out from food and beverage containers [188]. BPA has been measured in numerous human tissues and body fluids, including serum, urine, amniotic fluid, follicular fluid, placental tissue, and umbilical cord blood [187].

BPA is well known to have estrogenic effects. Specifically, it is known to bind to both types of nuclear estrogen receptors (ER), ERα and ERβ, though it binds with greater affinity to ERβ. BPA has historically been classified as a weak environmental estrogen, since its affinity for these classic nuclear estrogen receptors is typically weaker than that of estradiol [189]. Nevertheless, in some circumstances, BPA has been found to be just as potent as estradiol [189]. In addition to binding estrogen receptors, BPA also alters endogenous hormone synthesis, hormone metabolism, and hormone concentrations in blood [190]. Exposure to BPA results in changes to tissue enzymes and hormone receptors; it also alters thyroid and androgen hormone response systems [190].

In males, the endocrine-disrupting activity of BPA has been found to have several significant consequences, including changes in sex steroid concentrations, sexual function, and spermatogenesis [190]. In regard to sex steroid concentrations, animals exposed to BPA in the laboratory setting were noted to have dramatically decreased free testosterone levels after exposure [191]. In humans, several studies also observed inverse relationships between urinary BPA levels and free androgen index [192, 193].

Sexual function may also be impaired by BPA exposure. One study of Chinese men with workplace exposures to BPA found a dose-response relationship between an increasing level of cumulative BPA exposure and sexual dysfunction. BPA-exposed workers had a significantly greater risk of reduced sexual desire, increased erectile and ejaculatory dysfunction, and reduced satisfaction with sex life [194].

The association between BPA concentrations and spermatogenesis, however, is less clear. In one study of 375 men, there was no relationship between urinary BPA concentration and semen parameters [193]. In another study of 190 men recruited through an infertility clinic, urinary BPA concentration was associated with an increased, though not statistically significant, odds for decreased sperm concentration, motility, and morphology and increased sperm DNA damage [195]. However, in another study of 218 men in China with and without BPA exposure in the workplace, elevated urinary BPA levels were significantly associated with decreased sperm concentration, count, vitality, and motility [196].

Pesticides

Exposure to pesticides can occur in a variety of ways, including skin contact; inhalation; and, most commonly, dietary consumption [1]. Regardless of the mechanism of exposure, these chemicals can act as potent endocrine disruptors and exert a wide range of negative effects on human health. In men specifically, pesticides can damage spermatozoa, alter Sertoli cell or Leydig cell function, disrupt endocrine function at any stage of hormonal regulation, and impair fertility [197].

Several pesticides are particularly well known for their toxic effects, including dibromochloropropane (DBCP), ethylene dibromide (EDB), and vinclozolin.

Dibromochloropropane

DBCP is the best-known pesticide associated with impaired fertility [1]. One of the first reports of adverse effects of DBCP on semen quality was in 1977, when 14 of 25 workers in a California DBCP factory were found to have azoospermia or oligospermia and elevated FSH and LH levels [198]. A dose-response relationship between exposure and fertility was apparent, as men with sperm concentrations <1 million/mL had been exposed for at least 3 years, while men with sperm concentrations >40 million/mL had been exposed for fewer than 3 months [198]. In a larger follow-up study, 107 workers in a DBCP factory were found to have a duration-dependent relationship between DBCP exposure and semen parameters [199]. Of those exposed, 13.1% were azoospermic, 16.8% were severely oligospermatic, and 15.8% were mildly oligospermatic [199]. Testicular biopsies performed on 10 exposed men demonstrated significant damage to the seminiferous tubules and complete lack of spermatogonia in those with long-standing exposure [199]. These histological findings of seminiferous tubule damage were replicated in a similar, very small study of DBCP factory workers with azoospermia [200].

Several other large-scale studies of produce workers have also demonstrated the negative effect of DBCP exposure on semen parameters and fertility. In one study of 1500 male banana plantation workers in Costa Rica, 20%–25% of men exposed were diagnosed as infertile [201]. In another study of 26,400 males across 12 countries exposed to DBCP through their work on banana and pineapple plantations, the prevalence of azoospermia and oligospermia was striking. After a median exposure to DBCP of 3 years, 64.3% of all men had azoospermia or oligospermia; among men from the Philippines, this number was increased to 90.1% [202].

Fortunately, there has been some evidence that reproductive function can improve once exposure has ceased. In several studies, spermatogenesis was found to recover after anywhere between 18 and 45 months of nonexposure [203–205]; however, there was no improvement beyond 45 months [205]. Additionally, in one study by Eaton et al., the majority of azoospermic workers with a history of DBCP remained azoospermic after 5–8 years of nonexposure [206]. This evidence suggests that DBCP has significant and occasionally permanent effects on semen parameter and fertility in exposed men.

Ethylene Dibromide (EDB)
Another pesticide with clearly demonstrated male reproductive toxicity is the fumigant ethylene dibromide (EDB) [197]. In one study of 46 Hawaiian men employed in the papaya fumigation industry and exposed to airborne EDB, exposed men were found to have statistically significant decreases in sperm count, decreases in percentage of viable and motile sperm, and increases in morphological abnormalities compared with controls [207]. Another study of 10 EDB-exposed forestry workers also found decreased sperm velocity and decreased sperm volume compared with unexposed men [208], suggesting that EDB does indeed influence semen parameters. The effect of EDB on fertility outcomes remains under investigation.

Vinclozolin
Vinclozolin is a dicarboximide fungicide that has been widely used in Europe and the United States to protect grapes, fruits, vegetables, hops, ornamental plants, and grass from fungal damage [209]. Several studies have shown that vinclozolin has potent antiandrogenic effects that greatly influence normal development and fertility. In one study of male rats exposed to vinclozolin from gestational day 14 to postnatal day 3, these animals were found to have several developmental defects, including hypospadias, cleft phallus, supraingu-inal ectopic testes, a vaginal pouch, epididymal and testicular granulomas, atrophic seminal vesicles, and ventral prostate glands [210]. When the molecular mechanisms behind these developmental alterations were examined, it was discovered that while vinclozolin itself has a weak affinity for the androgen receptor, several of vinclozolin's metabolites act as effective antagonists of androgen receptor (AR) binding [209]. Similarly, Wong et al. found that these metabolites of vinclozolin exhibit antiandrogenic activity by blocking receptor binding

in a dose-dependent manner [211]. Remarkably, at very high concentrations, several metabolites of vinclozolin were also found to act as androgen receptor agonists in the absence of androgens [211], further emphasizing the potent endocrine-disrupting activity of vinclozolin.

This potent endocrine-disrupting activity can negatively affect spermatogenesis and male fertility. In one study of rats exposed to vinclozolin during embryonic days 8–14, the exposed male offspring were found to have a reduced spermatogenic capacity, which was transmitted to subsequent generations through the male germ cell line [212]. These results suggest that the endocrine-disrupting activity of vinclozolin has effects not only on the exposed generation but also on subsequent generations by inducing epigenetic alterations [212].

CONCLUSIONS

While environmental exposures and endocrine-disrupting chemicals remain the subject of active investigation, there is convincing evidence that these exposures have significant effects on male health and fertility. Various exposures are ubiquitous and nearly impossible to avoid. With certain health risks increasing and semen parameters declining, further research into mechanisms of action and preventative measures are warranted in order to limit exposures and optimize men's health.

References

[1] Gabrielsen JS, Tanrikut C. Chronic exposures and male fertility: the impacts of environment, diet, and drug use on spermatogenesis. Andrology 2016;4:648–61.

[2] Carlsen E, Giwercman A, Keiding N, Skakkebaek NE. Evidence for decreasing quality of semen during past 50 years. BMJ 1992;305:609–13.

[3] Geoffroy-Siraudin C, Loundou AD, Romain F, Achard V, Courbiere B, Perrard MH, Durand P, Guichaoua MR. Decline in semen quality among 10,932 males consulting for couple infertility over a 20-year period in Marseille, France. Asian J Androl 2012;14:584–90.

[4] Swan SH, Elkin EP, Fenster L. Have sperm densities declined? A reanalysis of global trend data. Environ Health Perspect 1997;105:1228–32.

[5] Swan SH, Elkin EP, Fenster L. The question of declining sperm density revisited: an analysis of 101 studies published 1934-1996. Environ Health Perspect 2000;108:961–6.

[6] Coalo A, Muscogiuri G, Piscitelli P. Environment and health: not only cancer. Int J Environ Res Public Health 2016;13:724.

[7] World Health Organization. WHO air quality guidelines for particulate matter, ozone, nitrogen dioxide and sulfur dioxide – Global update 2005, Available from: http://apps.who.int/iris/bitstream/handle/10665/69477/WHO_SDE_PHE_OEH_06.02_eng.pdf;jsessionid=15902BE60A3B5543D6166DFDE0F8F00D?sequence=1; 2006. Accessed 3 July 2018.

[8] Chen H, Goldberg MS, Villeneuve PJ. A systematic review of the relation between long-term exposure to ambient air pollution and chronic diseases. Rev Environ Health 2008;23:243–97.

[9] Pope CA, Burnett RT, Thun MJ, Calle EE, Krewski D, Ito K, Thurston GF. JAMA 2002;287:1132–41.

[10] Di Q, Wang Y, Zanobetti A, Wang Y, Koutrakis P, Choirat C, Dominici F, Schwartz J. N Engl J Med 2017;376:2513–22.

[11] Ostro B, Broadwin R, Green S, Feng W-Y, Lipsett M. Environ Health Perspect 2006;114:29–33.

[12] Watanabe N, Oonuki Y. Inhalation of diesel exhaust affects spermatogenesis in growing male rats. Environ Health Perspect 1999;107:539–44.

[13] Watanabe N. Decreased number of sperms and Sertoli cells in mature rats exposed to diesel exhaust as fetuses. Toxicol Lett 2005;155:55–8.

[14] Yoshida S, Sagai M, Oshio S, Umeda T, Ihara T, Sugamata M, Sugawara I, Takeda K. Exposure to diesel exhaust affects the male reproductive system in mice. Int J Androl 1999;22:307–15.

[15] Ieradi LA, Cristaldi M, Mascanzoni D, Cardarelli E, Grossi R, Campanella L. Genetic damage in urban mice exposed to traffic pollution. Environ Pollut 1996;92:323–8.

[16] Ono N, Oshio S, Niwata Y, Yoshia S, Tsukue N, Sugawara I, Takano H, Takeda K. Prenatal exposure to diesel exhaust impairs mouse spermatogenesis. Inhal Toxicol 2007;19:275–81.

[17] Pires A, de Melo EN, Mauad T, Nascimento Salvida PH, de Siqueria Bueno HM. Pre- and postnatal exposure to ambient levels of urban particular matter (PM(2.5)) affects mice spermatogenesis. Inhal Toxicol 2011;23:237–45.

[18] Somers CM, Yauk CL, White PA, Parfett CJ, Quinn JS. Air pollution induces heritable DNA mutations. Proc Natl Acad Sci U S A 2002;99:15904–7.

[19] Yauk C, Polyzos A, Rowan-Carroll A, Somers C, Godschalk RW, Van Schooten FJ, Berndt ML, Pogribny IP, Koturbash I, Williams A, Douglas GR, Kovalchuk O. Germ-line mutations, DNA damage, and global hypermethylation in mice exposed to particulate air pollution in an urban/industrial location. Proc Natl Acad Sci U S A 2008;105:605–10.

[20] De Rosa M, Zarrilli S, Paesano L, Carbone U, Boggia B, Petretta M, Maisto A, Cimmino F, Puca G, Colao A, et al. Traffic pollutants affect fertility in men. Hum Reprod 2003;18:1055–61.

[21] Guven A, Kayicki A, Cam K, Arbak P, Balbay O, Cam M. Alterations in semen parameters of toll collectors working at motorways: does diesel exposure induce detrimental effects on semen? Andrologia 2008;40:346–51.

[22] Hammoud A, Carrell DT, Gibson M, Sanderson M, Parker-Jones K, Peterson CM. Decreased sperm motility is associated with air pollution in Salt Lake City. Fertil Steril 2010;93:1875–9.

[23] Selevan SG, Borovec L, Slott VL, Zudova Z, Rubes J, Evenson DP, Perreault SD. Semen quality and reproductive health of young Czech men exposed to seasonal air pollution. Environ Health Perspect 2000;108:887–94.

[24] Radwan M, Jurewicz J, Polanska K, Sobala W, Radwan P, Bochenek M, Hanke W. Exposure to ambient air pollution – does it affect semen quality and the level of reproductive hormones? Ann Hum Biol 2016;43:50–6.

[25] Hansen C, Luben TJ, Sacks JD, Olshan A, Jeffay S, Strader L, Perreault SD. The effect of ambient air pollution on sperm quality. Environ Health Perspect 2010;118:203–9.

[26] Rubes J, Selevan SG, Evenson DP, Zudova D, Vozdova M, Zudova Z, Robbins WA, Perreault SD. Episodic air pollution is associated with increased DNA fragmentation in human sperm without other changes in semen quality. Hum Reprod 2005;20:2776–83.

[27] Carre J, Gatimel N, Moreau J, Parinaud J, Leandri R. Does air pollution play a role in infertility?: a systematic review. Environ Health 2017;16:82.

[28] Richardson SD, Plewa MJ, Wagner ED, Schoeny R, Demarini DM. Occurrence, genotoxicity, and carcinogenicity of regulated and emerging disinfection by-products in drinking water: a review and roadmap for research. Mutat Res 2007;636:178–242.

[29] Linder RE, Klinefelter GR, Strader LF, Suarez JD, Roberts NL. Spermatotoxicity of dichloroacetic acid. Reprod Toxicol 1997;11:681–8.

[30] Linder RE, Klinefelter GR, Strader LF, Suarez JD, Roberts NL, Dyer CJ. Spermatotoxicity of dibromoacetic acid in rats after 14 daily exposures. Reprod Toxicol 1994;8:251–9.

[31] Klinefelter GR, Strader LF, Suarez JD, Roberts NL, Goldman JM, Murr AS. Continuous exposure to dibromoacetic acid delays pubertal development and compromises sperm quality in the rat. Toxicol Sci 2004;81:419–29.

[32] Linder RE, Klinefelter GR, Strader LF, Veeramachaneni DNR, Roberts NL, Suarez JD. Histopathologic changes in the testis of rats exposed to dibromoacetic acid. Reprod Toxicol 1997;11:47–56.

[33] Zeng Q, Wang YX, Xie SH, Xu L, Chen YZ, Li M, Yue J, Li YF, Liu AL, Lu WQ. Drinking water disinfection by-products and semen quality: a cross sectional study in China. Environ Health Perspect 2014;122:741–6.

[34] Xie SH, Li TF, Tan YF, Zheng D, Liu AL, Xie H, Lu WQ. Urinary trichloroacetic acid levels and semen quality: a hospital-based cross-sectional study in Wuhun, China. Environ Res 2011;111:295–300.

[35] Zeng Q, Li M, Xie SH, Gu LJ, Yue J, Cao WC, Zheng D, Liu AL, Li YF, Lu WQ. Baseline blood trihalomethanes, semen parameters and serum total testosterone: a cross-sectional study in China. Environ Int 2013;54:134–40.

[36] Zeng Q, Chen YZ, Xu L, Chen HX, Luo Y, Li M, Yue J, Liu AL, Li YF, Lu WQ. Evaluation of exposure to trihalomethanes in tap water and semen quality: a prospective study in Wuhan, China. Reprod Toxicol 2014;46:56–63.

[37] Fenster L, Waller K, Windham G, Henneman T, Anderson M, Mendola P, Overstreet JW, Swan SH. Trihalomethane levels in home tap water and semen quality. Epidemiology 2003;14:650–8.

[38] Iszatt N, Nieuwenhuijsen MJ, Bennet J, Best N, Povey AC, Pacey AA, Moore H, Cherry N, Toledano MB. Chlorination by-products in tap water and semen quality in England and Wales. Occup Environ Med 2013;70:754–60.

[39] Luben TJ, Olshan AF, Herring AH, Jeffay S, Strader L, Buus RM, Chan RL, Savitz DA, Singer PC, Weinberg HS, Perreault SD. The Healthy Men Study: An Evaluation of Exposure to Disinfection By-Products in Tap Water and Sperm Quality. Environ Health Perspect 2007;115:1169–76.

[40] Stackelberg PE, Furlong ET, Meyer MT, Zaugg SD, Henderson AK, Reissman DB. Persistence of pharmaceutical compounds and other organic wastewater contaminants in a conventional drinking-water-treatment plant. Sci Total Environ 2004;329:99–113.

[41] Ivell R. Lifestyle impact and the biology of the human scrotum. Reprod Biol Endocrinol 2007;5:15.

[42] Durairajanayagam D, Agarwal A, Ong C. Causes, effects, and molecular mechanisms of testicular heat stress. Reprod BioMed Online 2015;30:14–27.

[43] Setchell BP. The Parkes Lecture. Heat and the testis. J Reprod Fertil 1998;144:179–94.

[44] Kim B, Park K, Rhee K. Heat stress response of male germ cells. Cell Mol Life Sci 2013;70:2623–36.

[45] Lue YH, Hiken AP, Swerdloff RS, Im P, Taing KS, Bui T, Leung A, Wang C. Single exposure to heat induces stage-specific germ cell apoptosis in rats: role of intratesticular testosterone on stage specificity. Endocrinology 1999;140:1709–17.

[46] Rockett JC, Mapp FL, Garges JB, Luft JC, Mori C, Dix DJ. Effects of hyperthermia on spermatogenesis, apoptosis, gene expression, and fertility in adult male mice. Biol Reprod 2001;65:229–39.

[47] Setchell BP, Ploen L, Ritzen EM. Effect of local heating on rat testes after suppression of spermatogenesis by pretreatment with a GnRH agonist and an anti-androgen. Reproduction 2002;124:133–40.

[48] Izu H, Inouye S, Fujimoto M, Shiraishi K, Naito K, Nakai A. Heat shock transcription factor 1 is involved in quality-control mechanisms in male germ cells. Biol Reprod 2004;70:18–24.

[49] Widlak W, Winiarski B, Krawczyk A, Vydra N, Malusecka E, Krawczyk Z. Inducible 70kDA heat shock protein does not protect spermatogenic cells from damage induced by cryptorchidism. Int J Androl 2007;30:80–7.

[50] Paul C, Murray AA, Spears N, Saunders PT. A single, mild, transient scrotal heat stress causes DNA damage, subfertility, and impairs formation of blastocysts in mice. Reproduction 2008;136:73–84.

[51] Agoulnik AI, Huang Z, Ferguson L. Spermatogenesis in cryptorchidism. Methods Mol Biol 2012;825:127–47.

[52] Taran I, Elder JS. Results of orchiopexy for the undescended testis. World J Urol 2006;24:231–9.

[53] Hadziselimovic F, Thommen L, Girard J, Herzog B. The significance of postnatal gonadotropin surge for testicular development in normal and cryptorchid testes. J Urol 1986;136:274–6.

[54] Rogers E, Teahan S, Gallagher H, Butler MR, Grainger R, McDermott TE, Thornhill JA. The role of orchiectomy in the management of postpubertal cryptorchidism. J Urol 1998;159:851–4.

[55] Feyles F, Peiretti V, Mussa A, Maneti M, Canavese F, Cortese MG, Lala R. Improved sperm count and motility in young men surgically treated for cryptorchidism in the first year of life. Eur J Pediatr Surg 2014;24:376–80.

[56] Pottern LM, Brown LM, Hoover RN, Javadpour N, O'Connell KJ, Stutzman RE, Blattner WA. Testicular cancer risk among young men: role of cryptorchidism and inguinal hernia. J Natl Cancer Inst 1985;74:377–81.

[57] Batata MA, Whitmore WF, Chu FC, Hilaris BS, Loh J, Grabstald H, Golbey R. Cryptorchidism and testicular cancer. J Urol 1980;124:382–7.

[58] Halme A, Kellokumpu-Lehtinen P, Lehtonene T, Teppo L. Morphology of testicular germ cell tumours in treated and untreated cryptorchidism. Br J Urol 1989;64:78–83.

[59] Herrington LJ, Zhao W, Husson G. Management of cryptorchidism and risk of testicular cancer. Am J Epidemol 2003;157:602–5.

[60] Kolon TF, Herndon CD, Baker LA, Baskin LS, Baxter CG, Cheng EY, Diaz M, Lee PA, Seashore CJ, Tasian GE, Barthold JS, American Urological Association. Evaluation and treatment of cryptorchidism: AUA guideline. J Urol 2014;192:337–45.

[61] Schoor RA, Elhanbly SM, Niederberger C. The pathophysiology of varicocele-associated male infertility. Curr Urol Rep 2001;2:432–6.

[62] Cozzolino DJ, Lipshultz LI. Varicocele as a progressive lesion: positive effect of varicocele repair. Hum Reprod Update 2001;7:55–8.

[63] Chiba K, Fujisawa M. Clinical outcomes of varicocele repair in infertile men: a review. World J Mens Health 2016;34:101–9.

[64] Green K, Turner T, Howards S. Varicocele: reversal of the testicular blood flow and temperature effects by varicocele repair. J Urol 1984;131:1208–11.

[65] Kay R, Alexander N, Baugham WL. Induced varicoceles in Rhesus monkeys. Fertil Steril 1979;31:195–9.

[66] Goldstein M, Eid J. Elevation of intratesticular and scrotal skin surface temperature in men with varicocele. J Urol 1989;142:743–5.

[67] Chehval MJ, Purcell MH. Deterioration of semen parameters in men with untreated varicoceles: evidence of progressive testicular damage. Fertil Steril 1992;57:174–7.

[68] Lipshultz LI, Corriere JN. Progressive testicular atrophy in the varicocele patient. J Urol 1977;117:175–6.

[69] Practice Committee of the American Society for Reproductive Medicine; Society for Male Reproduction and Urology. Report on varicocele and infertility: a committee opinion. Fertil Steril 2014;102:1556–60.

[70] Ding H, Tian J, Du W, Zhang L, Wang H, Wang Z. Open non-microsurgical, laparoscopic or open microsurgical varicocelectomy for male infertility: a meta-analysis of randomized controlled trials. BJU Int 2012;110:1536–42.

[71] Kovac JR, Fantus J, Lipshultz LI, Fischer MA, Klinghoffer Z. Cost-effectiveness analysis reveals microsurgical varicocele repair is superior to percutaneous embolization in the treatment of male infertility. Can Urol Assoc J 2014;8:619–25.

[72] Baazeem A, Belzile E, Ciampi A, Dohle G, Jarvi K, Salonia A, et al. Varicocele and male factor infertility treatment: a new meta-analysis and review of the role of varicocele repair. Eur Urol 2011;60:796–808.

[73] Abdel-Meguid TA, Al-Sayyad A, Tayib A, Farsi HM. Does varicocele repair improve male infertility? An evidence-based perspective from a randomized, controlled trial. Eur Urol 2011;59:455–61.

[74] Madgar I, Weissenberg R, Lunenfeld B, Karasik A, Goldwasser B. Controlled trial of high spermatic vein ligation for varicocele in infertile men. Fertil Steril 1995;63:120–4.

[75] Tanrikut C, Goldstein M, Rosoff JS, Lee RK, Nelson CJ, Mulhall JP. Varicocele as a risk factor for androgen deficiency and effect of repair. BJU Int 2011;108:1480–4.

[76] Hsiao W, Rosoff JS, Pale JR, Greenwood EA, Goldstein M. Older age is associated with similar improvements in semen parameters and testosterone after subinguinal microsurgical varicocelectomy. J Urol 2011;185:620–5.

[77] Su L, Goldstein M, Schlegel PN. The effect of varicocelectomy on serum testosterone levels in infertile men with varicoceles. J Urol 1995;154:1752–5.

[78] Carlsen E, Andersson AM, Petersen JH, Skakkebaek NE. History of febrile illness and variation in semen quality. Hum Reprod 2003;18:2089–92.

[79] Evenson DP, Jost LK, Corzett M, Balhorn R. Characteristics of human sperm chromatin structure following an episode of influenza and high fever: a case study. J Androl 2000;21:739–46.

[80] Garolla A, Torino M, Miola P, Caretta N, Pizzol D, Menegazzo M, Bertoldo A, Foresta C. Twenty-four-hour monitoring of scrotal temperature in obese men and men with a varicocele as a mirror of spermatogenic function. Hum Reprod 2015;39:1006–13.

[81] Shafik A, Olfat S. Scrotal lipomatosis. Br J Urol 1981;53:50–4.

[82] Shafik A, Olfat S. Lipectomy in the treatment of scrotal lipomatosis. Br J Urol 1981;53:55–61.

[83] Zorgniotti A, Reiss H, Toth A, Sealfon A. Effect of clothing on scrotal temperature in normal men and patients with poor semen. Urology 1982;19:176–8.

[84] Jung A, Leonhardt F, Schill WB, Schuppe HC. Influence of the type of undertrousers and physical activity on scrotal temperature. Hum Reprod 2005;20:1022–7.

[85] Minguez-Alacon L, Gaskins AJ, Chiu YH, Messerlian C, Williams PL, Ford JB, Souter I, Hauser R, Chavarro JE. Type of underwear worn and markers of testicular function among men attending a fertility center. Hum Reprod 2018;33:1749–56.

[86] Munkelwitz R, Gilbert BR. Are boxer shorts really better? A critical analysis of the role of underwear type in male subfertility. J Urol 1998;160:1329–33.

[87] Bujan L, Daudin M, Charlet JP, Thonneau P, Mieusset R. Increase in scrotal temperature in car drivers. Hum Reprod 2000;15:1355–7.

[88] Hjollund NHI, Storgaard L, Ernst E, Bonde JP, Olsen J. The relation between daily activities and scrotal temperature. Reprod Toxicol 2002;16:209–14.

[89] Stoy J, Hjollund NH, Mortensen JT, Vurr H, Bonde JP. Semen quality and sedentary work position. Int J Androl 2004;27:5–11.

[90] Sheynkin Y, Jung M, Yoo P, Schulsinger D, Komaroff E. Increase in scrotal temperature in laptop computer users. Hum Reprod 2005;20:452–5.

[91] Bonde JP. Semen quality in welders exposed to radiant heat. Br J Ind Med 1992;49:5–10.

[92] Thonneau P, Ducot B, Bujan L, Mieusset R, Spira A. Effect of male occupational heat exposure on time to pregnancy. Int J Androl 1997;20:274–8.

[93] Saikhun J, Kitiyanant Y, Vanadurongwan V, Pavasuthipaisit K. Effects of sauna on sperm movement characteristics of normal men measured by computer-assisted sperm analysis. Int J Androl 1998;21:358–63.

[94] Garolla A, Torino M, Sartini B, Cosci I, Patassini C, Carraro U, Foresta C. Seminal and molecular evidence that sauna exposure affects human spermatogenesis. Hum Reprod 2013;28:877–85.

[95] Brown-Woodman PD, Post EJ, Gass GC, White IG. The effect of a single sauna exposure on spermatozoa. Arch Androl 1984;12:9–15.

[96] Fukunaga H, Butterworth KT, Yokoya A, Ogawa T, Prise KM. Low-dose radiation-induced risk in spermatogenesis. Int J Radiat Biol 2017;93:1291–8.

[97] Azzam EI, Jay-Gerin J-P, Pain D. Ionizing radiation-induced metabolic oxidative stress and prolonged cell injury. Cancer Lett 2012;327:48–60.

[98] Clifton DK, Bremner WJ. The effect of testicular x-irradiation on spermatogenesis in man. A comparison with the mouse. J Androl 1983;4:387–92.

[99] Birioukov A, Meurer M, Peter RU, Braun-Falco O, Plewig G. Male reproductive system in patients exposed to ionizing irradiation in the Chernobyl accident. Arch Androl 1993;30:99–104.

[100] Fischbein A, Zabludovsky N, Eltes F, Grischenko V, Bartoov B. Ultramorphological sperm characteristics in the risk assessment of health effects after radiation exposure among salvage workers in Chernobyl. Environ Health Perspect 1997;105(Suppl. 6):1445–9.

[101] Zhou DD, Hao JL, Guo KM, Lu CW, Liu XD. Sperm quality and DNA damage in men from Jilin Province, China, who are occupationally exposed to ionizing radiation. Genet Mol Res 2016;15(1).

[102] Premi S, Srivastava J, Chandy SP, Ali S. Unique signatures of natural background radiation on human Y chromosomes from Kerala, India. PLoS ONE 2009;4.

[103] Meo SA, Arif M, Rashied S, Khan MM, Vohra MS, Usmani AM, Imran MB, Al-Drees AM. Hypospermatogenesis and spermatozoa maturation arrest in rats induced by mobile phone radiation. J Coll Physicians Surg Pak 2011;21:262–5.

[104] Shahin S, Singh SP, Chaturvedi CM. 1800 MHz mobile phone irradiation induced oxidative and nitrosative stress leads to p53 dependent Bax mediated testicular apoptosis in mice, Mus musculus. J Cell Physiol 2018;233:7253–67.

[105] Tas M, Dasdag S, Akdag MZ, Cirit U, Yegin K, Seker U, Ozmen MF, Eren LB. Long-term effects of 900 MHz radiofrequency radiation emitted from mobile phone on testicular tissue and epididymal semen quality. Electromagn Biol Med 2014;33:216–22.

[106] Dama MS, Bhat MN. Mobile phones affect multiple sperm quality traits: a meta-analysis. F1000Research 2013;2:40.

[107] La Vignera S, Condorelli RA, Vicari E, D'Agata R, Calogero AE. Effects of the exposure to mobile phones on male reproduction: a review of the literature. J Androl 2012;33:350–6.

[108] Adams JA, Galloway TS, Mondal D, Esteves SC, Mathews F. Effect of mobile telephones on sperm quality: a systematic review and meta-analysis. Environ Int 2014;70:106–12.

[109] Liu K, Li Y, Zhang G, Liu J, Cao J, Ao L, Zhang S. Association between mobile phone use and semen quality: a systematic review and meta-analysis. Andrology 2014;2:491–501.

[110] Kamiya K, Ozasa K, Akiba S, Niwa O, Kodama K, Takamura N, Zaharieva EK, Kimura Y, Wakeford R. Long-term effects of radiation exposure on health. Lancet 2015;386:469–78.

[111] McGlynn KA, Trabert B. Adolescent and adult risk factors for testicular cancer. Nat Rev Urol 2012;9:339–49.

[112] Yousif L, Blettner M, Hammer GP, Zeeb H. Testicular cancer risk associated with occupational radiation exposure: a systematic literature review. J Radiol Prot 2010;30:389–406.

[113] Sigurdson AJ, Doody MM, Rao RS, Freedman DM, Alexander BH, Hauptmann M, Mohan AK, Yoshinaga S, Hill DA, Tarone R, Mabuchi K, Ron E, Linet MS. Cancer incidence in the US radiologic technologists health study, 1983-1998. Cancer 2003;97:3080–9.

[114] Floderus B, Stenlund C, Persson T. Occupational magnetic field exposure and site-specific cancer incidence: a Swedish cohort study. Cancer Causes Control 1999;10:323–32.

[115] Visser AJ, Heyns CF. Testicular function after torsion of the spermatic cord. BJU Int 2003;92:200–3.

[116] Turner TT, Tung KS, Tomomasa H, Wilson LW. Acute testicular ischemia results in germ cell-specific apoptosis in the rat. Biol Reprod 1997;57:1267–74.

[117] Thomas WE, Cooper MJ, Crane GA, Lee G, Williamson RC. Testicular exocrine malfunction after torsion. Lancet 1984;2:1357–60.

[118] Farias JG, Bustos-Obregon E, Orellana R, Bucarey JL, Quiroz E, Reyes JG. Effects of chronic hypobaric hypoxia on testis histology and round spermatid oxidative metabolism. Andrologia 2005;37:47–52.

[119] Liao W, Cai M, Chen J, Huang J, Liu F, Jiang C, Gao Y. Hypobaric hypoxia causes deleterious effects on spermatogenesis in rats. Reproduction 2010;139:1031–8.

[120] Farias JG, Bustos-Obregon E, Reyes JG. Increase in testicular temperature and vascularization induced by hypobaric hypoxia in rats. J Androl 2005;26:693–7.

[121] Donayre J, Guerra-Garcia R, Moncloa F, Sobrevilla LA. Endocrine studies at high altitude. IV Changes in the semen of men. J Reprod Fertil 1968;16:55–8.

[122] Okumura A, Fuse H, Kawauchi Y, Mizuno I, Akashi T. Changes in male reproductive function after high altitude mountaineering. High Alt Med Biol 2003;4:349–53.

[123] Verratti V, Berardinelli F, Di Giulio C, Bosco G, Cacchio M, Pellicciotta M, Nicolai M, Martinotti S, Tenaglia R. Evidence that chronic hypoxia causes reversible impairment on male fertility. Asian J Androl 2008;10:602–6.

[124] Gonzales GF. Peruvian contributions to the study on human reproduction at high altitude: from the chronicles of the Spanish conquest to the present. Respir Physiol Neurobiol 2007;158:172–9.

[125] Eisenberg ML, Chen Z, Ye A, Buck Louis GM. Relationship between physical occupational exposures and health on semen quality: data from the Longitudinal Investigation of Fertility and the Environment (LIFE) study. Fertil Steril 2015;103:1271–7.

[126] Sheiner EK, Sheiner E, Carel R, Potashnik G, Shoham-Vardi I. Potential association between male infertility and occupational psychological stress. J Occup Environ Med 2002;44:1093–9.

[127] El-Helaly M, Awadalla N, Mansour M, El-Biomy Y. Workplace exposures and male infertility – a case-control study. Int J Occup Med Environ Health 2010;23:331–8.

[128] Safarinejad MR, Azma A, Kolahi AA. The effects of intensive, long-term treadmill running on reproductive hormones, hypothalamus-pituitary-testis axis, and semen quality: a randomized controlled study. J Endocrinol 2009;200:259–71.

[129] Tartibian B, Maleki BH. Correlation between seminal oxidative stress biomarkers and antioxidants with sperm DNA damage in elite athletes and recreationally active men. Clin J Sport Med 2012;22:132–9.

[130] Gaskins AJ, Mendiola J, Afeiche M, Jorgensen N, Swan SH, Chavarro JE. Physical activity and television watching in relation to semen quality in young men. Br J Sports Med 2015;49:265–70.

[131] Jozkow P, Rossato M. The impact of intense exercise on semen quality. Am J Mens Health 2017;11:654–62.

[132] Kavlock RJ, Daston GP, DeRosa C, Fenner-Crisp P, Gray LE, Kaattari S, Lucier G, Luster M, Mac MJ, Maczka C, Miller R, Moore J, Rolland R, Scott G, Sheehan DM, Sinks T, Tilson HA. Research needs for the risk assessment of health and environmental effects of endocrine disruptors: a report of the U.S. EPA-sponsored workshop. Environ Health Perspect 1996;104 (Suppl. 4):715–40.

[133] Kabir ER, Rahman MS, Rahman I. A review on endocrine disruptors and their possible impacts on human health. Environ Toxicol Pharmacol 2015;40:241–58.

[134] The Endocrine Disruption Exchange. List of potential endocrine disruptors, https://endocrinedisruption.org/interactive-tools/tedx-list-of-potential-endocrine-disruptors/search-the-tedx-list. Accessed 11 July 2018.

[135] Järup L, Berglund M, Elinder CG, Nordberg G, Vahter M. Health effects of cadmium exposure–a review of the literature and a risk estimate. Scand J Work Environ Health 1998;24(Suppl. 1):1–51. Review. Erratum in: *Scand J Work Environ Health* 1998;24(3):240.

[136] Thompson J, Bannigan J. Cadmium: toxic effects on the reproductive system and the embryo. Reprod Toxicol 2008;25:304–15.

[137] Hartwig A. Cadmium and cancer. Met Ions Life Sci 2013;11:491–507.

[138] Niewenhuis RJ. Effects of cadmium upon regenerated testicular vessels in the rat. Biol Reprod 1980;23:171–9.

[139] Laskey JW, Rehnberg GL, Laws SC, Hein JF. Reproductive effects of low acute doses of cadmium chloride in adult male rats. Toxicol Appl Pharmacol 1984;73:250–5.

[140] El-Ashmawy IM, Youssef SA. The antagonistic effect of chlorpromazine on cadmium toxicity. Toxicol Appl Pharmacol 1999;161:34–9.

[141] Pant N, Kumar G, Upadhyay AD, Gupta YK, Chaturvedi PK. Correlation between lead and cadmium concentration and semen quality. Andrologia 2015;47:887–91.

[142] Pant N, Upadhyay G, Pandey S, Mathur N, Saxena DK, Srivastava SP. Lead and cadmium concentration in the seminal plasma of men in the general population: correlation with sperm quality. Reprod Toxicol 2003;17:447–50.

[143] Xu DX, Shen HM, Zhu QX, Chua L, Wang QN, Chia SE, Ong CN. The associations among semen quality, oxidative DNA damage in human spermatozoa and concentrations of cadmium, lead and selenium in seminal plasma. Mutat Res 2003;534:155–63.

[144] Benoff S, Hauser R, Marmar JL, Hurley IR, Napolitano B, Centola GM. Cadmium concentrations in blood and seminal plasma: correlations with sperm number and motility in three male populations (infertility patients, artificial insemination donors, and unselected volunteers). Mol Med 2009;15:248–62.

[145] Hovatta O, Venäläinen ER, Kuusimäki L, Heikkilä J, Hirvi T, Reima I. Aluminium, lead and cadmium concentrations in seminal plasma and spermatozoa, and semen quality in Finnish men. Hum Reprod 1998;13:115–9.

[146] Jurasović J, Cvitković P, Pizent A, Colak B, Telisman S. Semen quality and reproductive endocrine function with regard to blood cadmium in Croatian male subjects. Biometals 2004;17:735–43.

[147] De Angelis C, Galdiero M, Pivonello C, Salzano C, Gianfrilli D, Piscitelli P, Lenzi A, Colao A, Pivonello R. The environment and male reproduction: the effect of cadmium exposure on reproductive function and its implication in fertility. Reprod Toxicol 2017;73:105–27.

[148] Lafuente A. The hypothalamic-pituitary-gonadal axis is target of cadmium toxicity. An update of recent studies and potential therapeutic approaches. Food Chem Toxicol 2013;59:395–404.

[149] Martin MB, Voeller HJ, Gelmann EP, Lu J, Stoica EG, Hebert EJ, Reiter R, Singh B, Danielsen M, Pentecost E, Stoica A. Role of cadmium in the regulation of AR gene expression and activity. Endocrinology 2002;143:263–75.

[150] Wang X, Wang M, Dong W, Li Y, Zheng X, Piao F, Li S. Subchronic exposure to lead acetate inhibits spermatogenesis and downregulates the expression of Ddx3y in testis of mice. Reprod Toxicol 2013;42:242–50.

[151] Alexander BH, Checkoway H, van Netten C, Muller CH, Ewers TG, Kaufman JD, Mueller BA, Vaughan TL, Faustman EM. Semen quality of men employed at a lead smelter. Occup Environ Med 1996;53:411–6.

[152] Telisman S, Cvitković P, Jurasović J, Pizent A, Gavella M, Rocić B. Semen quality and reproductive endocrine function in relation to biomarkers of lead, cadmium, zinc, and copper in men. Environ Health Perspect 2000;108:45–53.

[153] Benoff S, Centola GM, Millan C, Napolitano B, Marmar JL, Hurley IR. Increased seminal plasma lead levels adversely affect the fertility potential of sperm in IVF. Hum Reprod 2003;18:374–83.

[154] Anis TH, El Karaksy A, Mostafa T, Gadalla A, Imam H, Hamdy L, Abu el-Alla O. Chronic lead exposure may be associated with erectile dysfunction. J Sex Med 2007;4:1428–36.

[155] Gonulalan U, Hayırlı A, Kosan M, Ozkan O, Yılmaz H. Erectile dysfunction and depression in patients with chronic lead poisoning. Andrologia 2013;45:397–401.

[156] Ross AE, Marchionni L, Phillips TM, Miller RM, Hurley PJ, Simons BW, Salmasi AH, Schaeffer AJ, Gearhart JP, Schaeffer EM. Molecular effects of genistein on male urethral development. J Urol 2011;185:1894–8.

[157] Hashem NM, Abo-Elsoud MA, Nour El-Din ANM, Kamel KI, Hassan GA. Prolonged exposure of dietary phytoestrogens on semen characteristics and reproductive performance of rabbit bucks. Domest Anim Endocrinol 2018;64:84–92.

[158] Pan L, Xia X, Feng Y, Jiang C, Cui Y, Huang Y. Exposure of juvenile rats to the phytoestrogen daidzein impairs erectile function in a dose-related manner in adulthood. J Androl 2008;29:55–62.

[159] Meena R, Supriya C, Pratap Reddy K, Sreenivasula Reddy P. Altered spermatogenesis, steroidogenesis and suppressed fertility in adult male rats exposed to genistein, a non-steroidal phytoestrogen during embryonic development. Food Chem Toxicol 2017;99:70–7.

[160] Chavarro JE, Toth TL, Sadio SM, Hauser R. Soy food and isoflavone intake in relation to semen quality parameters among men from an infertility clinic. Hum Reprod 2008;23:2584–90.

[161] Mitchell JH, Cawood E, Kinniburgh D, Provan A, Collins AR, Irvine DS. Effect of phytoestrogen food supplement on reproductive health in normal males. Clin Sci (Lond) 2001;100:613–8.

[162] Song G, Kochman L, Andolina E, Herko RC, Brewer KJ, Lewis V. O-115: beneficial effects of dietary intake of plant phytoestrogens on semen parameters and sperm DNA integrity in infertile men. Fertil Steril 2006;86:S49.

[163] Wilcox AJ, Baird DD, Weinberg CR, Hornsby PP, Herbst AL. Fertility in men exposed prenatally to diethylstilbestrol. NEJM 1995;332:1411–6.

[164] Bibbo M, Gill WB, Azizi F, Blough R, Fang VS, Rosenfield RL, Schumacher GF, Sleeper K, Sonek MG, Wied GL. Follow-up study of male and female offspring of DES-exposed mothers. Obstet Gynecol 1977;49:1–8.

[165] Glaze GM. Diethylstilbestrol exposure in utero: review of literature. J Am Osteopath Assoc 1984;83:435–8.

[166] Stillman RJ. In utero exposure to diethylstilbestrol: adverse effects on the reproductive tract and reproductive performance and male and female offspring. Am J Obstet Gynecol 1982;142:905–21.

[167] Whitehead ED, Leiter E. Genital abnormalities and abnormal semen analyses in male patients exposed to diethylstilbestrol in utero. J Urol 1981;125:47–50.

[168] Henderson BE, Benton B, Cosgrove M, Baptista J, Aldrich J, Townsend D, Hart W, Mack TM. Urogenital tract abnormalities in sons of women treated with diethylstilbestrol. Pediatrics 1976;58:505–7.

[169] Klip H, Verloop J, van Gool JD, Koster ME, Burger CW, van Leeuwen FE, OMEGA Project Group. Hypospadias in sons of women exposed to diethylstilbestrol in utero: a cohort study. Lancet 2002;359:1102–7.

[170] Newbold RR, Bullock BC, McLachlan JA. Testicular tumors in mice exposed in utero to diethylstilbestrol. J Urol 1987;138:1446–50.

[171] Strohsnitter WC, Noller KL, Hoover RN, Robboy SJ, Palmer JR, Titus Ernstoff L, Kaufman RH, Adam E, Herbst AL, Hatch EE. Cancer risk in men exposed in utero to diethylstilbestrol. J Natl Cancer Inst 2001;93:545–51.

[172] Leary FJ, Resseguie LJ, Kurland LT, O'Brien PC, Emslander RF, Noller KL. Males exposed in utero to diethylstilbestrol. JAMA 1984;252:2984–9.

[173] Sharpe RM, Atanassova N, McKinnell C, Parte P, Turner KJ, Fisher JS, Kerr JB, Groome NP, Macpherson S, Millar MR, Saunders PT. Abnormalities in functional development of the Sertoli cells in rats treated neonatally with diethylstilbestrol: a possible role for estrogens in Sertoli cell development. Biol Reprod 1998;59:1084–94.

[174] Miyaso H, Naito M, Hirai S, Matsuno Y, Komiyama M, Itoh M, Mori C. Neonatal exposure to diethylstilbestrol causes granulomatous orchitis via epididymal inflammation. Anat Sci Int 2014;89:215–23.

[175] Newbold RR, Padilla-Banks E, Jefferson WN. Adverse effects of the model environmental estrogen diethylstilbestrol are transmitted to subsequent generations. Endocrinology 2006;147(6 Suppl):11–7.

[176] Gore AC, Chappell VA, Fenton SE, Flaws JA, Nadal A, Prins GS, Toppari J, Zoeller RT. EDC-2: The Endocrine Society's second scientific statement on endocrine-disrupting chemicals. Endocr Rev 2015;36:E1–E150.

[177] Silva MJ, Barr DB, Reidy JA, Malek NA, Hodge CC, Caudill SP, Brock JW, Needham LL, Calafat AM. Urinary levels of seven phthalate metabolites in the U.S. population from the National Health and Nutrition Examination Survey (NHANES) 1999-2000. Environ Health Perspect 2004;112:331–8.

[178] Li L, Bu T, Su H, Chen Z, Liang Y, Zhang G, Zhu D, Shan Y, Xu R, Hu Y, Li J, Hu G, Lian Q, Ge RS. In utero exposure to diisononyl phthalate caused testicular dysgenesis of rat fetal testis. Toxicol Lett 2015;232:466–74.

[179] Parks LG, Ostby JS, Lambright CR, Abbott BD, Klinefelter GR, Barlow NJ, Gray LE. The plasticizer diethylhexyl phthalate induces malformations by decreasing fetal testosterone synthesis during sexual differentiation in the male rat. Toxicol Sci 2000;58:339–49.

[180] Mahood IK, Hallmark N, McKinnell C, Walker M, Fisher JS, Sharpe RM. Abnormal Leydig Cell aggregation in the fetal testis of rats exposed to di (n-butyl) phthalate and its possible role in testicular dysgenesis. Endocrinology 2005;146:613–23.

[181] Aly HA, Hassan MH, El-Beshbishy HA, Alahdal AM, Osman AM. Dibutyl phthalate induces oxidative stress and impairs spermatogenesis in adult rats. Toxicol Ind Health 2016;32:1467–77.

[182] Pant N, Pant A, Shukla M, Mathur N, Gupta Y, Saxena D. Environmental and experimental exposure of phthalate esters: the toxicological consequence on human sperm. Hum Exp Toxicol 2011;30:507–14.

[183] Hauser R, Meeker JD, Duty S, Silva MJ, Calafat AM. Altered semen quality in relation to urinary concentrations of phthalate monoester and oxidative metabolites. Epidemiology 2006;17:682–91.

[184] Pant N, Kumar G, Upadhyay AD, Patel DK, Gupta YK, Chaturvedi PK. Reproductive toxicity of lead, cadmium, and phthalate exposure in men. Environ Sci Pollut Res Int 2014;21:11066–74.

[185] Specht IO, Toft G, Hougaard KS, Lindh CH, Lenters V, Jönsson BA, Heederik D, Giwercman A, Bonde JP. Associations between serum phthalates and biomarkers of reproductive function in 589 adult men. Environ Int 2014;66:146–56.

[186] Fong JP, Lee FJ, Lu IS, Uang SN, Lee CC. Relationship between urinary concentrations of di(2-ethylhexyl) phthalate (DEHP) metabolites and reproductive hormones in polyvinyl chloride production workers. Occup Environ Med 2015;72:346–53.

[187] Vandenberg LN, Hauser R, Marcus M, Olea N, Welshons WV. Human exposure to bisphenol A (BPA). Reprod Toxicol 2007;24:139–77.

[188] Benachour N, Aris A. Toxic effects of low doses of Bisphenol-A on human placental cells. Toxicol Appl Pharmacol 2009;241:322–8.

[189] Wetherill YB, Akingbemi BT, Kanno J, McLachlan JA, Nadal A, Sonnenschein C, Watson CS, Zoeller RT, Belcher SM. In vitro molecular mechanisms of bisphenol A action. Reprod Toxicol 2007;24:178–98.

[190] vom Saal FS, Akingbemi BT, Belcher SM, Birnbaum LS, Crain DA, Eriksen M, Farabollini F, Guillette LJ, Hauser R, Heindel JJ, Ho SM, Hunt PA, Iguchi T, Jobling S, Kanno J, Keri RA, Knudsen KE, Laufer H, LeBlanc GA, Marcus M, McLachlan JA, Myers JP, Nadal A, Newbold RR, Olea N, Prins GS, Richter CA, Rubin BS, Sonnenschein C, Soto AM, Talsness CE, Vandenbergh JG, Vandenberg LN, Walser-Kuntz DR, Watson CS, Welshons WV, Wetherill Y, Zoeller RT. Chapel Hill bisphenol A expert panel consensus statement: Integration of mechanisms, effects in animals and potential to impact human health at current levels of exposure. Reprod Toxicol 2007;24:131–8.

[191] Takao T, Nanamiya W, Nagano I, Asaba K, Kawabata K, Hashimoto K. Exposure with the environmental estrogen bisphenol A disrupts the male reproductive tract in young mice. Life Sci 1999;65:2351–7.

[192] Meeker JD, Calafat AM, Hauser R. Urinary bisphenol A concentrations in relation to serum thyroid and reproductive hormone levels in men from an infertility clinic. Environ Sci Technol 2010;44:1458–63.

[193] Mendiola J, Jørgensen N, Andersson AM, Calafat AM, Ye X, Redmon JB, Drobnis EZ, Wang C, Sparks A, Thurston SW, Liu F, Swan SH. Are environmental levels of bisphenol A associated with reproductive function in fertile men? Environ Health Perspect 2010;118:1286–91.

[194] Li D, Zhou Z, Qing D, He Y, Wu T, Miao M, Wang J, Weng X, Ferber JR, Herrinton LJ, Zhu Q, Gao E, Checkoway H, Yuan W. Occupational exposure to bisphenol-A (BPA) and the risk of self-reported male sexual dysfunction. Hum Reprod 2010;25:519–27.

[195] Meeker JD, Ehrlich S, Toth TL, Wright DL, Calafat AM, Trisini AT, Ye X, Hauser R. Semen quality and sperm DNA damage in relation to urinary bisphenol A among men from an infertility clinic. Reprod Toxicol 2010;30:532–9.

[196] Li DK, Zhou Z, Miao M, He Y, Wang J, Ferber J, Herrinton LJ, Gao E, Yuan W. Urine bisphenol-A (BPA) level in relation to semen quality. Fertil Steril 2011;95:625.

[197] Bretveld R, Brouwers M, Ebisch I, Roeleveld N. Influence of pesticides on male fertility. Scand J Work Environ Health 2007;33:13–28.

[198] Whorton D, Kraus RM, Marshall S, Milby TH. Infertility in male pesticide workers. Lancet 1977;2:1259–61.

[199] Whorton D, Milby TH, Krauss RM, Stubbs HA. Testicular function in DBCP exposed pesticide workers. J Occup Med 1979;21:161–6.

[200] Potashnik G, Ben-Aderet N, Israeli R, Yanai-Inbar I, Sober I. Suppressive effect of 1,2-dibromo-3-chloropropane on human spermatogenesis. Fertil Steril 1978;30:444–7.

[201] Thrupp LA. Sterilization of workers from pesticide exposure: the causes and consequences of DBCP-induced damage in Costa Rica and beyond. Int J Health Serv 1991;21:731–57.

[202] Slutsky M, Levin JL, Levy BS. Azoospermia and oligospermia among a large cohort of DBCP applicators in 12 countries. Int J Occup Environ Health 1999;5:116–22.

[203] Lantz GD, Cunningham GR, Huckins C, Lipshultz LI. Recovery from severe oligospermia after exposure to dibromochloropropane. Fertil Steril 1981;35:46–53.

[204] Olsen GW, Lanham JM, Bodner KM, Hylton DB, Bond GG. Determinants of spermatogenesis recovery among workers exposed to 1,2-dibromo-3-chloropropane. J Occup Med 1990;32:979–84.

[205] Potashnik G, Porath A. Dibromochloropropane (DBCP): a 17-year reassessment of testicular function and reproductive performance. J Occup Environ Med 1995;37:1287–92.

[206] Eaton M, Schenker M, Whorton MD, Samuels S, Perkins C, Overstreet J. Seven-year follow-up of workers exposed to 1,2-dibromo-3-chloropropane. J Occup Med 1986;28:1145–50.

[207] Ratcliffe JM, Schrader SM, Steenland K, Clapp DE, Turner T, Hornung RW. Semen quality in papaya workers with long term exposure to ethylene dibromide. Br J Ind Med 1987;44:317–26.

[208] Schrader SM, Turner TW, Ratcliffe JM. The effects of ethylene dibromide on semen quality: a comparison of short-term and chronic exposure. Reprod Toxicol 1988;2:191–8.

[209] Kelce WR, Monosson E, Gamcsik MP, Laws SC, Gray LE. Environmental hormone disruptors: evidence that vinclozolin developmental toxicity is mediated by antiandrogenic metabolites. Toxicol Appl Pharmacol 1994;126:276–85.

[210] Gray LE, Ostby JS, Kelce WR. Developmental effects of an environmental antiandrogen: the fungicide vinclozolin alters sex differentiation of the male rat. Toxicol Appl Pharmacol 1994;129:46–52.

[211] Wong C, Kelce WR, Sar M, Wilson EM. Androgen receptor antagonist versus agonist activities of the fungicide vinclozolin relative to hydroxyflutamide. J Biol Chem 1995;270:1998–2003.

[212] Anway MD, Memon MA, Uzumcu M, Skinner MK. Transgenerational effect of the endocrine disruptor vinclozolin on male spermatogenesis. J Androl 2006;27:868–79.

Further Reading

[213] Gebreegziabher Y, Marcos E, McKinon W, Rogers G. Sperm characteristics of endurance trained cyclists. Int J Sports Med 2004;25:247–51.

[214] Gould JC, Leonard SC, Maness SC, Wagner BL, Conner K, Zacharewski T. Bisphenol A interacts with the estrogen receptor alpha in a distinct manner from estradiol. Mol Cell Endocrinol 1998;142:203–14.

[215] Oldereid NB, Thomassen Y, Attramadal A, Olaisen B, Purvis K. Concentrations of lead, cadmium and zinc in the tissues of reproductive organs of men. J Reprod Fertil 1993;99:421–5.

[216] Perez KM, Titus-Ernstoff L, Hatch EE, Troisi R, Wactawski-Wende J, Palmer JR, Noller K, Hoover RN, National Cancer Institute's DES Follow-up Study Group. Reproductive outcomes in men with prenatal exposure to diethylstilbestrol. Fertil Steril 2005;84:1649–56.

[217] Zeng X, Jin T, Kong Q, Zhou Y. Changes of serum sex hormone levels in male workers exposed to cadmium. Zhonghua yu fang yi xue za zhi [Chin J Prev Med] 2002;36:258–60.

CHAPTER 7.2

Endocrine Disruptors and Men's Health

Christian Fuglesang S. Jensen*, Ulla N. Joensen[†], Zainab G. Nagras*, Dana A. Ohl[‡], Jens Sønksen*

*Department of Urology, Herlev and Gentofte Hospital, University of Copenhagen, Copenhagen, Denmark, [†]Department of Urology, Rigshospitalet, University of Copenhagen, Copenhagen, Denmark, [‡]Department of Urology, University of Michigan, Ann Arbor, MI, United States

ENDOCRINE DISRUPTORS

Endocrine disruptors are defined as exogenous substances that alter the function of the endocrine system. This includes a myriad of substances with a plethora of possible effects on human endocrine function, regulation, and development. The suspected specific endocrine disruptors are many, and knowledge in the field is rapidly expanding. Male reproductive function has been the focus of investigation, and the development of the male reproductive tract may be especially sensitive. This chapter will describe some of the proposed general mechanisms of endocrine disruptors relevant to male reproductive function, with focus on prenatal development.

Many suspected endocrine disruptors are used for industrial purposes and are present in consumer products. Among the most well-known examples of endocrine disruptors are pesticides, brominated flame retardants, polychlorinated biphenyls (PCBs), dichlorodiphenyl trichloroethane (DDT), parabens, bisphenol A, phytoestrogens, perfluorinated compounds (PFCs), and phthalates. Endocrine disruptors can be persistent or nonpersistent, and we are all exposed to many of them.

The exact mechanism of action of endocrine disruptors is still debated, and theories on mechanism of action have changed over time. The original "estrogen hypothesis" suggested that male reproductive developmental disorders were related to prenatal exposure to estrogens [1]. Today, the proposed mechanisms for several endocrine disruptors relevant to male reproductive function and sexual function have changed to include various antiandrogenic effects on hormone-dependent pathways [2]. Generally, however, endocrine disruptors are thought to induce subnormal levels of androgens during the development of the male urogenital tract in sensitive periods in fetal life.

The widely used analgesics, acetaminophen and ibuprofen, have also been proved to have endocrine-disrupting effects due to antiandrogenic effects in animal studies and decreased androgen production demonstrated in studies of human fetal testis tissue after exposure to clinically relevant doses of acetaminophen [3–5]. There are also studies associating cryptorchidism and paracetamol use in pregnant women, but as intervention studies cannot be carried out for obvious ethical reasons, clinical recommendations have not yet been decided.

A great deal of the evidence concerning the action and effects of endocrine disruptors originates from animal studies that are not easy to extrapolate to humans. This limitation is important to keep in mind when studying the topic of endocrine disruptors. Furthermore, in observational population studies, an unexposed control group can rarely be found, as virtually everyone is exposed. This further limits the value of epidemiological association studies that nevertheless are important as they are the only feasible studies in humans.

GONADAL DEVELOPMENT AND MALE DIFFERENTIATION

To better understand the action of endocrine disruptors and the possible detrimental effects on men's health, a brief review of gonadal development and male differentiation is needed.

One important concept is that the development of the male urogenital tract begins very early in fetal life. In fact, the mother might not be aware that she is pregnant at the time of important developmental events in the male fetus.

After activation of the sex-determining region Y (SRY) gene on the Y chromosome, the testes are formed and begin to produce androgens, stimulated by maternal HCG, and this early androgen production is vital to male sexual differentiation. Similarly, Sertoli cells are formed and aid in the development of the primordial spermatogonia, which have populated the primordial gonad from yolk sac migration. As testicular development and production of androgens drive the sex differentiation from the indifferent stage, the genital tubercle elongates by week 7 to begin form the future penis (Fig. 1). The final form of the external genitalia is seen in week 12 where the urethral groove is closed, and the urethral opening is at the tip of the glans. Finally, testicular descent occurs from about 27 to 35 weeks of gestation, and full-term boys should have both testicles in the scrotum [6].

During this process of androgen-induced masculinization, it is becoming clear that there are certain critical phases. These critical phases are referred to as

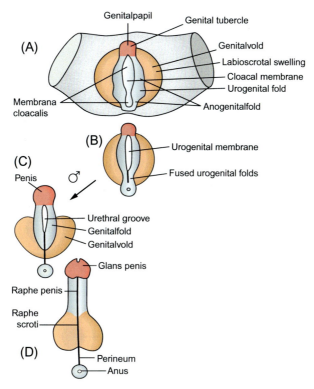

FIG. 1
Male differentiation of the external male genitalia from week 3–12 of gestation age. (A and B) The urogenital folds and the indifferent stage of the future external genitalia are present from around week 3 of gestation. (C) Male differentiation occurs as testicular development, and the production of androgens causes a progression from the indifferent stage with elongation of the genital tubercle by week 7 to begin form the future penis. (D) The final form of the external genitalia is seen in week 12 where the urethral groove is closed, and the urethral opening is at the tip of the glans.

"masculinization programming windows." If testosterone production is disturbed within these windows, the reproductive organs can exhibit irreversibly altered function or maldevelopment. Although the identification of masculinization programming windows was done in rats, the same mechanisms are relevant to humans, and based on the timings in rats, the masculinization programming windows in humans are likely to be in the period 8–14 weeks of gestation [7]. Endocrine disruptors may affect the production of circulating androgens in the fetus at these critical periods of gonadal development and male differentiation. Furthermore, it is possible that testicular dysgenesis of varying severity may render individuals more susceptible to later exposures with detrimental effects on male reproductive function. This chapter will focus on the effect of endocrine disruptors associated with fetal development.

ENDOCRINE DISRUPTORS AND ASSOCIATED CLINICAL CONDITIONS—THE "TESTICULAR DYSGENESIS SYNDROME"

In the late 1900s, epidemiological evidence accumulated to suggest adverse trends in male reproductive health. It was suggested in a landmark paper by Carlsen and colleagues [8] that semen quality in the general population was steadily decreasing in the study period from around 1940 to 1990. At the same time, in the same geographic areas, a rise in the incidence of cryptorchidism, hypospadias, and testis cancer was observed. This prompted researchers to look for common underlying mechanisms. In 2001, Skakkebæk and colleagues first described the "testicular dysgenesis syndrome" [9]. The theory explains how underlying genetic predispositions, intrauterine growth disorders, lifestyle factors including lifestyle factors of the mother, and exposure to endocrine disruptors may result in an abnormal development of the male urogenital tract resulting in certain clinical conditions in boys and men (Fig. 2). Thus, abnormal development is likely

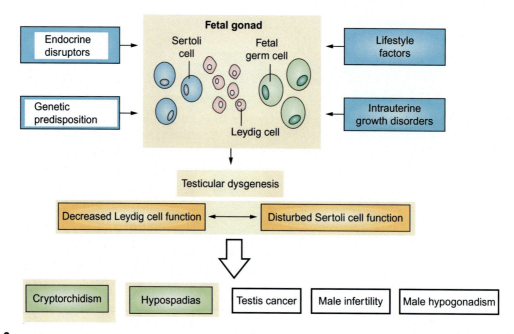

FIG. 2

The testicular dysgenesis syndrome. The combination of an underlying genetic predisposition, intrauterine growth disorders, lifestyle factors, and exposure to endocrine disruptors influencing the in utero environment causes a reduction of the levels of circulating androgens in the fetus, disrupting the normal embryonic programming and gonadal development. This testicular dysgenesis involves decreased Leydig cell function and disturbed Sertoli cell function that can result in clinical conditions including cryptorchidism; hypospadias; and long-term consequences such as testis cancer, male infertility, and male hypogonadism.

multifactorial, and some cases of testicular dysgenesis are potentially induced by exposure to lifestyle and environmental factors including endocrine disruptors. These cause a reduction of the levels of circulating androgens in the fetus resulting in disruption of the normal embryonic programming and gonadal development. The associated clinical conditions include cryptorchidism, testis cancer, hypogonadism, hypospadias, and infertility. A large amount of evidence points to fetal life as the sensitive period. However, evidence from animal studies suggests that some regeneration of prenatal damage is possible and that continued postnatal exposure could be required for the damage to be permanent.

Cryptorchidism

Cryptorchidism is the failure of descent of one or both testicles into the scrotum at birth and is one of the most common urogenital birth defects seen in newborn males. Full-term boys should have both testicles in their scrotum. Boisen and coworkers [10] examined full-term Danish boys at birth and showed an incidence of cryptorchidism of 9%. However, the incidence is substantially lower in most other locations, usually reported around 3%.

It is important to recognize that because the prenatal disruption of testicular development affects both testes, cryptorchidism should be viewed as a bilateral disease, even in cases where only one testicle is absent from the scrotum. Men with a history of cryptorchidism have 4–5 times increased risk of testis cancer [11], and the risk is there in both testicles. Most men with cryptorchidism have impaired semen quality in spite of early orchidopexy [12], and the risk of reduced spermatogenesis is present in both testes highlighting the fact that cryptorchidism is a bilateral disease. Looking at the impact of this on future fertility, men with a history of one undescended testis have a lower fertility rate (number of children fathered per man) but the same paternity rate (rate of men who have fathered at least one child) as those with bilateral descended testes, while men with a history of bilateral undescended testes have both lower fertility and paternity rates [13]. There are cases of complete azoospermia among men with a history of unilateral undescended testis, and unfortunately, there are no reliable markers to predict future fertility. It is believed that early orchidopexy may be associated with some catch-up growth, and it is clear that untreated bilateral undescended testes have a poor prognosis for fertility. The reduced fertility of cryptorchid men is likely due to germ cell loss and impaired germ cell maturation as demonstrated by Hadziselimovic and coworkers [14]. This might be caused by testicular dysgenesis, and cryptorchidism is considered to be a clinical condition associated with the testicular dysgenesis syndrome.

Hypospadias

Hypospadias is a urogenital birth defect in which the urethral opening is located more proximal than its usual location at the tip of the glans.

The incidence of hypospadias has increased during the last decades, and it is hypothesized that endocrine disruptors might be partially responsible [15]. The incidence of hypospadias is reported around 1%. Hypospadias ranges in severity from glandular hypospadias where the urethral opening is on the underside of the glans to more severe cases of proximal hypospadias with sometimes ambiguous genitalia. From Fig. 1, it is easy to understand how a failure of progress of male differentiation from the indifferent stage can cause hypospadias when the ventral groove is not closed properly.

To investigate the relationship of hypospadias to other possible clinical conditions of the testicular dysgenesis syndrome, Asklund and colleagues evaluated semen quality and reproductive hormones among 92 men with isolated hypospadias and 20 men with hypospadias and an additional genital disorder (cryptorchidism and/or germ cell tumors) and compared the results with a population of young healthy men [16]. There was no difference in semen quality between men with isolated hypospadias and controls. However, reproductive hormone levels did indicate a subtle impairment of testicular function in men with isolated hypospadias as a slight increase in follicle-stimulating hormone and luteinizing hormone was observed. When looking at semen quality of the 20 men with hypospadias and an additional genital disorder, it was found that all semen parameters were significantly lower compared with controls. Significant increases in follicle-stimulating hormone and luteinizing hormone were also observed for this group. Although these data do not demonstrate that hypospadias is a result of testicular dysgenesis, it seems that, at least in some cases, hypospadias is a part of a larger condition involving impaired testicular function. This is challenged in a recent publication by Ollivier and colleagues [17] arguing that hypospadias is largely due to inheritance caused by unknown genetic defects. In conclusion, hypospadias is a result of an abnormal gonadal development and male differentiation, and some cases might be due to testicular dysgenesis.

Testis Cancer

Testis cancer is the most common cancer in men 15–34 years of age [18], and the most common type of testis cancer is germ cell tumors. Testis cancer has been found to be associated with other clinical conditions of testicular dysgenesis, and it is well known that testis cancer and male infertility are interconnected. Male infertility is associated with an increased risk of testis cancer, and a diagnosis of testis cancer poses a significant risk to the reproductive health of a man, both from the associated dysgenesis of spermatogenesis and due to damage from surgical and medical treatment of the cancer.

The increased risk of testis cancer among infertile men has been shown in several retrospective cohort and case-control studies. A large analysis of US claim data including almost 1 million men found a hazard ratio of approximately 2 for testicular cancer in known infertile men versus control individuals [19]. This estimate seems valid, and it is believed that there is a twofold increased risk for testis cancer in infertile men compared with fertile men.

Male infertility is a complex and heterogeneous condition, and the etiology is often unknown. Among the idiopathic cases of male infertility, many are thought to be due to genetic defects. Thousands of genes contribute to spermatogenesis, and it might be that alterations in these genes also play a role in the development of testicular cancer. As previously described, the connection between the clinical conditions of the testicular dysgenesis syndrome is likely multifactorial, and the mechanisms responsible are overlapping. Acting together, these factors result in disruptions of testicular development and function, and the disrupted embryonal programming might act as a common underlying cause for impaired semen quality, testis cancer, cryptorchidism, and hypospadias. This also explains the 4–5 times increased risk of testis cancer in men with cryptorchidism. Other possible explanations include environmental and occupational factors that are not directly related to the testicular dysgenesis syndrome and endocrine disruptors [12].

Male Infertility

Infertility is defined as a failure to conceive after 12 months or more of regular unprotected sexual intercourse. The condition is a considerable problem today affecting up to 15% of couples. In some countries, 1 of 10 babies is born after assisted reproduction. It is estimated that a male factor contributes to the couples' infertility in approximately 50% of cases.

Male infertility is thought to be associated with other conditions of testicular dysgenesis, and for that reason, male infertility has been mentioned several times earlier in this chapter. Among the most well-established associations is the increased risk of testis cancer in infertile men possibly due to testicular dysgenesis.

Male infertility is complex and multifactorial, and knowledge regarding etiologic factors for male infertility is still developing. Up to 50% of infertile men are classified as idiopathic meaning that a likely cause for the infertility could not be established. When looking at testis tissue from infertile men, including idiopathic infertile men, many will have sections of tissue demonstrating Sertoli cell-only patterns, immature Sertoli cells, Leydig cell nodules, and microliths, which are all histological signs of testicular dysgenesis [20]. It has been hypothesized that some of the idiopathic cases of male infertility might be a result of testicular dysgenesis and can be considered long-term consequences of the abnormal development that took place in fetal life.

Hypogonadism

Male hypogonadism is the clinical condition caused by androgen deficiency [21]. The diagnosis requires persistent symptoms of testosterone deficiency and a decreased serum testosterone level. Among the symptoms of androgen deficiency are reduced libido, the loss of vigor, mood changes, and erectile dysfunction [22]. It is apparent that these symptoms can have an impact on the man's health, but is hypogonadism linked to exposure to endocrine disruptors and testicular dysgenesis and thus other clinical conditions potentially harmful to men's health?

In the theory of testicular dysgenesis syndrome, the collective effect of an underlying genetic predisposition, intrauterine growth disorders, lifestyle factors, and exposure to endocrine disruptors cause a decline in circulating androgens in the fetus resulting in an abnormal gonadal development that includes a decreased Leydig cell function resulting in a decreased production of testosterone later in life. As such, it is a lack of androgens in fetal life that in theory causes a lack of androgens in adult life. A recent study by Hauser and coworkers [23] did suggest that low testosterone concentrations in men aged 55–64 years were associated with exposure to phthalates. Similarly, male infertility was found to be linked to phthalate exposure. A substantial number of infertile men have low testosterone concentrations, even if not many have overt testosterone deficiency. Like other mentioned conditions, hypogonadism exhibits associations to other clinical conditions of testicular dysgenesis, and it might be that endocrine disruptors play a role in the development of male hypogonadism. It seems that there are several factors in adult life that are important for developing hypogonadism and endocrine disruptors may play a less important role in most cases of hypogonadism.

EVIDENCE SYNTHESIS AND CONCLUSION

The theory of endocrine disruptors and the testicular dysgenesis syndrome seems well established, but there are important limitations and missing links to most clinical cases of male reproductive disorders. Epidemiological studies, animal studies, and observational studies all point to a possible detrimental role of exposure to endocrine disruptors causing testicular dysgenesis involving clinical conditions such as cryptorchidism, hypospadias, infertility, testis cancer, and hypogonadism. A recent systematic review and meta-analysis on the effect of exposure to endocrine disruptors on male reproductive disorders (including cryptorchidism, hypospadias, and low sperm counts) found no overall effect with an OR of 1.11 (95% confidence interval 0.91–1.35) [24].

A lot of the evidence on which the theory is based stem from animal studies and animal models has shown that we cannot necessarily extrapolate to humans.

Data from observational studies are potentially confounded by several factors including the lack of a nonexposed control group, and a causal relationship is not possible to establish. We cannot and should of course not make intervention studies in pregnant women, which would be the only reliable way to establish causal relationships. True effect sizes of exposure to endocrine disruptors are impossible to get, and the topic is still theoretical and largely based on associations. However, it seems sensible that health policy makers should do an effort to limit the exposure of endocrine disruptors on pregnant women and educate about possible associated risks.

References

[1] Sharpe RM, Skakkebaek NE. Are oestrogens involved in falling sperm counts and disorders of the male reproductive tract? Lancet 1993;341(8857):1392–5.

[2] Kristensen DM, Skalkam ML, Audouze K, Lesne L, Desdoits-Lethimonier C, Frederiksen H, et al. Many putative endocrine disruptors inhibit prostaglandin synthesis. Environ Health Perspect 2011;119(4):534–41.

[3] Kristensen DM, Hass U, Lesne L, Lottrup G, Jacobsen PR, Desdoits-Lethimonier C, et al. Intrauterine exposure to mild analgesics is a risk factor for development of male reproductive disorders in human and rat. Hum Reprod 2011;26(1):235–44.

[4] van den Driesche S, Macdonald J, Anderson RA, Johnston ZC, Chetty T, Smith LB, et al. Prolonged exposure to acetaminophen reduces testosterone production by the human fetal testis in a xenograft model. Sci Transl Med 2015;7(288):288ra80.

[5] Kristensen DM, Desdoits-Lethimonier C, Mackey AL, Dalgaard MD, De Masi F, Munkbol CH, et al. Ibuprofen alters human testicular physiology to produce a state of compensated hypogonadism. Proc Natl Acad Sci U S A 2018;115(4):E715–e24.

[6] Pansky B. Review of medical embryology. MacMillan Publishing Company; 1982.

[7] Welsh M, Saunders PT, Fisken M, Scott HM, Hutchison GR, Smith LB, et al. Identification in rats of a programming window for reproductive tract masculinization, disruption of which leads to hypospadias and cryptorchidism. J Clin Invest 2008;118(4):1479–90.

[8] Carlsen E, Giwercman A, Keiding N, Skakkebaek NE. Evidence for decreasing quality of semen during past 50 years. BMJ 1992;305(6854):609–13.

[9] Skakkebaek NE, Rajpert-De Meyts E, Main KM. Testicular dysgenesis syndrome: an increasingly common developmental disorder with environmental aspects. Hum Reprod 2001;16(5):972–8.

[10] Boisen KA, Kaleva M, Main KM, Virtanen HE, Haavisto AM, Schmidt IM, et al. Difference in prevalence of congenital cryptorchidism in infants between two Nordic countries. Lancet 2004;363(9417):1264–9.

[11] Venn A, Healy D, McLachlan R. Cancer risks associated with the diagnosis of infertility. Best Pract Res Clin Obstet Gynaecol 2003;17(2):343–67.

[12] Hanson BM, Eisenberg ML, Hotaling JM. Male infertility: a biomarker of individual and familial cancer risk. Fertil Steril 2018;109(1):6–19.

[13] Trussell JC, Lee PA. The relationship of cryptorchidism to fertility. Curr Urol Rep 2004;5(2):142–8.

[14] Hadziselimovic F, Hoecht B. Testicular histology related to fertility outcome and postpubertal hormone status in cryptorchidism. Klin Padiatr 2008;220(5):302–7.

[15] Baskin LS, Ebbers MB. Hypospadias: anatomy, etiology, and technique. J Pediatr Surg 2006;41(3):463–72.

[16] Asklund C, Jensen TK, Main KM, Sobotka T, Skakkebaek NE, Jorgensen N. Semen quality, reproductive hormones and fertility of men operated for hypospadias. Int J Androl 2010;33(1):80–7.

[17] Ollivier M, Paris F, Philibert P, Garnier S, Coffy A, Fauconnet-Servant N, et al. Family history is underestimated in children with isolated hypospadias: a French multicenter report of 88 families. J Urol 2018;200(4):890–4.

[18] Shanmugalingam T, Soultati A, Chowdhury S, Rudman S, Van Hemelrijck M. Global incidence and outcome of testicular cancer. Clin Epidemiol 2013;5:417–27.

[19] Eisenberg ML, Li S, Brooks JD, Cullen MR, Baker LC. Increased risk of cancer in infertile men: analysis of U.S. claims data. J Urol 2015;193(5):1596–601.

[20] Juul A, Almstrup K, Andersson AM, Jensen TK, Jorgensen N, Main KM, et al. Possible fetal determinants of male infertility. Nat Rev Endocrinol 2014;10(9):553–62.

[21] Dohle GR, Arver S, Bettocchi C, Jones TH, Kliesch S. EAU guidelines on male hypogonadism. Arnhem, The Netherlands: EAU Guidelines Office; 2017.

[22] Lunenfeld B, Mskhalaya G, Zitzmann M, Arver S, Kalinchenko S, Tishova Y, et al. Recommendations on the diagnosis, treatment and monitoring of hypogonadism in men. Aging Male 2015;18(1):5–15.

[23] Hauser R, Skakkebaek NE, Hass U, Toppari J, Juul A, Andersson AM, et al. Male reproductive disorders, diseases, and costs of exposure to endocrine-disrupting chemicals in the European Union. J Clin Endocrinol Metab 2015;100(4):1267–77.

[24] Bonde JP, Flachs EM, Rimborg S, Glazer CH, Giwercman A, Ramlau-Hansen CH, et al. The epidemiologic evidence linking prenatal and postnatal exposure to endocrine disrupting chemicals with male reproductive disorders: a systematic review and meta-analysis. Hum Reprod Update 2016;23(1):104–25.

Index

Note: Page numbers followed by *f* indicate figures and *t* indicate tables.

A

Accessory pudendal artery (APA), 122–123
Acetaminophen, 404
Acupuncture, 103–104, 271, 282
Adipokines, 158–159
Adipose tissue, 152
Adult-onset diabetes, 127
Adult-onset hypogonadism (AOH), 254, 260
Advanced glycation end products (AGEs), 127–128
Aerobic exercise, 352, 356
AGEs. *See* Advanced glycation end products (AGEs)
AHI. *See* Apnea hypopnea index (AHI)
Air pollution, 363–366
AIs. *See* Aromatase inhibitors (AIs)
Alcohol, 11–12, 131
  acute effects
    ejaculatory dysfunction, 335
    erectile dysfunction, 333–335
  chronic effects
    ejaculatory dysfunction, 341
    erectile dysfunction, 336–341
    hypogonadism, 342
  confounding factors, 342–343
  cyclic AMP (cAMP) levels, 334
  disease states, 333
  oxidized LDL, 340–341
  PON1 gene, 340–341
  societal difference, 343
Alprostadil, 132–133
Amazonian diet, 6
American Association of Naturopathic Physicians (AANP), 282
American Medical Association, 186
American Society for Reproductive Medicine (ASRM) guidelines, 304, 306, 371
American Urological Association (AUA) guidelines, 12, 141–142, 237, 239
American Urological Association Symptom Index (AUA-SI), 35
AMPs. *See* Antimicrobial peptides (AMPs)
Anastrozole, 263–264, 328
Androgen(s), 125, 174
Androgen deficiency (AD), 235
Androstenedione, 289
Anejaculation, 11–12
  evaluation, 136
  treatment of, 136
Anorgasmia, 137
Antimicrobial peptides (AMPs), 64–66
Antioxidants, 44–46
Anxiety, 227
  cardiovascular patients, 226–227
  sexual dysfunction, 225
APA. *See* Accessory pudendal artery (APA)
Apnea hypopnea index (AHI), 191
Aromatase inhibitors (AIs), 143, 255
  benefits, 263
  dosing, 262
  indications, 262
  mechanism of action, 261, 262*f*
  safety concerns, 262–263
  side effects, 262–263
ART. *See* Assisted reproductive technology (ART)
Arteriosclerosis, 127–128
Assisted reproductive technology (ART), 41–42
  inflammation effects, 111
Athleticism, 213–214
AUA. *See* American Urological Association (AUA) guidelines
AUA-SI. *See* American Urological Association Symptom Index (AUA-SI)
Autoimmune factors, 108
Avanafil, 130–131
Azoospermia, 388–389

B

BACH. *See* Boston Area Community Health Survey (BACH)
Bacteroides, 113
  B. massiliensis, 70
Bacteroidetes phyla, 71
BC. *See* Bladder cancer (BC)
Benign prostatic hyperplasia (BPH), 27–36, 291–292
  obesity and, 149–152
Benzo(a)pyrene, 304–305
Bimix, 133
Bipolar disorder, 211–212
Bisphenol A (BPA), 387–388
Bladder cancer (BC)
  inflammation and, 113–114
  risk factors, 311
Blood alcohol content (BAC), 333–335
Blood vessels, 122–123
Blueberries, 12–13

Index

Body mass index (BMI), 7–8, 149, 261
Boston Area Community Health Survey (BACH), 27, 35
Boston Scientific devices, 134
BPH. *See* Benign prostatic hyperplasia (BPH)
Brief Male Sexual Function Inventory (BMSFI), 342–343
Buccal application, of testosterone, 142
Butyrate, 65, 69

C

Cabergoline, 288
Cadmium, 381–382
Calcium-D-glucarate, 291
Caloric load, 32
Cardiac death rates, 172
Cardiovascular diseases (CVDs), 153, 226–227
　and erectile dysfunction (ED), 169–171
　and gut microbiome, 156–157
　hypogonadism and, 174–175
　patients, sexual health in, 171–174
Cavernous nerves, 122
CC. *See* Clomiphene citrate (CC)
Cellular events, 129
Childhood diet, 19
Chinese medicine. *See* Traditional Chinese medicine (TCM)
Chlamydia trachomatis, 109–110
Chronic illness, 183–186
Chronic pelvic pain, 103–104
Chronic pelvic pain syndrome (CPPS), 293–294
Chronic prostatitis (CP), 103–104, 293
Cigarette smoking (CS)
　erectile dysfunction, 307–310
　male fertility, 304–306
　morbidity and mortality, 303
　phases, 303
　testosterone, 306–307
　urological cancers
　　BC, 311
　　kidney cancer, 312
　　PC, 311–312
Circumflex veins, 122
Citrus fruits, 13
Class, 208–209
Clomiphene, 143, 258

Clomiphene citrate (CC), 258, 263–264, 328
COBRA. *See* Cost and Outcome of Behavioural Activation (COBRA)
Coffee, 17–18
Colon health, 68–69
Coloplast devices, 134
Colorectal cancer, 68–69
Comfort food, 226
Commensal microorganisms, 65–66, 68
Constitutively produced HSPs, 369
Continuous positive airway pressure (CPAP), 186–187, 191, 195
Coping, 213–214
Cornus officinalis, 287
Corpora cavernosa, 121, 124, 253
Corynebacterium Prevotella, 101
Cost and Outcome of Behavioural Activation (COBRA), 171
CPAP. *See* Continuous positive airway pressure (CPAP)
C-reactive protein (CRP), 14, 35
Cryptorchidism, 107–108, 369–370, 407
CS. *See* Cigarette smoking (CS)
Cupping, 271
CVD. *See* Cardiovascular diseases (CVDs)
Cycling-induced ED, 355

D

Daidzein, 383
Dairy products, 31, 48–49
DBCP. *See* Dibromochloropropane (DBCP)
DBPs. *See* Disinfection by-products (DBPs)
DE. *See* Delayed ejaculation (DE)
Dehydroepiandrosterone (DHEA), 289, 323
Dehydroepiandrosterone sulfate (DHEAS), 323
Delayed ejaculation (DE), 9–10, 335
Delayed orgasm (DO), 137
Deoxycytidine, 110–111
Depression, 210–212, 227
　cardiovascular patients, 226–227
　sexual dysfunction, 225
Derogatis Interview for Sexual Functioning in Men-II (DISF-M-II) score, 240

Detumescence, 124
DHA. *See* Docosahexaenoic acid (DHA)
Diabetes mellitus (DM), 127
　erectile dysfunction, 127–128, 128*f*
　and sleep disorders, 184, 185*f*
Diabetic neuropathy, 128–129
Dibromochloropropane (DBCP), 388–389
Diet, 27–36
　fats, 42–44
　and sexual health
　　childhood diet, 19
　　and dietary patterns, 15–16
　　and ejaculatory/orgasmic dysfunction, 9–12
　　and erectile dysfunction (ED), 4–9
　　foods affect male hormones, 16–18, 16*t*
　　and hypogonadism, 14–16
　　ketogenic diet, 19
　　organic food, 19
　　overview of, 3–4, 16–18
　　and Peyronie's disease, 12–13
　　vs. weight loss, 8–9
Dietary patterns, 15–16
Dietary proteins, 28, 51–52
Diethylstilbestrol (DES), 384–385
Diets supplements, 44
Di(2-ethylhexyl) phthalate (DEHP), 386–387
Disinfection by-products (DBPs), 366–367
DM. *See* Diabetes mellitus (DM)
DNA approaches, gut microbiome, 61–62
DO. *See* Delayed orgasm (DO)
Docosahexaenoic acid (DHA), 42–43
Dorsal penile nerves, 121
Dyslipidemia, 33, 150–151

E

ED. *See* Erectile dysfunction (ED)
EDB. *See* Ethylene dibromide (EDB)
Ejaculation, physiology of, 124–125
Ejaculatory apparatus, 123
Ejaculatory dysfunction, 135–136
　alcohol
　　acute effects, 335
　　chronic effects, 341
　anejaculation evaluation, 136
　diet and, 9–12
　mechanism, 341

Index   415

OTC supplements
  cabergoline, 288
  Cornus officinalis, 287
  effects, 288t
  Epimedium koreanum, 286–287
  Lepidium meyenii, 287
  Tribulus terrestris, 287
overview, 135–136
treatment of anejaculation, 136
Ejaculatory reflex center, 125
Electromagnetic field (EMF) radiation exposure, 376–377
Endocrine-disrupting chemicals (EDCs)
  definition, 380
  heavy metals
    cadmium, 381–382
    lead, 382–383
  pesticides
    DBCP, 388–389
    EDB, 389
    vinclozolin, 389–390
  phytoestrogens, 383–384
  synthetic chemicals and solvents
    BPA, 387–388
    DES, 384–385
    phthalates, 385–387
  types, 380–390
The Endocrine Disruption Exchange, 380–381
Endocrine disruptors
  animal studies, 410–411
  definition, 403
  mechanism of action, 403
  observational studies, 410–411
Endocrinopathy, 156–157
Endothelial dysfunction, 128
Endothelial NO synthase (eNOS), 128, 309
Endurance aerobic training, 353, 356
Environmental exposures and male fertility, 110–111
Environmental toxins, 47–51
  air pollution, 363–366
  EDCs (*see* Endocrine-disrupting chemicals (EDCs))
  heat
    endogenous sources, 369–373
    exogenous sources, 373–375
  hypoxia, 378–379
  male fertility, 363
  physical and mental stress, 379–380

radiation
  spermatogenesis, 375–377
  testicular cancer, 377–378
  water pollution, 366–368
Epididymis, 123
Epimedium koreanum, 286–287
Equol, 16t, 17
Erectile dysfunction (ED), 104, 126–135, 186–188, 274
  alcohol
    acute effects, 333–335
    chronic effects, 336–341
  cardiovascular disease and, 169–171
  CS, 307–310
  cycling, 355
  definition, 282–283, 307, 349
  in diabetics, pathophysiology of, 127–129, 128f
  diet and, 4–9
  etiologies, 282–283
  exercise
    benefits, 349–350
    fertility, 352–353
    intervention studies, 351
    libido, 354
    pelvic floor, 352
    prostate cancer survivors, 351–352
  food groups and, 4–6
  lead exposure, 383
  management of, 130–135
  obesity and, 152–153, 153f
  OTC supplements
    effects, 286t
    Ginkgo biloba, 284
    ginseng, 283
    Korean mountain ginseng, 283–284
    KRG, 283
    Maca, 285
    Rubus coreanus, 284
    Schisandra chinesis, 284–285
    yohimbine, 285–286
    zallouh, 285
  overview, 126–127
  patient evaluation, 129
  prevalence, 239, 282–283
  testosterone therapy, 239–240
Erectile physiology, 123–124
Erectile tissue, 121
Erythrocytosis, 241–242
Estradiol, 261

Estrogens, 125, 154, 174–175, 290–291
Ethanol, 333, 334f
Ethylene dibromide (EDB), 389
Exercise, 213–214
  ED
    benefits, 349–350
    fertility, 352–353
    libido, 354
    pelvic floor, 352
    prostate cancer survivors, 351–352
    sperm concentration, 380
Exertion, 379

F

Failure of emission (FOE), 135–136
Fat, 5
Fatty acid, 43–44
FDA. *See* U.S. Food and Drug Administration (FDA)
Fertility, 225–226
  CS, 304–306
  exercise, 352–353
Ferula harmonis, 285
FFA. *See* Free fatty acids (FFA)
Fiber include fructooligosaccharides (FOS), 76
Firmicutes, 68
Fish, 49–50
Flax seed, 16–17
FOE. *See* Failure of emission (FOE)
Folates, 10, 46
Folic acid, 10, 47
Follicle-stimulating hormone (FSH), 47, 125–126, 190f, 253–254
Food
  and erectile function, 4–6
  as vehicle for environmental toxins, 47–51
FOS. *See* Fiber include fructooligosaccharides (FOS)
Free fatty acids (FFA), 76
Free testosterone (FT), 306
Fruits, 29, 50–51
FSH. *See* Follicle-stimulating hormone (FSH)
Functional anatomy, 121–123
  blood vessels, 122–123
  ejaculatory apparatus, 123
  erectile tissue, 121
  nerves, 121–122

Functional physiology
  erectile physiology, 123–124
  physiology of ejaculation, 124–125
  physiology of testosterone production, 125–126

G

Galactooligosaccharides (GOS), 75–76, 77t
GELDING theory. *See* Gut Endotoxin Leading to a Decline IN Gonadal function (GELDING) theory
Genistein, 383
Genital tract infections, 109
Genitourinary conditions, 103–105
Genitourinary infections, 109–110
  and infertility, 109–110
Genitourinary trauma, 354–355
Ginger, 17
Ginkgo biloba, 284, 286t
Ginseng, 283
L-Glutamine, 80, 83–84t
Glycyrrhizic acid, 16t, 17
Gonadotropin-releasing hormone (GnRH), 125–126, 143, 154
GOS. *See* Galactooligosaccharides (GOS)
Grains, 32
Gut Endotoxin Leading to a Decline IN Gonadal function (GELDING) theory, 104–105
Gut microbiome
  additional beneficial nutrients, 80
  colon health, 68–69
  colorectal cancer, 68–69
  dietary influence, 71–80
  differentiation of intestinal microbiome, 62–63
  DNA approaches, 61–62
  immunity, 63–66
  men's health and, 69–71
  microbes, 63–66
  microbial diversity, 66–68
  RNA approaches, 62

H

hCG. *See* Human chorionic gonadotropin (hCG)
Healthcare providers, 214
Health Professionals Follow-up Study, 127

Heat
  endogenous sources, 369–373
  exogenous sources, 373–375
Heat-shock factor 1 (HSF1), 369
Heat-shock proteins (HSPs), 369
Heavy alcohol consumption, 337
Heavy metals
  cadmium, 381–382
  lead, 382–383
Hegemonic masculinity, 213
Helicobacter pylori, 72, 78–79t
Help-seeking, coping behavior, 213–214
Herbaspirillum, 113
HG. *See* Hypogonadism (HG)
Homeostasis, 64–66
Homeostasis Model Assessment of Insulin Resistance (HOMA-IR) index, 257
Hormone residues, 48–49
Hormone secretion patterns, 189–191, 190f
Human chorionic gonadotropin (hCG)
  benefits, 257–258
  dosing, 256
  indications, 256
  mechanism of action, 256, 257f
  physiological role, 256
  safety concerns, 256–257
  side effects, 256–257
Human Microbiome Project, 61
Hypergonadotropic hypogonadism, 140–141
Hypogonadism (HG), 140–141, 188–191, 189f, 342, 410
  and cardiovascular disease, 174–175
  definition, 235
  diagnosis, 236–237
  diet and, 14–16
  and inflammation, 104–105
  obesity and, 154–155, 154f
  prevalence, 235
  symptoms, 235–236, 236t
  treatment
    follow-up, 237–239
    formulations, 237–239, 238t
    thorough counseling, 237
    (*see also* Testosterone therapy (TTh))
  weight loss and, 15

Hypogonadotropic hypogonadism (HH), 140–141, 256
Hypospadias, 383, 406f, 407–408
Hypothalamic-pituitary-gonadal axis (HPGA), 253–254, 254f

I

Iatrogenic etiologies, 136
Ibuprofen, 404
ICI. *See* Intracavernosal injections (ICI)
ICSM. *See* International Consultation on Sexual Medicine (ICSM)
IELT. *See* Intravaginal ejaculatory latency time (IELT)
IGF-1. *See* Insulin/insulin growth factor-1 (IGF-1)
IIEF. *See* International Index of Erectile Function (IIEF)
Immunity, 63–66
Immunomodulation, 69
Inducible HSPs, 369
Infertility, 41, 44–46
  cryptorchidism, 48–49
  definition, 304
  genitourinary infections and, 109–110
  obesity and, 155–157, 156f
  sleep disorders and, 195–198
  varicocele, 47–51
Inflammation, 101, 103–104, 106–108, 110–111
  and bladder cancer, 113–114
  and effects on LUTS/BPH, 34
  erectile dysfunction (ED) and, 104
  hypogonadism and, 104–105
  and prostate cancer, 111–113, 112f
  and testicular torsion, 108–109
  urogenital conditions associated with, 102t
Inflatable penile implant, 134
Insomnia, 182
Insulin/insulin growth factor-1 (IGF-1), 158
Insulin molecules, 158
Insulin resistance, 157
Integumentary microbiome, 62
International Classification of Sleep Disorders, 181
International Consultation on Sexual Medicine (ICSM), 130

Index

International Index of Erectile Function (IIEF), 127, 187–188, 239–240, 342–343
International Index of Erectile Function (IIEF-5), 349–351, 354–355
International Prostate Symptom Score (IPSS), 29, 33, 293–294
International Society for Sexual Medicine (ISSM), 342–343
Intestinal microbiome, 63
  broad functions, 64t
  differentiation, 62–63
Intracavernosal injections (ICI), 133–134
Intramuscular injections, 142–143
Intraurethral alprostadil, 132–133
Intravaginal ejaculatory latency time (IELT), 10, 137
IPSS. *See* International Prostate Symptom Score (IPSS)
Isoflavones, 16t, 17, 51

J

Joint Commission Journal on Quality and Patient Safety, 182

K

Kegel exercises, 352
Ketogenic diet, 19
Keystone species, 67
Korean mountain ginseng, 283–284
Korean red ginseng (KRG), 276–278, 283

L

Lactobacillus, 101
LCMS. *See* Liquid chromatography mass spectrometry (LCMS)
Lead, 382–383
Lepidium meyenii, 285, 287
LH. *See* Luteinizing hormone (LH)
Licorice, 17
Lipopolysaccharide (LPS), 64, 66
Lipoteichoic acid (LTA), 64, 66
Liquid chromatography mass spectrometry (LCMS), 141
Longitudinal Investigation of Fertility and the Environment (LIFE) Study, 195–197
Long sleep, 198–199
Long-term exhaustive exercise, 353

Lower urinary tract symptoms (LUTS), 27–36, 240, 291
  obesity and, 149–152
  sleep disorders and, 192–195, 193f
LPS. *See* Lipopolysaccharide (LPS)
LTA. *See* Lipoteichoic acid (LTA)
Luteinizing hormone (LH), 48, 125–126, 154
LUTS. *See* Lower urinary tract symptoms (LUTS)
Lycopene, 30, 159

M

Maca, 285
Mainstream smoke, 303
Male hormones, 16–18, 16t
Male infertility, 105–111, 408–409
  CS, 304–306
  dietary fats, sperm cell membrane and, 42–44
  sleep disorders in, 196f
Male obesity, 69
Male reproductive function, 403–405
Male silkworm extract, 289–290
Masculinity, 212
  help-seeking behaviors, 225
  integrative model, 215f
  society standards, 228
  stress, 224
Masculinization programming windows, 404–405
Massachusetts Male Aging Study (MMAS), 7–8, 139–140, 169–170, 307
Mast cells, 108
Meats, 48–49
MEDITA. *See* Mediterranean diet and type 2 diabetes (MEDITA)
Meditative practices
  definition, 224
  health benefits
    anxiety, 227
    cardiovascular disease, 226–227
    depression, 227
    diabetes, 226
    fertility, 225–226
    sexual health, 225
    weight, 226
  recommendations, 227–228
Mediterranean diet, 3, 6, 8–9
Mediterranean diet and type 2 diabetes (MEDITA), 8–9
Mental illness, 210–212

Metabolic crossfeeding, 67–68
Metabolic syndrome (MetS), 33, 34f, 155, 240–241
Metabolites, 64–65, 71, 309
Metaproteomics, 62
Metatranscriptomics, 62
Metformin, 158
Methylenetetrahydrofolate reductase (MTHFR), 46–47
Methylmercury, 49–50
Methyltestosterone, 142
MetS. *See* Metabolic syndrome (MetS)
Microbes, 63–66
Microbial diversity, 66–68
Microorganisms, 63
Mindful eating, 226
Mindfulness, 224
Mindfulness-based stress reduction (MBSR), 226
Moderate alcohol consumption, 336
MTHFR. *See* Methylenetetrahydrofolate reductase (MTHFR)
Multivariate analysis, 193, 308, 355
Muscle-invasive BC (MIBC), 311
Mycoplasma genitalium, 110

N

Naloxone, 328
Nanke, 271–272
National Commission on Sleep Disorders Research, 181
National Health and Nutrition Examination Survey (NHANES), 192, 198, 385
National Health and Nutrition Examination Survey (NHANES III) database, 35
National Institutes of Health, 61
Naturopathic medicine, 282
Neisseria gonorrhoeae, 109
Nephrolithiasis (NL), 310
Nerves, 121–122
Netherlands
  fertility clinic in, 48–49
  healthier dietary patterns, 51–52
Neuronal NO synthase (nNOS), 309
Neutrophil-to-lymphocyte ratio (NLR), 104
NHANES. *See* National Health and Nutrition Examination Survey (NHANES)
Nicotiana tabacum, 303

Nicotine, 303
Nitric oxide (NO), 123–124
NL. *See* Nephrolithiasis (NL)
NLR. *See* Neutrophil-to-lymphocyte ratio (NLR)
NO. *See* Nitric oxide (NO)
Noninsulin-dependent diabetes, 127
Nonmuscle-invasive BC (NMIBC), 311
Nonprescription medicine. *See* Over-the-counter medicine (OTC)
Nonsteroidal antiinflammatory medications (NSAIDs), 104
Nontestosterone treatment methods, 143
Norepinephrine, 124
NSAIDs. *See* Nonsteroidal antiinflammatory medications (NSAIDs)

O

Obesity, 32, 41–42, 104–105, 151f, 186, 226
  and BPH/LUTS, 149–152
  and erectile dysfunction (ED), 7–8, 152–153, 153f
  and hypogonadism, 14–15, 154–155, 154f
  and infertility, 155–157, 156f
  and prostate cancer, 158–159
Obstructive sleep apnea (OSA), 186–188, 191, 194–195, 199–200
Occupational exposures, 195–198
Occupational health, 182
Occupational stress, 379
Oligospermia, 388–389
Omega-3 fatty acids, 49–50, 52, 83–84t
One-carbon metabolism, 46–47
Ontario Farm Family Health Study, 305
Opioid epidemic, 321
Opioid-induced androgen deficiency (OPIAD)
  diagnosis, 323–325
  epidemiology, 321–322
  pathophysiology, 322–323
  treatment, 325–326
  treatment options, 326–328
Oral testosterone, 142
Organic food, 19
Orgasmic dysfunction, 137–139
  diet and, 9–12
  pathophysiology, 137
  patient evaluation, 137–138
  treatment, 138–139
OSA. *See* Obstructive sleep apnea (OSA)
Osteoporotic Fractures in Men Study, 149–150
Over-the-counter medicine (OTC)
  hormones
    estrogen, 290–291
    testosterone, 288–290
  market size, 281
  risks, 295
  scope, 281
  sexual function
    ED, 282–286
    ejaculatory and orgasmic function, 286–288
  urinary function
    BPH, 291–292
    prostatitis, 293–294
Overweight, 41–42
Oxidative injury, 304–305
Oxidative stress, 4–5, 30, 44–46
Oxytocin, 125

P

Pain syndromes, 103–105
Paleo diet, 6–7
Panax ginseng. *See* Korean red ginseng
Papaverine, 133
Parasympathetic nervous system (PNS), 121, 123
Particulate matter (PM), 363–365
Pathobionts, 63
Pathogens, 63, 65–66
PC. *See* Prostate cancer (PC)
PCPT. *See* Prostate Cancer Prevention Trial (PCPT)
PD. *See* Peyronie's disease (PD)
PE. *See* Premature ejaculation (PE)
Pelvic floor exercises, 352
Penile implants, 134–135
Penile vibratory stimulation (PVS), 139
Pesticides
  DBCP, 388–389
  EDB, 389
  residues, 50–51
  vinclozolin, 389–390
Peyronie's disease (PD), 12–13, 309–310

Phenolic compounds, 76–80, 78–79t, 81–82t
Phentolamine, 133
Phosphodiesterase type 5 inhibitors (PDE5i), 124, 130–132, 130t, 151–152, 283
Phthalates, 385–387
Physical activity. *See* Exercise
Physical health, 210
Physiology
  of ejaculation, 124–125
  of testosterone production, 125–126
Phytoestrogens, 383–384
PNS. *See* Parasympathetic nervous system (PNS)
Pollen extract, 293–294
Polyunsaturated fats, 5–6
Polyunsaturated fatty acids (PUFAs), 32, 43–44
POME. *See* Pulmonary oil microembolism (POME)
Pomegranates, 18
Porphyrobacter, 113
Prebiotics
  fiber variety and diversity, 75–76, 77t
  phenolic compounds, 76–80
Premature ejaculation (PE), 9–10, 335
Prescription medicines, 281
Prescription opioid abuse
  health-care issue, 321
  incidence, 321
Probiotics
  foods, 71–72
  microorganisms, 73–74t
  supplements, 72–75
Prostate cancer (PC), 198–200, 311–312
  inflammation and, 111–113, 112f
  obesity and, 158–159
  TTh, 242
Prostate Cancer Prevention Trial (PCPT), 28, 170
Prostate health, 70
Prostate-specific antigen (PSA), 30, 158, 198
Prostatitis, 293–294
Protein
  and metabolite approaches, 62–63
  sources, 27
Protodioscin (PTN), 289
Psychiatrists, 211–212

PUFAs. *See* Polyunsaturated fatty acids (PUFAs)
Pulmonary oil microembolism (POME), 143
PVS. *See* Penile vibratory stimulation (PVS)
Pygeum africanum, 292

Q

Quality of life (QOL), 195
Quercetin, 294

R

Race/ethnicity, 208–209
Radiation
  spermatogenesis, 375–377
  testicular cancer, 377–378
RCC. *See* Renal cell carcinoma (RCC)
RE. *See* Retrograde ejaculation (RE)
Reactive oxygen species (ROS), 105–106
Reduction by dutasteride of prostate cancer events (REDUCE), 34
Renal cell carcinoma (RCC), 312
Resistance training, 353
Resistant starch (RS), 76
Retrograde ejaculation (RE), 135–136
Rho-associated coiled-coil kinase (ROCK), 309
RNA approaches, gut microbiome, 62
ROS. *See* Reactive oxygen species (ROS)
RS. *See* Resistant starch (RS)
Rubus coreanus, 284
Rye-grass pollen extract, 292

S

Saw palmetto extract (SPE), 291–292, 294
SCFAs. *See* Short-chain fatty acids (SCFAs)
Schisandra chinesis, 284–285
Schistosoma haematobium, 113
Selective estrogen receptor modulator (SERMs), 143
Selective estrogen receptor modulators (SERMs)
  benefits, 260–261
  definition, 258
  dosing, 259
  indications, 259

mechanism of action, 258–259, 259f
safety concerns, 259–260
side effects, 259–260
Selective serotonin reuptake inhibitor (SSRI), 137
Semen analysis, 41–42
Sensate focus, 225
SERMs. *See* Selective estrogen receptor modulators (SERMs)
Serotonin, 10–11
Serotonin reuptake inhibitors (SSRIs), 10, 137
Sertoli cells, 43, 306, 384–385
Serum trihalomethane (THM) levels, 367
Severance secret (SS) cream, 287
Sex hormone-binding globulin (SHBG), 126, 156–157, 253–255
Sex positivity, 225
Sexual dysfunction, 137
  in diabetic men, prevalence of, 126t
  genitourinary conditions, pain syndromes and, 103–105
Sexual health, 3–4, 212, 225
  in CVD patients, 171–174
Sexual Health Inventory for Men (SHIM) questionnaire, 355
Sexually transmitted diseases (STDs), 213
Sexual stimulation, 123–124
SHBG. *See* Sex hormone-binding globulin (SHBG)
Shift work, 182–200, 189f, 193f
Shift work sleep disorder (SWSD), 181–182
Short-chain fatty acids (SCFAs), 63, 65f
Short sleep, 198–199
Sildenafil, 131
Sleep, 182–186
Sleep disorders, 183f, 185f, 186–200, 189f, 193f
  in male infertility, 196f
Smoking. *See* Cigarette smoking (CS)
Smooth muscle alterations, 129
Sneathia, 101
SNS. *See* Sympathetic nervous system (SNS)
Somatosensory evoked potentials (SSEP), 137–138

Southern Community Cohort Study, 149–150
Soy, 17
Soy products, 51
Spain, 49
Sperm, 123
  cell membrane, 42–44
  cells, structural changes in, 157
  diet and building blocks for, 42–47
  effects of inflammation on, 106f
Spermatogenesis, 46–47, 157
  hypoxia, 378–379
  mobile phone exposure, 377
  radiation exposure, 375–377
  testicular temperatures, 368
Sports
  athletic performance, sexual activity on, 355–356
  genitourinary trauma, 354–355
SSEP. *See* Somatosensory evoked potentials (SSEP)
SSRIs. *See* Serotonin reuptake inhibitors (SSRIs)
Starch, 32
STDs. *See* Sexually transmitted diseases (STDs)
Streptococcus, 101
Stress
  and contemporary male role, 208
  definition, 223
  meditative practice, 223
    (*see also* Meditative practices)
  and physical health, 210
  and sexual minority status to men's experience of, 208–209
  socialized for, 224
  unmanaged, 223
Subcutaneous pellets, 143
Superficial venous drainage, 122
SWSD. *See* Shift work sleep disorder (SWSD)
Sympathetic fibers, 122
Sympathetic hyperactivity, 27–36
Sympathetic nervous system (SNS), 124
Synthetic chemicals and solvents
  BPA, 387–388
  DES, 384–385
  phthalates, 385–387

T

TAC. *See* Total antioxidant capacity (TAC)
Tadalafil, 131
Tamoxifen, 258
T-cell differentiation, 63
T deficiency (TD), 140–141
   risk of, 139–140
Testicular cancer, radiation exposure, 377–378
Testicular dysgenesis syndrome, 406–410, 406f
Testicular heating
   endogenous sources
      cryptorchidism, 369–370
      varicocele, 370–372
   exogenous sources
      clothing, 373–374
      laptops, 374
      occupational hazard, 374
      sauna exposure, 375
      sitting, 374
Testicular torsion, inflammation and, 108–109
Testis cancer, 408–409
Testosterone (T), 69, 174–175, 288–290
   AIs
      benefits, 263
      dosing, 262
      indications, 262
      mechanism of action, 261, 262f
      safety concerns, 262–263
      side effects, 262–263
   CS, 306–307
   deficiency, 139–144, 235–237, 240–241, 254–255
      complications of, 144
      and impact on DM, treatment of, 141
      overview of, 139–140
      pathophysiology, 140–141
      patient evaluation, 141
      treatment of, 141–143
   gels, 326
   hCG
      benefits, 257–258
      dosing, 256
      indications, 256
      mechanism of action, 256, 257f
      physiological role, 256
      safety concerns, 256–257
      side effects, 256–257
   hypogonadism, 342
   patches, 326
   pellets, 143, 326–328
   production, 253–254
   production, physiology of, 125–126
   roles, 253
   SERMs
      benefits, 260–261
      definition, 258
      dosing, 259
      indications, 259
      mechanism of action, 258–259, 259f
      safety concerns, 259–260
      side effects, 259–260
Testosterone replacement therapy (TRT), 142t, 155, 174, 325–328
Testosterone therapy (TTh), 255
   benefits, 237
   cardiovascular events, 241–242
   efficacy
      erectile dysfunction, 239–240
      MetS, 240–241
      urinary function, 240
   prostate cancer, 242
   safety, 241–242
TMAO. *See* Trimethylamine N-oxide (TMAO)
TMC. *See* Total motile count (TMC)
Tobacco, 303–306, 311
Tomatoes, 30
Topical therapies, 143
Total antioxidant capacity (TAC), 110
Total motile count (TMC), 195–197
Total testosterone (TT), 306
Traditional Chinese medicine (TCM)
   andrological diseases, 274–276, 275t
   biomedical equivalents, 274–276, 275t
   diagnoses, 274–276, 275t
   network vessel-based body concept, 273
   percutaneous procedural interventions, 271
   theory and practice, 271
   treatment patterns and diagnoses, 276, 277t
   Western biomedicine, 272
Tribulus terrestris, 287
Trimethylamine N-oxide (TMAO), 70–71
Trimix, 134
TRT. *See* Testosterone replacement therapy (TRT)
Tryptophan, 10–11
TTh. *See* Testosterone therapy (TTh)
Tunica albuginea, 121, 309–310
2010 National Health Interview Survey, 184
Type II diabetes, 349–350

U

Unmanaged stress, 223
Ureaplasma, 101
Urinary microbiome, 101, 113
Urinary trichloroacetic acid (TCAA) levels, 367
Urological health, 27–36
Urologic malignancy, 111–114
U.S. Food and Drug Administration (FDA), 44
   OTC, 281
   TTh, 237–239, 241–242

V

Vacuum erection device (VED), 132
Vardenafil, 131
Varicocele, 106–107, 370–372
Vascular endothelial growth factor (VEGF), 158
Vegetables, 29, 50–51
VEGF. *See* Vascular endothelial growth factor (VEGF)
Veillonella, 101
Vinclozolin, 389–390
VITAL. *See* Vitamins and Lifestyle (VITAL)
Vitamin A, 83–84t, 106–107
Vitamin C, 45, 52
Vitamin D, 15–16, 83–84t
Vitamin E, 12, 45, 52
Vitamins and Lifestyle (VITAL), 171

W

Water pollution
   DBP, 366–367
   pharmaceuticals, pesticides and chemicals, 367–368
Weight, 226
Weight loss, 15, 157

diet *vs.*, 8–9
and hypogonadism, 15
Western diet, 3, 5–6
World Health Organization (WHO), 50–51
  air pollution guidelines, 363–364

alcohol effects, 333
obesity, 149

X
Xenoestrogens, 51
Xylooligosaccharides (XOS), 75–76, 77t

Y
Yang wilt, 274
Yoga, 224
Yohimbine, 285–286

Z
Zallouh, 285

Printed in the United States
By Bookmasters